FOURTH EDITION

Technical Communication

SITUATIONS AND STRATEGIES

Mike Markel

Boise State University

St. Martin's Press

New York

To my parents

EDITOR: Nancy Lyman
DEVELOPMENT EDITOR: Carla Kay Samodulski
MANAGER, PUBLISHING SERVICES: Emily Berleth
PUBLISHING SERVICES ASSOCIATE: Kalea Chapman
PROJECT MANAGEMENT: Books By Design, Inc.
PRODUCTION SUPERVISOR: Scott Lavelle
TEXT DESIGN: Books By Design, Inc.
COVER DESIGN: Lucy Krikorian
COVER ART: C-Squared Inc.

Library of Congress Catalog Card Number: 95-67058

Manufactured in the United States of America.

0 9 8 7 6
f e d c b a

For information, write:
St. Martin's Press, Inc.
175 Fifth Avenue
New York, NY 10010

ISBN: 0-312-11519-9

ACKNOWLEDGMENTS
Acknowledgments and copyrights are continued at the back of the book on page 698, which constitutes an extension of the copyright page.

Brief Contents

Contents

PART THREE. **Drafting the Text**

PART FOUR. Integrating Graphics and Design Elements

CHAPTER 12. Integrating Graphics

CHAPTER 13. Integrating Design Elements

PART FIVE. Revising the Document

PART SIX. Writing Common Kinds of Applications

Preface

The Fourth Edition of *Technical Communication: Situations and Strategies* has two goals: to be a clear and comprehensive teaching tool for technical-communication instructors and to prepare students for the various writing tasks they will face in the working world. Behind these goals is the belief that the best way to learn to write is to write and revise. Accordingly, the text includes numerous samples of technical communication along with dozens of writing and revising exercises that let students apply what they have learned in realistic technical-communication situations.

The difficult part of technical communication is to solve the simple, overriding problem involved in every kind of fact-based communication. This problem might be called "How to Write about Your Subject, Not about Yourself," or "How to Write to Your Readers, Not to Yourself." When technical communication fails, the most common cause is that writers did not consider *why* and *for whom* they were writing. Too many people write as though they are still in school and their readers are their instructors. The results, of course, are all too predictable: long, complicated documents that few people have the time, interest, or knowledge to read.

Technical Communication: Situations and Strategies is designed to help students analyze the writing situation—their audience and purpose in communicating—and then make decisions about the document's organization and content so that it meets the needs of its readers.

New to the Fourth Edition The changes in the field of technical communication reflect the changes in high-tech businesses and industry and in society as a whole. One major development in technical communication might be called the professionalization of the field. Technical communication is simply being taken more seriously as business and industry realize the enormous role that it plays in the success—or failure—of products in the marketplace. Technical communicators are more visible because they contribute to high-profile documents such as large manuals. They collaborate more often with technical professionals.

A second major development in technical communication is the increased presence of women, both as technical communicators and technical professionals. A rich vein of scholarship from feminist studies is changing the way we think about workplace communication, making it more flexible, more subtle, and more responsive to readers' needs.

A third major development in technical communication is internationalization. To participate in the global economy, manufacturers must pay attention to cultural and linguistic diversity, both within their own workforce and within their markets. A result of internationalization is a new awareness of—and a new sensitivity to—the needs and interests of people from all cultures.

A fourth major development is related to internationalization: an increased emphasis on graphics of all sorts. Fueled partly by increasingly sophisticated computer tools and partly by the need to decrease costs in addressing international readers, technical communication is more than ever a merging of the verbal and the visual. For this reason, I have changed the title of this book from *Technical Writing* to *Technical Communication* to more accurately reflect the importance of graphical means of communication.

A fifth major change in the field is a new appreciation for the importance of ethics. Technical communicators are often the link between the product manufacturer and the consumer, and for this reason technical communicators have sometimes been asked—or required—to misrepresent the product to the public. In an age of increasing lawsuits, technical communicators and technical professionals need a basic understanding of ethical thinking and of selected areas of the law.

To reflect these and other changes in technical communication, I have revised the text substantially, adding or expanding coverage of numerous topics.

New Chapters and New Elements

- *Collaborative writing.* To reflect the importance of collaboration in the workplace, I have added Chapter 3, "Writing Collaboratively," which includes guidance on how to organize and maintain a group, reduce conflict, and improve listening skills.

- *Multicultural audiences.* To help students understand the techniques of communicating effectively in a multicultural world, I have added a discussion of the needs of multicultural audiences to Chapter 4, "Analyzing Your Audience." Chapter 3 now covers different ways that people from other cultures participate in collaborative projects. Chapter 12 now discusses graphics for international readers.

- *Gender issues.* Because of the increasing importance of women in the workplace, Chapter 3 now includes a discussion of gender issues.

- *Usability testing.* To reflect the growing importance of usability testing as a means of quality control, the text includes a new chapter describing its theory and practice. Chapter 19, "Usability Testing of Instructions and Manuals," covers this important new aspect of technical communication.

- *Interviews with technical communicators and technical professionals.* Students need to know how technical communication will fit into their working lives. Therefore, throughout the book, technical communicators and technical professionals provide their own insights on different aspects of technical communication.

- *Appendix for ESL students.* The text includes a new appendix on the essentials of English grammar, focusing on those points that are most troublesome for speakers of English as a second language, such as the use of verbs and articles.

- *Commonly misused words.* To help students avoid common problems of usage, I have added Appendix C, "Commonly Misused Words."

Expanded Coverage

- *Argument.* Communication in the workplace needs to be persuasive; therefore, Chapter 8 now focuses on argument, with expanded coverage of the role of audience and of the nature of evidence.

- *High-tech tools.* More than ever, students need to know how to function in an electronic environment. Coverage of this topic begins in Chapter 1 and continues throughout the text, including discussions of the roles of word processors (Chapter 15), graphics software and desktop publishing (Chapter 12), the Internet (Chapter 6), CD-ROM databases (Chapter 6), hypertext (Chapter 18), and e-mail (Chapter 16).

- *Graphics and page design.* With the availability of high-tech tools, including color printers, technical communicators now have more options than ever before. Therefore, I have expanded and reorganized Chapter 12, "Integrating Graphics," to link the different kinds of graphics to the writer's goals. In addition, I have added color graphics and expanded the discussion of the role of color. Chapter 13, "Integrating Design Elements," now reflects our current understanding of the importance of design and the sophistication of modern desktop-publishing software. The chapter now includes a discussion of using graphics to signal the organization of an oral presentation, as well as the basics of presentation graphics.

- *Legal issues.* Chapter 2, "Ethical and Legal Considerations," now includes added coverage of four areas of law that all communicators need to understand: copyrights, trademarks, contracts, and liability.

- *Research methods.* Technical communicators need an array of information-gathering skills. Chapter 6, "Performing Secondary Research," now includes discussions of using electronic media such as CD-ROM and the Internet as information-gathering tools. And Chapter 7, "Performing Primary Research," includes expanded discussions of questionnaires, interviews, and other techniques of discovering and creating information.

- *Front and back matter.* Because technical communicators are involved more than ever in creating large documents, especially documentation sets, I have expanded the coverage of front and back matter and included information on how to write indexes. Chapter 11, "Drafting the Front and Back Matter," now covers both the APA and MLA documentation styles.

The Fourth Edition is completely reorganized to provide a clearer emphasis on the writing process and to highlight the most important elements of technical communication.

Because I feel strongly that technical communication is a humanistic discipline whose purpose is to help people live and work constructively with other people in a high-tech world, Part One now includes three chapters that describe the social context in which technical communication takes place. Chapter 1, "Introduction to Technical Communication," focuses on technical communication as a problem-solving medium that meets readers' needs and that is created collaboratively by people using high-tech tools. Chapter 2, "Ethical and Legal Considerations," provides a detailed introduction to ethical principles that technical communicators need to understand. Chapter 3, "Writing Collaboratively," discusses techniques for working effectively with others in creating documents.

Part Two, "Planning the Document," describes the planning process used for both individual and collaborative projects, then discusses techniques of finding and using information. Chapter 4, "Analyzing Your Audience," describes different kinds of audiences, including multicultural readers. Chapter 5, "Determining Your Purpose and Strategy," explains how to work within constraints (such as time and money) and how to propose a strategy to a supervisor. Chapter 6, "Performing Secondary Research," describes how to use the library and the Internet to gain an understanding of what is already known about a topic, and to assess the value of that information. Chapter 7, "Performing Primary Research," explains how to create new information about a topic by conducting inspections, experiments, field research, and interviews, and by sending questionnaires.

Part Three, "Drafting the Text," concentrates on the skills and tasks involved in drafting, beginning with the basic rhetorical patterns used in the argument (Chapter 8), then focusing on definitions (Chapter 9), descriptions (Chapter 10), and front and back matter (Chapter 11).

Part Four, "Integrating Graphics and Design Elements," discusses the important graphic elements of technical communication. Although many instructors will want to cover graphics and design after they have covered planning and drafting, others will want to treat these subjects earlier. For this reason, I have made both chapters as flexible as possible.

Part Five, "Revising the Document," consists of two chapters that cover strategies for making technical communication clearer and easier to follow. Chapter 14, "Revising for Coherence," describes techniques for revising the document as a whole, down to the paragraph level. Chapter 15, "Revising for Sentence Effectiveness," focuses on sentence-level concerns, including new discussions of sentence emphasis, formality, euphemisms, stating the obvious, meaningless modifiers, unnecessary prepositions, writing about people with disabilities, Simplified English, and structured writing. Although these issues are the focus of revising, some instructors may wish to cover them as part of the drafting phase.

Part Six, "Writing Common Kinds of Applications," describes how to prepare typical documents and oral presentations. Chapter 16, "Electronic Mail, Memos, and Letters," now reflects the importance of electronic mail. Chapter 17, "Job-Application Materials," covers a critically important communication task for all our students. Chapter 18, "Instructions and Manuals," focuses on an increasingly important application for technical communicators and technical professionals alike. Chapter 19, "Usability Testing of Instructions and Manuals," discusses this technique for determining whether documents are easy to use. Chapter 20, "Oral Presentations," relates oral presentations to written documents, emphasizing how to construct a clear and compelling argument.

Part Seven, "Writing Technical Reports," covers the process of proposing a project (Chapter 21), reporting on its progress (Chapter 22), and reporting on the completion of the project (Chapter 23).

Ancillary Materials

- *Expanded Instructor's Resource Manual.* The Instructor's Resource Manual now includes four articles on the teaching of technical communication.
- *Expanded Instructor's Resource Disk.* The disk now contains selected exercises from the text, as well as supplementary exercises.
- *Electronic and hard-copy transparencies.* The transparencies are now also included on the Instructor's Resource Disk, enabling you to modify them if you wish.
- *St. Martin's Technical Communication Hotline.*

Acknowledgments   All the examples in the book—from single sentences to complete documents—are real. Some were written by my students at Drexel University and Boise State University. Some were written by engineers, scientists, health-care providers, and businesspersons with whom I have worked as a consultant for over 20 years. Because much of the information in these documents is proprietary, I have silently changed brand names and other identifying information. I sincerely thank these dozens of individuals—students and professionals alike—who have graciously allowed me to reprint their writing. They have been my best teachers.

The Fourth Edition of *Technical Communication: Situations and Strategies* has benefited greatly from the perceptive observations and helpful suggestions of my fellow instructors throughout the country. Some completed extensive questionnaires about the previous edition; others reviewed the current edition in its draft form. I thank Beatrice C. Birchak, University of Houston–Downtown; James H. Brownlee, University of Minnesota; Stanley Coberly, West Virginia

University at Parkersburg; Bruce Shields Henderson, Santa Clara University; Robert T. Knighton, University of the Pacific; Wayne Losano, University of Florida; Kathy J. Russell, University of Pittsburgh at Johnstown; Gregory Saraceno, Broome Community College; Brenda Sims, University of North Texas; Thomas Stuckert, University of Findlay; Chester E. Tillman, University of Florida; and Marilyn J. Valentino, Lorain County Community College.

For sharing information on the use of computers in their classrooms and schools, I would like to thank David F. Beer, the University of Texas at Austin; Alma G. Bryant, University of South Florida; David Dyal, University of Florida; and Dirk Remley, Kent State University.

I would like to thank four other readers. Thomas R. Williams of the University of Washington and Elizabeth Keyes of Rensselaer Polytechnic Institute have generously shared their expertise in graphics and document design. Jocelyn Steer of the California School of Professional Psychology offered valuable suggestions to improve Appendix B, "Guidelines for Speakers of English as a Second Language." And librarian Adrienne M. Gardner helped me improve Chapter 6, "Performing Secondary Research."

Two other readers deserve special mention. Stephen C. Brennan of Louisiana State University, a gifted writer and rigorous editor, has helped me clarify my thinking and eliminate numerous errors. And Kevin S. Wilson of Boise State University, an extraordinary editor, has taught me more about writing and technical communication than he can imagine. The enormous contributions of Professors Brennan and Wilson have improved this book immeasurably. Every page of this text has benefited from their counsel.

I have been fortunate, too, to work with a superb team at St. Martin's Press, led by Carla Samodulski, Nancy Lyman, Emily Berleth, and Anne Dempsey. For me, St. Martin's continues to exemplify the highest standards of professionalism in publishing. They have been demanding but endlessly encouraging and helpful. I hope they realize the value of their contributions to this book.

I want to thank, too, my colleagues at Boise State University—Carol Martin, Chaman Sahni, Phil Eastman, and Daryl Jones—who have created a congenial, productive atmosphere for teaching and studying technical communication.

My greatest debt, however, is to my wife, Rita, who over the course of many months and, now, four editions, has helped me say what I mean.

A Final Word   I am more aware than ever before of how much I learn from my students, my fellow instructors, and my colleagues in industry and academia. If you have comments or suggestions for making this a better book, please get in touch with me at the Department of English at Boise State University, Boise, ID 83725. My phone number is (208) 385-3088, and my e-mail address is renmarke@idbsu.idbsu.edu. I hope to hear from you.

Mike Markel

CHAPTER 1

Introduction to Technical Communication

What Is Technical Communication?

Technical communication is the process of creating, designing, and transmitting technical information so that people can understand it easily and use it safely, effectively, and efficiently. Most technical communication is written by people working in or for organizations. Technical communication is read by people who need to carry out procedures and solve problems.

Much of what you read every day—textbooks, phone books, procedures manuals at the office, environmental impact statements, journal articles, the owner's manual for your car, cookbooks—is technical communication. The words and graphics of such documents are meant to be practical; that is, to communicate a body of factual information that will help an audience understand a subject or carry out a task. For example, an introductory biology text helps students understand the fundamentals of plant and animal biology and perform basic experiments. A user's manual for a software program describes how to use the program effectively.

Who Creates Technical Communication?

Most technical communication is created by one of two different categories of people:

- *Technical professionals.* Engineers, scientists, businesspeople, and other technically trained individuals are sometimes surprised to learn how much writing they do on the job. According to one survey, technical professionals can expect to devote at least one-fifth of their time to writing (Anderson, 1985). Another survey shows that 42 percent of engineers spent about one-third of their time writing, and that more than 28 percent spent more than 40 percent of their time writing (Barnum & Fischer, 1984). And, the percentage of time that technical professionals devote to communication increases as they advance: supervisors spend 40 percent of their time reading and writing; managers, 50 percent (MIT, 1984). A technical professional is a writer.

- *Technical communicators.* The job of a technical communicator is to write documents, including manuals, proposals, reports, sales literature, letters, journal articles, and speeches. Many technical communicators still call themselves *technical writers* (or *tech writers*), even though the term *technical communicator* more accurately reflects the increasing importance of graphics and the use of other media, such as on-line documentation. Other terms you will see include *learning product developer, information developer, documentation writer, information architect,* and *information engineer.* In small organizations, one technical communicator might be responsible for all aspects of creating the document, from drafting through production. In large organizations, a technical communicator might be a specialist, working full-time as a writer, editor, graphic artist, designer, or production specialist.

Although most technical communicators work as employees in organizations, many are independent contractors—sometimes called *freelancers*—who work on individual projects for organizations. These projects can last a few weeks, a few months, or a few years; the independent contractor can work at home, at the client's facility, or both.

The Evolution of Technical Communication

The earliest samples of writing are technical communication: invoices, receipts, and deeds. The English poet Geoffrey Chaucer (ca. 1342-1400) wrote some technical communication about the astrolabe, a navigational instrument. *The Origin of Species,* Charles Darwin's famous work on evolution, is another example of technical communication. But technical communication didn't become a field until around the time of World War II, when the military needed people to write user's manuals and maintenance manuals for hardware and weapons systems. In the past twenty years, the number of technical communicators has grown tremendously, largely because of the growth of the computer industry and related high-technology fields.

Before the 1970s, technical communication was often an afterthought. A computer manufacturer would market large systems without any instruction manuals at all. The company would send out technicians to install the system, and while they were there they tried to explain how to use it. Naturally, this wasn't a very effective or efficient way to educate the purchaser.

As more companies entered the richly profitable field, manufacturers realized that customers were frustrated when the product came without a user's manual, or when the manual arrived six months after the system, or when the manual was useless because of its poor quality. To keep their customers, high-tech companies started to pay more attention to their *documentation,* as these documents are called.

In many high-tech companies today, technical communicators work closely with design engineers and the legal and marketing staffs in creating a new product. The emphasis today is on "user friendly" products, and no product is friendly if the user can't figure out how to use it. Because technical communicators are valued members of professional organizations, their salaries and prestige have grown substantially. At many companies, such as IBM, technical communicators receive the same salary as hardware and software engineers.

The Role of Technical Communication in Business and Industry

The working world depends on written communication. Virtually every action taken in a modern organization is documented in writing (whether on paper or on-line). Here are a few examples of writing within an organization:

- Memos and electronic mail (e-mail) messages are written to request information and communicate new policies.
- Travel reports are written after technical professionals return from a business trip, such as attending a conference, inspecting a site, or teaching a training session.
- Detailed step-by-step instructions and policy and procedure statements are written when new processes and procedures are introduced in an organization.
- When a major project is being contemplated within an organization, a proposal is written to persuade management that the project would be not only feasible but also in the best interests of the organization.
- Once the project is approved and under way, progress reports are written periodically to inform the supervisors and managers about the current status of the project and about any unexpected developments.
- When the project has ended, a completion report is written to document the work and enable the organization to implement any recommendations.

In addition to in-house writing, every organization must communicate with other organizations and the public. Following are some examples:

- Inquiry letters, sales letters, goodwill letters, and claim and adjustment letters are written to customers, clients, and suppliers.
- Sales and marketing literature is written for distribution to potential customers.
- Research reports are written for external organizations.
- Articles are written for trade and professional journals.

The world of business and industry is a world of communication.

Technical Communication and Your Career

If you are taking a technical-communication course now, it was probably created to meet the needs of the working world. You have undoubtedly seen articles in newspapers and magazines about the importance of writing and other communication skills in a professional's career.

The first step in obtaining a professional-level position, in fact, is to write an application letter and a résumé. Your writing will help determine whether an organization decides to interview you. At the interview, your oral communication will be evaluated along with your other qualifications.

Once you start work, you will write e-mail, memos, letters, marketing materials, and short reports, and you might be asked to contribute to larger projects, such as proposals. During your first few months on the job, your supervisors will be looking at both your technical abilities and your communication skills.

A survey of engineering managers found that nearly two-thirds felt that technical communication was "extremely important" as a management tool (Spretnak, 1982). In another survey, 80 percent of engineers replied that their technical-communication skills helped them advance professionally (Barnum & Fischer, 1984).

Evidence that the working world values good communication skills is easy to find. Just look at the job ads in the newspaper and in the professional journals. Here are some examples:

From an organization that manufactures medical instruments:

> Senior Design Assurance Engineer. Duties include performing electronic/mechanical product, component, and material qualifications. Requires 3 years medical device industry experience, spreadsheet/word-processing abilities, and excellent written/oral communication skills. BSEE or biology degree preferred.

From a corporation that supplies agricultural products:

> Compensation Administrator. The successful candidate will administer the company-wide design, development, implementation, and monitoring of the compensation program. Minimum qualifications include 4–5 years of compensation analysis and salary administration experience. Experience writing job descriptions, developing salary structures, conducting and participating in salary surveys. Broad-based background in the general human resource function with specific knowledge of federal and state regulations affecting compensation systems. Demonstrated ability to operate a personal computer and spreadsheet software. Microsoft Excel preferred.

From a corporation that manufactures custom-molded plastic components:

> Custom injection molder has need for a process engineer. Manufacturing knowledge and experience with statistical process control required. Experience with injection molding and ASQC certification a plus. Excellent writing and speaking skills required, as well as word processing and spreadsheet skills.

Notice that these job ads mention not only communication skills but also computer skills.

The steady growth in continuing-education courses in technical and business communication also reflects this need for effective communicators. Organizations are paying to send their professionals back to the classroom to improve their technical-communication skills.

NURSING PROFESSOR ANNE PAYNE ON TECHNICAL COMMUNICATION AND NURSING

on the importance of communication for nurses

The most important thing is the ability to synthesize information and to record it and communicate it to other individuals succinctly and accurately. I'm talking about any sort of electronic or written medical record. We have to be able to communicate our conclusions and do it rapidly and to the appropriate individuals. We have to be accurate, but we also have to be brief; if you write three pages' worth, then nobody's going to read it.

on the importance of the electronic tools for nursing students

Right now St. Alphonsus Hospital is almost completely computerized; the nursing students will walk around with the notebook computer just like the nurses that they're working with. Our nursing students have to be comfortable with the technology.

on the kinds of writing that nurses do

Unit managers have to manage budgets, defend budgets, make a budget proposal. They have to do performance appraisals of other personnel. They have to be able to create a case for discussion in a team, maybe even have to communicate between agencies about a client's condition and make recommendations and referrals. It's very important for them to be able to analyze that information and get it across quickly, clearly, and with as much brevity as possible.

on the need to communicate with patients and families of patients

In the future, more and more of this will be in a written form; right now a lot of it is oral communication. But either way, nurses have to be able to gather appropriate information from a patient or from his or her family. If the patient is a child, the nurse must teach someone else how to care for the child. To do this, the nurse has to be able to communicate how to take care of someone at home.

on the role of communication in the nursing curriculum

I have wonderful faculty who have really taken it upon themselves to do a lot of writing across the curriculum. Right from the beginning, students are doing those micro themes and journals and a lot of different projects that make them express themselves correctly and completely in writing, and we also have students doing presentations.

on the importance of communication skills for a nurse entering the job market

One of the things that we do is have someone come over from the placement center; students work on writing a résumé and presenting themselves in writing and in interview situations. They have to learn to present themselves professionally. Written and oral communication skills can have a large impact on whether or not you're hired for a job.

Anne Payne, Ph.D., is chair of the Department of Nursing at Boise State University and a nurse practitioner.

The facts of corporate life today are simple. If you can communicate well, you are valuable; if you cannot, you are much less valuable.

Characteristics of Technical Communication

This section discusses seven characteristics of technical communication, as shown in Figure 1.1.

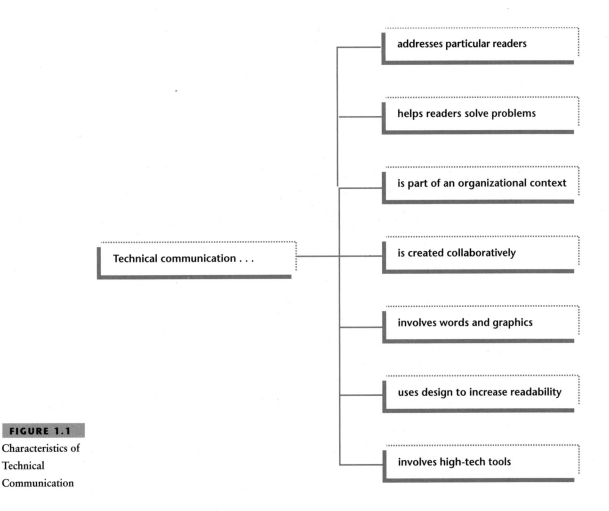

FIGURE 1.1
Characteristics of
Technical
Communication

**Technical
Communication
Addresses
Particular Readers**

Perhaps the most significant characteristic of technical communication is that it addresses particular readers. Although a journalist might have a profile of a typical reader in mind, you start with a much more specific profile when you create technical documents.

Sometimes you know the reader. For instance, if you are planning to write a proposal to your supervisor, you will think about that person's background, job responsibilities, history with the organization, attitudes toward your idea, and so forth. You will consider these factors as you decide what kind of document to write, how to structure it, what kinds of paragraphs to create, and what types of sentences and words to use. If you are writing to several different people with various backgrounds and needs, you might structure the document so that they can locate and understand the information they seek.

Sometimes you do not know the readers personally, such as when you write a brochure describing one of your company's products; but you create a profile. For example, your readers are police officers who are responsible for capital purchases. Even though you don't know their gender, their age, or other personal characteristics, at least you know that if they are reading the brochure, they share a police background and the responsibility for approving capital purchases.

Even though technical communication is addressed to particular readers, sometimes it is read by people who were never intended to be the audience. Other workers at your organization—especially managers and executives—might read sensitive documents; in addition, members of the public or the press might see documents that were never addressed to them. For these reasons, you must be careful that all your writing reflects the highest standards of professionalism.

Technical Communication Helps Readers Solve Problems

Technical communication is meant not to express the writer's creativity, or to entertain readers, but to help readers learn something or do something.

For instance, you look at your college's catalog to figure out the procedures for registration, to understand the requirements for a particular major, and to learn what is covered in a particular course. You read the owner's manual for your car to find out whether the manufacturer recommends that you rotate the tires front-to-back or diagonally. You read your company's employee-benefits manual to help you decide which benefits package you should select. You read technical communication because you need the information to help you analyze a situation and decide how to proceed.

People read technical communication to help them solve problems; people also write and distribute technical communication to help solve problems. If there is a parking problem on your campus, university officials educate the college community about the parking regulations and recommend alternative means of transportation. If one of your company's new products is not selling well, the company creates marketing materials and places ads in trade publications.

Technical Communication Is Part of an Organizational Context

Technical communication is created by people working within or for an organization to further its goals. Consider, for example, a department in a state government that oversees the state's programs in vocational education. Every activity undertaken by that department involves technical communication.

The department submits an annual report to the state legislature. This report covers each program offered by the department, indicating what need it was intended to address, who delivered it, who enrolled in it, where and when it was offered, how much it cost, and how much money it generated. The report also explains the successes and failures of each program and offers recommendations on how to make it more effective next time.

The department produces a vast quantity of technical information for the public as well. This information includes flyers, brochures, pamphlets, and even radio and television advertisements to publicize its offerings. In addition, the department produces materials used in the courses themselves, including written course texts and workbooks, and audio and video instructional material.

All the activities that support these services are technical communication. Staff members conduct interviews, distribute surveys, and perform database research to determine what programs should be offered; they make oral presentations to potential students at high schools and factories; they retain, train, and evaluate instructors; they write and produce publicity materials and monitor their effectiveness; they coordinate the programs; and they write letters to their students.

Technical Communication Is Created Collaboratively

You will often work alone in writing e-mail, memos, and letters, but you will work as part of a team in writing most of the larger, more sophisticated documents. Collaboration can take many forms, from having a colleague read a two-page memo you have written to working with a team of a dozen technical professionals and technical communicators to produce a 200-page catalog.

Collaboration is common in technical communication because no one person has all the information, skills, and time to put together a big document. Writers, editors, designers, production specialists, lawyers, and subject-matter experts—the various technical professionals—work together to create a better document than any one of them could have made working alone.

Much of the time spent on projects in organizations is devoted to communicating with collaborators. Large projects often begin with brainstorming sessions, at which the participants contribute ideas about the content, organization, and style of the document and try to reach consensus. Often, the different team members go off to work on different portions of the project, meeting periodically for briefings. Team members write progress reports and, at the end of a project, a completion report.

Because of the collaborative nature of technical communication, interpersonal skills are essential. You have to be able to listen to people with other views, express yourself clearly and diplomatically, and compromise.

Collaboration is discussed in more detail in Chapter 3.

Technical Communication Involves Words and Graphics

Virtually every technical document is a combination of words and graphics. In technical communication, graphics help the writer perform five main functions:

- communicate and reinforce difficult concepts
- communicate how-to instructions and descriptions of objects and processes

- communicate large amounts of quantifiable data
- communicate with nonnative speakers of English
- make the document more interesting and appealing to readers

Graphics are discussed in Chapter 12.

Technical Communication Uses Design to Increase Readability

Technical communicators use design features to make their documents more effective. Design features have three basic purposes:

- *To make the document attractive.* If the document looks attractive and professional, the reader is more likely to read it and more likely to form a positive impression of it and of you. Therefore, you are more likely to accomplish your purpose in writing.
- *To help the reader navigate the document.* Because a technical document can be long and complicated, and because most readers want to read only parts of it, design features help readers see where they are and get where they want to be. Some design features used commonly in technical communication are running headers and footers, tabs, and color coding.
- *To help the reader understand the document.* Design features help the reader see the organization of the document. For instance, if all first-level headings have one design and all second-level headings another, the reader will recognize this pattern and be better able to understand the discussion. You might also choose to display safety warnings in a color and size different from the rest of the text.

Design is discussed in Chapter 13.

Technical Communication Involves High-Tech Tools

Technical communication is produced on high-tech tools. The personal computer—the essential tool, along with the printer—is used in every phase of the production of documents. Everyone uses word-processing software; graphics software and desktop-publishing software are common as well. The personal computer is also the basic tool for using e-mail and on-line services such as the Internet.

The evolution from paper to on-line information is occurring not only on the input end (using tools to make the document) but also on the output end (using tools to publish it). The development of CD-ROM, with its massive storage capabilities, has made it practical to deliver large quantities of information—including text, video, and animation—in an inexpensive medium.

As information technology develops, becoming more powerful, easier to use, and less expensive, technical communicators and technical professionals alike are continuously upgrading their skills. We are all lifelong learners now.

Figure 1.2 shows a page from a user's manual for a computer printer (*Laser-Jet*, 1993, pp. 5–7). This page displays the characteristics of technical communication discussed in this section.

Printing Transparencies

The page is part of the user's manual, which is addressed to purchasers of this printer.

The page helps readers perform a particular task: print transparencies.

The manual was written collaboratively by employees at Hewlett-Packard and its subcontractors.

Design features include
- integrated words and graphics, with each graphic keyed to a numbered instruction
- clear diagrams and icons
- a footer—"Printing Tasks"—that helps readers see where they are
- rules (lines) and boxes
- plenty of white space

In addition, the writing is simple and clear, addressed to the reader's needs.

According to a note at the beginning, the manual "was created using text formatting software on an HP Vectra personal computer."

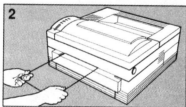

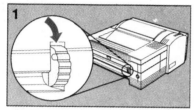

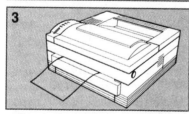

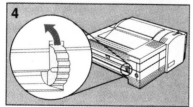

Selecting the Correct Media

CAUTION

To avoid damaging your printer, use only transparencies designed for use with laser printers. See Appendix B for detailed specifications.

Setting Up the Printer

First, select "manual feed" or "single sheet feed" through your software and send your job to print. Then set up the printer:

1 Set the paper path knob to the down position. This allows the transparencies to feed straight out the rear of the printer, reducing film curl.

2 Gently insert the transparency film into the manual feed slot until you feel some resistance on the upper left corner. The printer grabs the end of the transparency sheet.

3 Let the other end of the transparency hang down from the manual feed slot and rest on the table. The printer then prints the page.

4 After printing the transparencies, reset the paper path knob to the up position.

To Avoid Transparency Curl

To avoid transparency curl, grab the transparency as it comes out of the printer and set it on a flat surface to cool.

FIGURE 1.2

An Example of
Technical
Communication

Measures of Excellence in Technical Communication

Figure 1.3 shows the eight basic measures of excellence of technical communication.

Technical Communication Is Honest

The most important measure of excellence in technical communication is that it is honest. You have to tell the truth and not mislead the reader.

There are three reasons to be honest as a communicator:

- *It is the right thing to do.* Technical communication is not about using words and pictures to mislead or lie to people; technical communication

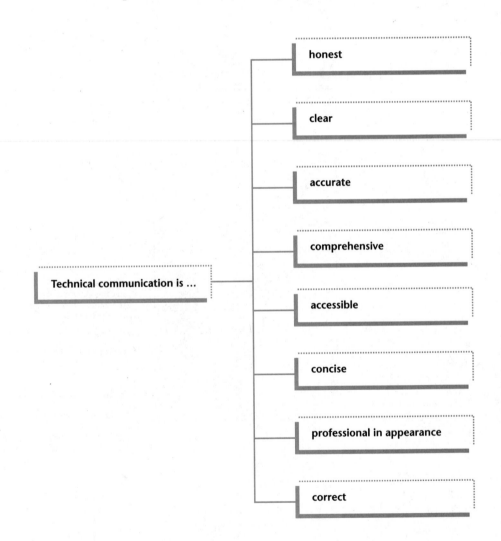

FIGURE 1.3

Measures of Excellence in Technical Communication

is about helping people understand how to make wise choices as they use the artifacts of a high-tech culture.

- *Readers can get hurt if you are not.* If you lie or mislead in writing about medication dosages, people could become seriously ill or could die. If you fail to warn readers about the safety hazards involved in operating a piece of equipment, people could be seriously injured or killed.

- *You and your organization could get in serious legal trouble if you are not.* If a plaintiff can convince a judge or jury that your document failed to provide honest, appropriate information, and that this failure led to a substantial injury or loss, you and your organization are likely to have to pay millions of dollars.

Chapter 2 discusses the ethical and legal aspects of technical communication.

Technical Communication Is Clear

An important characteristic of effective technical communication is clarity. That is, your goal is to produce a document that conveys a single meaning that the reader can understand easily.

The following directive, written by the British navy (*Technical Communication*, 1990), is an example of what you *don't* want to do:

> It is necessary for technical reasons that these warheads should be stored upside down, that is, with the top at the bottom and the bottom at the top. In order that there may be no doubt as to which is the top and which is the bottom, for storage purposes, it will be seen that the bottom of each warhead has been labeled with the word TOP.

Technical communication must be clear for two reasons:

- *Unclear technical communication can be dangerous.* A carelessly drafted building code tempts contractors to save money by using inferior materials or techniques. The 1986 space-shuttle tragedy might have been prevented had the officials responsible for deciding whether to launch received clear reports of the safety risks involved that day.

- *Unclear technical communication is expensive.* The cost of a typical business letter today—counting the writer's time and the cost of equipment, stationery, and postage—is more than ten dollars. In addition, every time an unclear letter is sent out, the reader has to write or phone to ask for clarification. The letter often must be rewritten and sent again to confirm the correct information. On a larger scale, clarity in technical communication is essential because of the collaborative nature of most projects. One employee investigates one aspect of the problem, while other employees work on other aspects of it. Unclear communication can jeopardize the entire project.

Technical Communication Is Accurate

Inaccurate writing causes as many problems as unclear writing. In one sense, accuracy is a simple concept: you must record your facts carefully. If you mean to write *4,000*, don't write *40,000*. The slightest inaccuracy will at least confuse and annoy your readers. A major inaccuracy, naturally, can be dangerous and expensive.

In another sense, however, accuracy is a question of ethics. Technical communication must be as objective and free of bias as you can make it. If readers suspect that you are slanting information—by overstating the significance of a particular fact or by omitting an important point—they have every right to doubt the validity of the entire document. Technical communication must be effective by virtue of its clarity and organization, but it also must be reasonable, fair, and honest.

Technical Communication Is Comprehensive

A good technical document is comprehensive; that is, it provides all the information that readers will need. It describes the background so that readers who are unfamiliar with the subject can understand it. It contains sufficient detail so that readers can follow the discussion and carry out any tasks called for. It refers clearly to supporting materials or includes them as attachments.

Comprehensiveness is crucial because the people who will act on the document need a complete, self-contained discussion in order to apply the information safely, effectively, and efficiently. In addition, they need an official company record of the project, from its inception to its completion.

Here are two examples of the importance of being comprehensive:

- A scientific article reporting on an experiment comparing the reaction of a new strain of bacterium to two different compounds will not be considered for publication unless the writer has fully described the methods used in the experiment. Because other scientists should be able to replicate the researcher's methods, every detail must be provided, including the names of the companies from which the researcher obtained all the materials.
- A report recommending that a company open a plant in a new location will be analyzed in detail before management commits itself to such an expensive and important project. The team charged with studying the report will need all the details. If the recommendations are implemented, the company will need a single, complete record in case changes have to be made several months or years later.

Technical Communication Is Accessible

Accessibility refers to the ease with which readers can locate the information they seek. Most technical documents are made up of small, relatively independent sections. Some readers are interested in only one or two sections; others might want to read several or most sections. Because few people will pick up the document and start reading from the first page all the way through, your job is to make the various parts of the document accessible. That is, readers should not be forced to flip through the pages to find the appropriate section.

Chapter 11 describes the major elements that make documents accessible, such as tables of contents and indexes. Chapter 13 discusses how to use design elements to improve accessibility.

Technical Communication Is Concise

To be useful, technical communication must be concise. The longer the document, the more difficult it is to use, for the obvious reason that it takes more of the reader's time. In a sense, conciseness works against clarity and comprehensiveness; for a technical explanation to be absolutely clear, it must describe every aspect of the subject in great detail. To balance the claims of clarity, conciseness, and comprehensiveness, you must make the document just long enough to be clear—given the audience, purpose, and subject—but not a word longer. You can shorten most writing by 10 to 20 percent simply by eliminating unnecessary phrases, choosing short words rather than long ones, and using economical grammatical forms.

The battle for concise writing, however, is often more a matter of psychology than of grammar. Some writers produce long documents to show their readers that they are trying hard. But if the document needs to be short to fulfill its purpose, your job is to figure out how to convey a lot of information in a little space. That is one of the challenges of writing technical communication.

Technical Communication Is Professional in Appearance

Documents start to communicate before anyone reads the first word. If the document looks neat and professional, readers will form a positive impression of both the content and the authors.

Your documents should adhere to the format standards that apply in your organization or your professional field. In addition, your documents should be well designed and neatly printed. For example, a letter should follow one of the traditional letter formats and have generous margins. It should be balanced both vertically and horizontally on the page.

Technical Communication Is Correct

Good technical communication correctly observes the conventions of grammar, punctuation, spelling, and usage.

Many of the rules of correctness are clearly important. If you mean to write "The three inspectors—Bill, Harry, and I—attended the session" but you use commas instead of dashes, your readers might think six people (not three) attended. If you write "While feeding on the algae, the researchers captured the fish," some of your readers might have a little trouble following you—at least for a few moments as they consider the image of researchers feeding on algae.

Most of the rules, however, make a difference primarily because readers will judge your thinking on how your writing looks and sounds. Carelessness and grammar errors make your readers doubt the accuracy of your information, or at least lose their concentration. You will still be communicating, but the message won't be the one you had intended. As a result, the document will not achieve its purpose, and readers could well judge you incompetent and unprofessional.

Technical communication is meant to fulfill a mission: to convey information to a particular audience or to persuade that audience to a particular point of view. To accomplish these goals, it must be honest, clear, accurate, comprehensive, accessible, concise, professional in appearance, and correct.

The rest of this book describes ways to help you say what you want to say.

EXERCISES

The following exercises call for you to write memos. See Chapter 16 for a detailed discussion of memos.

1. Locate a document that you believe to be an example of technical communication. Describe the aspects of the document that correspond to or exemplify the characteristics of technical communication discussed in this chapter. Then evaluate the effectiveness of the document. To what extent does it meet the measures of excellence discussed in this chapter? In what ways does it fall short of those measures? Write your response in a memo to your instructor. Submit a photocopy of the document (or a representative portion of it) with your assignment.

2. Locate an owner's manual for a consumer product, such as a coffeemaker, bicycle, or hair dryer. In a memo to your instructor, describe and evaluate the manual, using the strategy described in exercise 1. Submit a photocopy of the manual (or a portion of it) with your assignment.

3. Study the job ads in your field from a large newspaper or a professional journal. To what extent do they suggest that the employers value technical-communication skills? What percentage of the ads explicitly mention communication skills? What kinds of skills are mentioned most often? Are the ads themselves examples of effective technical communication? Write your response in a memo to your instructor.

4. Interview a faculty member in your major to learn his or her thoughts about the role of technical communication in that particular field and about the communication skills of recent graduates. How does the instructor try to help students improve their communication skills? Summarize the interview in a memo to your instructor.

5. An excerpt from a brochure from the Concord Coalition (1994), a non-profit political lobbying organization, appears on pages 17–18. Using the strategy explained in exercise 1, write a memo to your instructor explaining the characteristics of technical communication that you see in this excerpt and evaluating the effectiveness of the excerpt.

6. Evaluate the example of technical communication on page 19, an excerpt from a description of medical plans available to state employees. In a memo to your instructor, explain the strengths and weaknesses of the excerpt by referring to the measures of excellence described in this chapter.

REFERENCES

Anderson, P. (1985). What survey research tells us about writing at work. In L. Odell & D. Goswami (Eds.), *Writing in nonacademic settings*. New York: Guilford.

—————— **The Problem** ——————
A TOWERING MOUNTAIN OF DEBT

Amerida is at war — an economic war. Our nation's wealth is being drained, drop by drop, by federal budget deficits. Each year, the government spends $200 billion more than it has. These deficits are never paid off. They are added to the huge mountain of debt that now stands at over $4.5 trillion.

This debt mortgages our future to pay for current consumption. We're squandering our fiscal integrity and threatening America's prosperity. Without quick action to eliminate budget deficits, today's adults will be the first generation in American history to leave the nation worse off than it was when they came of age.

The Facts

◆ The federal government has run a budget deficit every year since 1969.

◆ Between 1980 and 1992, the national debt tripled from **$1 trillion** to **$4 trillion.** If unchecked, the debt will increase by **$1 trillion** more by 1996.

◆ Annual interest on the federal debt equals the income taxes paid by all those living west of the Mississippi River.

Why Worry about the Deficit?

Budget deficits strangle economic growth and hold down our standard of living. If the U.S. economy had grown as fast in the last 20 years as it did in the 1950s and 1960s, the median family income today would be **$50,000** instead of **$35,000.** In part, that's because large deficits use up the savings that would otherwise be available for private investment. Without that investment, it is harder for the economy to grow and provide more and better jobs.

Deficits cause cuts in important programs. Interest paid on the debt is the third largest program in the budget. Interest payments squeeze out programs we care about. This year, we'll spend more on interest than on agriculture, crime, education, environment, medical research, science and technology, and transportation — **combined.**

Deficits rob future generations. Instead of leaving a stronger, more competitive economy for our children, we will leave them ever-rising interest payments and stagnant economic growth. This is not the American Dream.

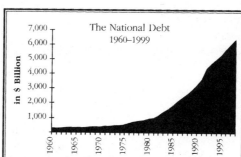

The National Debt 1960–1999

in $ Billion: 7,000 / 6,000 / 5,000 / 4,000 / 3,000 / 2,000 / 1,000

1960 1965 1970 1975 1980 1985 1990 1995

Barnum, C., & Fischer, R. (1984). Engineering technologists as writers: Results of a survey. *Technical Communication, 31*(2), 9–11.

Concord Coalition. (1994). The Concord Coalition. [Brochure]. Washington, DC: The Concord Coalition.

LaserJet 4L User's Manual. (1993). Boise, ID: Hewlett-Packard Company.

MIT Industrial Liaison Program. (1984). *Communication skills: A top priority for engineers and scientists.* (Report No. 32929).

Spretnak, C. M. (1982). A survey of the frequency and importance of technical communication in an engineering career. *The Technical Writing Teacher, 9*(3), 133–136.

Technical Communication. (1990). *37*(4), 385.

--------------------- **The Solution** ---------------------

JOIN THE CONCORD COALITION

The Concord Coalition is the citizens' voice against federal budget deficits. Even though most of the nation's elected officials say they want to eliminate the huge deficits that sap our economy and darken our future, too often they listen to the wishes of special interests when deciding budget issues.

The Concord Coalition is a nonpartisan grassroots movement to eliminate budget deficits. We are educating and mobilizing citizens and members of Congress to do something before it is too late.

How Will We Achieve Our Goals?

Large and vocal membership. Coalition chapters are building a broad base of participation through membership drives, highly visible meetings and activities, and local media attention. Our goal is to build large chapters across the country that provide a countervailing voice against special interest groups.

Education. State and local chapters take economic issues back from the economists and explain them in terms that everyone can understand. Coalition chapters host teach-ins, discussion groups, and budget-balancing exercises. A scorecard showing how senators and representatives voted on deficit issues is published annually.

Empowerment. Local and state Concord chapters hold well-publicized activities such as town meetings, teach-ins, deficit reduction exercises, candidates' forums,

Bring Home Your Congressman meetings, and other educational events.

Local volunteers help fundraise, organize events, put together newsletters, arrange phone trees, write letters to members of Congress, staff information booths, and network with local media.

No matter how you wish to get involved, the most important thing is to **join** The Concord Coalition and add your voice to the rising chorus of Americans demanding an end to federal deficits.

To find out the name of your state coordinator, call 202-737-1077.

What Will I Receive as a Concord Coalition Member?

- ◆ Automatic membership in your local chapter.
- ◆ Quarterly newsletter on budget developments and the activities of Coalition chapters nationwide.
- ◆ Invitation to annual national policy forum.
- ◆ Invitations to local candidates' forums, educational meetings, and other activities.
- ◆ Opportunities to order issue papers, booklets, and other educational materials.
- ◆ An effective way to make your voice heard.

The former Lincoln National/Peak Health HMO plan which was purchased by Idaho Preferred Healthcare in September, 1992, **WILL NOT** be offered this year. **Current members of the Peak Health plan will AUTOMATICALLY BE ENROLLED IN THE IDAHO PREFERRED HEALTHCARE PLAN unless they choose to enroll in another plan during Open Enrollment.**

If you would like to add yourself/dependents to medical coverage or change medical plans you must complete the appropriate enrollment application form and return it to your human resources/payroll office by May 14. **Late applications will not be accepted.**

Although there is no Open Enrollment for dental coverage at this time, if you are adding dependents to medical coverage, and **HAVE NOT PREVIOUSLY** declined dependent dental coverage, your dependents will automatically be added to dental coverage unless you complete a Dental Declination Form.

If you are moving from one medical plan to another, you and your dependents **currently covered by a dental plan** will be enrolled in the dental plan which corresponds to your new medical plan. For instance, if you are enrolled in Blue Shield Module 1/Dental Plan 1 and you enroll in Blue Shield Module 2, 3 or one of the HMO plans you will be covered under Dental Plan 2. If you are enrolled in Blue Shield Module 2, 3 or one of the HMO plans and you enroll in Blue Shield Module 1 you will be covered under Dental Plan 1.

The waiting periods for pre-existing and/or chronic conditions, and certain other conditions specifically listed under the Blue Shield, Group Health Northwest and Idaho Preferred Healthcare **MEDICAL** plans will be waived for all members **CHANGING** plans during Open Enrollment WHO ARE NOT CURRENTLY SUBJECT TO A WAITING PERIOD.

Any members WHO ARE SUBJECT TO THE WAITING PERIODS in their current **MEDICAL** plan WILL BE SUBJECT TO THE WAITING PERIODS in the plan they are enrolled in as of July 1. However, any months of coverage under a member's current medical plan will be credited toward the waiting periods in the plan he/she is enrolled in as of July 1.

All members changing from Dental Plan 1 to Dental Plan 2 **WILL BE SUBJECT** to the waiting periods and limitations in Dental Plan 2.

ALL NEW ENROLLEES, NEW HIRES, AND NEWLY ACQUIRED DEPENDENTS (EXCEPT NEWBORNS) WILL BE SUBJECT TO THE WAITING PERIODS IN EACH MEDICAL/DENTAL PLAN.

If you are adding dependents to medical coverage, or are changing plans, be sure to list all members of your family to be covered, not just the new enrollees. Members not listed on your new enrollment application form will not be covered, even if they were previously enrolled.

ALL CHANGES YOU MAKE WILL BE EFFECTIVE JULY 1, 1993. NECESSARY ENROLLMENT APPLICATION FORMS, HMO ENROLLMENT PACKETS AND DENTAL DECLINATION FORMS CAN BE OBTAINED FROM YOUR HUMAN RESOURCES/PAYROLL OFFICE.

Consult your Employee Group Insurance Handbook and HMO enrollment packets to compare the services of the HMOs and the Blue Shield modules.

FY94 MONTHLY EMPLOYEE PAID MEDICAL/DENTAL PLAN RATES

	Blue Shield Module 1/ Dental Plan 1	Blue Shield Module 2/ Dental Plan 2	Blue Shield Module 3/ Dental Plan 2
Single	$ 6.00	$ 6.00	$ 24.65
2-Party with Dependent Dental	34.90	34.90	72.00
2-Party **without** Dependent Dental	30.93	30.93	59.63
Family with Dependent Dental	60.80	60.80	113.55
Family **without** Dependent Dental	50.22	50.22	83.90
*Split with Dependent Dental	6.00	6.00	40.10
*Split **without** Dependent Dental	6.00	6.00	29.33

	Group Health Northwest/ Dental Plan 2	Idaho Preferred Healthcare/ Dental Plan 2
Single	$ 35.40	$ 44.98
2-Party with Dependent Dental	93.51	112.65
2-Party **without** Dependent Dental	81.14	100.28
Family with Dependent Dental	148.50	179.61
Family **without** Dependent Dental	118.85	149.96
*Split with Dependent Dental	64.29	85.83
*Split **without** Dependent Dental	53.52	75.06

*A Split membership applies when both spouses in a family, with dependent children covered, are employed by the state and are both enrolled in the **same** plan. One spouse enrolls as a single, the spouse covering the children enrolls as a split.

Ethical and Legal Considerations

A power company in a big American city faced a problem recently: its obsolete generating equipment produced illegal levels of air pollution. Unfortunately, the company didn't have the millions of dollars needed to replace the old equipment or even to add scrubbers that would lower the pollution to legal levels. The Environmental Protection Agency, fully aware of the pollution levels, routinely fined the utility—$250, four times a year. Why were the penalties so low? Because the courts recognized that older power companies would not be able to meet modern pollution standards and continue to supply power to the city, as well as employ thousands of people. The utility industry and the lawmakers in effect had worked out a plan to replace the older technology with newer equipment over a period of years, thus enabling the companies to stay in business.

What does this have to do with you? As a technical communicator or a technical professional, you will likely be involved with problems like this— problems that have an ethical dimension. Everyone knows why pollution is undesirable, but getting rid of it, or lowering it to acceptable levels, is often difficult and expensive. In business, government, and industry, people work every day to solve ethical problems such as this in ways that address the needs of different constituencies: in this case, the home owners and businesspeople who need the power, the company employees who need paychecks, and everyone who breathes the city's air.

As a professional, you will have to make decisions about how to treat clients, customers, and other organizations; how your organization deals with the government regulatory bodies that oversee your industry; and how the actions of your organization affect the environment. All these decisions will be communicated in your writing and speaking.

This chapter provides an introduction to professional ethics. Throughout the rest of this book, you will also encounter other discussions of ethics. The chapter on proposals, for example, offers a discussion of some of the ethical issues involved in writing proposals.

This chapter briefly introduces ethics and explains the standards commonly used to decide ethical questions. Then it discusses legal considerations for communicators: copyrights, trademarks, contract law, and liability. Next, the chapter discusses codes of conduct and whistleblowing. Finally, the chapter provides a set of guidelines for communicating technical information ethically.

A Brief Introduction to Ethics

According to Claudia Huff and Saul Carliner (personal communication, June 7, 1993), 35 percent of the software used in the United States is pirated, that is, copied illegally. In Spain, the figure is 86 percent; in Thailand, 99 percent.

A recent survey of two thousand faculty members and two thousand doctoral students in the United States found that 22 percent of the faculty know of department members who have overlooked sloppy use of data, over 30 percent report cases of inappropriate credit given for authorship of articles, and 40 percent know of colleagues who have used college resources for personal business (Swazey, Louis, & Anderson, 1994). Other widespread types of unethical conduct reported in the survey include violations of safety and animal-research guidelines, sexual harassment of students, discrimination, and cheating by students.

Computer-monitoring software tells an employer how long a networked computer was used, what files the user accessed, whether any files were modified or deleted, how quickly the user worked, and how long the user was away from the desk. This software is used to monitor some 26 million American workers; the pay of 10 million workers is based on computer monitoring (Forester & Morrison, 1994, p. 157). Does the use of this software invade the worker's privacy and degrade the work environment intolerably, or does it provide a fair, objective way to measure productivity and award good performance? Does the company fulfill its ethical obligation by informing workers that it is using, or might use, the software? Questions such as these are being debated across the country.

Virtually every day you see an article in a newspaper or journal about ethics. But what exactly is ethics? Some people equate ethical conduct and legal conduct; if an act is legal, it is ethical. Most people, however, believe that ethical standards are more demanding than legal standards. It is perfectly legal, for example, to try to sell an expensive life insurance policy to an impoverished elderly person who has no dependents and therefore no need for such a policy. Yet many people would consider the attempt unethical.

When businesspeople were asked to give their definition of ethics, half of them answered, "What my feelings tell me is right." One-quarter of the respondents replied, "What is in accord with my religious beliefs." And most of the rest said, "What conforms to 'the golden rule'" (Baumhart, 1968, pp. 11–12). Although philosophers cannot agree on what constitutes ethical conduct, most would agree to the following definition: *Ethics is the study of the principles of conduct that apply to an individual or a group.*

What are the basic principles used to think through an ethical problem? Ethicist Manuel G. Velasquez (1992) outlines three kinds of moral standards useful in confronting ethical dilemmas:

- *Rights.* The standard of rights concerns the basic needs and welfare of particular individuals. If a company has agreed to provide continuing employment to its workers, the standard of rights requires that the company either keep the plant open or provide adequate job training and placement services.

- *Justice.* The standard of justice concerns how the positive and negative effects of an action or policy can be distributed fairly among a group. For example, it would seem reasonable that the expense of maintaining a highway be borne primarily by people who use that highway. However, since everyone benefits from the highway, it is just that general funds also be used for maintenance.
- *Utility.* The standard of utility concerns the effects, both positive and negative, of a particular action or policy on the general public. If, for example, a company is considering closing a plant, the standard of utility requires that the company's leaders consider not only their savings from shutting down the plant but also the financial hardships on the unemployed workers and the economic impact on the rest of the community.

Although it is best to think about the implications of any serious act in terms of all three standards, often they conflict. For instance, from the point of view of utility, no-fault car insurance laws—under which people are prohibited from suing for damages below a certain dollar amount—are good because they reduce the number of nuisance suits that clog up the court system and increase everyone's insurance costs. However, from the point of view of justice, these laws are unfair because they force innocent drivers to pay for the repairs.

In conflicts among the three standards, the standard of rights is generally considered the most important, and the standard of justice the second most important. Simply ranking the three standards cannot solve all ethical problems, however. If an action or policy were to greatly affect the general public, the standard of utility—the "least important" standard—might outweigh the standard of rights. For instance, if the power company has to cross your property to repair a transformer on a utility pole, the standard of utility (the need to get power to all the people affected by the problem) takes precedence over the standard of rights (your right to private property). Of course, the power company is obligated to respect your rights, insofar as possible, by explaining what it wants to do, trying to accommodate your schedule, and repairing any damage to your property.

Ethical problems are difficult to resolve precisely because no rules exist to determine when one standard outweighs another. In the example of the transformer, how many customers have to be deprived of their power before the power company is morally entitled to violate the property owner's right of privacy? Is it a thousand? Five hundred? Ten? Only one?

Most people do not debate the conflict among rights, justice, and utility when they confront a serious ethical dilemma; instead, they simply do what they think is right. Perhaps this is good news. However, the depth of ethical thinking varies dramatically from one person to another, and the consequences of superficial ethical thinking can be profound. For these reasons, ethicists have described a general set of principles that can help people organize their thinking about the role of ethics within an organizational context. These principles form a web of rights and obligations that connect an employee, an organization, and the world in which the organization is situated.

For example, in exchange for their labor, employees enjoy three basic rights:

- fair wages
- safe and healthy working conditions
- due process in the handling of such matters as promotions, salary increases, and firing

Although there is still serious debate about employee rights, such as the freedom from surreptitious surveillance and unreasonable searches in the case of drug investigations, the question almost always concerns the extent of employees' rights, not the existence of the basic rights themselves. For instance, ethicists disagree about whether hiring undercover investigators to discover drug users at a job site is an unwarranted intrusion on the employees' rights, but there is no debate about the right of exemption from unwarranted intrusion.

Your Ethical Obligations

In addition to enjoying rights, an employee assumes obligations, which can form a clear and reasonable framework for discussing the ethics of technical communication. The following discussion outlines three sets of obligations:

- to your employer
- to the public
- to the environment

Your Obligations to Your Employer You will be hired to further your employer's legitimate aims and to refrain from any activities that run counter to those aims. Specifically, you will have the four obligations shown in Figure 2.1.

- *Competence and diligence. Competence* refers to your skills; you should have the training and experience to do the job adequately. *Diligence* simply means hard work.
- *Honesty and candor.* You should not steal from your employer. Stealing involves such obvious practices as embezzlement as well as the common occurrences of "borrowing" office supplies and padding expense accounts. *Candor* means truthfulness; report problems to your employer that might threaten the quality or safety of the organization's product or service. If, for instance, you have learned that a chemical your company wants to make could pollute the drinking water, you are obligated to inform your supervisor, even though the news may be unwelcome.

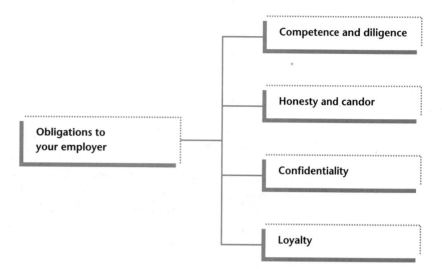

FIGURE 2.1

Obligations to Your
Employer

Another kind of problem involving honesty and candor concerns what Sigma Xi, the Scientific Research Society, calls trimming, cooking, and forging (*Honor,* 1986, p. 11). *Trimming* is the smoothing of irregularities to make the data look extremely accurate and precise. *Cooking* is retaining only those results that fit the theory and discarding the others. And *forging* is inventing some or all of the research data, and even reporting experiments that were never performed. In carrying out research, an employee might feel some pressure to report only positive, statistically significant findings but must resist the pressure.

- *Confidentiality.* You should not divulge company business outside of the company. If a competitor knew that your company is planning to introduce a new product, it might introduce its own version of that product, robbing you of your competitive advantage. Many other kinds of privileged information—such as quality-control problems, personnel matters, relocation or expansion plans, and financial restructuring—also could be used against the company. A well-known problem of confidentiality involves *insider information;* an employee who knows about a development that will increase the value of the company's stock buys the stock before the information is made public, thus reaping an unfair (and illegal) profit.

- *Loyalty.* You should act in the employer's interest, not in your own. Therefore, it is unethical to invest heavily in a competitor's stock, because that could jeopardize your objectivity and judgment. For the same reason, it is unethical to accept bribes or kickbacks. It is unethical to devote considerable time to moonlighting (performing an outside job, such as private consulting) because the outside job could lead to a conflict of interest and because the heavy workload could make you less productive in your primary position.

Every organization that offers products or provides services is obligated to treat its customers fairly. As a representative of an organization, and especially as an employee communicating technical information, you will frequently confront ethical questions.

In general, an organization is acting ethically if its product or service is both *safe* and *effective*. It must not injure or harm the consumer, and it must fulfill its promised function. However, these commonsense principles provide little guidance in dealing with the complicated ethical problems that arise routinely.

Every year, some 34 million Americans are hurt in accidents related to consumer items; 28,000 of these people die, and 100,000 are permanently disabled. The annual cost of these injuries is $12 billion (Velasquez, 1992, p. 271). Even more commonplace, of course, are product and service failures: the items are difficult to assemble or operate, they don't do what they are supposed to do, they break down, or they require more expensive maintenance than indicated in the product information.

Who is responsible for the injuries and product failures: the company that produces the product or service, or the consumer who purchases it? In individual cases, blame is sometimes easy enough to fix. A person who operates a chain saw without reading the safety warnings and without seeking any instruction in how to use it is surely to blame for any injuries caused by the normal operation of the saw. But a manufacturer who knows that the chain on the saw is liable to break when used under certain circumstances, and who failed to remedy this problem or warn the consumer, is responsible for any resulting accidents.

However, such cases do not outline a rational theory that can help people understand how to act ethically in fulfilling their obligations to the public. There is no clear consensus on this issue. Today three main theories describe a company's obligations to the public, as shown in Figure 2.2.

- *The contract theory.* The contract theory holds that a person who buys a product or service is entering into a contract with the manufacturer. The manufacturer and, by implication, the employee representing the manufacturer have four main obligations:

  – to ensure that the product or service complies with the contract in several respects: it does what its advertisements say it does, it operates a certain period of time before needing service or maintenance, and it is at least as safe as the product information states and the advertising suggests

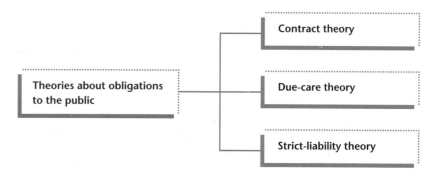

FIGURE 2.2

Theories about Obligations to the Public

– to disclose all pertinent information about the product or service, so the potential consumer can make an informed decision about whether to purchase it

– to avoid misrepresenting the product or service

– to avoid coercion, like that used by an unethical funeral director who takes advantage of a consumer's grief to sell an expensive funeral

Critics of the contract theory argue that the typical consumer cannot understand the product as well as the manufacturer does, and that consumer ignorance invalidates the "contract."

• *The due-care theory.* The due-care theory places somewhat more responsibility on the manufacturer than the contract theory does. This theory holds that the manufacturer knows more about the product or service than the consumer does and therefore has a greater responsibility to make sure it complies with all manufacturer's claims and is safe. Therefore, in designing, manufacturing, testing, and communicating about a product, the manufacturer has to make sure the product will be safe and effective when used according to the instructions. However, the manufacturer is not liable when something goes wrong that it could not foresee or prevent.

Critics of the due-care theory argue that because it is almost impossible to determine whether the manufacturer has in fact exercised due care, the theory offers little of practical value.

• *The strict-liability theory.* The strict-liability theory, which goes one step further than the due-care theory, holds that the manufacturer is at fault when any injury or harm occurs from any product defect, even if it has exercised due care and could not possibly have predicted or prevented the failure or injury. This theory assumes that the only way to assess blame is to hold the manufacturer guilty and thereby force it to assume all liability costs. In so doing, society subsidizes the product when the manufacturer builds these costs into the price of the product.

Critics of the strict-liability theory hold that, although it might be practical, it is unjust, because no person or organization should be held liable for something that is not its fault.

Your Obligations to the Environment

One of the most important lessons we have learned in the last two decades is that we are polluting and depleting our limited natural resources at an unacceptably high rate. Our overreliance on fossil fuels not only deprives future generations of their use but also causes possibly irreversible pollution problems, such as global warming. Everyone—government, business, and individuals—must work to preserve the fragile ecosystem, to ensure the survival not only of our own species but also of the other species with which we share the planet.

But what does this have to do with you? In your daily work, you probably do not cause pollution or deplete the environment in any extraordinary way. Yet because of the nature of your work, you will often know how your organization's actions affect the environment. For example, if you work for a manufacturing company, you might be aware of the environmental effects of making or using your company's products. Or you might contribute to the creation of the environmental impact statement.

Communicators should treat every actual or potential occurrence of environmental damage seriously. We should alert our supervisors to the situation and work with them to try to reduce the damage. The difficulty, of course, is that protecting the environment is expensive. Clean fuels cost more than dirty ones. Disposing of hazardous waste properly costs more (in the short run) than merely dumping it. Organizations that want to cut costs may be tempted to cut corners on environmental protection.

Your Legal Obligations

This chapter mentioned earlier that most people distinguish between ethical and legal obligations, and that they feel that ethical obligations are more comprehensive and more important. Still, the two are related very clearly in our culture. We attempt to use the law to enforce many of our ethical guidelines. Through the legal system, we try to punish people and organizations that violate our most fundamental ethical beliefs.

As a person involved in technical communication, you should know the basics of the four different bodies of law shown in Figure 2.3.

Copyright Law

The Copyright Act of 1976 provides legal protection to the author of any work—such as printed material, software, photographs—whether it is published

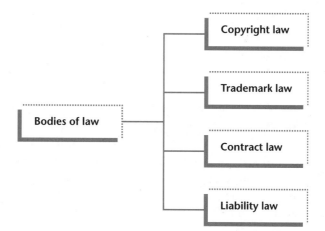

FIGURE 2.3
Bodies of Law Related
to Technical
Communication

or unpublished, and whether the author is an individual or a corporation. The author is entitled to profit from the sale and distribution of the work, in exchange for the obligation to make the work accessible.

Fair Use and Competitive Pressures

The concept by which work is made accessible is called *fair use*. According to Kozak (1993, p. 2), fair use gives you the right to use material, without getting permission, for "purposes such as criticism, comment, news reporting, teaching . . . , scholarship, or research." Unfortunately, fair use is not a specific legal term but a convention that is interpreted differently in different situations.

Kozak (1993) points out that courts consider four factors in determining whether a use of material is "fair use":

- *The purpose and character of the use, focusing on whether the use is for profit.* Profit-making organizations are scrutinized more carefully.
- *The nature and purpose of the copyrighted work.* When information is essential to the public good, such as medical information, the concept of fair use is applied more liberally.
- *The amount and substantiality of the portion of the work used.* A 200-word passage would be a small portion of a book but a large portion of a 500-word brochure. Although you will frequently see guidelines stating that 400 words are the maximum you may use, the courts have not always ruled in favor of that number.
- *The effect of the use on the potential market for the copyrighted work.* Your use of the work cannot hurt the author's potential to profit from the original work. This stipulation is deliberately vague.

Fair use does not apply to graphics: you must obtain written permission to use any graphics.

Although writers usually adhere to fair-use guidelines in reports and journal articles, the situation is different in the case of manuals and product information. For these applications, unfortunately, some companies have been known to take whole sections of another company's product information or manual, make minor changes, and publish it themselves. This practice is unethical, of course, even if it has been done so skillfully that it is technically legal.

The difficulty in thinking about the ethics of this problem is that the difference between stealing and learning from your competitors can be subtle. Words are protected by copyright, but ideas are not. For instance, hundreds of companies produce personal computers, all of which are sold with user's guides. Rare is the manufacturer who doesn't study the competitors' user's guides to see how a feature or topic is described. Inevitably, a good idea spreads from one document to another. If one manual contains a particularly useful kind of troubleshooting guide, soon many others contain similar ones. Even though this process of imitation tends to produce a dull uniformity, it can improve the overall quality of the documents.

Many organizations treat internal and external documents differently.

- *Internal documents.* If your organization asks you to update an internal procedures manual, it probably expects you to use any material from the existing manual, even if you cannot determine the original author. This is perfectly legal, under the concept known as *work for hire*. Anything written on the job by an employee being paid by your organization is the property of the organization. Any revisions you make will also be the property of the organization.

- *External documents.* Many organizations credit the authors of the external documents they publish, such as external manuals or articles. Many user's guides, as well as many software programs, acknowledge the authors and illustrators.

  However, the question of what constitutes authorship can be complicated because most large technical documents are produced collaboratively, with several persons contributing text, another doing the graphics, a third reviewing for technical accuracy, and a fourth reviewing for legal concerns. Problems are compounded when a document is revised, and parts of original text or graphics are combined with new material.

Guidelines for Dealing with Copyright Questions

To deal with copyright questions, follow these five guidelines:

- *Abide by the fair-use concept.* The fair-use concept protects everyone; try not to use excessive amounts of another source's work.

- *Seek permission.* Write to the source, seeking permission to use the work. Describe accurately what portion of the work you wish to use and what kind of publication or other work you wish to use it in. The source is likely to charge you for permission.
- *Cite your sources accurately.* Not only is citing your sources an ethical matter; it strengthens your writing by showing the reader the range of your research. In addition, it protects you in case some of the facts or interpretations in your work turn out to be inaccurate: you have not claimed that you were solely responsible for them.
- *Discuss authorship questions openly.* The best way to determine the authorship of a document is to discuss the issue openly with everyone who contributed to it. Some persons might deserve to be listed as authors; others, only credited in an acknowledgment section.
- *Seek legal counsel if you have questions.* Copyright questions can be very complicated. There are questions of authorship, especially when you are dealing with electronic documents found on the Internet. Exactly who created a document, or whether the document has been altered without the author's knowledge or permission, is sometimes difficult to determine. There are also questions related to permissions, particularly in the case of a multimedia product such as a laser disc, which can require dozens of permissions for a minute of running time. Work with your organization's legal counsel. Not only can legal counsel answer many of your questions, but the mere fact that you sought counsel is evidence of good faith, an important factor should the copyright question ever have to be resolved in court.

For a more detailed discussion of plagiarism, see the discussion of references in Chapter 11.

Trademark Law   Companies use *trademarks* and *registered trademarks* to ensure that the public recognizes the name or logo of their products; therefore, you need to know the definitions of these two legal terms. The following discussion is based on excellent articles by Branscomb (1993) and Jim Grey (personal communication, January 31, 1994).

- A *trademark* is a word, phrase, name, or symbol that is identified with a company. The company uses the ™ symbol after the product name to signify that it is claiming the design or device as a trademark. For instance, Aldus Corporation claims Pagemaker™ as a trademark. However, claiming a trademark does not ensure significant legal rights.
- A *registered trademark* refers to a word, phrase, name, or symbol that has been registered with the U.S. Patent and Trademark Office. The company uses the ® symbol after the product name to indicate that the name

is a registered trademark. Registering a trademark, a process that can take years, ensures much more legal protection in the United States, as well as protection in other nations. The application for registration gives other companies an opportunity to challenge trademark ownership. For instance, Mead Data Central, owner of the database service Lexis®, sued Toyota after it introduced the Lexus® automobile, although it is hard to imagine someone confusing a database service with a Toyota. (Both Cadillac Pet Foods, Inc., and General Motors have registered the trademark Cadillac, apparently confident that consumers can distinguish a can of dog food from a luxury automobile.)

As a communicator, you are responsible for using the trademark and registered trademark symbols accurately when you refer to the items. Unfortunately, doing so is not always easy, because, as Branscomb (1993, p. 98) points out, companies often inappropriately claim trademark status for their products. He recommends that you or your legal staff consult sources such as the following for the most accurate and up-to-date information:

Official Gazette of the United States Patent and Trademark Office: a weekly periodical published by the U.S. Patent and Trademark Office.

Trademark Register of the United States: an annual directory published by the U.S. Patent and Trademark Office. This directory is also available online.

To protect your own company's trademarks, Jim Grey (personal communication, January 31, 1994) recommends the following four guidelines:

- *Distinguish your trademarks from other material.* Use boldface, italics, different size type, or a different color to distinguish the trademarked term.

- *Use the trademark symbol.* At least once in each document—preferably, the first time—use the appropriate symbol after the name or logo, followed by an asterisk. At the bottom of the page, include a statement such as the following: "*COKE is a registered trademark of the Coca-Cola Company."

- *Use the trademarked item as an adjective, not a noun or verb.* Trademarks can get confused with the generic term they refer to; the people at Xerox spend a lot of money running ads explaining that you cannot "xerox" anything, even on a Xerox® photocopier; you can only photocopy something. Therefore, use the trademarked item along with the generic term: LaserJet® printer.

- *Do not use the plural form or the possessive form of the term.* Doing so reduces the uniqueness of the item and encourages the public to think of the term as generic.

INCORRECT:	take some Kodacolors®
CORRECT:	take some photographs using Kodacolor® film
INCORRECT:	Kodacolor's® fine quality
CORRECT:	the fine quality of Kodacolor® film

Contract Law Contract law deals with agreements between two parties. In most cases, disputes concern whether a product lives up to the manufacturer's claims. These claims are communicated as express warranties or implied warranties.

An *express warranty* is an explicit statement in a piece of product documentation that the product has a particular feature or can perform a particular function. An *implied warranty* is a warranty that is not written explicitly but rather inferred by the purchaser.

What exactly does *implied* mean? If a user's manual for a printer shows sophisticated graphics contained on sample pages, the company is implying that the printer can print those graphics—and the purchaser has a right to infer that the printer can—even if the manual never explicitly makes that claim. Implied warranties also occur in more casual communication, such as memos or letters to customers or potential customers. Therefore, it is important to be careful about what you write when you describe your company's product.

To protect themselves, most companies include a disclaimer with their product information. A *disclaimer* is a statement that limits the company's claims about the product. Manuals usually include disclaimers such as the following:

> Every effort has been made to supply complete and accurate information. However, Ottorino Publishers assumes no responsibility for the information contained herein or its use.

Software disclaimers often contain statements such as the following:

> Ottorino Software warranties this product to be free of defects at the time of purchase and agrees to replace the product at no cost if it is found to be defective under normal conditions of use, for a period of ninety (90) days from purchase. However, Ottorino Software assumes no responsibility for incidental or consequential damages.

Disclaimers such as this are often parodied, as in the following example (quoted in Forester & Morrison, 1990, p. 119):

> We don't claim Interactive EasyFlow is good for anything—if you think it is, great, but it's up to you to decide. If Interactive EasyFlow doesn't work: tough. If you lose a million because Interactive EasyFlow messes up, it's you that's out the million, not us. If you don't like this disclaimer: tough. We reserve the right to do the absolute minimum provided by law, up to and including nothing.
>
> This is basically the same disclaimer that comes with all software packages, but ours is in plain English and theirs is in legalese.
>
> We didn't really want to include a disclaimer at all, but our lawyers insisted.

Or:

> Omniverse Software warrants the diskettes to be of black color and of square shape under normal use for a period of ninety (90) days from the date of purchase.

These parodies are funny, but they make a serious point: many companies try as hard as they can to avoid taking responsibility for the quality of their products.

Liability Law   Liability law is a serious subject. A product-liability action is "a lawsuit for personal injury, death, property damage, or financial loss caused by a defective product" (Helyar, 1992, p. 126). In the years 1974–1992, the number of product-liability cases increased 800 percent. Liability is a concern for communicators because courts routinely decide that manufacturers have a legal obligation to provide adequate operating instructions and to warn consumers about the risks of using their products. In other words, product documentation in the form of instructions and manuals is treated just like any other component of the product itself.

Helyar (1992) summarizes the communicator's obligations, as reflected in recent court rulings. Helyar offers ten guidelines for abiding by liability laws:

- *Understand the product and its likely users.* Before writing the product information, perform research to understand the product thoroughly and to learn everything possible about the users and how they will use the product.

- *Describe the product's functions and limitations.* Describe the product's function to enable potential users to determine whether it is the appropriate product to buy. You must also describe the product's shortcomings. In one case, the court found the manufacturer liable for not stating that its electric smoke alarm does not work during a power outage.

- *Instruct users on all aspects of ownership.* The instructions should cover the following seven topics:
  - assembly
  - installation
  - use and storage
  - testing
  - maintenance
  - first aid and emergencies
  - disposal

- *Use appropriate words and graphics in the instructions.* The text must use common terms, simple sentences, and brief paragraphs; it must be structured logically; and it must contain explicit directions. The graphics must be simple, unified, and easy to understand; where appropriate, they should show people performing tasks.

The words and graphics must be appropriate for the intended users in terms of their education, mechanical ability, manual dexterity, and intelligence. For products to be used by children or nonnative speakers of English, the instructions should include symbols, graphics, and pictographs.

- *Warn about risks of using or misusing the product.* The instructions must warn not only about the obvious dangers of using the product (such as electrical shock or chemical poisoning) but also about less obvious dangers. For example, an automobile manufacturer was found guilty for not having warned that parking a car on grass, leaves, or other combustible material can cause a fire.

  The instructions should also describe the cause, extent, or seriousness of the danger. One judge noted in a ruling that a sign warning "Keep off the grass" is inadequate if there are deadly snakes in the grass.

  For particularly dangerous products, the instructions should include the following information:

  – the nature of the danger
  – the uses or misuses that can lead to the danger
  – the extent of the harm caused by the uses or misuses
  – a clear statement warning of that harm
  – clearly written steps on how to use the product safely

  In warning about danger, the instructions should use mandatory language, such as *must* and *shall,* rather than *might, could,* or *should.* The instructions should also use the terms *warning* and *caution* appropriately. See Chapter 18 for a discussion of *danger, warning,* and *caution.*

- *Include warnings along with assertions of safety.* When users read in the product information that a product is safe, they tend to pay less attention to warnings. Therefore, the product information must include detailed safety warnings to balance the claims of safety.

- *Make the directions and warnings conspicuous.* As discussed in Chapter 18, warnings and safety information must be

  – large
  – easily visible
  – in the appropriate location
  – durable

- *Make sure the instructions comply with applicable company standards and any applicable local, state, or federal statutes.*

- *Test the product and instructions to make sure they are accurate and easy to understand.* For more information on usability testing, see Chapter 19.
- *Make sure the users get the information.* Obviously, the instructions must accompany the product. But if the manufacturer discovers a problem later, it is obligated to get in touch with the users, using direct mail if possible or newspaper advertising if not. Automobile recall notices are examples of how manufacturers contact their users.

Codes of Conduct

Virtually all professional organizations, as well as three-quarters of the country's 1,500 largest corporations, have codes of conduct: written statements of the ethical standards employees are to follow (Brockmann & Rook, 1989, p. 91).

Codes of conduct vary greatly from organization to organization. Some are brief and general, offering only guidelines for proper behavior. The Code for Communicators written by the Society for Technical Communication, for example, is reprinted in Figure 2.4.

Other organizations' codes describe in great detail proper and improper behavior, and actually stipulate penalties for violating the principles. These codes could be thought of more as sets of rules than as general guidelines. The American Society of Mechanical Engineers' code, for instance, specifies procedures that the society is to follow in cases of complaints about ethical violations. These procedures are quite specific about the proper functioning of the society's Professional Affairs and Ethics Committee, even indicating, for instance, that the complaint must be acknowledged "by Certified Mail" (Chalk, 1980, p. 193).

Do codes of conduct really encourage ethical behavior? This is not an easy question to answer, of course; no statistics exist on how many unethical acts were prevented because people were inspired or frightened by a code of conduct. Many ethicists and officers in professional societies are skeptical. Because codes must be flexible enough to cover a wide variety of situations, they tend to be so vague that whether a person in fact violated one of their principles is virtually impossible to say.

A study conducted by the American Association for the Advancement of Science (Chalk, 1980) found that whereas most professional organizations in the sciences and engineering have codes that provide for hearing cases of unethical conduct, relatively few such cases have ever been brought before the organizations. Many societies reported, for example, that only three or four allegations were ever lodged against individuals, and most of the societies have never taken disciplinary action—censure or expulsion—against any of their members for ethical violations. And a survey about how chemical engineers make decisions about ethical conflicts reveals that of the 4,318 respondents, fewer than a half

Society for Technical Communication
Code for Communicators

As a technical communicator, I am the bridge between those who create ideas and those who use them. Because I recognize that the quality of my services directly affects how well ideas are understood, I am committed to excellence in performance and the highest standards of ethical behavior.

I value the worth of the ideas I am transmitting and the cost of developing and communicating those ideas. I also value the time and effort spent by those who read or see or hear my communication.

I therefore recognize my responsibility to communicate technical information truthfully, clearly, and economically.

My commitment to professional excellence and ethical behavior means that I will

- Use language and visuals with precision.
- Prefer simple, direct expression of ideas.
- Satisfy the audience's need for information, not my own need for self-expression.
- Hold myself responsible for how well my audience understands my message.
- Respect the work of colleagues, knowing that a communication problem may have more than one solution.
- Strive continually to improve my professional competence.
- Promote a climate that encourages the exercise of professional judgment and that attracts talented individuals to careers in technical communication.

FIGURE 2.4

STC Code for Communicators

Used with permission from *Code for Communicators,* published by the Society for Technical Communication, Arlington, Virginia.

dozen even mentioned codes of conduct in describing their thinking (Bryan, 1992, p. 81).

If codes of conduct are not often systematically enforced, do they have any real value? Critics point out that almost no organization is willing to support someone who brings a charge. An accuser within an organization is likely to face overt or covert punishment; thus self-interest compels many people to remain silent, regardless of the high-sounding statements in the organization's code of conduct. Few professional organizations have ever come to the financial aid of an accuser who has lost a job because of a justified allegation.

For this reason, codes of conduct sometimes seem to be public-relations tools intended to persuade employees, organization members, the general public, and the government that the organization polices its own members.

Ethicist Jack N. Behrman finds the greatest value in merely writing a code: an organization thereby clarifies its own position on ethical behavior (1988, p. 156). Of course, distributing the code might also foster an increased awareness of ethical issues, in itself a positive result.

ATTORNEY BRENT GARDNER ON ETHICS AND HEWLETT-PACKARD'S STANDARDS OF BUSINESS CONDUCT

on Hewlett-Packard's long-term commitment to ethics

The founders of the company set an expectation that employees act with the highest standards of ethics. They felt it was good business: if you act ethically with your customers there will be a bond of trust established and they will come back.

on Hewlett-Packard's approach to ethics training for its employees

Every employee gets a copy of our Standards of Business Conduct, and then there is training on a periodic basis, beginning when the employee first starts at HP. We also have prepared a training tape that can be used by any supervisor. The latest edition of the Standards is simpler and more straightforward than past editions.

on Hewlett-Packard's other efforts to maintain high ethical standards

We have internal auditors who review all of the entities of the company to see whether there are any standards-of-conduct violations. Also, HP has a post office box where anyone concerned with an ethical issue can anonymously raise that issue with the company. Every manager is responsible for seeing that his or her employees follow the standards that we have established. Our efforts have been quite effective.

on the major areas covered by the Standards of Business Conduct

We cover dealings by the employee with the company, conflicts of interest, gifts to or from customers and suppliers, legal issues in dealing with customers and competitors and suppliers, and dealing with the federal government.

on the major impediments to ethical behavior in the workplace

One problem is a general lack of understanding of what ethical standards are expected. Few employees have studied ethics in school or given the subject much thought. This means that they often just adopt the standards of those with whom they work closely. Another problem is that economic pressures sometimes cause people to do things they think make business sense but which are really shortsighted.

on how ethics should be taught in colleges and universities

My oldest son graduated a couple of years ago in computer science; one of his required classes was on the ethics of computer science. I think that is a good idea. Ethics is actually a legitimate issue for every course of study. I would include ethics in teaching technical communication.

Brent Gardner is assistant general counsel for Hewlett-Packard Company in Boise, Idaho.

Whistleblowing

Whistleblowing is the practice of going public with information about serious unethical conduct within the organization. For example, an engineer is blowing the whistle in telling a government regulatory agency or a newspaper that quality-control tests on a product the company sells have been faked.

In deciding whether to blow the whistle, you must choose between loyalty to your employer, on one hand, and to your own standards of ethical behavior, on the other. Some people believe that an employee owes complete loyalty to the employer. James M. Roche, former president of General Motors, has written: "Some of the enemies of business now encourage an employee to be *disloyal* to the enterprise. They want to create disharmony, and pry into the proprietary interests of the business. However this is labeled—industrial espionage, whistle blowing, or professional responsibility—it is another tactic for spreading disunity and creating conflict" (Beauchamp & Bowie, 1988, p. 262).

Yet most people believe that workers should not be asked to steal or lie or take actions that could physically harm others. Where does loyalty to the employer end and the employee's right to take action begin? And what should an employee do before blowing the whistle?

Ethicist Manuel Velasquez (1992, p. 403) argues that whistleblowing is justified if four conditions are satisfied:

- There is clear and compelling evidence that the organization is doing something that is hurting or will hurt other parties.
- The employee has made a serious but unsuccessful attempt to prevent the wrongdoing by going through channels internally.
- External whistleblowing is reasonably certain to prevent or stop the wrongdoing.
- The wrongdoing is serious enough to warrant the injuries that the whistleblowing will probably cause for the employee, his or her family, and any other parties.

But saying that you are *justified* in blowing the whistle is not the same as saying that you are *obligated* to do so. Velasquez suggests you are obligated to blow the whistle when you are both justified to and when at least one of the following two conditions is satisfied:

- Your position or professional responsibility calls for you to prevent the wrong. For instance, if you are a professional engineer, your professional code of ethics stipulates that you are to prevent certain kinds of wrongdoing, such as the building of unsafe structures.

- The wrongdoing is sufficiently serious, for society, a group, or an individual. Wrongdoing that involves health, safety, or basic economic welfare falls into this category.

A number of organizations have instituted procedures that encourage employees to bring ethical questions to management rather than blow the whistle. Among the more common means that companies use are anonymous questionnaires and ombudspersons. An *ombudsperson* is an employee whose job includes bringing ethical grievances to management's attention. An ombudsperson who feels that management has not responded satisfactorily to the situation is empowered to report the information freely.

Most companies still have no formal procedures for handling serious ethical questions, however, and whistleblowing thus remains risky. Although the federal government and about half the states have laws intended to protect whistleblowers, these laws are not highly effective. It is simply too easy for the organization to penalize the whistleblower—subtly or unsubtly—through negative performance appraisals, transfers to undesirable locations, or isolation within the company. For this reason, many people feel that an employee who has unsuccessfully tried every method of alerting management to a serious ethical problem would be wise simply to resign rather than face the professional risks of whistleblowing. Of course, resigning quietly is much less likely to force the organization to remedy the situation.

As many ethicists say, doing the ethical thing does not always advance a person's career.

Guidelines for Ethical Communication

Ethical problems are particularly common in two different kinds of writing you will do:

- *Proposals.* You might be asked to exaggerate or lie about your organization's past accomplishments, pad the résumés of the project personnel, or list as project personnel workers who in fact will not be contributing to the project.
- *Product information.* This category includes everything from descriptions in sales catalogs to specification sheets and operating instructions and manuals.

As Bryan (1992) points out, most ethical dilemmas in any kind of product information arise because you are doing two things at once: describing and advertising. These two functions are not only different, but often in conflict. (Ironically, most advertisers maintain they are merely providing product information, even though most ads are light on facts.)

The two guidelines presented here cannot solve all the ethical problems and dilemmas you will face as a working professional. However, they can help you prevent many kinds of ethical problems.

- Tell the truth.
- Do not mislead.

Tell the Truth   Perhaps the simplest and strongest guideline is to tell the facts honestly and accurately. Employees report that sometimes they are asked to lie. Obviously, lying—knowingly providing inaccurate information—is unethical. If your company's own tests of the electric screwdriver it produces show that a charge lasts 50 minutes, but the competitor's model lasts 75 minutes, you might be pressured by a supervisor to simply lie: to say that your product's charge lasts 75 minutes. Your responsibility is to resist this pressure, by going over the supervisor's head, if necessary.

Do Not Mislead   Providing misleading information is a little more complicated than lying, but ethically it amounts to the same thing. A misleading statement or graphic enables or even encourages the reader to believe false information. For instance, a product-information sheet for a set of skis is misleading if the accompanying photograph shows bindings and poles that are not included with the skis. Some stores advertise computer systems at low prices, with the small print stating "monitor optional." The monitor isn't optional if you actually want to use the computer.

Information can mislead the reader in a number of other ways:

- *Scare tactics.* A document can wrongly suggest that a product—or a particular brand of the product—is necessary. For example, a flashlight manufacturer misleads in suggesting that only its own brand of batteries will power the flashlight.
- *Euphemisms.* If you need to fire someone, use a word such as *fire* or *release,* not *alternative employment facilitation.* See Chapter 15 for more information on euphemisms.
- *Clichés.* Communicators sometimes use such terms as *user friendly, ergonomically designed,* and *state of the art* to make the product sound better than it is. If you make a judgment about the product, back it up with accurate, specific information. Do not write "We carried out extensive market research" if all you did was make a few phone calls.
- *Ignoring negative features.* For instance, if an information sheet for a portable compact-disc player suggests that it can be used by joggers without mentioning that the bouncing will probably make it skip, the information sheet is misleading.

- *Legalistic constructions.* Avoid phrases that say one thing but actually mean another. For instance, it is unethical to write "The 3000X was designed to operate in extreme temperatures, from −40 degrees to 120 degrees Fahrenheit" if in fact the unit cannot reliably operate in those temperatures. The fact that the statement might actually be accurate—the unit was *designed* to operate in those temperatures—doesn't make it any less misleading.

- *Failing to acknowledge assistance from others.* Don't suggest that you did all the work yourself if you didn't. Cite your sources and your collaborators accurately and graciously.

EXERCISES

The following exercises call for you to write memos. See Chapter 16 for a discussion of memos.

1. Research an event that involved ethical matters, such as the Chernobyl nuclear accident, the *Exxon Valdez* oil spill, the *Challenger* tragedy, or an incident in which a news organization manipulated a film, video, or photograph. Write a memo to your instructor analyzing the ways in which elements of the incident represented a violation of ethical standards, and the ways in which the organization (or organizations) applied (or didn't apply) ethical standards in its public statements and actions in the aftermath of the incident.

2. Research your college or university's code of conduct for students, and write a memo to your instructor describing and evaluating it. Consider such questions as the following: How long is it? How comprehensive? Does it provide detailed guidelines or merely general statements? Where does the code appear? From your experience, does it appear to be widely publicized, enforced, and adhered to? Are there sources of information on campus that could provide information on how the code is applied on your campus?

3. Research the code of conduct that applies in the field you are studying. Write a memo to your instructor describing and evaluating the code. How effective do you think it would be in preventing ethical lapses and in punishing people who have knowingly violated its precepts? Can you obtain any information on how the code is in fact used within your field?

4. Find a newspaper or magazine ad that you feel contains untrue or misleading information. Write a memo to your instructor describing the ad and analyzing the unethical techniques. How might the information have been presented more honestly? Include a photocopy of the ad with your memo.

5. Find a newspaper or magazine article that you feel contains untrue or misleading information. Write a memo to your instructor describing the article and analyzing the unethical techniques. How might the information have been presented more honestly? Include a photocopy of the article with your memo.

6. Read the following case. In a memo to your instructor, respond to the questions at the end.

   The town of Acton, Ohio (population 6,500), like many other small communities in the Rust Belt, has suffered economically during the last decade. Much of its infrastructure is old and in need of repair, and the town has a shrinking tax base. Young people routinely leave the area after high school in search of better jobs.

   The main employer in Acton is Diversified Construction Materials, which employs over 1,000 people from Acton and surrounding communities. Like Acton, Diversified has known better times. Its products are known for their high quality, but foreign manufacturers and domestic manufacturers who have moved their production facilities to third-world countries are undercutting Diversified's prices and gaining market share.

   However, the Research and Development Department at Diversified has just formulated a new type of blown insulation that the company thinks will perform as well as fiberglass but beat its price. This new substance promises to be a major part of Diversified's highly regarded insulation products. A number of retailers have ordered large quantities of the new insulation based on Diversified's exhibits at trade shows and some preliminary advertisements in industrial catalogs.

As the head technical writer at Diversified, Susan Taggert oversees the creation of all the product information for the insulation. As she normally does in such cases, she gathers all the documentation from R&D and any other materials available in the company, which she will study before mapping out a strategy.

About one week into the project, Taggert discovers from laboratory notebooks that three of the seven technicians participating in the project experienced abnormally high rates of absence from work during the four months they spent developing the insulation. One of the three technicians requested to be transferred from the project at the end of the first month. His request was granted.

Calling Diversified's Personnel Department, Taggert learns that all three of the technicians complained of the same condition, bronchial irritation of varying degrees of severity, but that the irritation ceased two to three days after the last exposure to the insulation. Apparently some compound in the insulation, which the company physician could not immediately identify, affected some of the technicians who worked closely on it.

Taggert goes to the vice president of operations, Bill Mondale, who is in charge of introducing all new products. Taggert presents her information to Mondale and suggests that the company find out what is causing the bronchial irritation before it ships any of the product. Although the irritation does not appear to be serious, there are no data on the potential effects of long-term exposure to the insulation when used in houses or offices.

Mondale points out Diversified's tight deadline; delivery is scheduled in less than two weeks. Determining the cause of the irritation could take weeks or months and cost many thousands of dollars. Taggert points out the financial risks involved in selling a product that poses a health risk. Mondale responds that it is a risk the company will have to take, and adds that the product is in compliance with all applicable federal guidelines. The company has staked its reputation—and its third-quarter profits—on the insulation. He directs Taggert to proceed with the product literature as quickly as possible and not to spend any more time worrying about the health hazard.

What should Susan Taggert do? Why is this course of action preferable to other courses of action? Explain carefully the ethical implications of the course of action you recommend.

7. Read the following case. In a memo to your instructor, respond to the question at the end.

John Boorman manages the Technical Publications Department at Santa Barbara Equipment, a company that manufactures equipment used in the semiconductor industry. John supervises four technical communicators who produce the internal and external communication at Santa Barbara Equipment.

Santa Barbara Equipment, only three years old, was founded by Eugene Froom, who invented the two major products the company produces. Froom has molded the company in his image. He doesn't delegate authority, and he has no patience with standard business practices. After three years, the company still does not have a policies-and-procedures manual for new employees, nor is there a code of conduct. If Mr. Froom thinks you are very bright, he will hire you, despite—or even because of—your lack of business experience and sophistication.

The newest member of the Technical Publications Department, Lynn Stone, was hired largely because of her computer skills. She knows more about computers than anyone else in her department, and in her free time she loves to "surf the Internet," as she describes it.

Lynn is contributing to the department's latest project: online documentation to accompany Santa Barbara's new product. Her job is to help the other technical communicators put the information online. A week before the documentation is to be pressed onto CDs, John Boorman learns through a conversation with Lynn that the software she used in coding the information was downloaded off the Internet. Concerned about copyright questions, he asks her to find out everything she can about the software.

She learns the following: the software was made by a company called Visionary Software, which has no idea how it got onto the Internet. After checking the coding, Visionary's president tells Lynn that the software has been altered. He also tells her that he will refer the matter to the Visionary's legal counsel.

Lynn reports this information to John Boorman, who communicates it to Eugene Froom. Froom is furious.

What should John Boorman do? What is his obligation regarding Lynn's conduct? What is his responsibility to Eugene Froom, the company president? To Visionary Software? To Santa Barbara's customers? In writing your response, be sure to explain your reasoning and to evaluate the ethical and practical implications of each recommendation you make.

REFERENCES

Baumhart, R. (1968). *An honest profit: What business-men say about ethics in business.* New York: Holt, Rinehart and Winston.

Beauchamp, T. L., & Bowie, N. E. (1988). *Ethical theory and business* (3rd ed.). Englewood Cliffs, NJ: Prentice-Hall.

Behrman, J. N. (1988). *Essays on ethics in business and the professions.* Englewood Cliffs, NJ: Prentice-Hall.

Branscomb, E. S. (1993). Trademarks: Caveat scriptor. *Technical Communication, 40*(1), 97–99.

Brockmann, R. J., & Rook, F. (Eds.). (1989). *Technical communication and ethics.* Washington, DC: Society for Technical Communication.

Bryan, J. (1992). Down the slippery slope: Ethics and the technical writer as marketer. *Technical Communication Quarterly, 1*(1), 73–88.

Chalk, R. (1980). *AAAS professional ethics project: Professional ethics activities in the scientific and engineering societies.* Washington, DC: American Association for the Advancement of Science.

Forester, T., & Morrison, P. (1994). *Computer ethics: Cautionary tales and ethical dilemmas in computing.* (2nd ed.). Cambridge, MA: MIT Press.

Helyar, P. S. (1992). Products liability: Meeting legal standards for adequate instructions. *Journal of Technical Writing and Communication, 22*(2), 125–147.

Honor in science. (1986). New Haven: Sigma Xi. The Scientific Research Society.

Kozak, E. M. (1993). Copyright and fair use: A legal tug of war. *Editorial Eye, 16*(3), 2–3.

Swazey, J. P., Louis, K. S., & Anderson, M. S. (1994). The ethical training of graduate students requires serious and continuing attention. *Chronicle of Higher Education, 40*(27), B1–B2.

Velasquez, M. G. (1992). *Business ethics: Concepts and cases* (3rd ed.). Englewood Cliffs, NJ: Prentice Hall.

CHAPTER 3 Writing Collaboratively

You have probably noticed that many professors require collaborative writing in their courses. The popularity of collaboration on campus reflects a similar trend in the working world. This chapter focuses on the techniques of collaborative writing.

The chapter begins by describing different patterns of collaboration and the strengths and weaknesses of the technique. Next, the chapter discusses reducing conflict, improving listening skills, and understanding the role of gender differences and multiculturalism in collaboration.

In one sense, virtually every document is collaborative, for every writer uses information created, discovered, or interpreted by someone else. However, a useful description (Allen, Atkinson, Morgan, Moore, & Snow, 1987) is that collaborative writing

- leads to the production of a document
- involves substantive interaction among group members
- involves shared decision-making power over and responsibility for the document

Collaborative writing is common in organizations, and as documents and the techniques used to produce them become more complex, the amount of collaboration is likely to continue to increase. One survey found that 73.5 percent of 200 college-educated businesspeople collaborate in order to produce about a quarter of their documents (Faigley & Miller, 1982, p. 567). Another survey found that 87 percent of 520 professionals collaborate at least some of the time (Ede & Lunsford, 1990, p. 20). A third study of more than 400 professionals found that they often write collaboratively (Couture & Rymer, 1989, p. 78).

What kinds of documents do people write collaboratively? Everything from brief notes to books. However, the three kinds of documents that appear to be written collaboratively most often are proposals, memos, and reports (Ede & Lunsford, 1990, p. 63).

Collaboration exists in highly structured bureaucracies and more flexible organizations. Some organizations follow the *division-of-labor model,* in which the manager assigns roles, and then employees work separately and submit their products to the manager. Other organizations follow the *integrative model,* in which people work together cooperatively, sometimes forming and reforming their own groups (Killingsworth & Jones, 1989).

Patterns of Collaboration

Regardless of whether the organization uses the division-of-labor model or the integrative model, there are three different patterns of collaboration, as shown in Figure 3.1:

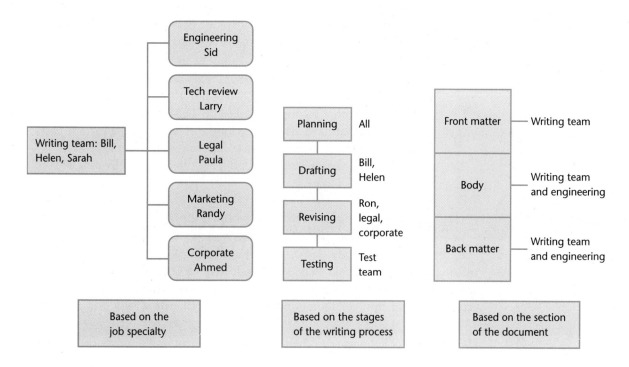

FIGURE 3.1

Patterns of Collaboration

- *Collaboration based on job specialty.* Perhaps the most common pattern is to divide up the work according to the expertise of the group members. For instance, an engineer is the subject-matter expert in charge of contributing all the technical information; a graphic artist is responsible for the graphics and the design; a technical communicator drafts the document.

- *Collaboration based on the stages of the writing process.* In many organizations, a great deal of collaboration occurs during the *prewriting* stage. Different technical professionals and technical communicators come together to plan a big document, such as a manual. They share ideas about the content, organization, and style of the document, and they establish a production schedule and an evaluation program.

  Collaboration during the *drafting* stage is much less common, largely because working together is much more time consuming than working individually. However, advances in groupware are making it more convenient to draft collaboratively. *Groupware* is computer software that lets several people at different locations work on the same document at the same time.

Collaboration during the *revising* phase is very common. Often the people who helped launch the project return to review the draft and make suggestions.

- *Collaboration based on the section of the document.* Sometimes one person takes responsibility for one section of the document, another person does another section, and so forth. This pattern is common for large projects that have discrete sections. For example, the various parts of proposals, which have standard components, such as the project description and the credentials section, are often written by different people.

Advantages and Disadvantages of Collaboration

Ede and Lunsford's study (1990, p. 50) found that 58 percent of writers considered collaborative writing to be very productive or productive, whereas 42 percent found it not very productive or not at all productive. As a technique for creating documents, collaboration has both strengths and weaknesses.

Advantages of Collaboration

Writers who collaborate can create a better document and improve the functioning of an organization:

- *Collaboration draws on a greater knowledge base.* Two heads are better than one, and three are better yet. The more people involved in planning and writing the document, the larger the knowledge base. This principle translates to a document that can be more comprehensive and more accurate than the product of a single author.
- *Collaboration draws on a greater skills base.* No one person can be an expert manager, writer, editor, graphic artist, and production person.
- *Collaboration provides a better sense of how the audience will read the document.* Working collaboratively involves having different people read drafts. Each person who reads and comments on a draft acts as an audience, raising questions and suggesting improvements that one person writing alone would not likely generate.
- *Collaboration enables people to share the responsibility for the document.* No one person should have to shoulder the responsibility for a 400-page manual or proposal.
- *Collaboration improves communication among employees.* Collaboration breaks down physical, organizational, and emotional barriers. People who work together for a long time get to know one another and learn about their jobs, responsibilities, and frustrations. A shared goal—putting together a professional-quality document that will reflect posi-

tively on everyone involved—can enable people to work together more productively.

- *Collaboration improves the socialization of new employees.* Collaboration helps acclimate new employees to an organization. They get to know people from different parts and different levels of the company. Working collaboratively helps new employees learn the procedures used in the organization: how to get things done, which people to see, what forms to fill out, and so forth. In addition, collaboration teaches the values of the company, such as the importance of shared goals and the willingness to work hard and sacrifice for an important initiative.

Disadvantages of Collaboration

Six principal problems are associated with collaboration:

- *Collaboration takes more time than individual writing.* It takes longer to create a document collaboratively than individually simply because of the need for the collaborators to communicate.
- *Collaboration can lead to groupthink.* When members of a group place a higher value on getting along with each other than on thinking clearly and critically about the subject, *groupthink* can be the result. Groupthink can lead to an inferior document because no one wanted to cause a scene by asking tough questions about the group's work.
- *Collaboration can yield a disunified document.* The document can end up disunified in content or style. For example, sections written by two people can contradict each other or contain unintended duplication. Or one section can refer to another section that never gets written. And the more people involved in the collaboration, the greater the variation of style, in everything from design to spelling. To prevent this sort of disunity, writers must plan and edit the document with extreme care.
- *Collaboration can lead to inequitable workloads.* No matter how hard the project leader tries to prevent it, some people will end up working much harder than others.
- *Collaboration can reduce the collaborators' sense of pride in the document.* If the project calls for many people to contribute relatively minor effort, they might miss the sense of satisfaction that writers generally feel. They can start to think of themselves as very small cogs in a very large wheel; as a result, the quality of the document can suffer.
- *Collaboration can lead to interpersonal conflict.* People can disagree about the best way to create the document or about the document itself. Such disagreements can poison the working relationship not only during the project but long after it is complete. Techniques for reducing conflict are discussed in more detail later in this chapter.

Organizing and Maintaining a Collaborative Group

Although collaborative groups take many shapes and function in many different ways, the following seven guidelines will help you set up and maintain a successful group:

- *Define your task.* In some groups, the first and most important guideline is to define your task. Your group members must agree on what you are supposed to accomplish. You have to reach consensus, for instance, that your task is to revise your company's policies-and-procedures manual, and that it has to be finished by April 10. If there are other requirements, such as that the page count not exceed 300, you should articulate them as well.

- *Choose a group leader.* In some groups, the first task is to choose a group leader. This person serves as the link between the group and management. In addition, the group leader is responsible for keeping the group on track, leading the meetings, and coordinating the communication among group members.

  Sometimes group leaders are not chosen by the group members but are assigned by management. For instance, companies that bid for contracts often have project managers whose job is to organize and supervise proposal-writing teams.

- *Define tasks for the group members.* In most groups, work is divided among the group members. As shown earlier in Figure 3.1, the division of labor can follow different patterns.

- *Establish working procedures.* To prevent problems later, establish working procedures to cover meetings, communication, and conflict resolution. Members need answers to the following questions—in writing, if possible—before the work of the group proceeds:

  – When do we meet?
  – Where do we meet?
  – What procedures will we follow in the meetings?
  – How are we to communicate with other members of the group, including the group leader?
  – How often are we to communicate?
  – What access will we have to people outside the group and to other resources?
  – What procedures are we to follow in resolving conflict?

- *Establish a work schedule.* The first thing you did was define your task, focusing on the "deliverable": the document you are to produce. If you know the date on which it is due, you have to work backward and create a schedule for the entire project: for a document to be shipped on April

10, it might have to be at Shipping by April 5 and at the printer by March 19. Each of these dates is called a *milestone*. Figure 3.2 shows a work-schedule form that can be used to assist in planning the schedule.

- *Create evaluation materials.* Group members have a right to know how they will be evaluated for their participation in the project. In some hierarchical organizations, the project manager evaluates the group members without communicating to them either the criteria or the results of the evaluation. If it is possible to make the criteria explicit, however, do so, for a more open process creates a better atmosphere and increases the chances that the group members will perform satisfactorily. Figure 3.3 on page 52 is a sample evaluation form that can be used by each group member to evaluate the contributions of other members of the group.

Some organizations and college instructors also use self-evaluation to help people assess their own performance in group projects. Figure 3.4 on page 53 is a self-evaluation form.

Notice that milestones are presented in reverse chronological order; the delivery-date milestone, for instance, comes first. In other forms, items are presented in normal chronological order.

The form includes spaces to list the person responsible for each of the milestones and progress reports, as well as spaces to state the progress of each milestone and progress report.

Name of Project:
Principal Reader:
Other Readers:
Group Members:

Milestones	Responsible Member	Status	Date
Deliver Document			
Proofread Document			
Send Document to Print Shop			
Complete Revision			
Review Draft Elements			
Assemble Draft			
Establish Tasks			

Progress Reports	Responsible Member	Status	Date
Progress Report 3			
Progress Report 2			
Progress Report 1			

Meetings	Agenda	Location	Date	Time
Meeting 3				
Meeting 2				
Meeting 1				

Notes

FIGURE 3.2

Work-Schedule Form

Your name_____   Date_____

Title of the project_____

Instructions

Use this form to evaluate the other members of your group. Write the name of each group member in one of the columns, then assign a grade of 0 to 10 (0 being the lowest grade, 10 the highest) for each group member for each criterion. Then total the grades for each member. Because each group member has different strengths and weaknesses, the scores you write will differ. On the back of this sheet, write any comments you wish to make.

Criterion	Group Members			
1. Regularly attends meetings	1._____	1._____	1._____	1._____
2. Is prepared at the meetings	2._____	2._____	2._____	2._____
3. Meets deadlines	3._____	3._____	3._____	3._____
4. Contributes good ideas in meetings	4._____	4._____	4._____	4._____
5. Contributes ideas diplomatically	5._____	5._____	5._____	5._____
6. Submits good work	6._____	6._____	6._____	6._____
7. Listens to other members	7._____	7._____	7._____	7._____
8. Shows respect for other members	8._____	8._____	8._____	8._____
9. Helps to reduce conflict	9._____	9._____	9._____	9._____
10. Your overall assessment of this person's contribution	10._____	10._____	10._____	10._____
Total Points				

FIGURE 3.3

Evaluation Form

- *Involve outsiders.* Groups benefit from the involvement of outsiders for the same reasons that individuals benefit from groups: the more perspectives the group has, the more successful the project will be. For this reason, groups often call in persons or organizations to assess the project periodically. For example, a group commonly submits its project plan to management for review and suggestions before beginning the project.

Your name _____   Date _____
Title of the project _____

Instructions

On this form, record and evaluate your activities on this project. In the Log
section, record the activities you performed as an individual; then record the
activities you performed as part of the group. For all activities, record the date
and the number of hours you spent. In the Evaluation section, write two brief
statements: one about the aspects of your contribution to the group that you
think were successful, and one about the aspects that you want to improve.

Log

Individual Activities	Date	Number of Hours

Activities as Part of Group	Date	Number of Hours

Evaluation

Aspects of My Participation That Were Successful
_____
_____
_____
_____
_____

Aspects of My Participation That I Want to Improve
_____
_____
_____
_____
_____

FIGURE 3.4

Self-Evaluation Form

Then, periodically, the group consults with outside subject-matter experts. At the end of the project, the group usually presents its work to subject-matter experts and management for review and approval.

Of course, you should also consult with other members of your group as much as possible. Before you articulate an idea in a group meeting, try it out on as many group members as you can privately; you will see their reactions to the idea and be able to refine it.

Reducing Conflict

Because collaborating on an important project is stressful, it can lead to interpersonal conflict. People can get frustrated and angry with each other because of personality clashes or disputes about the process of collaborating or about the substance of the project itself. But for the project to succeed, the group members must be able to work together productively. One technical communicator, quoted in Ede and Lunsford (1990, p. 66), sums it up this way: "No one here loses a job because of incompetence; they lose jobs if they can't work with others." Following are nine guidelines for establishing and maintaining good relationships while collaborating:

- *Arrive on time and be prepared to work.* Show up on time, have all your papers and other materials ready, and know what your goal is for the meeting. Everyone is late or unprepared once in a while, but conscientious professionals seem to get caught in fewer traffic problems than other people.

- *Meet your deadlines.* Another aspect of professionalism is to have your work ready on time. If you know you will not be able to, give as much warning as possible so that the group leader can plan around the problem and not waste the time of the other members.

- *Listen carefully.* It's a fact of life: most people would rather talk than listen. But in any group you ought to listen far more than you talk. Listen carefully, trying to understand and interpret what the speaker is saying. Listening will be discussed in more detail in the next section.

- *Ask pertinent questions.* To make sure you understand what is being said, ask questions. Don't be afraid that you will appear ignorant; the brightest people constantly try to understand what they hear and to make connections to other ideas. Asking pertinent questions helps not only you but also the other members of the group: it forces them to examine speakers' assumptions, evidence, and logic.

- *State your views diplomatically.* Experienced professionals earn respect by thinking well, not by bulldozing other people. You want to appear self-confident but not overbearing. Avoid personal remarks and insults, and avoid overstating your position. A modest qualifier such as "I think" or "it seems to me" is an effective signal to your listeners.

OVERBEARING: My plan is a sure thing; there's no way we're not going to kill Allied next quarter.

DIPLOMATIC: I think this plan has a good chance of success: we're playing off our strengths and Allied's weaknesses.

Notice, in the diplomatic version, the speaker's decision to call it "this plan" rather than "my plan."

PERSONAL: John, you're wrong about the export figures for 1994; we did $32 million, not $22 million.

BETTER: John, are you sure those export figures for 1994 are accurate? I thought I read in the February summary that we did $32 million, not $22 million.

The second version of this sentence is better than the first because it does not accuse John of being wrong; it merely questions his facts. Another advantage is that the second version states the source where the correct figures might be found, thus suggesting how the discrepancy can be resolved without making it a personal issue.

- *Don't get emotionally attached to your own ideas.* You cannot look at your own ideas completely objectively, but it is smart to try. When you meet opposition, try to understand why other members of the group don't share your views. Digging in is usually unwise—unless it is a serious matter of principle—because although you may be right and everyone else wrong, it's not likely.

- *Be supportive, not critical.* Be tolerant of other people's views and working methods. Show them respect and trust them. Doing so is right, both ethically and practically. It is ethical to treat people respectfully, as you want them to treat you. And it is practical: if you anger people, they will go out of their way to oppose you.

- *Don't hog the shared space.* In most groups, there is a shared work space: a workstation, a chalkboard, or a flip chart. Don't monopolize the shared space; let everyone have an opportunity to control it.

- *Share your resources with other members of the group.* The more resources everyone has, the more productive the group will be. Make sure you communicate to the other group members the resources—article, books, software, and so on—you have; they might be able to use those resources in ways that had not occurred to you.

Improving Listening Skills

Listening is not the same as hearing. *Hearing* consists of the physiological, sensory processes by which auditory sensations are received by the ear and transmitted to the brain. *Listening* is the psychological process of interpreting and understanding the content of the sensory experience.

Regardless of whether we are working in a collaborative group, listening is a critically important skill. According to Hulbert (1989, p. 3), 45 percent of the time we spend in communicating is devoted to listening, yet our listening skills are very poor. Right after hearing a 10-minute presentation, the average person has understood and remembered about half of what the speaker said; after 48 hours, that figure drops to about 25 percent.

The following discussion presents guidelines to improve your listening skills. The first four guidelines relate to the physical environment and the listener's physical presence:

- *Reduce distractions.* If the temperature is uncomfortable and the room is noisy, or if you're hungry or tired, you won't be able to listen effectively.
- *Face the speaker.* You will concentrate better if you face the speaker. Doing so will also help the speaker, for you will seem more alert and interested in listening.
- *Maintain eye contact.* Looking at the speaker's eyes helps you concentrate on the message. In addition, it gives the speaker more confidence and thus can improve the quality of the information.
- *Reduce irrelevant movements.* Pay attention physically. Don't tap a pencil on the table or play with your keys.

Here are three guidelines to follow as you listen to the speaker:

- *Set aside your preconceptions.* Too often we make up our minds before the speaker says a word. We react to the speaker's gender, age, race, personal appearance, or clothing. Or we decide that we already know what the speaker will say because we know that he or she is advocating environmental interests, say, or handgun interests.
- *Listen to the ideas, not the words.* One barrier to effective listening is to focus on the words rather than the meaning. Don't focus on pronunciation or accents; don't worry about the speaker's grammar errors. Instead, concentrate on the points the speaker is making.
- *Use the time productively.* Although people speak at about 150 words per minute, they think at about 500 words per minute. Speakers use the phrase "cruising down Route 350" to describe a listener drifting off because the speaker is talking so much more slowly than the listener is thinking. Use your time well: think about what the speaker is saying.

Examine the evidence and the logic. Compare the speaker's message with what you already know about the subject. Think of questions you might ask later.

Finally, follow these three guidelines if the speaker is talking with you alone or in a small group and you are permitted or expected to respond:

- *Don't interrupt.* If someone is speaking, listen.
- *Paraphrase what the speaker has said.* Use phrases such as the following: "So what you are saying is. . . ." Paraphrasing demonstrates to the speaker that you are listening carefully; in addition, it reduces the chances of misunderstanding.
- *Ask open questions.* If it is appropriate for you to ask questions, include open questions as well as closed ones. A *closed question* calls for a brief response, either yes/no or a simple fact. An *open question* encourages the speaker to elaborate.

CLOSED QUESTIONS:	"Did we mail the original or fax it?"
	"When did we do the last preventive maintenance on the roller?"
OPEN QUESTIONS:	"Why do you think we're getting these results?"
	"Where do you see the flat-panel market in five years?"

Gender and Collaboration

As this discussion has suggested, effective collaboration involves two related challenges: maintaining the group as an effective, friendly working unit and accomplishing the task. Scholars of gender and collaboration see these two challenges as representing the feminine and masculine perspectives.

Any discussion of gender studies must begin with a qualifier: when we talk about gender, we are generalizing, not talking about particular people; the differences in behavior between two men or between two women are likely to be greater than the difference between men and women.

The differences in how the sexes communicate and work in groups are traced to our traditional family structure. As the primary caregivers, women learn to value nurturing, connection, growth, and cooperation; as the primary breadwinners, men learn to value separateness, competition, debate, and even conflict (Chodorow, 1978).

For decades, scholars have studied the differences in speech between women and men. Women use more qualifiers and tag questions (such as "Don't you think?") (Tannen, 1990). However, some scholars suggest that women might be using these patterns because it is expected of them, and that they use these patterns mainly in groups that include men (McMillan, Clifton, McGrath, & Gale,

TECHNICAL WRITER RONDA WILSON ON COLLABORATION

on the value of collaboration in making documents

Collaboration is important, just to ensure technical accuracy. I try to talk to as many people as possible, and sometimes I get different information. Maybe the electrical person told you one thing in the beginning, and the mechanical person told you another, and they clearly contradict, so you have to get to the bottom of it. Sometimes collaboration actually exposes a real problem with the equipment that has to be fixed for it to go out.

on the challenge of working collaboratively

I think the biggest challenge is probably the writer's lack of technical knowledge. You're approaching people who are talking another language. In general, I do as much research as I possibly can before talking with subject-matter experts. A mechanical engineer is looking at you and saying, "I've got to explain something that's very technical and very complicated, and this person just isn't a technical person." The engineer either wants to give you the short easy answer that doesn't get you what you want, or wants to get too complex.

on the irony of this communication problem

The irony is, sometimes the person who is not an expert is closer in profile to the actual user of the equipment than the engineer is. The person who designed it is just not going to see things from a user's perspective, so writers without a technical background provide a useful point of view.

on ways to solve the communication problem

I've been to several training sessions on teamwork and collaboration, dealing with ways for different disciplines to see the other perspectives around them. But I really think common sense and politeness will take you most of the way. You try to establish a relationship with the different technical people.

on the value of politeness

These are busy people, so I try not to waste their time. I plan out what I'm going to ask them: this is the first question I'm going to ask, then this one, and we're going to go either from specific to general or from general to specific. Or I'm going to go over there with this diagram, or I'm going to give them a rough draft. Sometimes I will give them a list of questions and then set up an interview time. I think they appreciate a little advance notice, because sometimes you're asking them things that maybe aren't right at the tips of their tongues and they have to research a little bit, and it gives them a chance to do so. Also, they can do the research when they want, work that in whenever, take it home and look at it, and then set up an interview a day or two later.

on gender and collaboration

Most of the technical people are men, so there are times when I think that maybe I would do better if I weren't a woman. I guess, in general, gender does get involved, but once you've established a relationship with someone, gender becomes less important.

Ronda Wilson, a technical writer at Santa Clara Plastics in Boise, Idaho, has ten years' experience as a writer and editor for several engineering and high-tech companies.

1977). Many experts caution against using qualifiers and tag questions; for some listeners, qualifiers and tags suggest subservience and powerlessness.

The study of women and men in collaborative groups is a newer research subject, but as expected, women appear to value consensus and relationships, show more empathy, and demonstrate superior listening skills (Borisoff & Merrill, 1987); men appear to be more competitive and more likely to assume leadership roles.

Although many questions remain unanswered, it seems clear that women's communication patterns are more focused on maintaining the group, and men's on completing the task. For instance, women talk more about topics unrelated to the task (Duin, Jorn, & DeBower, 1991). But this talk is central to maintaining group coherence, which is necessary for the group to accomplish its task.

Scholars of gender recommend that all professionals strive to achieve an androgynous mix of the skills and aptitudes commonly associated with women and men. According to William Eddy (1983):

> Traditional male traits of task focus, objectivity, confrontation, and control are clearly important in many situations. . . . But to build and lead groups that attain effectiveness and viability by fully utilizing their human resources, you also need some of the traits traditionally thought of as female. . . . It is not surprising, when you think about it, why an androgynous combination of skills is best.

For an excellent review of the literature, see Lay (1994).

Multiculturalism and Collaboration

Most collaborative groups in industry and in the classroom include people from other cultures. The challenge for all people in collaborative groups is to understand the ways that cultural differences can affect the behavior of the group.

Some 70 percent of the world's cultures value the family, the community, and the corporation more highly than the individual (Thiederman, 1991). As Bosley (1993) points out, this fact would suggest that people from these cultures would be excellent participants in collaborative projects. But this inference is not necessarily true.

Collaborative groups in the United States are based on the premise that the individual should speak up. But people from other cultures often find it difficult to assert themselves in collaborative groups in the United States. In a study of college students from other cultures in collaborative groups, Bosley (1993, p. 57) lists several behaviors that differ from those of students from the United States. Students from other cultures

- may be unwilling to respond with a definite "no"
- may be reluctant to admit when they are confused or to ask for clarification
- may avoid criticizing others
- may avoid initiating new tasks or performing creatively

Even the most benign gestures of friendship from a U.S. student can cause confusion. If a U.S. student casually asks a Japanese student about her major and the courses she is taking, the Japanese student might find the question too personal—but find it perfectly appropriate to talk about her family and her religious beliefs (Lustig & Koester, 1993, p. 234).

This brief discussion of multiculturalism is meant only to point out the range and variety of cultural differences you are likely to encounter—or have encountered already. There is no substitute for immersing yourself in the culture you wish to learn about, but that is usually impractical. The best alternative is to try to remain open to encounters with people from other cultures, without jumping to conclusions about what their behaviors might mean.

A good first step is to read a full-length discussion of multiculturalism. The most respected scholar of multiculturalism in the United States is Edward T. Hall, whose major books include the following:

Beyond Culture. Garden City, NY: Anchor, 1976.

The Hidden Dimension. Garden City, NY: Doubleday, 1969.

Understanding Cultural Differences. Yarmouth, ME: Intercultural Press, 1990 (with Mildred Reed Hall).

For a detailed discussion of multiculturalism, see Chapter 4.

WRITER'S CHECKLIST

1. Has your group
   - ❏ defined your task?
   - ❏ chosen a group leader?
   - ❏ defined tasks for the group members?
   - ❏ established working procedures?
   - ❏ established a work schedule?
   - ❏ created evaluation materials?
   - ❏ involved outsiders?

2. To reduce conflict in collaborating, do you
   - ❏ arrive on time and prepared to work?
   - ❏ meet your deadlines?
   - ❏ listen carefully?
   - ❏ ask pertinent questions?
   - ❏ state your views diplomatically?
   - ❏ avoid digging in when the group rejects your idea?
   - ❏ act supportively, not critically?
   - ❏ share the work space?
   - ❏ share your resources?

3. To improve your listening skills, do you
   - ❏ reduce distractions?
   - ❏ face the speaker?
   - ❏ maintain eye contact?
   - ❏ reduce irrelevant movements?
   - ❏ set aside your preconceptions?
   - ❏ listen to the ideas, not the words?
   - ❏ use your listening time productively?
   - ❏ refrain from interrupting?
   - ❏ ask open questions?

EXERCISES

Several of the following exercises call for you to write a memo. See Chapter 16 for a discussion of memos.

1. Interview a professor in your department who has published a cowritten article. Ask this professor to describe the collaborative techniques used in writing the article and to evaluate the strengths and weaknesses of these techniques. Write a memo to your instructor that briefly summarizes the professor's main points and then analyzes the information you gathered. In what ways, and to what extent, do the professor's insights confirm the discussion in this chapter about collaboration?

2. Interview a technical communicator or technical professional in your community who has participated in a collaborative project. Ask this person to describe the collaborative process and to evaluate the strengths and weaknesses of the process. Write a memo to your instructor that briefly summarizes the technical communicator's main points and then analyzes the information you gathered. In what ways, and to what extent, do the technical communicator's insights confirm the discussion in this chapter about collaboration?

3. Keep a log of your experiences during a collaborative project. In a memo to your instructor, describe what techniques worked well, what problems the group encountered, how the group tried to solve them, and how successful the collaboration was.

4. Your local Chamber of Commerce, which is revising its tourism brochure, feels that your college or university should be described in more detail. It has asked that your team draft a 2,000-word description of the sites and activities on your campus that might attract tourists. (You may use three or four photographs to supplement the description.) With your collaborators, brainstorm, do the necessary research, outline, draft, and revise the description.

REFERENCES

Allen, N., Atkinson, D., Morgan, M., Moore, T., & Snow, C. (1987). What experienced collaborators say about collaborative writing. *Iowa State Journal of Business and Technical Communication, 1*(2), 70–90.

Bolton, R. (1979). *People skills.* New York: Simon & Schuster.

Borisoff, D., & Merrill, L. (1987). Teaching the college course in gender differences as barriers to conflict resolution. In L. B. Nadler, Nadler, M. K., & Todd-Mancillas, W. R. (Eds.), *Advances in gender and communication research* (pp. 351–361). Lanham, MD: University Press of America.

Bosley, D. (1993). Cross-cultural collaboration: Whose culture is it, anyway? *Technical Communication Quarterly, 2*(1), 51–62.

Chodorow, N. (1978). *The reproduction of mothering: Psychoanalysis and the sociology of gender.* Berkeley: University of California Press.

Couture, B., & Rymer, J. (1989). Interactive writing on the job: Definitions and implications of collaboration. In M. Kogen (Ed.), *Writing in the business professions* (pp. 73–93). Urbana, IL: National Council of Teachers of English.

Duin, A. H., Jorn, L. A., & DeBower, M. S. (1991). Collaborative writing—Courseware and telecommunications. In M. M. Lay & W. M. Karis (Eds.), *Collaborative writing in industry: Investigations in theory and practice* (pp. 146–169). Amityville, NY: Baywood.

Eddy, W. (1983). Qtd. in A. G. Sargent (1983), *The androgynous manager,* in J. Stewart, (1990). *Bridges not walls: A book about interpersonal communication* (5th ed.) (pp. 274–282). New York: McGraw-Hill.

Ede, L., & Lunsford, A. (1990). *Singular texts/plural authors: Perspectives on collaborative writing.* Carbondale: Southern Illinois University Press.

Faigley, L., & Miller, T. P. (1982). What we learn from writing on the job. *College English, 44,* 557–569.

Hulbert, J. (1989). Barriers to effective listening. *Bulletin of the Association of Business Communication, 52*(2), 3–5.

Killingsworth, M. J., & Jones, B. G. (1989). Division of labor or integrated teams: A crux in the management of technical communication? *Technical Communication, 36*(3), 210–221.

Lay, M. M. (1994). The value of gender studies to professional communication research. *Journal of Business and Technical Communication, 8*(1), 58–90.

Lustig, M. W., & Koester, J. (1993). *Intercultural competence.* New York: HarperCollins.

McMillan, J. R., Clifton, A. K., McGrath, D., & Gale, W. S. (1977). Women's language: Uncertainty or interpersonal sensitivity and emotionality? *Sex Roles, 3,* 545–549.

Tannen, D. (1990). *You just don't understand.* New York: William Morrow.

Thiederman, S. (1991). *Profiting in America's multicultural marketplace.* New York: Macmillan.

| CHAPTER 4 | # Analyzing Your Audience |

The content and form of every technical document you write are determined by the situation that calls for that document: your audience and your purpose. Understanding the writing situation helps you devise a strategy to meet your readers' needs—and your own.

This chapter covers how to analyze your audience. The discussion begins with how to understand the basic categories of readers—experts, technicians, managers, and general readers. Then the chapter explains how to analyze the individual characteristics of your readers. Finally, the chapter discusses how to address multicultural audiences.

Understanding the Writing Situation

The concepts of audience and purpose are hardly unique to technical communication; most everyday communication is the product of the same two-part environment. When a classified advertisement describes a job opening for prospective applicants, the writing situation of the advertiser is clearly identifiable:

Audience: prospective applicants

Purpose: to describe the job opening so that qualified persons will apply

Once you have defined the two basic elements of your writing situation, you must analyze each before deciding on the document's content and form.

Although you might assume that purpose would be your primary consideration, it's better to start by analyzing your audience, for examining its characteristics often improves your understanding of your purpose. The separation of audience and purpose is to some extent artificial—you cannot think about one without thinking about the other—but it is nonetheless useful. The more thoroughly you can define first your audience and then your purpose, the more exactly you will be able to tailor your document to the situation.

Purpose is discussed in detail in Chapter 5.

Primary and Secondary Audiences

Identifying and analyzing the audience can be difficult. You have to put yourself in the position of people you might not know, reading something you haven't yet started to write. Most writers want to concentrate first on the content: what they have to say. Feeling most comfortable with content, they want to write something first and shape it later.

Resist this temptation. If you have written mostly for teachers, you probably haven't had to think much about the audience. In most cases, you have some idea of what your teachers want to read, and you have received explicit guidelines and expectations for the assignments. And you realize that most teachers usually know more about the subject than their students do.

In business and industry, however, you will often have to write to different audiences, some of whom you will know little about and some of whom will know little about your technical field.

In addition, these different audiences may well have very different purposes in reading what you have written. Readers are often classified into two categories:

- A *primary audience* that consists of people who have a direct role in responding to your document. They might be people who use your information in doing their jobs. They might evaluate and revise your document, or they might act on your recommendations. An executive who decides whether to authorize building a new production facility is a primary reader. So is the treasurer who has to determine whether the organization can pay for it.

- A *secondary audience* that consists of people who need to know what is being planned, such as salespeople who want to know where a new facility will be located, what products it will produce, and when it will be operational. A secondary audience does not have a direct role in responding to your document.

As shown in Figure 4.1, the first step in analyzing your audience is to make general assumptions about them by classifying them into one of several basic categories. The second and third steps, to determine the individual characteristics of each reader you can identify and to create audience-profile sheets, are discussed in the next sections.

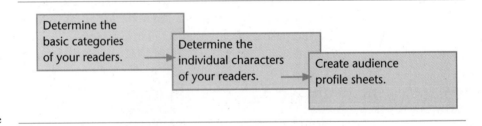

FIGURE 4.1

Analyzing Your Audience

Basic Categories of Readers

Although everyone is unique, a useful first step is to try to classify each reader on the basis of knowledge of the subject you are writing about. In general, every reader could be classified into one of four categories, as shown in Figure 4.2.

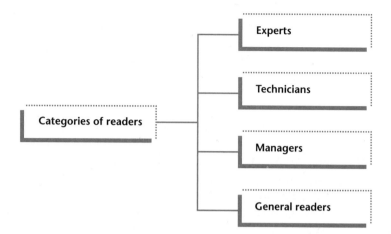

FIGURE 4.2

Basic Categories of
Readers

Of course, these categories are generalizations. For example, everyone is a general reader in almost every field outside of his or her own specialty. On the job, however, most people fit into one or perhaps two categories. Ellen DeSalvo, for example, a Ph.D. in materials engineering, would be an expert in that particular field, and perhaps she is also the manager of the Materials Group at her company. Off the job, she might be an expert on Native American folk art, having studied it seriously for many years as a hobby.

For this reason, classifying readers into categories is simply a first step in helping you understand how you will have to tailor a particular document to them. It is not a useful way to understand them as people. That part of audience analysis is discussed later in this chapter.

The Expert   The expert is a highly trained individual with an extensive theoretical and practical understanding of the field. Often the expert carries out basic or applied research and communicates those research findings. Here are some examples:

- a *physician* who is trying to understand how the AIDS virus works and who delivers papers at professional conferences and writes research articles for scholarly journals
- an *engineer* who is trying to devise a simpler and less expensive test for structural flaws in composite materials
- a *forester* who is trying to plan a strategy for dealing with the threat of forest fires

In short, almost everyone with a postgraduate degree—and many people with an undergraduate degree in a technical field—is an expert in one area. However, not everyone with a degree is an expert, and many experts have no formal advanced training.

Because experts share a curiosity about their subject and a more or less detailed understanding of the theory in their field, they usually have no trouble understanding technical vocabulary and formulas. Therefore, when you write to them, you can get right to the details of the technical subject, without spending time sketching in the fundamentals.

In addition, most experts are comfortable with long sentences, if the sentences are well constructed and no longer than necessary. Like all readers, experts appreciate graphics, but they can understand more sophisticated diagrams and graphs than most readers can.

Figure 4.3 is a passage from an article on how artificial intelligence can be used to produce engineering labor standards (Yazici, Benjamin, & McGlaughlin, 1994, p. 303). This passage illustrates the needs and interests of the expert reader.

The Technician

The technician has practical, hands-on skills. The technician takes the expert's ideas and turns them into real products and procedures. The technician fabricates, operates, maintains, and repairs mechanisms of all sorts, and sometimes teaches other people how to operate them. An engineer having a problem with an industrial laser will talk over the situation with a technician. After they agree on a possible cause of the problem and a way to try to fix it, the technician will go to work.

Like experts, technicians are very interested in their subject, but they know less about the theory. They work with their heads and their hands. Technicians have a wide variety of educational backgrounds; some have a high school education, while others have attended trade schools or earned an associate's degree or even a bachelor's degree. When you write to technicians, keep in mind that they do not need complex theoretical discussions. They want to finish a task safely, effectively, and quickly. Therefore, they need schematic diagrams, parts lists, and step-by-step instructions. Most technicians prefer short or medium-length sentences and common vocabulary, especially in documents such as step-by-step instructions.

Figure 4.4 on page 68, an excerpt from the installation instructions ("Tradeline," 1981) for a heating thermostat, illustrates the needs and interests of the technician.

The Manager

The manager is harder to define than the technical person, for the word *manager* describes what a person does more than what a person knows. A manager makes sure an organization operates smoothly and efficiently. For instance, the manager of the procurement department at a manufacturing plant sees that raw materials are purchased and delivered on time so that production will not be interrupted. The manager of the sales department of that same organization sees that salespeople are out in the field, creating interest in the products and follow-

The discussion assumes a strong background in the relationship between artificial intelligence (AI) and the manufacturing process. The general reader would be unable to follow this discussion because of its highly technical vocabulary ("automated knowledge acquisition tools," for instance).

The discussion refers frequently to the published literature on the subject; expert readers would be familiar with some—or all—of the literature cited here. (The bracketed numbers 1, 2, and 25 are citations.)

The discussion uses fairly long sentences, not because the writers want to show off but because the material is complicated.

The discussion refers to a fairly complex graphic.

Further implementation of AI-based tools can be found in manufacturing control. For instance, in a study done for the Department of Defense, an automated knowledge acquisition tool was designed to evaluate the hardness of weapon systems. The tool, called KNACK, interacts with the design engineers and program managers and automatically generates the knowledge base code to evaluate the performance of the weapon system. KNACK evaluates whether the weapon system is "hard" in relation to the effect of a specific nuclear weapon being investigated [2]. Examples of the knowledge acquisition tools to assist the task analysis can be found in [1] and [25].

These studies showed the importance of AI techniques for production planning and control. The transfer of knowledge can be facilitated by the use of knowledge-based systems that can result in improved production. The AI-based approach can enhance the knowledge transfer within the organization and provide better planning and control of products and resources.

Automated generation of shop-floor job descriptions and relevant labor time estimates can be a valuable alternative to the traditional methods of developing engineering labor standards. Fig. 1 presents a conceptual framework for incorporating AI-based techniques in this domain. Data from the shop floor on parts, machines and processes are processed to establish the method, to generate the necessary job elements, to compute times associated with the job elements, and to estimate standard job times for the application and provide a complete job detail sheet. An AI-based module composed of automated knowledge acquisition and codification can assist in better integrating the functions of methods engineers and computer programmers, and in reducing delays and conflicts between engineers and programmers.

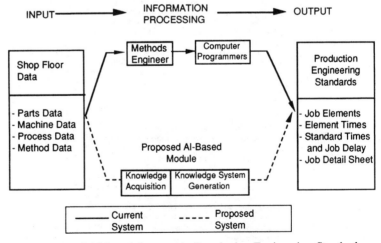

Fig. 1.   Role of AI-based Systems in Developing Engineering Standards.

FIGURE 4.3

Writing Addressed to an Expert Audience

TRADELINE

T822D HEATING THERMOSTAT

Notice that point 3 in the installation instructions explicitly states, "Installer must be a trained, experienced service technician."

The writers have used design features—for instance, caution boxes and numbered steps—to help the reader install the thermostat safely and effectively.

The instructions for mounting and adjusting the thermostat call for knowledge, experience, and even some tools that the general reader does not possess.

APPLICATION

The T822D 2-wire, mercury switch thermostat provides low voltage control of heating systems. The T822D features an adjustable heat anticipator.

INSTALLATION

WHEN INSTALLING THIS PRODUCT . . .

1. Read these instructions carefully. Failure to follow them could damage the product or cause a hazardous condition.

2. Check the ratings given in the instructions and on the product to make sure the product is suitable for your application.

3. Installer must be a trained, experienced service technician.

4. After installation is complete, check out product operation as provided in these instructions.

CAUTION

Disconnect power supply before beginning installation to prevent electrical shock or equipment damage.

LOCATION

Locate the thermostat on an inside wall about 5 ft [1.5 m] above the floor in an area with good air circulation at average temperature.

Do not mount the thermostat where it may be affected by—

—drafts, or dead spots behind doors and in corners.
—hot or cold air from ducts.
—radiant heat from the sun or appliances.
—concealed pipes and chimneys.
—unheated areas behind the thermostat.

WIRING

Disconnect power supply before beginning installation to prevent electrical shock or equipment damage. All wiring must comply with local codes and ordinances. See Figs. 2-5 for internal schematic and typical connection diagrams.

For new installations, run low voltage thermostat wire to the chosen location. For replacement applications, check the old thermostat wires for frayed or broken insulation. Replace any wires in poor condition.

MOUNTING

The T822D Heating Thermostat is designed to be mounted on a wall or vertical outlet box.

1. Grasp the thermostat cover at the top and bottom with one hand. Pull outward on the top of the thermostat cover until it snaps free of the base. Remove the red plastic shipping pin from the thermostat.

2. Pull about 4 in. [101.6 mm] of wire through the wall or into the outlet box. Attach the thermostat wires to the appropriate terminals on the back of the thermostat base (Fig. 1). Push any excess wires back through hole and plug any opening to prevent drafts that may affect thermostat performance.

3. Set the adjustable heat anticipator. See SETTING, Heat Anticipator.

4. Fasten the thermostat to the wall or outlet box with a screw through the top mounting hole (Fig. 1).

NOTE: It may be necessary to move the set point lever to uncover the mounting hole.

5. Place a spirit level across the top of the thermostat. Adjust the thermostat until it is leveled. Start a screw in the center of the bottom mounting hole.

IMPORTANT

These thermostats are calibrated at the factory mounted at true level. Any inaccuracy in leveling will cause control deviation.

6. Recheck thermostat leveling and tighten the mounting screws.

7. Replace the thermostat cover.

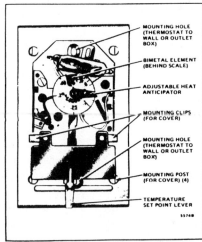

Fig. 1—Internal view of the T822D Heating Thermostat.

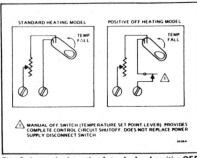

Fig. 2—Internal schematic of standard and positive OFF T822D.

L.M.
Rev. 8-81

Form Number
60-0313—5

FIGURE 4.4

Heating Thermostat
Installation Instructions

Reproduced by permission of Honeywell, Inc.

ing up leads. In other words, managers coordinate and supervise the day-to-day activities of the organization.

Upper-level managers, known as executives, address longer-range concerns. They foresee problems years ahead by considering questions such as the following:

- Is current technology at the company becoming obsolete?
- What are the newest technologies?
- How expensive are they?
- How much would they disrupt operations if they were adopted?
- What other plans would have to be postponed or dropped altogether?
- When would the conversion start to pay for itself?
- What has been the experience of other companies that have adopted the new technologies?

Executives are concerned with these and dozens of other broad questions that go beyond the day-to-day managerial concerns.

Because management is a popular college major, many managers today have studied general business, psychology, and sociology. Often, however, managers start out in a technical area. An experienced chemical engineer, for instance, might manage the engineering division of a consulting company. Although he has a solid background in chemical engineering, he earned his managerial position because of his broad knowledge of engineering and his ability to work well with colleagues. In his daily routine he might have little opportunity to use his specialized engineering skills.

Although generalizing about the average manager's background is difficult, identifying the manager's needs is easier. Managers want to know the bottom line. They have to get a job done on schedule; they don't have time to study a theory the way an expert does. Rather, managers must juggle constraints—financial, personnel, time, and informational—and make logical and reasonable decisions quickly. And they have to communicate with their own supervisors.

When you write to a manager, try to determine his or her technical background; then choose an appropriate vocabulary and sentence length. Regardless of the individual's background, however, focus on the practical information the manager will need. For example, if you are a research-and-development engineer who is describing a new product line to the sales manager, you might begin with some theoretical background so that the sales representatives can communicate effectively with potential clients. For the most part, however, you should concentrate on the product's capabilities and its advantages over the competition.

If you know that your reader will take your information and use it in a document that will be addressed to higher-level managers or executives, make your reader's job easier. Include an executive summary (see Chapter 11) and use frequent headings (Chapter 14) to help your reader see the major points you are making. Ask your reader if there is an organizational pattern, a format, or a strategy for writing the document that will help him or her in using your document as source material.

Figure 4.5, an executive summary (D'Ottavi, 1993) from a report entitled "Manufacturing Bidirectional Valves: A Feasibility Study," illustrates the interests and needs of the manager.

This executive summary provides little technical information about the technology—how the valves work—and concentrates on what the manager cares about: the demand for the product, the costs of researching and developing it, and the market value for it.

The writer uses a simple vocabulary; the only term that the general reader might not understand is "break-even analysis," which the intended reader would surely understand.

The writer concludes with a clear recommendation.

Executive Summary

Advances in biomechanical engineering have created a demand for small, quick-acting, bidirectional hydraulic valves. Presently, no valves exist commercially that can satisfy bidirectional flow requirements. Experts speculate that federal and private funding for biomechanical projects will exceed $2 million by 1996. Other specialty valves similar to bidirectional valves have proven to be highly patentable, resulting in fifteen years or more of exclusive rights to manufacture and sell. Comparable valves can cost between $100 and $300.

This report describes an investigation that examined the demand, the production costs, and the market value for bidirectional valves.

Presently, the demand for these valves is low, due to the fairly limited number of applications. In addition, their future use, although expected to be high, is uncertain at this time.

The initial investment to research and develop bidirectional valves is estimated to be $42,500. The cost to manufacture the valves is estimated at $81 each.

The estimated market value is $200 each. According to a break-even analysis, revenue from 360 valves would equal the initial investment costs plus the costs per value. These estimates are very conservative, since the actual production costs would be significantly lower if a high volume of valves were manufactured.

Because bidirectional valves are versatile (they can also be used and sold as unidirectional valves), investment in the research and development of bidirectional valves would be a relatively low risk. In addition, the initial investment costs are no higher than those for a unidirectional valve. Considering that technological development of devices that use bidirectional valves could increase dramatically, a valve prototype available for customer testing and evaluation could prove highly profitable.

I therefore recommend that we research and develop bidirectional valves. However, because of the uncertain future demand for them, I recommend delaying their manufacture.

The General Reader   Often you will have to address the general reader, sometimes called the layperson. A nuclear scientist reading about economics is a general reader, as is a homemaker reading about new drugs used to treat arthritis.

The layperson reads out of curiosity or self-interest. The typical article in the magazine supplement of the Sunday paper—for example, on attempts to increase the populations of endangered species in zoos—will attract the general reader's attention if it seems interesting and well written. The general reader may also seek specific information that will bring direct benefits: someone interested in buying a house might read articles on new methods of alternative financing.

In writing for a general audience, use a simple vocabulary and relatively short sentences when you are discussing subjects that might be confusing. Translate jargon into standard English idiom. Use analogies and examples to clarify your discussion. Discuss the human angle—how the situation affects people. Sketch in any special background—historical or ethical, for example—so that your reader can follow your discussion easily. Concentrate on the implications for the general reader. For example, in discussing a new substance that removes graffiti from buildings, focus on its effectiveness and cost, not on its chemical composition.

Figure 4.6 (Schwartz, 1994) on page 72, the beginning of a newspaper article addressed to the general reader, illustrates the interests and needs of the general reader.

Table 4.1 summarizes the previous discussion of how to write to the basic categories of readers.

TABLE 4.1	*Basic Categories of Readers*	
Audience	**Reasons for Reading**	**Guidelines for Writing**
EXPERT	To gain an understanding of the theory and its implications	Include theory, technical vocabulary, citations, and sophisticated graphics.
TECHNICIAN	To gain a hands-on understanding of how something works or how to carry out a task	Include graphics. Use common words, short sentences, and short paragraphs. Avoid excessive theory.
MANAGER	To learn the bottom-line facts to aid in making decisions	Focus on managerial implications, not technical details. Use short sentences and simple vocabulary. Put details in appendices.
GENERAL READER	To satisfy curiosity and for self-interest	Use short sentences and paragraphs, human appeal, and an informal tone.

Dawn Price is telling a joke about three men on a train—a Cuban, a Russian, and an American. The joke relies on timing, body language, and audience rapport.

The difference here is that Dawn, who is 12, is deaf, as are the three schoolmates she is entertaining. What's more, her schoolmates cannot see her directly. They are watching a personal computer screen showing an outlined image of Dawn performing sign language.

Dawn and her friends are communicating through a sign language telephone that is being developed at the A. I. duPont Institute, a children's hospital in Wilmington, Delaware.

The system is necessary because sign language requires fluid motion, so much so that deaf people have difficulty understanding the jerky pictures that are transmitted by even the most expensive video phones.

"We look not for the quality of the picture, but for the quality of the movement," said Richard A. Foulds, director of the University of Delaware's Applied Science and Engineering Laboratories, who heads the development of the project.

The sign language telephone operates with computers, not telephones. It is a custom-designed circuit board that fits into an IBM-compatible PC. Users need a video camera or camcorder, which can be plugged into the back of the system. As the camera takes in images of a person performing sign language, the circuit board strips out unimportant information.

The only elements retained are the "luminance valleys," which correspond to the edges of a person's fingers, clothing, and facial features.

The result is a black-and-white outlined drawing of the person. Mr. Foulds said transmitting only the essential elements of an image—so-called "edge detection"—reduces the number of bits, or pieces of information, in the original image 3,500 times.

The passage focuses on the human aspects of the phenomenon, not on the technical aspects. It concentrates on the girl, Dawn Price, and the researcher, Richard Foulds.

The passage uses an informal tone, including idiomatic phrases and contractions ("What's more" in paragraph 2).

The passage requires no previous knowledge about the subject.

The passage gradually introduces technical vocabulary, such as "luminance valleys" and "edge detection." A technical term—*bits*—is defined in a parenthetical definition.

FIGURE 4.6

Writing Addressed to the General Reader

Individual Characteristics of Readers

Classifying your readers into categories is a good first step, but it is a very imprecise process. Sometimes it's difficult to be more precise because you don't know who the readers will be. You might be writing a report for your supervisor, who you know is likely to distribute it, but you don't know to whom. Or you might be addressing an audience of several hundred or even several thousand people who share some characteristics but are not a unified group.

Still, it is a good idea to try to find out as much information as you can about the individual characteristics of your readers. Figure 4.7 shows nine

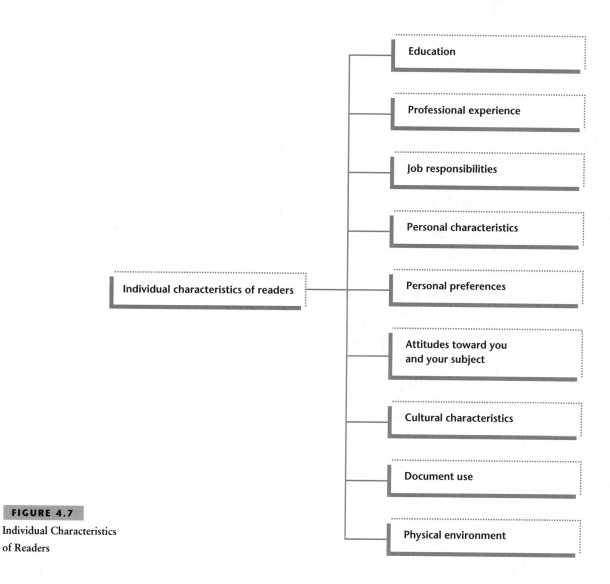

FIGURE 4.7
Individual Characteristics
of Readers

aspects to consider for each person who you know will be getting a copy of
your document.

Education   Think not only about the degree the person has earned, such as a bachelor of
science in nutrition; think also about what kinds of courses, and concentrations
of courses, the person has taken. For example, a metallurgical engineer who
earned a B.S. in 1971 has a much different technical background from the person who earned the same degree in 1995, because the field has changed so much
in recent years.

Don't forget, too, formal and informal course work on the job; a reader who has not studied business in college might have studied it in on-site workshops or training programs.

Knowing your readers' educational backgrounds will help you determine how much supporting material to provide, what level of vocabulary to use, what kind of sentence structure and length to use, what type and number of graphics to include, and whether to provide such formal elements as a glossary or an executive summary.

How do you discover your readers' educational backgrounds? It isn't easy. You cannot ask people to send you their current résumés. But you can try to learn as much as possible in conversation with your colleagues. The longer you stay at your organization, and the more you talk with people, the more you will learn.

Professional Experience

Although formal education in school and on the job is significant, a person's professional experience is equally important. As a person's professional career progresses, of course, work experience becomes even more critical than formal education. A reader who has been out of school for a decade or more likely has had a variety of professional experiences. A nurse might have represented her hospital on a community committee to encourage citizens to give blood, might have worked with the hospital administration to choose vehicles for the emergency medical staff, and might have contributed to the planning for the hospital's new delivery room or hospice. In short, her range of experience might have provided several areas of competence or even expertise.

If one of your readers has experience in a subject related to the one you are writing about, you might want to adjust the content and style of your document accordingly. In addition, you should ask that person about tactics to try or to avoid, both in carrying out the project and in writing about it.

Job Responsibilities

Learn as much you can about your readers' job responsibilities. If you are writing a document for internal distribution, take a copy of your company's organization chart and circle the boxes that represent your readers. This process will give you a clear idea of your readers' different technical and managerial areas of responsibility—useful information as you plan the content, organization, and style of the document. Write down in a sentence the major job responsibility for your primary reader. Then think about how your document will help that person accomplish it. For example, if you are writing a feasibility study of several means of cooling the air for a new office building, you know that your reader—an upper-level manager—will have to worry about electricity or other utility costs over the lifetime of the cooling system. Therefore, in your report you need to make clear how you are estimating utility costs in the future.

Personal Characteristics

Do not overlook your readers' personal characteristics. Do the age and gender of your readers tell you anything important about how they will read and interpret your document? Can you draw inferences about your readers' knowledge of and attitudes toward your topic? Do your readers have any other personal characteristics you should consider, such as vision impairment, that would affect the way you write and design your document?

Personal Preferences

In addition to personal characteristics, people have individual tastes and biases: some rational and reasonable, some not. One person hates to see the first-person pronoun *I*. Another finds the word *interface* distracting when the writer isn't discussing computers. A third reader appreciates a descriptive abstract on the title page of documents; a fourth relies on executive summaries. One good way to find out a person's preferences is to read that person's own documents. With careful study, you will find that everyone's writing is as distinct as a fingerprint.

Common sense dictates that you accommodate as many of your readers' preferences as you can. Sometimes you can't, of course, either because the preferences contradict one another or because the special demands of the subject won't permit you to accommodate them. But try to avoid alienating or distracting your readers.

Attitudes toward You and Your Subject

Try to learn as much as you can about your readers' attitudes toward you and your subject.

- *Attitudes toward you.* If things are going your way, most people will like you because you are hardworking, intelligent, and cooperative. However, some people will dislike you because you are hardworking, intelligent, and cooperative. Why is this? I don't know.

  If a reader's animosity toward you is irrational or unrelated to the current project, you can do little other than try to repair the damage. You can try to earn that person's respect and trust by meeting him or her on some neutral ground, perhaps by discussing other, less volatile projects or some shared interest: gardening, skiing, science-fiction novels.

- *Attitudes toward your subject.* If possible, discuss the subject thoroughly with your primary readers to determine whether they are positive, neutral, or negative. Figure 4.8 on page 76 explains basic strategies for responding to a particular type of reader attitude.

Cultural Characteristics

What you know about your readers' cultural characteristics can help you appeal to their interests and avoid confusing or offending them. As discussed later in this chapter, cultural characteristics can affect virtually every aspect of a reader's

If . . .	Do this . . .
Your reader is neutral or positively inclined toward your subject	• Write the document so that it responds to the needs of your readers; make sure the vocabulary, level of detail, organization, and style are appropriate for them.
Your readers are hostile to the subject or your approach to it	• Try to find out what the objections are and then answer them effectively. Explain clearly why the objections are either not valid or not as important as the benefits. • In addition, you might organize the document so that the actual recommendation follows the explanation of the benefits. This strategy encourages the hostile reader to understand your argument rather than reject it out of hand. • As you make your case, avoid describing the subject as a dispute. Seek areas of agreement and concede points. Avoid trying to persuade your readers; people don't like to be persuaded, because it threatens their ego. Instead, suggest that there are new facts that need to be expressed. People are less reluctant to change their minds when they realize that there are additional facts to be considered.
One of your readers was instrumental in creating the policy or procedure that you are arguing is ineffective	• Be diplomatic when you discuss the shortcomings of the present system, but be especially careful if there is a chance of offending one of your readers. People can be very defensive about their ideas, and sometimes they are most defensive about their worst ideas. You don't want someone to dig in and do everything to oppose you just because you were tactless. • When you address such a reader, don't write, "The present system for logging customer orders is completely ineffective." Instead, write, "While the present system has worked well for many years, new developments in electronic processing of orders might enable us to improve the speed and reduce the errors substantially."

FIGURE 4.8

Responding to Your
Readers' Attitudes
toward Your Subject

comprehension of a document and perception of the writer. If you are writing to a multicultural audience, you'll want to modify your vocabulary, sentence length and structure, and use of graphics.

Document Use   What will readers do with the document? Will they

- file it?
- skim it?
- read only a portion of it?
- study it carefully?
- modify it and submit it to another reader?
- attempt to implement recommendations?
- use it to perform a test or carry out a procedure?
- use it as a source document for another document?

Just as you have a purpose in writing the document, each reader will have a purpose in reading it. If only 1 of 15 readers needs detailed information, you must provide it, but you don't want to make the other 14 people wade through it. Determine some way to make the details available, but clearly set them off from the rest of the document; an appendix might do the job. If you know that your reader wants to use your status report as raw material for a report to a higher-level reader, try to write your report so that it requires little rewriting. You might use your reader's own writing style, and even submit the computer disk or put it out on the network so that your work can be merged with the new document without retyping.

Physical Environment

Technical documents are often formatted in a special way or constructed of special materials to improve their effectiveness in particular physical environments. For example, documents used in poorly lit places are printed with a type larger than normal. Some documents will be used on ships or on aircraft or in garages, where they might be exposed to wind, salt water, and grease. Special waterproof bindings, oil-resistant or laminated paper, coded colors, unusual-sized paper—these are just some of the special formats and materials that you might have to consider. A user's manual for a computer system should be bound so that it can lie flat on a table for a long period. See Chapter 13 for a more detailed discussion of how to design a document for use in different environments.

Figure 4.9 summarizes the discussion of the individual characteristics of readers.

FIGURE 4.9

Questions to Ask about the Individual Characteristics of Readers

Characteristic	Questions to Ask about Your Readers
Education	• What do the readers know about the subject from their formal or informal education?
Professional experience	• What do the readers know about the subject from their previous experience on the job?
Job responsibilities	• What are your readers' major areas of technical and managerial responsibility?
Personal characteristics	• What is the age and gender of your readers? • Do they have any disabilities that affect how they read?
Personal preferences	• Do your readers have any personal preferences or biases about writing that you should consider?
Attitudes toward you and your subject	• Do your readers have any attitudes toward you that you should consider? • Do your readers have any attitudes toward your subject or your approach to it that you should consider?
Cultural characteristics	• Do your readers have any cultural characteristics that you should consider as you create text or graphics?
Document use	• What do you know about how your readers will be using your document that will help you decide how to write or deliver it?
Physical environment	• What do you know about the physical environment in which your readers will use the document that will help you design or produce it?

The Audience Profile Sheet

To help you analyze your audience, you might create an audience profile sheet. You could then fill out the sheet for each primary reader and each secondary reader (but only if there are a few).

To understand how to use the audience profile sheet, assume you work in the drafting department of an architectural engineering firm. You know that the company's CAD equipment is out of date and that recent CAD technology would make it easier and faster for the draftspeople to do their work. You want to persuade your company to authorize the purchase of a CAD workstation that costs about $4,000.

Your primary reader is Harry Becker, the manager of the Drafting/Design Department. Your secondary reader is Tina Buterbaugh, manager of the Finance Department.

Figures 4.10 and 4.11 on pages 79 and 80 show two audience profile sheets: one for Harry Becker and one for Tina Buterbaugh.

Accommodating the Multiple Audience

You will be writing to a multiple audience far more often than writers did even a couple of decades ago, for two reasons:

- *The knowledge explosion.* Two decades ago, there was little difference between writing to a single reader and writing to multiple readers. Most readers, whether technical people or managers, shared a basic understanding of the organization's product or service. In fact, most managers had risen from the technical staff. Today, however, the knowledge explosion and the increasing complexity of business operations have created a fairly wide gap between what the technical staff knows and what the managers know.

- *New technology for communication.* The photocopy machine has made it simple and inexpensive to send copies of most documents to dozens of people. Electronic mail makes it even simpler and less expensive.

If you think your document will have a number of readers, consider making it *modular*. That is, break it up into different components addressed to different kinds of readers. A modular report, for example, might contain an executive summary for the managers who don't have the time, the knowledge, and the desire to read the whole report. It might also contain a full technical discussion for expert readers, an implementation schedule for technicians, and a financial plan in an appendix for budget officers.

Audience Profile Sheet

1. Reader's name and job title.

   Name <u>Harry Becker</u>   Job Title <u>Manager, Drafting/Design Dept.</u>

2. Kind of reader   Primary <u>x</u>   Secondary _____

3. Reader's educational background.

   Formal education <u>BS, Architectural Engineering, Northwestern, 1985</u>

   Training Courses and Workshops <u>CAD/CAM Short Course 1988; Motivating Your Employees</u> <u>Seminar, 1989; Writing on the Job Short Course, 1990.</u>

4. Reader's professional background (previous positions or work experience).
   <u>Worked for two years in a small architecture firm. Started here 10 years ago as a draftsperson.</u> <u>Worked his way up to Assistant Manager, then Manager. Instrumental in the Wilson project,</u> <u>particularly in coordinating personnel and equipment.</u>

5. Reader's chief responsibilities on the job. <u>Supervises a staff of 12 draftspersons. Approves or</u> <u>disapproves all requests for capital expenditures over $1,000 coming from his dept. Works with</u> <u>employees to help them make the best case for the purchase. After approving or disapproving</u> <u>the request, forwards it to Tina Buterbaugh, Manager, Finance Dept., who maintains all capital</u> <u>expenditure records.</u>

6. Reader's personal characteristics. <u>N/A</u>

7. Reader's likes. <u>Straightforward, simple documents, lots of evidence, clear structure.</u>

   Reader's dislikes. <u>Long, complicated documents full of technical terms.</u>

8. Reader's attitude toward you and the subject of the document.
   Positive <u>x</u>   Neutral _____   Negative _____
   Why? In what ways? <u>Strongly in favor of the request, but he knows that his department has</u> <u>authorized a lot of computer expenditures, and is skeptical because of an unwise purchase last</u> <u>year. His attitude is, if someone wants to buy it, let them make a good case for it.</u>

9. Reader's cultural characteristics. <u>Nothing unusual.</u>

10. How the reader will use the document.
    Skim it _____   Study it <u>x</u>
    Read a portion of it _____   Which portion? _____
    Modify it and submit it to another reader <u>x</u>
    Attempt to implement recommendations _____
    Use it to perform a task or carry out a procedure. _____
    Use it to create another document. _____
    Other? _____ Explain _____

11. Reader's physical environment. <u>Normal corporate environment.</u>

FIGURE 4.10
Audience Profile Sheet

79

Audience Profile Sheet

1. Reader's name and job title.

   Name <u>Tina Buterbaugh</u>   Job Title <u>Manager, Finance Dept.</u>

2. Kind of reader.   Primary _____   Secondary <u>x</u>

3. Reader's educational background.

   Formal education <u>B.S., Finance, USC, 1982; MBA, Wharton, 1987.</u>

   _____

   Training Courses and Workshops <u>Don't know, but she is on the training and development</u>
   <u>committee.</u>

4. Reader's professional background (previous positions or work experience).
   <u>Not sure, but I know she worked for 6 years as Director of Finance for a large retailer.</u>

5. Reader's chief responsibilities on the job. <u>In general, in charge of long-range financial planning. In</u>
   <u>this case, she records all capital expenditures and, at the end of the year, will decide the</u>
   <u>department's capital budget for next year.</u>

6. Reader's personal characteristics. <u>NA</u>

7. Reader's likes. <u>Same as Harry Becker.</u>

   Reader's dislikes. <u>Long, complicated documents full of technical terms.</u>

8. Reader's attitude toward you and the subject of the document.
   Positive _____   Neutral <u>x</u>   Negative _____

   Why? In what ways? <u>She has no feelings about proposals this small ($4,000), and she knows Harry</u>
   <u>Becker is good at forcing the writer to make a good case, but the company has spent a lot lately on</u>
   <u>computers. She expects Harry to ask for a bigger capital budget next year.</u>

9. Reader's cultural characteristics. <u>I know she lived in Paris for five years and has traveled extensively.</u>

10. How the reader will use the document.
    Skim it <u>x</u>   Study it _____
    Read a portion of it _____   Which portion? _____
    Modify it and submit it to another reader _____
    Attempt to implement recommendations _____
    Use it to perform a task or carry out a procedure _____
    Use it to create another document _____
    Other? _____ Explain <u>She will probably spend no more than 5 minutes on it.</u>

11. Reader's physical environment. <u>Normal.</u>

FIGURE 4.11

Audience Profile Sheet

Strategies for accommodating the multiple audience are discussed in Chapter 11, as well as in Chapters 16-23, which treat different kinds of technical documents.

Understanding Multiculturalism

Our culture and our workforce are becoming increasingly diversified culturally and linguistically, and our businesses are relying more on exports. Technical communicators and technical professionals need to communicate effectively with various groups:

- nonnative speakers of English in the United States
- nonnative speakers of English outside the United States
- speakers of other languages who read texts translated from English into their own languages

Thrush (1993) tells the story of a U.S. elevator company installing Korean-made escalators in Mexico. Even though the U.S. company had the Korean manufacturer translate the manuals into English, the style and organization of the documentation were so different from what the Mexican engineers and technicians expected that they couldn't understand it.

Consider these facts about our dependence on global trade (Lustig & Koester, 1993, pp. 5–6):

- One-third of U.S. corporate profits are earned in international trade.
- The 23 largest U.S. banks do almost half of their business overseas.
- Four of five new jobs in the United States are created to produce goods and services for foreign trade.
- Current U.S. investments overseas are worth more than $300 billion.

And consider these facts about the demographics of the United States (Lustig & Koester, 1993, pp. 9–10):

- In California in the 1980s, the Asian population increased by 1.5 million; the Latino population, by 3.1 million. Today, 43 percent of that state's population is nonwhite; by the year 2000, it is expected to be more than 50 percent nonwhite.
- Children in U.S. schools speak more than 150 languages; the figure in Los Angeles alone is 108.
- By the year 2000, white males are expected to be a minority in the U.S. workforce.

Effective communication requires an understanding of the different categories of cultural patterns.

**Categories of
Cultural Patterns** First, what are cultural patterns? *Cultural patterns* are the "shared beliefs, values, and norms that are stable over time and that lead to roughly similar behaviors across similar situations" (Lustig & Koester, 1993, p. 98). Unless you have lived in a different culture, you are probably unaware of how powerful and pervasive cultural patterns are. In fact, more than half of U.S. businesspeople on long overseas assignments return home early because of their inability to adapt to the local culture (Ferraro, 1990).

As shown in Figure 4.12, this section discusses five basic categories of cultural patterns, focusing on information of use to technical communicators and technical professionals.

Differences in Values and Beliefs

Our culture differs significantly from others in its gender roles and nonverbal communication patterns. Consider these differences carefully, especially when you create graphics. Cultures also vary in their attitudes toward time. Think about this difference when you create graphics or text; if you refer to the fast pace of today's business when addressing readers in India, they might not understand what you mean.

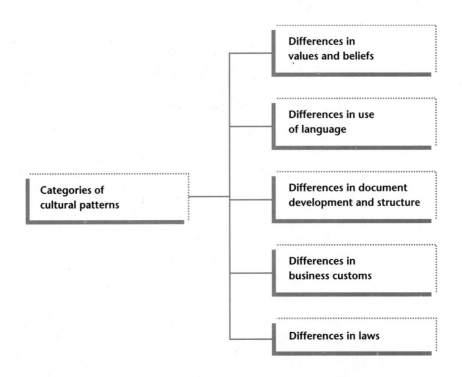

FIGURE 4.12

Categories of Cultural
Patterns

- *Gender roles.* In most Western cultures, women play a much greater role in the workplace than they do in many Middle Eastern and Asian cultures. For this reason, women appear more frequently in marketing and advertising materials in Western countries than in others.
- *Attitudes toward time.* Western cultures often view time as a linear concept; once the time has passed, it is lost forever. The focus is on the future and the past. Many other cultures see time as a cycle, with no beginning and no end; the focus is on the present, and a meeting that extends two hours beyond its scheduled length is no problem (Limaye & Victor, 1991). The Hopi language, for instance, does not have tenses, and the Hopi culture has no concept of past, present, and future. For the Hopi, time merely "gets later" (Littlejohn, 1992, p. 209).

Differences in Use of Language

Every culture is different in its use of language and the mechanics of language. Some of the most important differences are in the following three areas:

- *Vocabulary and sentence length.* In the United States, readers expect simple vocabulary and short sentences. In most European countries, readers expect more complicated vocabulary and long sentences.
- *Numbers.* The United States is the only major country that has not adopted the metric system. In addition, Americans use periods to separate whole numbers from decimals, and commas to separate the thousand from the hundreds. Much of the rest of the world reverses this usage.

United States: 3,425.6

Europe: 3.425,6

- *Dates.* We write out and abbreviate dates differently from most other cultures:

United States: March 2, 1996   3/2/96

Europe: 2 March 1996   2/3/96

Japan: 2 March 1996   96/2/3

The European style—2 March 1996—is gaining popularity in the United States.

Differences in Document Development and Structure

Many aspects of document development and structure are culturally determined. You should consider the following eight factors before you start to write for a multicultural audience:

- *Individual and group orientation.* The typical U.S. professional is highly individualistic in outlook; the average Asian professional is highly group oriented. Therefore, U.S. writers use *I* often and refer to themselves more than to the organization they represent; Asian writers do not express a personal viewpoint but refer to their organization frequently.

- *The nature of evidence.* In most Western countries, evidence is empirical and testable. In many other cultures, such as the Japanese, tradition, authority, and group consensus are more important than empirical evidence.

- *Organization.* People in the United States tend to structure documents deductively: from the general to the particular. In much of the rest of the world, the organization is inductive: from particulars to the general.

- *Directness and indirectness.* Readers in the United States expect explicit, direct documents with the structure clearly spelled out. Many European and Asian cultures value indirectness and digression.

- *Graphics.* Graphics that are perfectly clear in one culture can be confusing or meaningless in another. The garbage-can icon on the Macintosh computer has often been cited as a graphic that does not translate well, for garbage cans are shaped differently and made of different materials around the world.

- *Document design.* Different cultures have different attitudes toward space that affect their use of white space in their documents. Fisk (1992) discusses differences in page design in the United States, England, and Japan: the English use a denser, more crowded design than Americans, and the Japanese design documents with a focal point in the center of the page.

- *Format.* In business letters, many cultures use distinctive patterns for the salutation, the opening and closing paragraphs, and the complimentary close. In Japan, for example, business letters usually begin with a reference to the season, such as "The snow has now disappeared, and the buds are starting to appear on the branches."

- *Amount of detail.* Cultures are classified as high context and low context. In a *high-context culture,* such as the Japanese, the meaning is communicated more implicitly than explicitly. The reader is left to infer the details through the context and tone of the document; therefore, documents tend to be relatively short. In a *low-context culture,* such as that of the United States, the message is stated explicitly and fully supported by details; therefore, documents tend to be longer.

Differences in Business Customs

Business customs vary greatly from culture to culture. Differences include the following:

- *Business protocol.* This category includes greetings, business dress, and customs regarding gifts.
- *Formality and informality.* The level of formality used in addressing a person varies greatly from culture to culture.
- *Nonverbal communication.* Innocent hand gestures in one culture are highly offensive in another. The space that people maintain between one another also differs significantly from culture to culture. In many European cultures, the distance between two people talking is much smaller than in the United States. Seating arrangements in offices and at meetings also differ from one culture to another.
- *Punctuality.* Some cultures require absolute punctuality in business meetings. Other cultures expect you to be 10 or 20 minutes late.
- *Bribery.* Bribery is still required for doing business in many cultures. This reality has serious implications—involving ethics and U.S. law—that your organization has to consider (Martin & Chaney, 1992). For a good introduction to the topic, see Velasquez (1992).

Differences in Laws

When you distribute documents in foreign countries, you must adhere to their laws regarding safety, health, and the environment. In addition, you must be aware of local laws regarding timing of delivery, form and terms of payment, the role of consultants, censorship, and authority to make binding decisions. Recently, a group of American businesspeople trying to invest in a Russian company were surprised to learn that they could not meet in their hotel because they had not obtained the permit required for meetings of more than 10 people.

Writing to Multicultural Audiences

The preceding discussion of the major areas of difference among cultures cannot answer questions about how to write for any particular multicultural audience; all it can do is point out the areas you need to investigate. The most valuable piece of advice is to do your homework about the culture of your readers; read everything you can about the society. In addition, as you plan the document, enlist the help of someone from the culture of your readers. This person can help you prevent blunders that might confuse or offend your readers.

In general, the following eight guidelines will help you improve the success of your writing:

on translating by machine and by humans

In our company, translations are done exclusively by highly trained professional translators who are native speakers of the target language. We strongly believe that sophistication and perfection cannot be obtained through machine translation. Even though the world of machine translations is improving rapidly, a machine will never be able to grasp cultural issues and nuances. An alternative solution to bridging the bilingual and bicultural gap would be a machine translation accompanied by human editing. However, this process is very labor intensive and, in most instances, it would be easier and faster for the translator to translate a document from scratch.

on why translating is more than just translating individual words

A year or two ago, McDonald's printed take-out bags decorated with flags from around the world. On the bag was additional text translated into the different languages of the flags depicted. Had McDonald's taken the time to consult with a cultural expert, they would have avoided committing the major faux pas of printing bags that contained a Saudi flag with holy Koran scripture. This was extremely offensive to the Saudis. To print the Saudi flag, which contains holy scripture, on a disposal bag that would wind up in a garbage heap was sacrilegious. Another example that comes to mind is the Chevy Nova. When it was introduced to the Latin-American market, no one would buy it. No one at General Motors realized that "No va" means "doesn't go" in Spanish.

on internationalization

In the development process of a project, consideration should be given to the possibility of going global and, therefore, internationalizing the product. This is a fancy way of saying that at the development stage the product has to be designed for easy adaptation to international market requirements. If you haven't internationalized the product, then the localization to a specific geographic market can pose a problem. For example, one area of localization is the conversion of measurements (metric system, British and U.S. systems), colors, dates (day first followed by the month and vice versa, depending on the country), calendars (Gregorian used in Israel; Imperial used in Japan, etc.), currencies, temperatures (celsius versus fahrenheit), translation of icons, etc., to a specific market. Localization is an expensive process, and those expenses can be significantly reduced if the product has been internationalized from its inception. Taking all these issues into consideration at the beginning makes the globalization process much easier.

on domestic localization for products sold in the United States

If you look at U.S. demographics, we are really a melting pot. With so many non-English-speaking consumers, domestic localization can be a major selling point. We are living in a very competitive environment. If you do not speak English and need to purchase a certain product, you would most likely choose the product that has packaging instructions written in your language. By applying simple localization rules, the various ethnic groups can be easily reached.

on the aspects of culture that are vulnerable to miscommunication

Issues such as the use of icons in written text can be considered offensive or misleading to many cultures. For example, in Australia, Greece or Turkey, the familiar "thumbs up," both physical and printed, is considered very offensive. Similarly, the American gesture of "OK" means "zero" or "worthless" in France, while in Japan it means you want your change in coins. Pitfalls such as these must be carefully avoided to get the right message across.

Idiomatic expressions are not always translatable. One idiom can mean something quite different in another language; therefore, cultural equivalents must be used. Other aspects to consider are the level of formality, and foods that are acceptable or unacceptable in other cultures. Using this information in your favor gives you an edge over your competitors.

on the process of translating

The very first thing to consider before translating is the target audience and localization issues that need to be addressed. We then select a team of subject-specific translators/editors. The linguists chosen are not only proficient in the language pair combination, but are native speakers of the target language. They are degreed linguists who frequently hold dual degrees in the subjects they translate. After a document is translated, the material goes to an editor to ensure content and continuity and to guarantee that sensitive cultural issues are appropriately addressed. The final step is an in-depth quality control review by our project managers. This includes proofreading, formatting, and incorporation of any special requests the client may have.

George P. Rimalower is CEO of ISI Translation Services in Valley Village, California.

- *Limit your vocabulary.* Make it a goal that every word you use have only one meaning. See Chapter 15 for a discussion of Simplified English, a group of languages that rely on a limited vocabulary.
- *Define abbreviations and acronyms in a glossary.* Don't count on your readers to know what a GFI—ground fault interrupter—is, because the abbreviation is based on English vocabulary and word order. See Chapter 11 for a discussion of glossaries.
- *Avoid jargon unless you know your readers are familiar with it.* For instance, your readers might not know what a *graphical user interface* is.
- *Avoid idioms.* Don't write anything immediately after watching a game on television. You'll end up telling your Japanese readers that your company plans to put on a full-court press, and they will have no idea what you're talking about. One person new to American culture got in the carpool lane on the expressway because he wanted to get his car washed.
- *Keep sentences short.* There is no magic number, but try to stay within an average range of 20–25 words.
- *Use the active voice whenever possible.* The active voice—in which the actor is the grammatical subject of the sentence—is easier for nonnative speakers of English. For more information on voice and other aspects of sentence construction, see Chapter 15.
- *Be sensitive to the culture when creating graphics.* Keep in mind the culture's values, particularly regarding gender roles. See Chapters 3 and 12.
- *Have the document reviewed by someone from the culture.* Even if you had help planning the document, have it reviewed before you send it out or get it printed.

WRITER'S CHECKLIST

Following is a checklist for analyzing your audience. Remember that your document might be read by one person, several people, a large group, or several groups with various needs.

1. Have you filled out a reader's profile sheet for your primary and secondary audiences that considers
   - ❑ education?
   - ❑ professional experience?
   - ❑ personal preferences?
   - ❑ attitudes toward you and your subject?
   - ❑ cultural characteristics?
   - ❑ use of the document?
   - ❑ physical environment?

2. In planning to write to a multicultural audience, have you considered
   - ❑ differences in values and beliefs?
   - ❑ differences in use of language and in mechanics?
   - ❑ differences in document development and structure?
   - ❑ differences in business customs?
   - ❑ differences in laws?

3. In writing for a multicultural audience, did you
   - ❑ limit your vocabulary?
   - ❑ define abbreviations and acronyms in a glossary?
   - ❑ avoid jargon unless you know that your readers are familiar with it?
   - ❑ avoid idioms?
   - ❑ keep sentences short?
   - ❑ use the active voice whenever possible?
   - ❑ use sensitivity in creating graphics?
   - ❑ have the document reviewed by someone from the culture?

Several of the following exercises call for you to write a memo. For a discussion of memos, see Chapter 16.

1. Choose a 200-word passage from a technical article addressed to an expert audience. Rewrite the passage so that it is clear and interesting to the general reader. Submit the original passage along with your memo.

2. Audience is the primary consideration in many types of nontechnical writing. Choose a one- or two-page magazine advertisement for an economy car, such as a Subaru, and one for a luxury car, such as a Mercedes. In a memo to your instructor, contrast the audiences for the two ads by age, sex, economic means, hobbies, interests, and leisure activities. In contrasting the two audiences, consider the explicit information in the ads—the writing—as well as the implicit information—hidden persuaders such as background scenery, color, lighting, angles, and the situation portrayed by any people photographed. Submit the ad along with your memo. (Submit color photocopies or the original ads from the magazines.)

3. The letter on page 89 was written by a branch manager of a lawn service company to a home owner who inquired about the safety of the lawn-care service. The home owner mentioned in her letter that she has an infant who likes to play outside on the lawn. Assume that you have recently been hired by Greenlawn's Customer Service Department. Your supervisor has asked you to review some of the company's recent letters to customers. Write a memo to your supervisor, evaluating the way the letter responds to the concerns of its reader.

4. Fill out an audience profile sheet about yourself, as if you were the reader of your own writing. Then fill out an audience profile sheet about your instructor. What are the major differences between your profile of your instructor and your self-profile?

5. Interview a student from another culture. If you do not know someone, consult the International Student Office, the English Department, or the English as a Second Language Office. Choose one of the topics discussed in this chapter, such as gender roles, notions of time, or document standards, and write a memo to your instructor comparing and contrasting your culture's view with that of the other student.

(For a discussion of comparison and contrast, see Chapter 8.)

6. The student-exchange program at your college wishes to write information booklets about each of the countries in which students from your school study. Research one country and write a 2,000-word guide that explains the culture for students preparing to study there. Concentrate on how the expectations for students in that country differ from those for students at your own college.

7. The following passage is from an advertisement from a translation service. Revise the passage to make it more appropriate for a multicultural audience.

> If your technical documents have to meet the needs of a global market but you find that most translation houses can't handle the huge volume, accommodate the various languages you require, or make your deadlines, where do you turn?
>
> Well, your search is over. Translations, Inc., provides comprehensive translations in addition to full-service documentation publishing.
>
> We utilize ultra-sophisticated translation programs that can translate a page in a blink of an eye. Then our crack linguists comb each document to give it that personalized touch. No job too large! No schedule too tight! Give us a call today!

REFERENCES

D'Ottavi, J. (1993). *Manufacturing bidirectional values: A feasibility study.* Unpublished manuscript, Drexel University.

Ferraro, G. P. (1990). *The cultural dimensions of international business.* Englewood Cliffs, NJ: Prentice Hall.

Fisk, M. J. (1992). People, proxemics, and possibilities for technical writing. *IEEE Transactions on Professional Communication, 35*(3), 176–182.

Limaye, M. R., & Victor, D. A. (1991). Cross-cultural business communication research: State of the art and hypotheses for the 1990s. *Journal of Business Communication, 28*(3), 277–299.

Littlejohn, S. (1992). *Theories of human communication* (4th ed.). Belmont, CA: Wadsworth.

Lustig, M. W., & Koester, J. (1993). *Intercultural competence.* New York: HarperCollins.

Martin, J. S., & Chaney, L. H. (1992). Determina-

GREENLAWN   **April 5, 19XX**

Dear Mrs. Smith,

Thank you for inquiring about the safety of the Greenlawn program. The materials purchased and used by professional landscape companies are effective, nonpersistent products that have been extensively researched by the Environmental Protection Agency. Scientific tests have shown that dilute tank-mix solutions sprayed on customers' lawns are rated "practically nontoxic," which means that they have a toxicity rating equal to or lower than such common household products as cooking oils, modeling clays, and some baby creams. Greenlawn applications present little health risk to children and pets. A child would have to ingest almost 10 cupsful of treated lawn clippings to equal the toxicity of one baby aspirin. Research published in the *American Journal of Veterinary Research* in February 1984 demonstrated that a dog could not consume enough grass treated at the normal rate of application to ingest the amount of spray material required to produce toxic symptoms. The dog's stomach simply is not large enough.

A check at your local hardware or garden store will show that numerous lawn, ornamental, and tree-care pesticides are available for purchase by home owners either as a concentrate or combined with fertilizers as part of a weed and feed mix. Label information shows that these products contain generally the same pesticides as those programmed for use by professional lawn-care companies, but contain higher concentrations of these pesticides than found in the dilute tank-mix solutions applied to lawns and shrubs. By using a professional service, home owners can eliminate the need to store pesticide concentrates and avoid the problems of improper overapplication and illegal disposal of leftover products in sewers or household trash containers.

Greenlawn Services Corporation has a commitment to safety and to the protection of the environment. We have developed a modern delivery system using large droplets for lawn-care applications. Our applicators are trained professionals and are licensed by the state in the proper handling and use of pesticides. Our selection of materials is based on effectiveness and safety. We only apply materials that can be used safely, with all applications made according to label instructions.

On the basis of these facts, I am sure that you will be pleased to know that the Greenlawn program is a safe and effective way to protect your valued home landscape. I have also enclosed some additional safety information. I encourage you to contact me directly should you have any questions.

Sincerely,

Helen Lewis
Branch Manager

tion of content for a collegiate course in inter-cultural business communication by three Delphi panels. *Journal of Business Communication, 29*(3), 267–283.

Schwartz, E. I. (1994, August 21). A work in progress: Sign language telephones. *New York Times,* p. 3: 7.

Thrush, E. A. (1993). Bridging the gaps: Technical communication in an international and multicultural society. *Technical Communication Quarterly, 2*(3), 271–283.

"Tradeline T822D Heating Thermostat." (1981). Minneapolis: Honeywell.

Velasquez, M. (1992). *Business ethics: Concepts and cases* (3rd ed.). Englewood Cliffs, NJ: Prentice Hall.

Yazici, H., Benjamin, C., & McGlaughlin, J. (1994). AI-based generation of production engineering labor standards. *IEEE Transactions on Engineering Management, 41*(3), 302–309.

Determining Your Purpose and Strategy

Chapter 4 introduced the concept of the writing situation: the audience and the purpose of a document. This chapter begins with a discussion of how to determine your purpose. Next, the chapter provides guidelines for planning a strategy and proposing it to your managers.

Determining Your Purpose

Once you have identified and analyzed your audience, examine your general purpose in writing. Ask yourself this simple question: "What do I want this document to accomplish?" When your readers have finished reading what you have written, what do you want them to *know* or *believe*? What do you want them to *do*? Think of your writing not as an attempt to say something about the subject but as a way to help others understand it or act on it. Think of your writing as an attempt to persuade your readers: to convince them to hold a particular belief or take a particular action.

To define your purpose more clearly, think in terms of verbs. Isolate a single verb that represents your purpose, and keep it in mind throughout the writing process. (Of course, in some cases a technical document has several purposes, and therefore you might want to choose several verbs.) Following are examples of verbs that indicate typical purposes in technical documents. The list has been divided into two categories: verbs used when you primarily want to communicate information to your readers, and verbs used when you want to convince them to accept a particular point of view.

Communicating Verbs	*Convincing Verbs*
to explain	to assess
to inform	to request
to illustrate	to propose
to review	to recommend
to outline	to forecast
to authorize	to evaluate
to describe	
to define	
to instruct	
to summarize	

This classification is not absolute. For example, *to review* could in some cases be a *convincing verb* rather than a *communicating verb*: one writer's review of a complicated situation might be very different from another's review of the same situation. Following are a few examples of how you can use these verbs in clarifying the purpose of your document. (The verbs are italicized.)

- This report *describes* the research project that is intended to determine the effectiveness of the new waste-treatment filter.
- This report *reviews* the progress in the first six months of the heat-dissipation study.
- This letter *authorizes* the purchase of six new PCs for the Jenkintown facility.
- This memo *recommends* that we study new ways to distribute specification revisions to our sales staff.

As you devise your purpose statement, remember that your real purpose might differ from your expressed purpose. For instance, if your real purpose is to persuade your reader to lease a new computer system rather than purchase it, you might phrase the purpose this way: *to explain the advantages of leasing over purchasing.* As mentioned in Chapter 4, many readers don't want to be *persuaded* but are willing to learn new facts or ideas.

Determining Your Strategy

Once you have analyzed your audience and determined your purpose, the next step is to put the two together and create a strategy for writing the document. *Strategy* refers to your game plan: how you are going to create a document that accomplishes your goals. Specifically, planning a strategy involves determining the constraints you must work within.

As a student writer, you work within a number of limits or constraints: the amount of information you can gather for a paper, the length and format, the due date, and so forth. In business, industry, and government, you will be working within similar constraints. Figure 5.1 on page 94 shows seven sets of constraints.

Ethical Constraints   As discussed in Chapter 2, we all work within a web of rights and responsibilities. Our greatest responsibility is to our own sense of ethical behavior. Our ethical standards are challenged when we are asked to lie or mislead in documents.

In most cases, you have some options when you feel you are being asked to act unethically. Some organizations and professional communities have a published code of conduct. In addition, most large companies have ombudspersons, people whose job is to help employees resolve ethical and other conflicts within the organization. Don't feel that you are obligated to do whatever your supervisor tells you to. If you think that you are being asked to act unethically, think

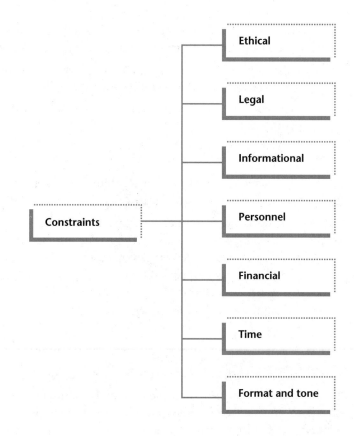

FIGURE 5.1

Constraints

about the issue carefully and then take advantage of the resources available to you.

Legal Constraints   Chapter 2 also discusses legal issues. You are obligated to abide by all applicable laws concerning labor practices, environmental issues, fair trade, consumer rights, and so forth. If you think that the action recommended by your supervisor has potential legal implications for you or your company, meet with your organization's legal counsel and, if necessary, attorneys outside the organization.

Informational Constraints   You might face constraints on the kind or amount of information you can use in the document. Most often, the information you need is not available. For example, you might want to recommend that your organization buy a piece of equipment, but you can't find any objective evidence that it will do the job. You have advertising brochures and testimonials from satisfied users, but apparently nobody has performed the kinds of controlled experiments or tests that would convince the most skeptical reader.

What do you do? You tell the truth. You state exactly what the situation is, weighing the available evidence and carefully noting what is missing. Your most important credential on the job is credibility; you will lose it if you unintentionally suggest that your evidence is better than it really is. In the same way, you don't want your readers to think that you don't realize your information is incomplete; they will doubt your technical knowledge.

Personnel Constraints   As discussed in Chapter 3, much technical communication is written collaboratively by teams of writers, editors, production specialists, and subject-matter experts. One constraint you will face in working on large projects is that you won't have access to as many collaborators as you need. Every organization hires as few people as it can, of course. For this reason, you must have a good idea of the kind of personnel you will need, and the sooner you can communicate your needs to management, the more likely you are to get what you need.

Many organizations hire technical communicators and other personnel project by project, not as regular employees. In such cases, present a clear and persuasive proposal to hire the personnel you need. However, you will probably have to make do with fewer people than you want, and everyone will have to work harder and do some kinds of tasks for which they are ill suited.

Financial Constraints   Related to personnel constraints are financial constraints, for if you had unlimited funds, you could hire all the personnel you need. But financial constraints can affect other categories of resources, too. For instance, a printing expert can tell you, almost to the penny, what the extra costs will be to move up a grade in paper, or to add color to a document, or to use a comb binding instead of saddle binding. Producing large, complex documents—especially in large quantities—is an exercise in precise estimating. If your organization has a technical-publications department, consult it for advice. Otherwise, work closely with print shops and service bureaus.

Time Constraints   People, especially writers, never think they have enough time to do their jobs. In fact, many managers don't allot enough time for their employees to do a good job putting together all the necessary memos, letters, reports, and manuals. Most working professionals resign themselves to taking work home in the evenings or on the weekends. Some people come in early to do their writing before things get hectic.

The first piece of information you need, of course, is the deadline for the document. Sometimes a document will have a number of deadlines—for intermediate reviews, progress and status reports, and so forth. Once you know when a particular document (or part of it) is due, work backward to devise a schedule. For example, if a major report is due on a Friday, you know that you must have it essentially done by the day before, because you want to let it sit overnight before you proofread it for the last time. If it has to be done by Thursday, you might want to give yourself at least four days for drafting and revising. That means you want to finish your outlining by the previous Friday.

Keep in mind that few people can accurately estimate the time a particular activity will take. As you probably know intuitively, tasks almost always take longer than estimated. People call in sick, information is lost or delayed in the mail—whatever can go wrong will go wrong, as Murphy's Law states. When you collaborate, the number of potential problems increases dramatically, because when one person is delayed, other people may lack the necessary information to proceed, causing a snowball effect. In addition, friction between people who are working hard can slow down a project substantially.

Creating a schedule is discussed in Chapter 4.

Format and Tone Constraints

You will be expected to work within one more set of constraints:

- *Format.* Format constraints are limitations placed on the size, shape, or style of the document, either implicitly or explicitly, by the organization for which you work or the organization to which you are writing. For example, many organizations have guidelines on how to present different kinds of information—all tables and figures are presented at the end of the report, or the names of the people receiving copies of your memo are listed in alphabetical order, or by hierarchy. You can best determine the organizational constraints by finding out whether a company style guide exists. If not, the company might use an externally prepared style guide, such as the *Chicago Manual of Style* or the *U.S. Government Printing Office Style Manual*. If your organization has no official style guide, check similar documents on file to see what other writers have done, and, of course, talk with more experienced coworkers.

  If you are writing to another company, learn what you can about its preferences. If you are responding to an invitation for a proposal, investigate whether the company specifies length, organization, and style. Many government agencies, for example, require that potential suppliers submit their proposals in two separate parts: the technical proposal and the budget.

- *Tone.* You already know that you write one way to an organizational superior and another way to a peer or a subordinate. When addressing superiors, use a formal, polite tone. When addressing peers or subordinates, use a less formal tone but be equally polite. Politeness is common courtesy, and it is common sense: you should not bully people. If you obnoxiously tell the janitor to change the bulb in the ceiling fixture above your desk, you might end up waiting a few weeks for it, or the next day you might find that you have a new bulb and a large footprint in the middle of your desk blotter.

Planning Your Tactics

You have analyzed your audience, determined your purpose, and figured out under what constraints you will be working. Now it is time to put all these factors together and to think about what the document will eventually look like and how you are going to get from here to there.

For illustration, let's go back to the writing situation described in Chapter 4: you want to persuade your two readers to authorize the purchase of some CAD equipment. Your primary reader, Harry Becker, is the head of your department. Your secondary reader is Tina Buterbaugh, the head of finance. Your audience analysis suggests that only Harry Becker knows much about CAD equipment. However, because Tina Buterbaugh has been at the company for a number of years, she probably understands the basics of the technology: what the equipment is used for, why it makes draftspeople much more efficient, and so on.

In this case, your purpose in writing is clear: to convince Harry Becker that your proposal is reasonable, so that he will authorize the purchase of the equipment. But as was mentioned earlier, you might want to phrase that purpose differently when it appears in the document: *to explain the advantages of acquiring new CAD equipment*, rather than *to persuade you to authorize me to purchase new CAD equipment.*

Think about the constraints you will be working under:

- *Ethical.* You can't think of any ethical constraints. The project is a straightforward proposal to upgrade equipment.
- *Legal.* Again, nothing.
- *Informational.* As far as you know, obtaining the necessary information should be no problem because the CAD system you are interested in has been on the market for some months and has been reviewed in several professional journals. In addition, you know draftspeople at other companies who have purchased it and will talk with you about their experience and let you try out the equipment.
- *Personnel.* The project is relatively simple and should not require any personnel. You can research, write, and produce the document by yourself. However, you realize that it is a good idea to talk with other draftspeople during the planning stage and hear their thoughts about your draft. In addition, you decide to show the final draft to Karen, one of the technical writers, who often critiques and proofreads documents for the draftspeople.
- *Financial.* There are no financial constraints for your proposal. Ordinarily, a recommendation to spend $4,000 would be routine. However, in light of an unwise recent project that wasted $50,000, you know that Harry has become careful about authorizing any capital expenditures, even small ones, because he knows that next year's budget will depend, to a great extent, on how successful this year's purchases turn out to be.

- *Time.* The only time constraint is the familiar problem: finding the time to write the document. There is no external deadline, because this is a self-generated project; however, you have not received authorization to spend company time on the proposal. Therefore, you will probably have to work on it for a few evenings at home.
- *Format and tone.* Although your company doesn't have a style manual that describes internal proposals, the format has remained fixed over the years, and there are plenty of models to consult.

What does your analysis of audience and your determination of purpose and constraints tell you about how to write the document? Because your readers will be particularly careful about large capital expenditures, you must clearly show that the type of equipment you want is necessary and that you have recommended the most effective and efficient model. In addition, your proposal will have to answer a number of questions that will be going through your readers' minds:

- What system is being used now?
- What is wrong with that system, or how would the new equipment improve our operations?
- On what basis should we evaluate the different kinds of available equipment?
- What is the best piece of equipment for the job?
- How much will it cost to purchase (or lease), maintain, and operate the equipment?
- Is the cost worth it? At what point will the equipment pay for itself?
- What benefits and problems have other purchasers of the equipment experienced?
- How long would it take to have the equipment in place and working?
- How would we go about getting it?

Your readers' personal preferences about writing style indicate that you should be straightforward, direct, and objective. Avoid technical vocabulary and unnecessary technical details, but make clear the practical advantages of the new equipment. In addition, it would be a good idea to include an executive summary for Tina Buterbaugh's convenience.

Proposing a Strategy

After you figure out what you want to do, consider gaining management's approval before proceeding.

You needn't seek approval for every kind of project, of course. Sometimes you simply do your research and write your document. But the larger and more complex the project and the document, the more sense it makes to be sure that you are on the right track before you invest too much time and effort. For the sake of simplicity, projects can be classified into two categories: small and large.

Small and *large* mean different things in different contexts. At a three-person software company, a $5,000 project might be large; at Microsoft, a $5,000,000 project would be relatively small. Despite these enormous differences of scale, these categories can help show the range of possibilities in proposing a strategy.

Proposing a Small Project

You are planning the CAD equipment project. You have a good understanding of your audience, purpose, and strategy, and a general outline is starting to take shape in your mind. But before you actually start to write an outline or gather all the information you will need, it's a good idea to spend another 10 or 15 minutes making sure your primary reader agrees with your thinking. You don't want to waste days or even weeks working on a document that won't fulfill its purpose.

Submitting to your primary reader a statement of your understanding of the audience, purpose, and strategy is a means of establishing an informal contract. Although readers can change their minds later—in response to new developments or after more thought—provisional approval is better than no approval at all. Then, if the project is subsequently revised or canceled, at least you can prove that you have been using your time well.

Writing such a statement has an additional benefit: it helps redefine the project. What was earlier an idea you were thinking about is now a project, and this redefinition makes it easier to receive authorization to use company resources.

Some writers are reluctant to submit such a statement to the primary reader, either because they don't want to be a nuisance or because they're worried that it might reveal a serious misunderstanding of the writing situation. There must be some people out there who don't want to be bothered, but I haven't met any. Why should they object? It doesn't take more than a minute to read your brief statement, and if there *has* been a misunderstanding, it is far easier to remedy at this early stage. Having the document come out right the first time around is in everyone's interest.

Your statement can serve a different purpose as well: if you want your reader's views on which of two strategies to pursue, you can describe each one and ask your reader to state a preference.

What should this statement look like? It doesn't matter. You can send an e-mail or write a memo, as long as you clearly and briefly state what you are trying to do. Figure 5.2 on page 100 is an example of the statement you might submit to your boss about the CAD equipment.

Harry:

   Tell me if you think this is a good approach for the proposal on CAD equipment.

   Outright purchase of the complete system will cost more than $1,000, so you would have to approve it and send it on for Tina's approval. (I'll provide leasing costs as well.) I want to show that our CAD hardware and software are badly out of date and need to be replaced. I'll be thorough in recommending new equipment, with independent evaluations in the literature, as well as product demonstrations. The proposal should specify what the current equipment is costing us and show how much we can save by buying the recommended system.

   I'll call you later today to get your reaction before I begin researching what's available.

   Renu

FIGURE 5.2

Proposal for a Small
Project

In composing this statement, the writer began with her audience profile sheets for the two principal readers (see Chapter 4, Figures 4.10 and 4.11). She describes a logical, rational plan for proposing the equipment purchase.

Once you have received your primary reader's approval, you can feel confident in starting to gather and interpret your information and then writing your document. These tasks will be discussed in the next three chapters.

Proposing a Large Project

Most organizations require that large, complex projects be planned more carefully than small, less complex ones. Within your organization, there might be written guidelines on what kinds of projects require what kinds of planning documents, or the guidelines might be unwritten.

Many kinds of planning documents exist. Complex projects usually call for a *proposal*, a formal statement that answers the following questions:

- What do you plan to do?
- Who will participate in the project?
- How do you plan to do the project?
- Why do you plan to do it that way?
- When do you plan to do the project?
- Where do you plan to do the project?
- What are your credentials to do the project?

See Chapter 21 for a discussion of proposals.

1. What is your purpose in writing? What is the document intended to accomplish?

2. Is your purpose consistent with your audience's needs?

3. How does your understanding of your audience and your purpose determine your strategy: the scope, structure, organization, tone, and vocabulary of the document?

4. Will you have to accommodate
   ❑ any ethical constraints?
   ❑ any legal constraints?
   ❑ any informational constraints?
   ❑ any personnel constraints?
   ❑ any financial constraints?
   ❑ any time constraints?
   ❑ any format or tone constraints?

5. Have you proposed your project to your primary readers to see if they approve of it?

EXERCISES

Several of the following exercises call for you to write a memo. For a discussion of memos, see Chapter 16.

1. Choose two articles on the same subject, one from a general-audience periodical, such as *Time* or *Newsweek,* and one from a technical journal, such as *Science, Journal of Visual Languages and Computing, Management Science,* or *Journal of Microcomputer Applications.* Write a memo to your instructor, comparing and contrasting the two articles from the point of view of the authors' assessment of the writing situation: the audience and the purpose. (For more information on comparison and contrast, see Chapter 8.) As you plan the memo, keep in mind the following questions:

   a. What is the background of each article's audience likely to be? Does either article require that the reader have specialized knowledge?

   b. What is the author's purpose in each article? In other words, what is each article intended to accomplish?

   c. How do the differences in audience and purpose affect the following elements in the two articles?
      1. scope and organization
      2. sentence length and structure
      3. vocabulary
      4. number and type of graphics
      5. references within the articles and at the end
      6. headings and white space

   Submit the two articles (or photocopies of them) along with your memo.

2. Locate a fund-raising letter from a charitable organization, such as the American Cancer Society, the March of Dimes, or the Special Olympics. Write a memo to your instructor, describing the different tactics used in the letter to persuade you to give money to the cause. Which tactics work well, and which do not? Do not limit your analysis to the argument made by the words themselves; consider also the design of the letter, the appearance of the type, and the type of paper used. If the letter includes any graphics or the envelope included any materials other than the letter itself, analyze them as well.

CHAPTER 6 | # Performing Secondary Research

The Role of Research in the Problem-Solving Process

Characteristics of Good Information

If You Can Choose Your Own Topic

Generating Ideas about the Topic
  Asking Journalistic Questions
  Brainstorming
  Freewriting
  Talking
  Sketching

Using the Library and the Internet
  Using the Library
    Reference Librarians
    Online Catalogs
    Reference Books
    Periodical Indexes
    Newspaper Indexes
    Abstract Services
    Government Publications
    Guides to Business and Industry
    Online Databases
    CD-ROM
  Using the Internet
    Electronic Mail
    Usenet Newsgroups
    File Transfer
    Remote Log-in

Understanding the Information
  Skimming
  Taking Notes
    Paraphrasing
    Quoting
  Summarizing

Evaluating the Information
Evaluating for Accuracy
Evaluating for Bias
Evaluating for Comprehensiveness
Evaluating for Timeliness
Evaluating for Clarity
The Special Case of the Internet

Writer's Checklist

Exercise

Chapters 4 and 5 discussed planning a writing project: analyzing the document's audience and determining its purpose, then planning a writing strategy that accommodates the audience's needs and your time and budgetary constraints. The next step in most writing projects is to find and evaluate the technical information that will go into the document.

This chapter covers the techniques of *secondary research,* which is the process of collecting information that other people have discovered or created. The most common way to perform secondary research is to read books and journals, but researchers also talk with colleagues, consult databases, attend conferences, and gather information from the Internet.

Primary research, which involves creating technical information yourself, is the subject of Chapter 7. This book presents secondary research first because rarely would you conduct primary research before having conducted secondary research. To design the experiments or the field research that goes into primary research, you need a thorough understanding of the information that already exists about the subject.

This chapter covers how to generate ideas, how to use a library, how to understand the basics of the Internet, and how to understand and evaluate the information you have gathered.

The Role of Research in the Problem-Solving Process

Sometimes students think they write research reports so they can show their teachers that they know how to write research reports. In the working world, you do research as part of the larger task of solving problems. Nobody has ever said, "I have an easy week coming up; I think I'll write a report."

In the working world, you don't have to go looking for problems and opportunities; they present themselves all the time. For instance, you are on a business trip, during which you use your laptop to make a slide presentation for a potential client. Presentations are on your mind, and during a free hour you visit an exhibition of office products and see a new kind of device for projecting slide presentations from a laptop. You bring back some product literature. After talking with your colleagues about the new product, you conclude that it might be worthwhile to investigate the need for this device at your company. You will want to answer questions such as these:

- How many presentations do we make from laptops?
- How do we currently handle projection?
- What are the advantages and disadvantages of the way we do it now?
- What are our technical and financial needs for a projection device?
- What are the available products?
- What are the advantages and disadvantages of each product?
- Which product would be best for us?

Each of these questions calls for research. Here is how you might research several of them:

- *How many presentations do we make from laptops?* You make a few phone calls within the company to find out how many business trips your company's employees go on, how many times they take a laptop, and how often they make presentations from the laptop.
- *What are the advantages and disadvantages of the way we do it now?* You ask an assistant to review some of the recent trip reports to see whether there is any mention of how your colleagues currently make presentations from a laptop. If that technique doesn't yield any useful information, you prepare a survey or phone three or four colleagues who are frequent travelers and who represent different groups within your company: females and males, more experienced and less experienced, more computer literate and less computer literate.
- *What are the available products?* You could visit dealers, but they will want to sell you what they carry, not necessarily what is best for you. You could review product literature, but, again, you're dealing with a biased source of information. So you review the literature in professional journals, focusing on articles appearing in the most prestigious ones. You look for articles written by unbiased sources who have no financial stake in the outcome. You look for articles that include some sort of *bench-*

mark testing—the same task is performed on each competing product, and the evaluators measure the results and evaluate the output.

You do all kinds of secondary and primary research: you conduct interviews, send out surveys, use online bibliographies to find journal articles, post questions to an Internet newsgroup on the subject. With more information than you can possibly use, you start to write—to try to figure out what all the information means. In the process, you realize that you still have some questions, that some of the information is incomplete, some contradictory, and some just unclear. You go back and do more research, then more writing.

Characteristics of Good Information

There is no shortage of information out there; the challenge is to find good information. Look for information that is

- *Accurate.* If you think that people in your company make 500 trips a year but the real figure is 50, you might end up buying equipment that you should be renting.

- *Unbiased.* Obviously, you want sources that don't have any financial interest in your decision. But you also want to watch out for other kinds of bias; some sources might have a bias in favor of one kind of computer operating system, a bias that prevents them from offering a helpful analysis.

- *Comprehensive.* You want to hear from different kinds of people—in terms of gender, cultural characteristics, and age—and people representing all views on the topic. The last person you interview might be the first one to point out something you never thought of: if the projection device you decide to buy weighs more than eight pounds, he would never take it on an airplane, although he would take it on a trip by car.

- *Current.* If your information about laptop presentation devices is from 1993, it might as well be from 1953.

- *Clear.* You want information that is easy to understand; otherwise, you'll waste time figuring it out, and you might misinterpret it.

Later, this chapter discusses how to evaluate the information according to these five criteria.

If You Can Choose Your Own Topic

Few professionals get to choose their topics, whereas students often do. This freedom is a mixed blessing. An instructor's request that you "come up with an appropriate topic" is for many people frustratingly vague; they would rather be told what to write about—how computers have changed the process of manufacturing, for instance—even though the topic might not interest them. If you don't spend weeks agonizing over the decision, however, the freedom to choose your own topic and approach is a real advantage: you're more likely to want to read and write about a topic that interests you, and therefore you'll likely do a better job.

Forget topics such as the legal drinking age, the draft, and abortion, which don't relate to most writing situations on the job. Ask yourself what you are *really* interested in: perhaps something you are studying at school, some aspect of your part-time job, something you do during your free time. Browse through three or four recent issues of a news magazine or professional journal, and you will find dozens of articles that suggest interesting, practical topics involving technical information.

Table 6.1 shows the titles of several articles from the June 20, 1994, *U.S. News and World Report* that suggest possible paper topics.

TABLE 6.1   *Possible Topics for a Research Paper*

Title of the Article	Possible Subject Areas for a Report
The gigantic costs of developing tiny chips	Advances in chip technology The economic battle for chip dominance
The most dangerous place on earth [North Korea]	Techniques of nuclear monitoring The changing nuclear threat in post–cold war times
The smoke next door [secondhand smoke]	The health effects of secondhand smoke The public-relations battle between health groups and the tobacco industry
The suffocating politics of pollution	The health effects of air pollution The economic effects of air pollution

Many of these subjects are too broad for a brief research report of 10 to 15 pages, but they are good starting points. Think about your local community. When you read about the remarkable advances in computer-chip technology, think about the computer labs on your campus. How is your school attempting to keep up with hardware advances? When you read about secondhand smoke, think about your school's policy on smoking in dormitories, classrooms, offices, and public areas. Or think about how local restaurants are responding to health concerns about secondhand smoke. Good topics are everywhere.

Generating Ideas about the Topic

Once you have a topic, it is time to start generating ideas about it. In this discussion, the task of generating ideas is explained before the task of doing research, but in fact you will be generating ideas throughout the writing process. Although you will generate ideas before you do the research, the research sources you read will force you to reevaluate your ideas and create new ones. Even after you are well under way drafting the document, you will revise and add ideas.

By *ideas* I don't mean formal, page-length explanations that are ready to be placed into the document, although if you have these in mind after choosing a topic, all the better. By *ideas* I mean subtopics of the overall topic you are writing about. For instance, if you plan to write about a new technology that mixes water with oil to increase the efficiency of and reduce pollution in power plants, one of your ideas probably will be the theory behind the new technology, because almost certainly your readers will need to know it. However, you might also think of other ideas—such as the history of this technology, its cost, its advantages and disadvantages—that might not make their way into the document.

At this stage, many ideas might be little more than phrases or questions; logic tells you that you ought to know about them, and you remember reading about them, but you are not prepared to write about them yet. Generating ideas is a way to start mapping out what will be included in the document, where it will go, and what additional information you will need to get.

How do people generate ideas? However they can. Many people don't even know what process they use. For some, the best ideas come when they aren't consciously thinking about their topic. A billboard, a television commercial, a conversation with a neighbor—anything can trigger an idea. The brain works on a problem when we are driving or taking a shower or sleeping. Some creative people actually keep a pad and pen on their nightstand because they frequently wake up in the middle of the night with a great idea.

Even though we can't completely understand or control the process of generating ideas, we can help it along. Figure 6.1 on page 108 shows five techniques that have proven useful for many people.

Note that these techniques are used throughout the writing process. Good writers rethink their documents at every stage.

Asking Journalistic Questions

The journalistic questions are the *who, what, when, where, why,* and *how* of the topic. If, for instance, you think you might be interested in writing about digital audio tape (DAT), ask yourself the following questions:

- Who invented DAT? Who uses it?

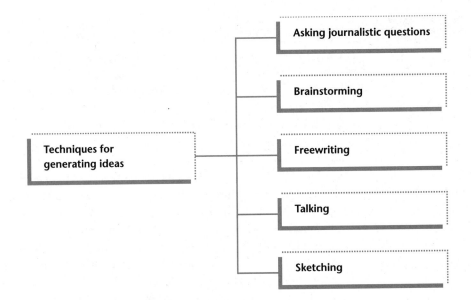

FIGURE 6.1

Techniques for
Generating Ideas

- What is DAT?
- When was DAT invented? When was it licensed? When will it become popular?
- Where was it invented? Where is it used now? Where will it be used in the future?
- Why is DAT used? What advantages does it offer over analog audio tape or compact discs?
- Why is it not widely used now?
- How does DAT work? What is its principle of operation?

Answering these basic questions will help you understand how much you know about the topic and what kinds of information you need to acquire.

Brainstorming   Brainstorming involves spending 10 or 15 minutes listing ideas about your subject. List them as quickly as you can, using short phrases, not sentences. For instance, if you are describing global warming, the first five items in your brainstorming list might look like this:

- definition of global warming
- role of chlorofluorocarbons
- political issues
- effect on smokestack industries
- international agreements to limit global warming

As the word *brainstorming* suggests, you are not trying to impose any order on your thinking. You will probably skip around from one idea to another. Some ideas will be subsets of larger ideas. Don't worry about it; you will straighten out the logic later.

Brainstorming has one purpose: to free your mind so that you can think of as many ideas as possible that relate to your subject. When you have done your initial research and start to construct an outline, you will probably find that some of the items you listed do not in fact belong in the document. Just toss them out. The advantage of brainstorming is that for many people it is the most effective and efficient way to catalog what *might* be important to the document. Using a more structured way of generating ideas, ironically, would cause you to miss more of them.

Brainstorming is actually a way of classifying everything you know into two categories:

- information that probably belongs in the document
- information that does not belong

Once you have decided what material probably belongs, you can start to write the outline.

Freewriting   You might have used freewriting in one of your previous writing classes. Freewriting consists of writing without any plans or restrictions. If your topic is the technology for mixing oil and water in power plants, you write whatever comes into your mind about it. You don't consult an outline, you don't stop to think about sentence construction, you don't consult a reference book. You just make the pen or the cursor move.

In most cases, the freewriting text never becomes a part of the final document. Your explanations might be only half-correct; they almost certainly will be expressed unclearly. And half of what you write might be merely phrases or questions that you think you might need to answer. But something valuable might be happening: the mere act of trying to make sense helps you determine what you do and do not understand. And one phrase or sentence might spark an important idea.

Figure 6.2 on page 110 is an example of a brief freewriting text created in about five minutes by a student in a technical-communication course. She was working for a company considering a health-promotion program.

This freewriting text is a mess: a series of incomplete thoughts, and more questions than answers. Yet the writer did start to work out some of the problems her company would face (keeping the cost down, motivating employees, possible liability for injuries during the program) and some possible solutions to research (such as asking the insurance company for assistance and finding out if there is an organization that oversees the different kinds of programs). By start-

> Insurance rates have gone up over 10% a year for last 5 years. Other problem: the loss of employees to preventable health problems. Check out health-promotion programs. How much do they cost? Are they administered by us or by a subcontractor? How about reduced-rate memberships at health clubs? Will we have further liability if someone gets hurt in the programs? How much would we save—in premiums and reduced health problems? Is there an organization that rates these programs? First we have to find out how many employees would be interested. Maybe the insurance company could help us with some of this information.

FIGURE 6.2

Freewriting Text

ing to think through the problems her company would face, the writer is of course starting to think about her own problems: what kind of information her readers will need. Not bad for five minutes' work.

Talking

Talking with someone about your topic is an excellent way to find out what you already know about it and to generate new ideas. Another student is a good choice, because you will be able to return the favor. Ask the person to throw questions at you as you speak. If you start with your main idea—"Our company is thinking about instituting some sort of health-promotion program for its employees"—the person could ask "Why?" You respond, "Because our health insurance premiums are too high and because too many of our employees suffer preventable health problems." The person might then ask, "How high are the premiums? What is the rate of increase?" or "What percentage of your employees get sick? What is an acceptable rate? Is the rate changing?" A careful listener doesn't have to be a detective, just someone trying to understand what you are saying and what it means.

As the person asks questions such as these, you will quickly get a sense of how clearly you understand your topic, what additional information you will need to get, and what questions your readers will want answered in your document. You will also find yourself making new connections between ideas; the mere act of forcing yourself to put your ideas into sentences for someone else helps you synthesize the information.

Sketching

The four techniques for generating ideas that have been discussed so far—asking journalistic questions, brainstorming, freewriting, and talking with someone—are all nonhierarchical. That is, when you use them, you do not worry about the relative importance of ideas; minor ideas are mixed in with major ideas. However, some people prefer a more structured way to generate ideas.

Sketching ideas—making a visual representation rather than a verbal one—

can help you start to structure the document. Sketching has an additional benefit: it can help you see what information is weak or missing altogether. Two kinds of sketches are used often in the professional world:

- *Clustering.* Write the main idea or a main question in the middle of the page. Then draw a circle around it. Write second-level ideas around the main idea. Then add third-level ideas around the second-level ideas.

  Figure 6.3 shows a cluster for the report on health-promotion programs. The writer has started with the journalistic questions. Notice that the writer doesn't yet know how to answer the question "Who?" She has to go back to her sources to find out who actually administers these kinds of programs.

- *Branching.* There is only one major difference between clustering and branching: the movement from larger to smaller ideas is from the center

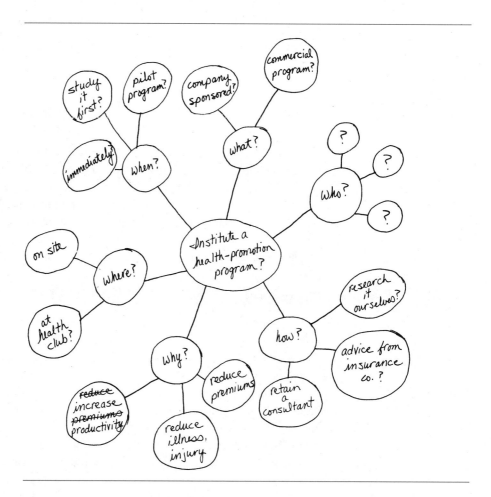

FIGURE 6.3

Clustering

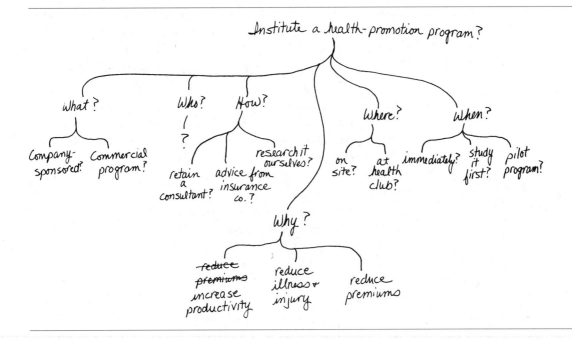

FIGURE 6.4

Branching

to the perimeter in the cluster format and from top to bottom in the branching format. Figure 6.4 shows branching.

As you think more about your topic, you will extend and revise your sketch. You might want to rethink the sketch if, for example, one of the branches contains many more subbranches than the others. A short branch might signal that you need more research; or it might just be a smaller idea that is fine as it is. A long branch could signal that you went into much more detail on that one idea than you did on the other ideas; or it could merely be a more complicated idea. Counting branches and clusters provides no sure answers.

Using the Library and the Internet

If as a student you realize that your topic would require much research and time—weeks to wait for an interlibrary loan, for example—consider changing to a topic that you know more about. Even though many writers like to have a precise topic in mind before they begin to search the literature, sometimes it is best to start with only a general topic and see whether information is available and easy to access. For example, you might want to write about computers but be unsure about what aspect to focus on. The basic research techniques described here will enable you to discover, quickly and easily, what aspects of computers will make the most promising topics. The advantage of being flexible about your topic is obvious: in 15 minutes you can discover dozens of possible topics from which to choose.

Using the Library   The best place to begin your search for additional information is a college or university library. Most college libraries have substantial reference collections and receive the major professional journals. Large universities have more comprehensive collections. Many large universities have specialized libraries that complement selected graduate programs, such as those in zoology or architecture. In addition, large cities often have special scientific or business libraries that are open to the public.

Reference Librarians

The most important information sources at any library are the reference librarians. They are invariably willing to suggest new ways to search for what you need—specialized directories, bibliographies, or collections that you didn't know existed. They can also assist you in using online databases. Perhaps most important, they will tell you if the library *doesn't* have the information you need and suggest other libraries to try. Reference librarians can save you much time, effort, and frustration. Don't be afraid to ask them questions when you run into problems.

Online Catalogs

In most libraries, the card catalog has been replaced by an online catalog that lists almost all the library's holdings: its books, microforms, films, phonograph records, tapes, and other materials. Most online catalogs work essentially the same way, even though their screens and commands differ. Study the instructions sheet for the system you will be using, and read the on-screen commands.

An online catalog does what the card catalog did—and much more. You can search for an item if you know its subject, title, or author, or any key term from one of these three. For instance, if you know that the title of the book you want contains the word *quantitative,* you can use an online catalog to locate the title *The Visual Display of Quantitative Information.* And the display indicates not only the standard bibliographic information about the title—author, publisher, call number, and so forth—but also whether the title is on the shelves or checked out.

Using an online catalog, you can modify the search to include or exclude different parameters. For instance, you can limit the search by a number of parameters:

- medium (such as books, reports, or videotapes)
- language (such as English)
- publication date (such as 1993 or later)
- publisher (such as St. Martin's Press)

The online catalog lists and describes the holdings at one particular library. Using the Internet, you can search the catalogs of hundreds of major libraries around the world. See the discussion of using the Internet later in this chapter.

Reference Books

Some books listed in the catalog have call numbers preceded by the abbreviation *Ref.* These books are part of the reference collection, a separate grouping of books that normally may not be checked out of the library.

The reference collection includes the general dictionaries and encyclopedias, biographic dictionaries (*International Who's Who*), almanacs (*Facts on File*), atlases (*Rand McNally Commercial Atlas and Marketing Guide*), and dozens of other general research tools. In addition, the reference collection contains subject encyclopedias (*Encyclopedia of Banking and Finance*), dictionaries (*Psychiatric Dictionary*), and handbooks (*Biology Data Book*). These specialized reference books are especially useful when you begin a writing project, for they provide an overview of the subject and often list the major works in the field.

How do you know if there is a dictionary of the terms used in a given field, such as nutrition? As you probably have guessed, you can find out in the reference collection. It would be impossible to list here even the major reference books in science, engineering, and business, but you should be familiar with the reference books that list the many others available. Among these guides-to-the-guides are the following:

> Harris, S. (1991). *The New York Public Library book of how and where to look it up*. New York: Prentice Hall.
>
> Mullay, M., & Schlicke, P. (Eds.). (1993). *Walford's guide to reference material* (6th ed.). 3 vols. London: Library Association.
>
> Sheehy, E. P. (Ed.). (1986). *Guide to reference books* (10th ed.). Chicago: American Library Association.

The most comprehensive guide-to-the-guides is Sheehy's *Guide to Reference Books* (also see its recent *Supplements*), an indispensable resource that lists bibliographies, indexes, abstract services, dictionaries, directories, handbooks, encyclopedias, and many other sources. However, Sheehy is not updated as frequently as other guides.

Periodical Indexes

Periodicals are the best source of information for most research projects because they offer recent, authoritative discussions of limited subjects. The biggest challenge in using periodicals is identifying and locating the dozens of relevant articles that are published each month. Although only half a dozen major journals may concentrate on your field, a useful article might appear in one of a hundred other publications. A periodical index, which is simply a listing of articles classified according to title, subject, and author, can help you determine which journals you want to locate.

More and more periodical indexes are available online—on a database or on CD-ROM—as well as in printed versions. The online versions offer three major advantages:

- They are available sooner than printed versions and are therefore more current.
- They are easier to search because they are cumulative; that is, they include entries for a number of years, not just the current year.
- They are easier to copy. You can highlight the entries you want and print them out or download them onto a floppy disk. You don't have to copy information by hand or use a photocopier.

Newspaper Indexes

Many major newspapers around the world are indexed by subject. The three most important indexed U.S. newspapers are the following:

- *The New York Times*
- *The Christian Science Monitor*
- *The Wall Street Journal*

The first two are highly reputable general newspapers; the third is the authoritative source on business, finance, and the economy.

Abstract Services

Abstract services are like indexes but also provide abstracts: brief summaries of the articles' important results and conclusions. (See Chapter 11 for a discussion of abstracts.) In most cases, reading the abstract will enable you to decide whether to search out the full article. The title of an article, alone, is often a misleading indicator of its contents.

Some abstract services, such as *Chemical Abstracts,* cover a broad field, but many are specialized rather than general. *Adverse Reaction Titles,* for instance, covers research on the subject of adverse reactions to drugs. Figure 6.5 on page 116, an excerpt from a guide included in *America: History and Life,* provides an excellent explanation of how to use that journal. Most abstract journals are organized in this way.

As is the case with periodical indexes, abstract services are going online.

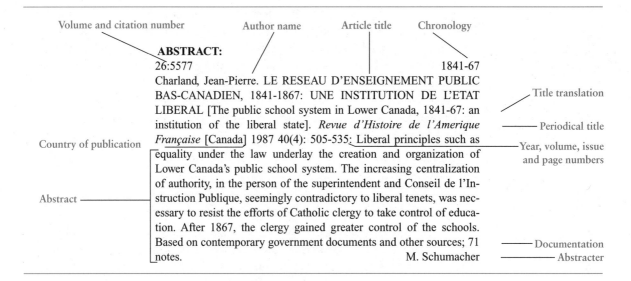

Volume and citation number   Author name   Article title   Chronology

ABSTRACT:
26:5577   1841-67
Charland, Jean-Pierre. LE RESEAU D'ENSEIGNEMENT PUBLIC
BAS-CANADIEN, 1841-1867: UNE INSTITUTION DE L'ETAT   Title translation
LIBERAL [The public school system in Lower Canada, 1841-67: an
institution of the liberal state]. *Revue d'Histoire de l'Amerique*
Française [Canada] 1987 40(4): 505-535: Liberal principles such as
equality under the law underlay the creation and organization of
Lower Canada's public school system. The increasing centralization
of authority, in the person of the superintendent and Conseil de l'Instruction Publique, seemingly contradictory to liberal tenets, was necessary to resist the efforts of Catholic clergy to take control of education. After 1867, the clergy gained greater control of the schools.
Based on contemporary government documents and other sources; 71   Documentation
notes.   M. Schumacher   Abstracter

Country of publication

Periodical title

Year, volume, issue
and page numbers

Abstract

FIGURE 6.5

Guide to the Use of an
Abstract Service

Government Publications

The U.S. government publishes thousands of documents annually. In researching
any field of science, engineering, or business, you are likely to find that a government agency has produced a relevant brochure, report, or book.

Government publications usually are not listed in the indexes and abstract
journals. The *Monthly Catalog of United States Government Publications,* put
out by the U.S. superintendent of documents, provides extensive access to these
materials. The *Monthly Catalog* is indexed by author, title, and subject. If your
research requires information published by the government before 1970, you
should know about the *Cumulative Subject Index to the Monthly Catalog,*
1900–1971 (Washington, DC: Carrollton Press, 1973–1976). This 15-volume
index eliminates the need to search through the early *Monthly Catalogs.*

Government publications are usually cataloged and shelved separately from
the others. They are classified according to the Superintendent of Documents
system, not the Library of Congress system. See the reference librarian or the
government documents specialist for information about finding government
publications in your library.

State government publications are indexed in the *Monthly Checklist of State*
Publications (1910 to date), published by the Library of Congress.

You can also access many databases on the Internet through a feature called
Telnet. For instance, the Environmental Protection Agency, NASA, the National

Science Foundation, and some 130 other government agencies offer numerous databases from which you can copy information. See the discussion of the Internet later in this chapter.

The federal government is also beginning to offer more information on CD-ROM. For additional information on government publications, consult the following three guides:

> Bailey, W. G. (1993). *Guide to popular U.S. government publications* (3rd ed.). Englewood, CO: Libraries Unlimited.
>
> Morehead, J., & Fetzer, M. K. (1992). *Introduction to United States government information sources* (4th ed.). Littleton, CO: Libraries Unlimited.
>
> Schwarzkopf, L. C. *Government reference books.* Littleton, CO: Libraries Unlimited. (A biennial publication.)

Guides to Business and Industry

Many kinds of research require access to information about regional, national, and international business and industry. If, for example, you are researching a new product for a report, you may need to contact its manufacturer and distributors to discover the range of the product's applications. You could consult one of the following two directories:

> *Thomas register of American manufacturers.* New York: Thomas. 1905 to date.
>
> *Poor's register of corporations, directors, and executives, United States and Canada.* New York: Standard and Poor's. 1928 to date.

These annual national directories provide information on products and services and on corporate officers and executives. In addition, many local and regional directories are published by state governments, local chambers of commerce, and business organizations. These resources are particularly useful for students and professionals who wish to communicate with potential vendors or clients in their own area.

Also valuable are two major investment services—publications that provide balance sheets, earnings data, and market prices for corporations.

> *Moody's investors service.* New York: Moody's.
>
> *Standard and Poor's corporation records.* New York: Standard and Poor's.

Two government periodicals contain valuable information about national business trends:

> *Federal Reserve bulletin.* Washington, DC: U.S. Board of Governors of the Federal Reserve System. (A monthly publication that outlines banking and monetary statistics.)

Survey of current business. Washington, DC: U.S. Office of Business Economics. (A monthly record of general business indicators.)

For more information on how to use business reference materials, consult the following guides:

Daniells, L. M. (1993). *Business information sources* (3rd ed.). Berkeley: University of California Press.

Lavin, M. R. (1992). *Business information: How to find it, how to use it* (2nd ed.). Phoenix: Oryx.

Woy, J. (Ed.). (1994). *Encyclopedia of business information sources* (10th ed.). Detroit: Gale.

Online Databases

Most college and university libraries—and even some public libraries—have facilities for online information searching. Technological improvements in computer science, along with continuing growth in the amount of scholarly literature produced, ensure that in the foreseeable future online searching will be the major way to access information.

Libraries today lease access to different database services, with which they communicate by computer. The largest database service is DIALOG Information Services, which in 1994 offered more than 450 different databases containing some 350 million bibliographic entries in a wide range of fields.

For more information on database services, see the *Directory of Online Databases* (New York/Cuadra/Elsevier), a quarterly publication that describes thousands of different databases. The journal *Online* is also devoted to the subject. For more information on how to use databases, see Kathleen Young Marcaccio's *Gale Directory of Databases* (Detroit: Gale, 1993, 2 vols.).

CD-ROM

CD-ROM is transforming the way large volumes of information are stored and accessed. Currently, CD-ROMs are used primarily to store research tools such as indexes, abstract services, and reference texts. However, full-text disks—holding the full texts of journal articles and books—are fast coming on the market. *Fulltext Sources Online* (Needham Heights, MA: BiblioData), a directory published twice a year, lists 4,500 journals, newspapers, newsletters, and television and radio transcripts available in full text online.

In 1994, there were 15 million CD-ROM systems in use in the United States, and by 1999, eight of ten personal computers will be sold with CD-ROM drives (Flynn, 1994, p. 10). CD-ROM is likely to become the dominant form of information storage in all research libraries; it has already become a standard

TABLE 6.2	*Using* And *and* Or *in a CD-ROM Search*	
Term	**Example**	**Explanation**
AND	nursing and technical communication	• Any item with both terms in the entry will be selected.
OR	nursing and technical communication or writing	• Any item with the term *nursing* and either the term *technical communication* or the word *writing* will be selected.

technology in business and industry, where its relatively low cost and small size make it the best storage medium.

Searching a bibliography on CD-ROM calls for the use of key words and the words *and* and *or*. Using *and* and *or*, you can expand or contract your search to find an appropriate number of items. Table 6.2 shows how to use the two words if you are searching for information about technical communication for nurses.

Because the same reference source often is available both online and on CD-ROM, the different online directories include the CD-ROM sources. For more information on how to use CD-ROM, see Kathleen Young Marcaccio's *Gale Directory of Databases* (Detroit: Gale, 1993, 2 vols.).

Using the Internet   The Internet is the information source likely to have the most profound effect on the daily lives not only of scientists, engineers, and scholars but of average citizens. The Internet (the Net) is the home to "virtual communities," electronic spaces where people discuss topics of interest. As of 1994 ("On the Internet," 1994, p. A26), there were some 27,000 mailing lists and newsgroups, where people discuss topics ranging from religion to home brewing to financial markets. People use the Internet to carry out business, too. They post their résumés, reserve airline flights, buy mail-order clothing, and advertise their products.

The Internet is a web of networks, connecting millions of computers and users. Accurate numbers are hard to find for two reasons:

• *The Net is growing too quickly.* Bollag (1994, p. A17) estimates that the Net links 20 to 30 million people using 35,000 different networks in 150 countries. With a growth rate of 10 percent per month, the Net will reach 100 million people by 1998.

• *The Net is highly decentralized.* The Net does not have a headquarters, nor are there any managers. It doesn't physically exist anywhere. Nobody knows for sure how big it is or who is connected to it.

As an information source, the Internet is unparalleled. Today, physicians in remote locations use the Net to transmit medical information to specialists for expert diagnosis. Scientists sit at their personal computers and log on to supercomputers; when it comes time to write their papers, they use electronic mail and file transfer protocol to collaborate with other scientists around the world. Political dissidents in totalitarian countries maintain contact with each other and with the outside world through the Net.

The Internet is also an excellent source of information for students. There are four basic applications on the Internet:

- electronic mail
- Usenet newsgroups
- file transfer
- remote log-in

Electronic Mail

E-mail is the simplest and most popular use of the Net. Essentially, the Net provides the same e-mail function as that carried out in local-area networks (LANs), but you can communicate with anyone in the world whose computer is connected to the Net. On the Internet, sending a message to the other side of the world usually takes less than a half hour.

E-mail on the Net is used for three basic functions:

- *Sending messages.* You can write to people in your class and to people in similar classes around the country and around the world. You can also write to companies, foundations, museums, and the president of the United States.
- *Delivering text files.* You can send a text file from your computer over the Net to another person.
- *Participating in mailing lists.* You can subscribe to thousands of mailing lists through e-mail. A mailing list is an e-mail address that automatically distributes all incoming mail to everyone on the list.

  Although thousands of mailing lists are devoted to entertainment and recreation, many are of professional interest as well. For instance, as of this writing there are mailing lists on computers and writing, on technical communication, on rhetoric, on editing, and on indexing.

For more information on e-mail, including e-mail etiquette, see Chapter 16.

Usenet Newsgroups

As with mailing lists, Usenet—a network that runs on the Internet—offers participants an opportunity to discuss issues, ask questions, and get answers. Usenet consists of thousands of different newsgroups, organized according to seven basic categories, including computer science, science, recreation, and network software. In a Usenet newsgroup, mail is not sent to individual computers, as with mailing lists, but stored on local databases, which you then access. Articles usually are posted within 24 hours (Tennant, Ober, & Lipow, 1993, p. 97). Another difference is that a newsgroup often posts copyrighted information, such as articles. In addition, newsgroups offer computer files for you to copy.

File Transfer

One of the major applications of the Internet is file transfer: moving a computer file from one computer to another. This function is called *file transfer protocol* (ftp).

In some cases, ftp requires authorization; that is, you have to have a password that lets you into the other computer. For example, two scientists collaborating on a research project would use this kind of ftp because they want to be able to transfer files privately. In other cases, however, ftp is open to anyone. This is called *anonymous ftp*.

Anonymous ftp lets you log in as an anonymous user, then search the host computer—usually a government, university, or research-center computer—to see if there are any files or software programs you want. One challenge in using anonymous ftp is that there are literally millions of files available, and therefore it is difficult to determine where a file you want is located, or whether it exists at all.

Remote Log-In

Via the Internet, you can connect your computer to another computer, using your terminal as if it were connected to the remote computer. This kind of remote log-in is called *Telnet*.

Why would you want to connect to another computer? You might want to check the collection at a library thousands of miles away. There are hundreds of OPACs—Online Public Access Catalogs—from which you can get detailed bibliographic information and, in some cases, abstracts of books or articles. Some libraries let you set up interlibrary loans through Telnet.

Because the tools used to search and create information on the Internet change month by month, nobody can predict what the Internet will look like in five years. Mosaic and Netscape, for instance, are two popular graphical software tools used for browsing the World-Wide Web, a network of multimedia sites on the Internet. Software companies are now incorporating navigational tools into their common applications, such as word processing, a development that promises to hasten the rate of change on the Internet.

The best source of information on how to use the Internet is, of course, the Internet, for only an electronic medium can keep up with the rapid change.

However, dozens of books about the Net are published each year. As of this date, some of the highly regarded guides to the Internet are the following:

Hahn, H., & Stout, R. (1994). *The Internet complete reference.* Berkeley: Osborne McGraw-Hill.

Kehoe, B. P. (1994). *Zen and the art of the Internet: A beginner's guide* (3rd ed.). Englewood Cliffs, NJ: Prentice Hall.

LaQuey, T. (1994). *The Internet companion: A beginner's guide to global networking* (2nd ed.). Reading, MA: Addison-Wesley.

Understanding the Information

Once you have gathered your books, articles, and electronic information, you need to assimilate them. When you have a good understanding of what you have, you can go on to the next stage—evaluating the information.

To understand the information you have gathered, you will want to skim it, take notes, and summarize some of it.

SCIENCE WRITER JOSEPH CHEW ON THE INTERNET

on starting to use the Internet
About 1988, while I was working here at Lawrence Berkeley Laboratory, it became obvious that I should use it, because that's how the international accelerator physics community talks to each other. It's a very geographically scattered bunch of people.

on why he uses the Internet
Partly for amusement and partly to fulfill our public information mission, to correct some of the more outrageous things people say about science and to point them toward various sources of information.

on the Internet as a teacher
I've probably learned a lot more on the Internet by making mistakes than by searching for information, because a lot of people who know something have to be provoked in some way into telling you.

on the excitement of the Internet
I think the most exciting thing about it is the presence of live experts. A FAQ [Frequently Asked Questions] file is really nothing more than an encyclopedia, and using it is nothing more than looking things up, but actually participating in the dynam-ics of the information building processes, that's useful and tremendously exciting. What you're really talking about is a professional conference without the cheap wine.

on the role of the Internet
What's the purpose of the Internet? Is it a peer-to-peer forum for experts trying to do work, or is it essentially a teaching forum? As more people get onto it, it gravitates to the latter, just by sheer force of numbers. It doesn't belong to the Unix wizards who originated it, it doesn't even belong to the scientists who have been using it for ten years. It belongs by sheer force to AOL [America On Line] and Prodigy and .Edu [educators].

on the future of the Internet
I really think that the general direction of the future is the World Wide Web and the graphical interfaces in particular, because you can just present so much more information so much more easily with text and graphics together.

Joseph Chew is the scientific coordinator of the Accelerator and Fusion Research Division at Lawrence Berkeley Laboratory in Berkeley, California.

Skimming Inexperienced writers sometimes make the mistake of trying to read every potential source. All too often, they get only halfway through one of their several books when they realize they must start writing immediately to submit their reports on time. Knowing how to skim is invaluable.

To skim a book, read the following parts:

- *the preface and introduction:* to get a basic idea of the writer's approach and methods
- *the acknowledgments section:* to learn about any assistance the author received from other experts in the field, or about the author's use of important primary research or other resources
- *the table of contents:* to get an idea of the scope and organization of the book
- *the notes at the ends of chapters or the end of the book:* to understand the nature and extent of the author's research
- *the index:* to get a sense of the coverage that the information you need receives
- *a few paragraphs from different portions of the text:* to gauge the quality and relevance of the information

To skim an article, focus on the following elements:

- *the abstract:* to get an overview of the content of the article
- *the introduction:* to get a broad understanding of the purpose, main ideas, and organization of the article
- *the notes and references:* to understand the nature and extent of the author's research
- *the headings and several of the paragraphs:* to get a sense of the organization of the article and the quality and relevance of the information

Skimming will not always tell you whether a book or article is going to be useful. A careful reading of a work that looks useful might prove disappointing. However, skimming can tell you if the work is *not* going to be useful: because it doesn't cover your subject, for example, or because its treatment is too superficial or too advanced. Eliminating the sources you don't need will give you more time to spend on the ones you do.

Taking Notes Note taking is often the first step in actually writing the document. Your notes will provide the vital link between what your sources have said and what you are going to say. You will refer to them over and over. For this reason, it is smart to take notes logically and systematically.

The best way to take notes is electronically. If you can download files from the Internet, download bibliographic references from a CD-ROM database, and do your note taking on a laptop computer, you will save a lot of time and prevent many errors.

But if you do not have access to these electronic tools, buy two packs of note cards: one 4 inches by 6 inches or 5 inches by 8 inches, the other 3 inches by 5 inches. (There is nothing special about ready-made note cards; you can make up your own out of scrap paper.) The major advantage of using cards is that they are easy to rearrange later, when you want to start outlining the report.

On the smaller cards, record the bibliographic information for each source from which you take notes. Figure 6.6 shows the information to record for books and articles.

On the large cards, write your notes. To simplify matters, write on one side of a card only and limit each card to a narrow subject or discrete concept so that you can easily reorder the information to suit the needs of your document.

Most note taking involves two different kinds of activities: paraphrasing and quoting.

Paraphrasing

A paraphrase is a restatement, in your own words, of someone else's words. *In your own words* is crucial: if you simply copy someone else's words—even a mere two or three in a row—you must use quotation marks.

What kind of material should you paraphrase? Any information that you think *might* be useful in writing the report: background data, descriptions of mechanisms or processes, test results, and so forth.

To paraphrase accurately, follow these four steps:

- *Study the original until you understand it thoroughly.*
- *Rewrite the relevant portions of the original.* If you find it easiest to rewrite in complete sentences, fine. If you want to use fragments or merely list information, make sure you haven't compressed the material

Information to Record for a Book	Information to Record for an Article
• Author	• Author
• Title	• Title of the article
• Publisher	• Periodical
• Place of publication	• Volume
• Year of publication	• Number
• Call number	• Date
	• Pages on which the article appears
	• Call number

FIGURE 6.6

Bibliographic Information
to Record

so much that you'll have trouble understanding it later. Remember, you might not be looking at the card again for a few weeks.

- *Title the card so that you'll be able to identify its subject at a glance.* The title should include the general subject and the author's attitude or approach to it, such as "Criticism of open-sea pollution-control devices."
- *Include the author's last name, a short title of the article or book, and the page number of the original.* You will need this information later when you cite your source.

Figure 6.7 on page 126 shows two paraphrased note cards based on the following discussion (adapted from Lovgren, 1994, pp. 87–88). The author is describing his business: creating metaphors to be used in graphical user interfaces on computers. The student has paraphrased each paragraph on a separate note card.

How do you find good metaphors? To be usable, a metaphor must be in the user's sphere of knowledge. The lack of understanding of this requirement frequently causes problems in interface design. Many of the interface metaphors I have seen are metaphors within the knowledge of the developer, but not within the knowledge of the user.

You have to talk to the users, not just to other developers. In my consultancy, we use a technique called *contextual interviews* to gather this information. I'll explain it by walking you through a session with a hypothetical potential user I'll call Sally.

The first thing we do is phone Sally and schedule a two-hour meeting at her place of employment. We always try to meet the person where he or she works. When we interview users outside their work environment, they give us an overly organized picture of their work, apparently forgetting that real work tends to be less structured and is often interrupted. Also, we have found that two hours is just about the right time. Longer sessions tend to distort results because the session is exhausting for both interviewer and the interviewee.

Before beginning the interview, we ask Sally if she has any comments, questions, or issues of a general nature. Our intent is to put her at her ease. As part of the preinterview, we inform Sally that the interview will be audiotaped.

We then ask Sally what she would be doing if we weren't here and ask her to start the task. While she is doing it, we ask her to describe what she is doing, why, and for whom. We also ask what kinds of difficulties she might run into that we don't see.

As follow-up, we prepare a complete report of the interview, including an annotated transcript, sketches, and copies of any forms, calendars, and other supplemental material. In annotating the transcript, I mark the nouns Sally used, the actions she mentioned, and any relationships among the nouns.

After repeating this process for several other potential users, we generally get a sense of the metaphors that might be appropriate. For example, if several people in addition to Sally refer to a "red book," and the system is to manage information in this book, then the red book becomes a likely metaphor. It works because when anybody in the organization mentions this term, everyone knows what they are talking about.

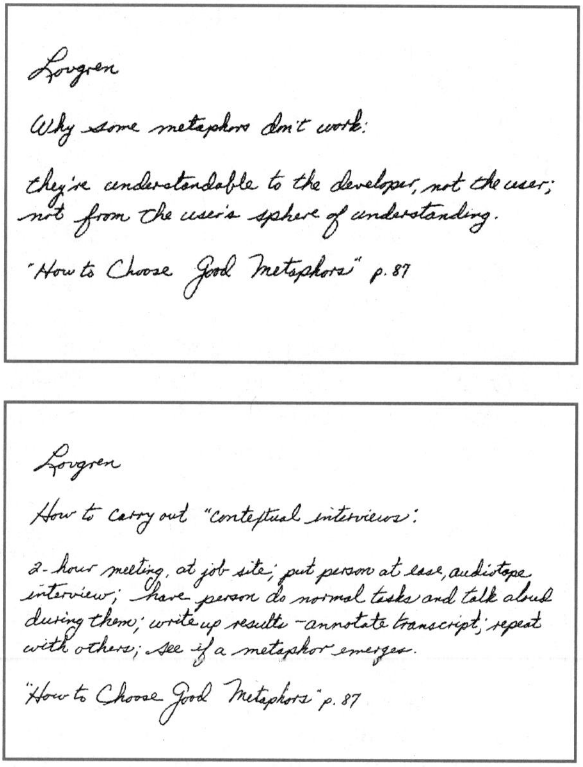

> Lovgren
>
> Why some metaphors don't work:
>
> they're understandable to the developer, not the user;
> not from the user's sphere of understanding.
>
> "How to Choose Good Metaphors" p. 87

> Lovgren
>
> How to carry out "contextual interviews:
>
> 2-hour meeting, at job site; put person at ease, audiotape
> interview; have person do normal tasks and talk aloud
> during them; write up results — annotate transcript; repeat
> with others; see if a metaphor emerges.
>
> "How to Choose Good Metaphors" p. 87

FIGURE 6.7

Paraphrased Notes

Notice how a heading provides a focus for each card. The student has omitted the information he does not need, but he has recorded the necessary bibliographic information so that he can document his source easily or return to it if he wants to reread it. There is no one way to paraphrase: you have to decide what to paraphrase—and how to do it—on the basis of your analysis of the audience and the purpose of your report.

Quoting

Sometimes you will want to quote a source, either to preserve the author's particularly well-expressed or emphatic phrasing or to lend authority to your discussion. In general, do not quote passages more than two or three sentences long, or your report will look like a mere compilation. Your job is to integrate an author's words into your own work, not merely to introduce a series of quotations.

The simplest form of quotation is an author's exact statement:

As Jones states, "Solar energy won't make much of a difference in this century."

To add an explanatory word or phrase to a quotation, use brackets:

> As Nelson states, "It [the oil glut] will disappear before we understand it."

Use ellipses (three spaced dots) to show that you are omitting part of an author's statement:

| ORIGINAL STATEMENT: | "The generator, which we purchased in May, has turned out to be one of our wisest investments." |
| ELLIPTICAL QUOTATION: | "The generator . . . has turned out to be one of our wisest investments." |

For more details on the mechanics of quoting, see the entries under "Quotation Marks," "Ellipses," and "Brackets" in Appendix A. For a discussion of different styles of documentation, see Chapter 11.

Summarizing   Summarizing, a more comprehensive form of taking notes, is the process of rewriting a passage to make it shorter while still retaining its essential message.

The two main reasons to summarize are:

- *To learn a body of information.* When you skim a source and find it to be useful, you might want to summarize it so that you can integrate it with other information in creating your own document.

- *To create one or more of the summaries that will go into your own document.* Most long technical documents contain several different kinds of summaries:

  – a letter of transmittal that provides an overview of the document
  – an abstract, a brief technical summary
  – an executive summary, a brief nontechnical summary directed to the manager
  – a summary at the end that draws together a complicated discussion

These summaries are discussed in Chapters 8 and 11.

The present discussion covers the process of summarizing printed information that you uncover in your research and might use in your own document. Although the technique described here will prove useful as you prepare the abstract and the different kinds of summaries in your own writing, the focus in this section is on extracting the essence of a passage by summarizing it.

The process of summarizing information consists of the following seven steps:

1. *Read the passage carefully several times.* Before you proceed, you must thoroughly understand what the writer is saying.

2. *Underline the key ideas.* Most writers put their main ideas in a few obvious places: titles, headings, topic sentences, transitional paragraphs, concluding paragraphs. The bodies of the paragraphs usually explain and exemplify the main points, and therefore are not the most likely places to find the general ideas of the passage.

3. *Combine the key ideas.* Take a short break and then study what you have underlined. Paraphrase the underlined ideas. Don't worry about your grammar, punctuation, or style; if you will be submitting the summary to another reader, you can revise later. At this point, you just want to see if you can reproduce the essence of the original.

4. *Check your draft against the original for accuracy and emphasis.* Reread the original to make sure your summary is accurate and reflects the writer's emphasis. In checking for accuracy you need to pay attention to a wide variety of matters, from getting individual statistics and names correct to ensuring that your version of a complicated concept represents the original faithfully. Checking for proper emphasis means getting the proportions right; if the original devotes 20 percent of its space to a particular point, your draft should not devote 5 percent or 50 percent of its space to that point.

5. *If necessary, edit your summary for style.* If your purpose in summarizing is to learn the material yourself, you don't need to edit for style. However, if you will be submitting the summary to another reader, now is the time to revise it for grammar, punctuation, usage, style, and spelling. See Chapters 14 and 15 for more information on the revision process.

6. *Record the bibliographic information carefully.* Even though a summary might contain all your own words, you still must cite it, because the main ideas are someone else's. If you don't already have the bibliographic information in an electronic form, use bibliography cards, as described in the discussion of taking notes.

7. *If necessary, add key words.* If you are creating an abstract (see Chapter 11), or if for some other reason readers will want to be able to retrieve your summary by using key words, list the key words now.

Figure 6.8 (based on McComb, 1991, pp. 19–21) is a narrative history of television technology addressed to the general reader. Figure 6.9 on page 130 is a summary that includes the key terms. This summary is 10 percent of the length of the original.

A Brief History of Television

Although it seems as if television has been around for a long time, it's a relatively new science, younger than rocketry, internal medicine, and nuclear physics. In fact, some of the people that helped develop the first commercial TV sets and erect the first TV broadcast antennas are still living today.

The Early Years

The first electronic transmission of a picture was believed to be made by a Scotsman, John Logie Baird, in the cold month of February 1924. His subject was a Maltese Cross, transmitted through the air by the magic of television (also called "Televisor" or "Radiovision" in those days) the entire distance of ten feet.

To say that Baird's contraption was crude is an understatement. His Televisor was made from a cardboard scanning disk, some darning needles, a few discarded electric motors, piano wire, glue, and other assorted odds and ends. The picture reproduced by the original Baird Televisor was extremely difficult to see—a shadow, at best.

Until about 1928, other amateur radiovision enthusiasts toyed around with Baird's basic design, whiling long hours in the basement transmitting Maltese Crosses, model airplanes, flags, and anything else that would stay still long enough under the intense light required to produce an image. (As an interesting aside, Baird's lighting for his 1924 Maltese Cross transmission required 2,000 volts of power, produced by a roomful of batteries. So much heat was generated by the lighting equipment that Baird eventually burned his laboratory down.)

Baird's electromechanical approach to television led the way to future developments of transmitting and receiving pictures. The nature of the Baird Televisor, however, limited the clarity and stability of images. Most of the sets made and sold in those days required the viewer to peer through a glass lens to watch the screen, which was seldom over seven by ten inches in size. What's more, the majority of screens had an annoying orange glow that often marred reception and irritated the eyes.

Modern Television Technology

In the early 1930s, Vladimir Zworykin developed a device known as the iconoscope camera. About the same time, Philo T. Farnsworth was putting the finishing touches on the image dissector tube, a gizmo that proved to be the forerunner to the modern cathode ray tube or CRT—the everyday picture tube. These two devices paved the way to the TV sets we know and cherish today.

The first commercially available modern-day cathode ray tube televisions were available in about 1936. Tens of thousands of these sets were sold throughout the United States and Great Britain, even though there were no regular television broadcasts until 1939, when RCA started what was to become the first American television network, NBC. Incidentally, the first true network

FIGURE 6.8

Original Passage

(Figure 6.8 continued)

transmission was in early 1940, between NBC's sister stations WNBT in New York City (now WNBC-TV) and WRGB in Schenectady.

Postwar Growth

World War II greatly hampered the development of television, and during 1941–1945, no television sets were commercially produced (engineers were too busy perfecting radar, which, interestingly enough, contributed significantly to the development of conventional TV.) But after the war, the television industry boomed. Television sets were selling like hotcakes, even though they cost an average of $650 (based on average wage earnings, that's equivalent to about $4,000 today).

Progress took a giant step in 1948 and 1949 when the four American networks, NBC, CBS, ABC, and Dumont, introduced quality, "class-act" programming, which at the time included Kraft Television Theatre, Howdy Doody, and The Texaco Star Theatre with Milton Berle. These famous stars of the stage and radio made people want to own a television set.

Color and Beyond

Since the late 1940s, television technology has continued to improve and mature. Color came on December 17, 1953, when the FCC approved RCA's all-electronic system, thus ending a bitter, four-year bout between CBS and RCA over color transmission standards. Television images beamed via space satellite caught the public's fancy in July of 1962 when Telstar 1 relayed images of AT&T chairman Frederick R. Kappell from the U.S. to Great Britain. Pay-TV came and went several times in the 1950s, 1960s and 1970s; modern-day professional commercial videotape machines were demonstrated in 1956 by Ampex; and home video recorders appeared on retail shelves by early 1976.

The first electronic transmission of a picture was produced by Baird in 1924. The primitive equipment produced only a shadow. Although Baird's design was modified by others in 1920s, the viewer had to look through a glass lens at a small screen that gave off an orange glow.

Zworykin's iconoscopic camera and Farnsworth's image dissector tube—similar to the modern CRT—led in 1936 to the development of modern TV. Regular broadcasts began in 1939, on the first network, NBC. Research stopped during WWII, but after that, sales grew, even though sets cost approximately $650, the equivalent of $4,000 today.

Color broadcasts began in 1953; satellite broadcasting began in 1962; and home VCRs were introduced in 1976.

Key terms: television, history of television, NBC, color television, satellite broadcasting, video cassette recorders, Baird, Zworykin, Farnsworth.

FIGURE 6.9

Summary of the Original

Passage

Evaluating the Information

Earlier, this chapter made the point that good information is accurate, unbiased, comprehensive, current, and clear. Following are some ways to evaluate information.

Evaluating for Accuracy   There is no special technique for determining the accuracy of the information you have gathered. Stay alert, however, to three factors:

- *Reputation.* What is the reputation of the source, either the author or the publication? A book should be published by a reputable trade, academic, or scholarly publisher. A journal should be sponsored by a professional association or university. Read the list of editorial board members; they should be well-known names in the field. If you have any doubts about the authority of a book or journal, ask the reference librarian or a professor in the appropriate field.

  Of course, nobody believes supermarket tabloid reports of World War II bombers on the moon, but mainstream publications also have varying reputations for accuracy. Be careful about citing research from publications that repackage other people's information. For instance, the weekly news magazines such as *Time* and *Newsweek* base many of their articles on other articles in the professional literature, and sometimes the accuracy slips as the information is translated to a level appropriate for the general audience.

  Be alert to the reputations of individual authors. Many books and journals include biographical sketches. Does the author appear to have solid credentials in the subject? Look for academic credentials, other books or articles written, membership in professional associations, awards, and so forth. If no biographical information is provided, consult a "Who's Who" of the field.

- *Contradictions.* If you find contradictions in your information, that's a clue that you need to do more research.

- *Hunches.* If some information just doesn't sound right to you, keep searching.

Evaluating for Bias   Evaluate the source of the information. Trade publications—magazines about a particular industry or group—often seek to promote the interests of that industry or group. Don't automatically assume the accuracy of information in trade publications for loggers *or* environmentalists. And some authors, such as hosts of television and radio shows, are less scrupulous with their facts than they ought to be. Using them as sources is doubly dangerous: not only is their information likely to be tainted, but your readers are likely to assume that it is even if it isn't.

Evaluating for Comprehensiveness

Doing research is hard work; it's a natural human impulse to hope we have "enough information" for our report. But our purpose is not to gather just enough information to support a report. Our purpose is to find the best information. That means tracking down sources even if we are quite sure they won't have any useful information. Only with perseverance can we be certain that we have searched for an appropriate diversity of viewpoints.

Evaluating for Timeliness

Check the date of publication. In high-technology fields, in particular, a five-year-old article or book is likely to be of little value (except, of course, for historical studies).

Evaluating for Clarity

Unclear information is at least a nuisance; you don't want to have to reread an article five times to get the point just because the author and the editor didn't do their jobs well. But unclear information is often a symptom of a more serious problem. Authors who didn't take the time to make the presentation clear might not have taken the time to check their facts, or perhaps they feared that a clear presentation would reveal logical problems.

Of course, information that is difficult to understand is not necessarily unclear; it could be highly technical. You might have to do additional research in order to understand it.

The Special Case of the Internet

The most difficult kind of material to evaluate is information from the Internet. The Internet's greatest strengths—that it is open to anyone who is connected, and that one person's voice is heard as loudly as that of anyone else—pose a real challenge when it comes to evaluating sources of information. In most cases, information appears on the Internet without passing through the formal review procedure that is characteristic of books and professional journals. Therefore, you have to be alert to three possibilities:

- that the source is not reputable
- that the information has been revised by someone other than the source, without the source's knowledge or consent
- that the source is not really the person identified

Although this last situation is uncommon, some people use other people's identities to transmit fraudulent or malicious information through the Internet, a practice called *spoofing* (Seabrook, 1994, p. 75). For this reason, before you use any information acquired through the Net, it is a good idea to get in touch with the source by phone, mail, or fax to see if he or she really wrote it—in the version you have in your hands.

Once you have performed your basic primary and secondary research, you can turn your attention to shaping the document. Chapters 8–11 discuss techniques of drafting the text. Chapters 12 and 13 discuss graphics and document design and layout. Keep in mind, however, that because you will probably return to do more research, it is smart to keep all your research in order. Don't toss out any information you have gathered, even if you think you will never need it again.

WRITER'S CHECKLIST

1. Have you focused your topic?
2. Have you used techniques to generate ideas?
3. Have you devised a strategy for conducting secondary and primary research?
4. Have you consulted the appropriate reference books, including periodicals indexes, newspaper indexes, abstract services, government publications, guides to business and industry, online databases, CD-ROM, and the Internet?
5. Have you studied the information by skimming, taking notes, and summarizing?
6. Have you evaluated the information according to
   - ❏ accuracy?
   - ❏ bias?
   - ❏ comprehensiveness?
   - ❏ timeliness?
   - ❏ clarity?
7. In evaluating information from the Internet, have you been especially careful about questions of authorship?

EXERCISE

Check with your instructor about whether this assignment should be done individually or in a small group. Choose a topic on which to write a report for this course (see Chapters 21-23 for information on reports). Make sure the topic is sufficiently focused that you will be able to cover it in some detail. Subjects that can be focused into topics include the following:

- A campus problem, such as a shortage of parking space. What are the symptoms and causes of the problem? What effects has the problem had on life at the college and its surrounding area? Has the college or a community group attempted to investigate the problem? What have been the findings? Conduct your own analysis of the problem, draw conclusions, and recommend what should be done next.
- The evolving role of the Internet as an information source. Which aspects of the Internet will prove most useful for researchers? Which resources already in place are most likely to change over the next few years? What new services are likely to evolve? What possible problems—technical, political, economic, social, ethical—must be faced if the potential of the Internet is to be realized?
- Alternative fuels and energy sources for automobiles. What alternatives are being developed now? What are their strengths and weaknesses? What are the factors that encourage their development? What are the social or political barriers to their widespread introduction? Which is most likely to succeed in the marketplace?
- Job prospects in a particular field. What data exist on job prospects, both in your own geographic area and nationally? What are the trends? Does one aspect of the field appear to have better prospects than other aspects? What steps can a student take to ensure the best preparation for the job market? How well does the curriculum meet the needs of industry?
- The evolving role of the technical communicator. Determine the different skills and backgrounds of technical communicators in your area. What kinds of documents do they make? Do they use multimedia or hypertext? Where did they learn the skills they use? How well equipped is your college to prepare tomorrow's technical communicators?

What steps could be taken to improve their preparation?

a. Using one of the guides-to-the-guides, plan a strategy for researching your topic.

1. Which guides, handbooks, dictionaries, and encyclopedias contain the background information you should read first?

2. Which basic reference books discuss your topics?

3. Which major indexes and abstract journals cover your topic?

4. Which of these basic tools is available online in your library? In which medium (for instance, online bibliographies, CD-ROM) are they available?

5. If you have access to the Internet, does it provide additional resources: mailing lists, newsgroups, ftp, experts you can query through e-mail?

b. Write down the call numbers of the three indexes and abstract journals most relevant to your topic (if you are using hard copy).

c. Make up a preliminary bibliography of two books and five articles that relate to your topic.

d. Using the bibliography from c, above, find one of the works listed and write a brief assessment of its value.

e. Paraphrase any three paragraphs, each on a separate note card, from the first two pages of an article listed in your bibliography in c, above. Also note at least two quotations, each on a separate card: one should be a complete sentence, and one an excerpt from a sentence. Include a photocopy of the original when you submit the assignment, with the sentences you have quoted highlighted or underlined.

f. Write a 500-word summary of an article listed in your bibliography from c, above. Include a photocopy of the article when you submit the assignment.

REFERENCES

Bollag, B. (1994, June 29). The "great equalizer." *Chronicle of Higher Education, 40*(43), pp. A17, A19.

Flynn, L. (1994, April 24). CD-ROM's: They're not just for entertainment. *New York Times,* p. 3: 10.

Lovgren, J. (1994). How to choose good metaphors. *IEEE Software, 11*(3), pp. 86–88.

McComb, G. (1991). *Troubleshooting and repairing VCRs* (2nd ed.). Blue Ridge Summit, PA: TAB/McGraw-Hill.

On the Internet. (1994, March 23). *Chronicle of Higher Education, 40*(29), p. A26.

Seabrook, J. (1994, June 6). My first flame. *New Yorker, 70*(16), pp. 70–79.

Tennant, R, Ober, J., & Lipow, A. G. (1993). *Crossing the Internet threshold: An instructional handbook.* San Carlos, CA: Library Solutions Press.

Performing Primary Research

Chapter 6 began by explaining the difference between secondary and primary research: *secondary research* is the process of finding out information that other people have created or gathered, whereas *primary research* is the process of creating or gathering that research yourself. This chapter discusses basic techniques of primary research, including how to conduct inspections, experiments, field research, and interviews; how to write letters of inquiry; and how to administer questionnaires.

Conducting Inspections, Experiments, and Field Research

Although the library and the Internet offer a wealth of secondary sources, as explained in Chapter 6, often you will need to conduct primary research in addition to consulting the work of others. Inspections, experiments, and field research are three common primary-research techniques. To inspect something is, of course, to look at it, either with or without the use of specialized tools. To experiment is to test a theory about why something happened, or whether something would happen. For instance, an engineer could devise an experiment to test whether cars with antilock brakes handle better in the rain than do cars with conventional brakes. To conduct field research is to study people (or other animals) working in their natural environment.

Conducting Inspections

Regardless of the field you work in, you are likely to read many sentences that begin "An inspection was conducted to determine. . . ." A civil engineer can often determine what caused the cracking in a foundation by inspecting the site. An accountant can learn a lot about the financial health of an organization by inspecting the company's financial records.

These people are looking at a site, an object, or a document (for example, the financial records) and applying their knowledge and professional judgment to what they see. Sometimes the inspection techniques are more complicated. A civil engineer inspecting foundation cracking might want to test hunches by taking a soil sample back to the lab and testing it. An accountant checking the books might need to perform some computerized analyses on the information.

When you carry out an inspection, be sure to take good notes. Try to answer the appropriate journalistic questions—*who, what, when, where, why,* and *how*—as you inspect, or as soon after as possible. Where appropriate, photograph or sketch the site or print out the output from computer-assisted inspections. You will probably need the data later for the document.

Conducting Experiments

Learning how to conduct the many kinds of experiments used in a particular field can take months or even years. Therefore, this discussion can serve only as a brief introduction.

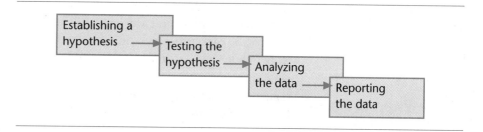

FIGURE 7.1

Conducting Experiments

In many cases, conducting an experiment involves four phases, as shown in Figure 7.1.

- *Establishing a hypothesis.* A hypothesis is an informed guess about the relationship between two factors. For instance, if you want to determine the relationship between gasoline octane and miles per gallon, you can set up an experiment with the following hypothesis: "A car will get better mileage with 89 octane gas than with 87 octane gas."

- *Testing the hypothesis.* In most cases, you need an *experimental group* and a *control group*—two groups identical except for the condition you are studying: in this case, the gasoline used. The control group would be a car running on 87 octane. The experimental group would be an identical car running on 89 octane. The experiment would consist of driving the two cars over an identical course at the same speed—preferably in some sort of controlled environment such as a laboratory—over a given distance, such as 1,000 miles. At the end of the 1,000 miles you would measure how much gasoline each car used and calculate the miles per gallon. The results would either support or refute your original hypothesis.

    An experiment as neat as this one is often difficult to set up. For instance, you probably wouldn't be able to get two identical cars and run them for 1,000 miles in a lab. So you do the best you can: you run your own car with two or three tanks of 87 octane and calculate your miles per gallon. Then you run it on the same amount of 89 octane and do the same calculations. Of course, you try to make sure that you are subjecting the car to the same driving conditions to control other variables that could affect the results. The important point is that you try to make the conditions for both groups as similar as possible.

- *Analyzing the data.* You must try to understand whether your data show merely a correlation—one factor changes along with another—or a real causal relationship. For instance, nonsmokers tend to live seven years longer than smokers. The tobacco industry claims this is only a statistical coincidence. Most scientists see the relationship as causal: smoking causes people to die sooner. Another example: we know that sports cars are involved in more fatal accidents than sedans, but we don't know whether the car has much to do with that fact, or whether driving habits are the important factor. Analyzing the data is a challenge because it is tempting to see what you want to see.

- *Reporting the data.* When researchers explain their findings in a report, they try to explain as clearly as they can what they did, why they did it, what they saw, what it means, and what ought to be done next. For more information on writing reports, see Chapters 21–23.

Conducting Field Research

Whereas an experiment is quantitative (it yields statistical data that can be measured), field research is usually qualitative; it yields data that cannot be measured or cannot be measured as precisely as experimental data. Often in field research you seek to understand the quality of an experience. For instance, you might want to understand how a new seating arrangement would affect group dynamics in a classroom. You could design a study in which you observed and recorded the classes and perhaps interviewed the students and the instructor about their reactions. Then you could do the same in a traditional classroom and compare the results.

Some kinds of studies have both quantitative and qualitative elements. In the case of the classroom seating arrangements, you could create some quantitative measures, such as the number of times students talked with each other or the length of the interchanges between them. In addition, you could create questionnaires to elicit the opinions of the students and the instructor. If you used these quantitative measures on a sufficiently large number of classrooms, you could gather valid quantitative information.

When you are doing quantitative or qualitative studies on the behavior of animals—from rats to monkeys to people—try to minimize two common problems:

- *The effect of the experiment on the behavior you are studying.* When a television camera crew arrives on the scene to cover a protest demonstration, the protest becomes more animated. Similarly, if you are studying the effects of the classroom seating arrangement, you have to try to minimize the effects of your being there. For instance, make sure that the camera is placed unobtrusively, and that it is set up before the students arrive, so they don't see the process. Still, anytime you bring in a videotape camera, you can never be sure that what you witness is typical. Even an outsider who sits quietly taking notes can disrupt typical behavior.
- *Bias in the recording and analysis of the data.* This problem can occur because researchers want to confirm their hypotheses. For instance, in an experiment to determine whether word processors help students write

better, the researcher is likely to see improvement where other people don't. For this reason, the experiment should be designed so that it is *double blind*. That is, the students doing the writing that is to be studied shouldn't know what the experiment is about, so they won't change their behavior to support or negate the hypothesis. And when the data are being analyzed, they should be disguised so that you don't know whether you are examining the results from the control group or the experimental. If the control group wrote in ink and the experimental group used word processors, the control papers should be typed on a word processor, so that all the papers look identical as you analyze the writing.

Conducting an experiment or field research is relatively simple; the hard part is designing your study so that it accurately measures what you want to measure.

Conducting Interviews

The previous section mentioned that researchers often conduct interviews in field research. Interviews are also extremely useful when you need information on subjects too new to have been discussed in the professional literature or inappropriate for widespread publication (such as local political questions). Most students are inexperienced at interviewing and hence are reluctant to do it. Interviewing, like any other communication skill, requires practice. The following discussion explains how to make interviewing less intimidating and more productive.

Choosing a Respondent

Follow these three steps in choosing a *respondent,* a person to interview:

1. *Determine what you want to find out.* Only then can you begin to search for a person who can provide the necessary information.

2. *Determine who could provide the information.* The ideal respondent is an expert willing to talk. Many times your topic will dictate whom you should ask. If you are writing about research at your university, for instance, the logical choice would be a researcher involved in the project. Sometimes, however, you might be interested in a topic about which a number of people could speak knowledgeably, such as the reliability of a particular kind of office equipment or the reasons behind the growth in the number of adult students. Use directories, such as local industrial guides, to locate the names and addresses of potential respondents.

3. *Determine whether the person is willing to be interviewed.* On the phone or in a letter, state what you want to ask about; the person might not be able to help you but might be willing to refer you to someone who can. And be

sure to explain why you have decided to ask the respondent; a well-chosen compliment works better than admitting that the person you really wanted to interview is out of town. Don't forget to mention what you plan to do with the information, such as write a report or give a talk. Then, if the person is willing to be interviewed, set up an appointment at his or her convenience.

Preparing for the Interview

Follow these two steps in preparing for the interview:

1. *Do your homework.* Never give the impression that you are conducting the interview to avoid doing other kinds of research. If you ask questions that are already answered in the professional literature, the respondent might become annoyed and uncooperative.

2. *Prepare good questions.* Good questions are clear, focused, and open.

   – *Clear.* The respondent should be able to understand what you are asking.

UNCLEAR:	Why do you sell Trane products?
CLEAR:	What are the characteristics of Trane products that led you to include them in your product line?

   The unclear question can be answered a number of unhelpful ways: "Because they're too expensive to give away" or "Because I'm a Trane dealer." The clear version of the question forces the respondent to think about specific reasons for selling Trane products as opposed to (or in addition to) other brands.

   – *Focused.* The question must be sufficiently narrow to be answered briefly. If you want additional information, you can ask a follow-up question.

UNFOCUSED:	What is the future of the computer industry?
FOCUSED:	What will the American chip industry look like in ten years?

   – *Open.* Your purpose is to get the respondent to talk. Don't ask a lot of questions that have yes or no answers.

CLOSED:	Do you think the federal government should create industrial partnerships?
OPEN:	What are the advantages and disadvantages of the federal government's creating industrial partnerships?

Conducting the Interview

Remember the following four points as you begin the interview:

- *Arrive on time.*
- *Thank the respondent for taking the time to talk with you.*
- *Repeat the subject and purpose of the interview and what you plan to do with the information.*

- *If you wish to tape-record the interview, ask permission.* Taping makes some people uncomfortable.

As you conduct the interview, keep in mind the following four guidelines:

- *Take notes.* Write down important concepts, facts, and numbers, but don't take such copious notes that you are still writing when the respondent finishes an answer. It is important to maintain eye contact with the respondent.
- *Start with prepared questions.* Because you are likely to be somewhat nervous at the start, you might leave out important questions. Therefore, it is wise to begin with prepared questions.
- *Be prepared to ask follow-up questions.* Listen carefully to the respondent's answer. Be ready to ask a follow-up question or request a clarification. Have your other prepared questions ready, but be willing to deviate from them. In a good interview, the respondent probably will lead you in directions you had not anticipated.
- *Be prepared to get the interview back on track.* Gently return to the point if the respondent begins straying unproductively, but don't interrupt rudely or show annoyance.

Follow these three guidelines as you conclude the interview:

- *Thank the respondent.* After all your questions have been answered (or you have run out of time), thank the respondent again.
- *Ask for permission for a follow-up interview.* If a second meeting would be useful—and you think the person would be willing to talk with you further—ask for the second meeting now.
- *Ask for permission to quote the respondent.* If you might want to quote the respondent by name, ask permission now.

Following Up after the Interview

Do two things after the interview:

- *Write up the important information.* Do it while the interview is still fresh in your mind. (This step is unnecessary, of course, if you have recorded the interview.) If you will be printing a transcript of an interview, make the transcript now.
- *Send a brief thank-you note.* Within a day or two of the interview, show the respondent that you appreciate the courtesy extended to you and that you value what you have learned. In the letter, confirm any previous offers you made, such as to send a copy of the report.

Figure 7.2 on page 142 is a portion of a transcript of an interview with an attorney specializing in information technology. The interviewer is a student who is writing about legal aspects of software ownership.

Q. Why is copyright ownership important in marketing software?
A. If you own the copyright, you can license and market the product and keep other people from doing so. It could be a matter of millions of dollars if the software is popular.

Q. Shouldn't the programmer automatically own the copyright?
A. If they wrote the program on their own time, they should and do own the copyright.

Q. So "on their own time" is the critical concept?
A. That's right. We're talking about the "work-for-hire" doctrine of copyright law. If I am working for you, anything I make under the terms of my employment is owned by you.

Q. What is the complication, then? If I make the software on my machine at home, I own it; if I'm working for someone, they own it.
A. Well, the devil is in the details. Often the terms of employment are casual, or there is no written job description or contract for the particular piece of software.

Q. Can you give me an example of what you're referring to?
A. Sure. There was a 1992 case, *Aymes* v. *Bonelli*. Bonelli owned a swimming pool and hired Aymes to write software to handle recordkeeping on the pool. This was not part of Bonelli's regular business; he just wanted a piece of software written. The terms of the employment were casual. Bonelli paid no health benefits, Aymes worked irregular hours, usually unsupervised— Bonelli wasn't a programmer. When the case was heard, the court ruled that even though Bonelli was paying Aymes, Aymes owned the copyright because of the lack of involvement and participation by Bonelli. The court found that the degree of skill required by Aymes to do the job was so great that, in effect, he was creating the software by himself, even though he was receiving compensation for it.

Q. How can such disagreements be prevented? By working out the details ahead of time?
A. Exactly. The employer should have the employee sign a statement that the project is being carried out as work-made-for-hire, and should register the copyright with the U.S. Copyright Office in Washington. Conversely, employees should try to have the employer sign a statement that the project is not work-for-hire, and should try to register the copyright themselves.

Q. And if agreement can't be reached ahead of time?
A. Then stop right there. Don't do any work.

FIGURE 7.2

Excerpt from an Interview

Notice how the student prompts the attorney to expand her answers. Also notice how the student responds to the attorney's answers, making the interview more of a discussion.

Sending Letters of Inquiry

A letter of inquiry is often a useful alternative to a personal interview. If you are lucky, the respondent will provide detailed and helpful answers. However, the person might not understand what information you are seeking or might not want to take the trouble to help you. Also, you can't ask follow-up questions in a letter, as you can in an interview. Although the strategy of the inquiry letter is essentially that of a personal interview—persuading the reader to cooperate and phrasing the questions carefully—inquiry letters in general are less successful, because the reader has not already agreed to provide information and therefore sometimes does not respond. In addition, an inquiry letter doesn't give you the opportunity to follow up, as an interview does.

For a full discussion of inquiry letters, see Chapter 16.

Administering Questionnaires

Questionnaires enable you to solicit information from a large group of people. Although they provide a useful and practical alternative to interviewing dozens of people spread out over a large geographical area, questionnaires rarely yield completely satisfactory results, for three reasons:

- *Some of the questions will misfire.* No matter how careful you are, some questions are not going to work: the respondents will misinterpret them or supply useless answers.
- *You won't obtain as many responses as you want.* The response rate will almost never exceed 50 percent; in most cases, it will be closer to 10 or 20 percent.
- *You cannot be sure the respondents are representative.* In general, people who feel strongly about an issue are much more likely to respond than are those who do not. For this reason, be careful in drawing conclusions based on a small number of responses to a questionnaire.

When you send a questionnaire, you are asking the recipient to do a favor. If the questionnaire requires only two or three minutes to complete, of course you are more likely to receive a response than if it requires an hour. Your goal, then, should be to construct questions that will elicit the information you need as simply and efficiently as possible.

Creating Effective Questions

Effective questions are unbiased and clearly phrased:

- *Use unbiased language.* Don't ask "Should we protect ourselves from unfair foreign competition?" Instead, ask "Are you in favor of imposing tariffs on men's clothing?"

- *Be specific.* If you ask "Do you favor improving the safety of automobiles?" only an eccentric would answer "No." You are more likely to get a useful response if you ask "Do you favor requiring automobile manufacturers to equip new cars with side-impact air bags, which would raise the price $300 per car?"

As you make up the questions, keep in mind that there are several formats from which to choose:

- *Multiple choice*

  Would you consider joining a company-sponsored sports team?
  Yes _____   No _____

  How do you get to work? (Check as many as apply.)
  my own car _____
  car/van pool _____
  bus _____
  train _____
  walk _____
  other _____ (please specify)

- *Likert scale.* Likert-scale questions consist of a statement to which the respondent registers the degree of agreement or disagreement.

  The flextime program has been a success in its first year.
  Strongly Disagree _____ _____ _____ _____ _____ _____ Strongly Agree

  Although opinions differ, most statisticians recommend using an even number of possible responses (six, in this case); with an odd number, too many respondents choose the middle response, which does not provide useful data.

- *Semantic differentials.* Semantic differential questions call for the respondent to register a response along a continuum between a pair of opposing adjectives. Usually, semantic differential questions are used to gauge a person's subjective response to a task, an experience, or an object.

simple	____	____	____	____	____	____	difficult
attractive	____	____	____	____	____	____	ugly
interesting	____	____	____	____	____	____	boring
neat	____	____	____	____	____	____	sloppy

- *Ranking.* Ranking questions call for the reader to indicate a priority among a number of alternatives.

Please rank the following work schedules in order of preference. Put a 1 next to the schedule you would most like to have, a 2 next to your second choice, and so on.
8:00–4:30 _____
8:30–5:00 _____
9:00–5:30 _____
flexible _____

- *Short answer.* Short-answer questions call for the reader to respond briefly using phrases or sentences.

    What do you feel are the major *advantages* of the new parts-requisitioning policy?
    1. _____
    2. _____
    3. _____

- *Short essay.* Short-essay questions call for a longer response.

    The new parts-requisitioning policy has been in effect for a year. How well do you think it is working?
    _____
    _____
    _____
    _____
    _____
    _____
    _____

The primary advantage of an essay question is that sometimes you will discover information you never would have found using closed-end questions. But keep in mind that you will receive fewer responses if you ask for essay answers; moreover, essays, unlike other kinds of answers, cannot be quantified precisely. A simple statement with a specific number in it—"Seventy-five percent of the respondents own two or more PCs"— helps you make a clear and convincing case. However thoughtful and persuasive an essay answer may be, it is subject to the interpretations of different readers.

After you have created the questions, write a letter or memo to accompany the questionnaire. A letter to someone outside your organization is basically an inquiry letter (sometimes with the questions themselves on a separate sheet); therefore, it must clearly indicate who you are, why you are writing, what you plan to do with the information, and when you will need it. (See Chapter 16 for a discussion of inquiry letters.) For people within your organization, a memo accompanying a questionnaire should provide the same information.

Testing the Questionnaire

Before you send *any* questionnaire, show it and the accompanying letter or memo to a few people who can help you see any problems in them. After you revise the materials, test them on people whose backgrounds are similar to those of your real respondents. Revise the materials a second time, and, if possible, test them again. Remember, after you send out the questionnaire, you cannot revise it and send it to the same people.

Sending the Questionnaire After drafting the questions and testing them, administer the questionnaire. Determining whom to send it to can be simple or difficult. If you want to know what the residents of a particular street think about a proposed construction project, your job is easy. But if you want to know what mechanical-engineering students in colleges across the country think about their curricula, you will need background in sampling techniques to isolate a representative sample.

Be sure also to include a self-addressed, stamped envelope with questionnaires sent to people outside your organization.

Figure 7.3 shows a sample questionnaire.

WRITER'S CHECKLIST

1. If appropriate, have you
   □ conducted inspections?
   □ conducted experiments?
   □ performed field research?
   □ conducted interviews?
   □ sent letters of inquiry?
   □ administered questionnaires?

EXERCISES

1. The exercise in Chapter 6 called for you to choose a topic on which to write a report for this course and then perform several kinds of secondary research. For the same assignment, perform the following kinds of primary research:
   a. Using a local industrial guide, list five persons who might have firsthand knowledge of your topic. Arrange for and carry out an interview with one of the five.
   b. Make up two sets of questions: one for an interview with one of the persons on your list from exercise 1a, and one for a questionnaire to be sent to a large group of people.

2. Revise the following questions from interviews to make them more effective. In a brief paragraph each, explain why you have revised the question as you have.
   a. What is the role of communication in your daily job?
   b. Do you think it is better to relocate your warehouse or go to just-in-time manufacturing?
   c. Isn't it true that it's almost impossible to train an engineer to write well?
   d. Where are your company's headquarters located?
   e. Is there anything else you think I should know?

3. Revise the following questions from questionnaires to make them more effective. In a brief paragraph each, explain why you have revised the question as you have.
   a. Does your company provide tuition reimbursement for its employees?
      Yes _____   No _____
   b. What do you see as the future of bioengineering?
      _____
      _____
      _____
   c. How satisfied are you with the computer support you receive?
      _____
      _____
   d. How many employees work at your company?
      5–10 _____   10–15 _____   15 or more _____
   e. What kinds of documents do you write most often?
      memos _____
      letters _____
      reports _____

September 6, 19XX

To: All employees
From: William Bonoff, Vice-President of Operations
Subject: Evaluation of the Lunches Unlimited food service

As you may know, every two years we evaluate the quality and cost of the food service that caters our lunchroom. We would like you to help by sharing your opinions about the food service. Your anonymous responses will help us in our evaluation. Please drop the completed questionnaires in the marked boxes near the main entrance to the lunchroom. Thank you for your cooperation.

1. Approximately how many days per week do you eat lunch in the lunchroom?
   0 _____   1 _____   2 _____   3 _____   4 _____   5 _____

2. At approximately what time do you eat in the lunchroom?
   11:30–12:30 _____   12:00–1:00 _____   12:30–1:30 _____   varies _____

3. It is always easy to find a clean table.
   Strongly Disagree _____   _____   _____   _____   _____   _____   Strongly Agree

4. The Lunches Unlimited personnel are polite and helpful.
   Strongly Disagree _____   _____   _____   _____   _____   _____   Strongly Agree

5. Please comment on the quality of the different kinds of food you have eaten in the lunchroom.
   a. Hot meals (daily specials)
      excellent _____   good _____   satisfactory _____   poor _____
   b. Hot dogs and hamburgers
      excellent _____   good _____   satisfactory _____   poor _____
   c. Sandwiches
      excellent _____   good _____   satisfactory _____   poor _____
   d. Salads
      excellent _____   good _____   satisfactory _____   poor _____
   e. Desserts
      excellent _____   good _____   satisfactory _____   poor _____

6. What *foods* would you like to see served that are not served now?
   _____

7. What *beverages* would you like to see served that are not served now?
   _____

8. Please comment on the prices of the foods and beverages served.
   a. Hot meals (daily specials)   d. Desserts
      too high _____ fair _____ a bargain _____   too high _____ fair _____ a bargain _____
   b. Hot dogs and hamburgers   e. Beverages
      too high _____ fair _____ a bargain _____   too high _____ fair _____ a bargain _____
   c. Sandwiches
      too high _____ fair _____ a bargain _____

9. Would you be willing to spend more money for a better-quality lunch, if you thought the price was reasonable?   yes, often _____   sometimes _____   not likely _____

10. Please provide whatever comments you think will help us evaluate the catering service.
    _____

FIGURE 7.3

Questionnaire

Developing the Argument

At this point in the writing process, you know to whom you are writing, and why, and you have performed most of the primary and secondary research. Although you will likely revise your analysis of your audience and purpose, and you will probably continue to gather information throughout the entire writing process, now it is time to start developing your argument.

This chapter begins with a brief review of outlining and guidelines for drafting the body, focusing on how to use the computer effectively as a drafting tool. The chapter then turns to a discussion of argument, explaining how to craft a convincing case for your point of view. Next the chapter discusses basic patterns for organizing the argument. The chapter ends with a discussion of ways to introduce and conclude the body.

A Brief Review of Outlining

Chapters 6 and 7 offered strategies for generating ideas and gathering more information, concentrating on using the library; performing experiments, field research, and inspections; conducting interviews; and sending out questionnaires. After gathering a mass of information, most writers like to create an outline, a process that is fundamental to organizing an effective argument. This section reviews the basic principles of outlining.

Creating an outline involves five main tasks, as shown in Figure 8.1.

Placing Similar Items Together in a Group
If you used branching or clustering to generate ideas, you have already grouped similar items. But if you used brainstorming, you have not. Start by looking at the first item on your brainstorming list and determining what major category it belongs to. For instance, it could be an item that belongs in the background section. Scan the rest of the list, noting other items that belong in the background section. Link them together. Then go back to the top of the list and determine the logical category of the next item on the list. Link it with other items that

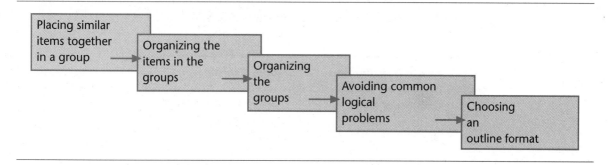

FIGURE 8.1

Steps in Outlining

belong in the same category. Repeat this process until all the items are either grouped into logical categories or discarded.

Even though the outline is incomplete, some writers like to start drafting the document at this point. They want to see what they have to say, and the only sure way to do that is to start to write. Once they have started drafting, they feel better able to sequence items. They juggle paragraphs instead of juggling outline items. Other writers feel more comfortable working from a more refined outline. For them, the next step is to sequence the items within each group.

Organizing the Items in the Groups   In some cases, sequencing the items in each group is a simple matter—one item might precede another in time, or perhaps one item is more important than another and therefore should come first. Likewise, one item in the group might represent the problem, another the methods for solving it, and the next the solution. Later, this chapter offers a detailed discussion of eight basic patterns of organizing an argument.

Organizing the Groups   To organize the groups, follow the same logical processes you used to organize the items within each group. Often your readers will expect a particular structure. For instance, the body of a report begins with an introduction, followed by the detailed discussion and the major findings. If your material does not lend itself to a traditional pattern of organization, you'll need to create one that your readers will find clear and easy to follow.

Avoiding Common Logical Problems   As you create and refine your outline, avoid two common logical problems:

- faulty coordination
- faulty subordination

Faulty coordination occurs when a writer equates items that are not of equal value or not of the same level of generality. Figure 8.2 shows a portion of an outline with faulty coordination and a corrected version.

Faulty subordination occurs when an item is made a subunit of a unit to which it does not belong, as shown in Figure 8.3.

Faulty Coordination	Proper Coordination

Common household tools
1. screwdrivers
2. drills
3. claw hammers
4. ball-peen hammers
	Common household tools
	1. screwdrivers
	2. drills
	3. hammers
	    3.1 claw hammer
	    3.2 ball-peen hammer

FIGURE 8.2

Faulty Coordination and Proper Coordination

Faulty Subordination	Proper Subordination
Power sources for lawnmowers 1. manual 2. gasoline 3. electric 4. riding mowers	Power sources for lawnmowers 1. manual 2. gasoline 3. electric

FIGURE 8.3

Faulty Subordination and Proper Subordination

In Figure 8.3, "riding mowers" is out of place because it is not a power source of a lawnmower. Whether it belongs in the outline at all is another question, but it certainly doesn't belong here.

A second kind of faulty subordination occurs when only one subunit is listed under a unit. The solution is to incorporate the single subunit into the unit: in Figure 8.4, item 1.1 is deleted and restated as item 1.

Single Subunit	Proper Subordination

Types of sound-reproduction systems
1. records
   1.1 phonograph records
2. tapes
   2.1 cassette
   2.2 open reel
   2.3 DAT
3. compact discs
	Types of sound-reproduction systems
	1. phonograph records
	2. tapes
	   2.1 cassette
	   2.2 open reel
	   2.3 DAT
	3. compact discs

FIGURE 8.4

Single Subunit and Proper Subordination

Choosing an Outline Format

If you are creating an outline purely for your own use, you don't have to worry about its format, as long as you can understand it and work effectively from it. However, you should use one of the standard outline formats if your outline will be read by someone else, or if your document will contain a table of contents, which is basically an outline of the document. (See Chapter 11 for a discussion of tables of contents.)

Two kinds of outlines are common:

- topic outlines
- sentence outlines

In a topic outline, the entries are phrases, such as "increased worker productivity." In a sentence outline, the entries are sentences, such as "A major advantage of health-promotion programs is that they increase worker productivity." Relatively few writers use sentence outlines, because they know they will have to revise the sentences later. The main advantage of the sentence outline is that it keeps you honest: if you don't know what to say about an item, you will realize your uncertainty clearly when you try to come up with a sentence.

If you will be submitting your outline to someone else, you will have to make one more decision: whether to use a traditional alphanumeric system or the decimal system. The two systems are shown in Figure 8.5.

An advantage of the decimal system is that it is simple to use and to understand. You don't have to remember how to represent the different heading levels. In the same way, your readers can easily see what level they are reading.

Use whichever system helps you stay organized—or is required by your organization.

Traditional Outline	Decimal Outline
I.	1.0
A.	1.1
1.	1.1.1
2.	1.1.2
B.	1.2
1.	1.2.1
a.	1.2.2
b.	2.0
(1)	
(2)	2.1 etc.
(a)	
(b)	
2.	
C. etc.	

FIGURE 8.5

Traditional and Decimal Outlines

Guidelines for Drafting

People use different methods for drafting. Some people start with a detailed outline; others work with only a very brief, general one. Still others work from sketches of graphics. Whatever technique you use, the following eight guidelines should help you efficiently create a draft that you can then revise into a final document.

- *Get comfortable.* Drafting is hard work, so you might as well get comfortable. Figure out the best conditions for you. Obviously, a good chair set at the right height for the keyboard is important, and so is good light that doesn't reflect off the screen. But everything else about your environment is a matter of personal preference. Some people work best in a quiet room; others concentrate better with music.

- *Draft right on the outline.* One advantage of drafting on a computer is that you can draft on the outline. By using the outline, you are less likely to lose sight of your overall plan. Before drafting, make a copy of your outline. Then draft on one copy of the outline. With this technique, you can easily refer to a clean copy of the outline. Many writers like to print out a paper copy of their outline to place next to their keyboard.

- *Start with the easiest topics.* There is no rule that says you have to start writing at the beginning of the document. Many writers like to start with a section from the middle, usually a technical section that they are comfortable writing about.

- *Draft quickly.* Many students make the mistake of trying to draft slowly so that their first draft will be good enough to serve as a final draft. It probably won't be. Expressing ideas on paper is so complicated that you shouldn't expect to say what you mean—and say it well—the first time around. Don't even try. Just try to make your fingers keep up with your brain. Turn the phrases from your outline into paragraphs.

- *Don't stop to get more information or to revise.* Set a timer and draft without stopping. Many writers prefer one hour; some, two. When you hit an item in your outline that you don't understand or that requires more research, just go on to the next item. Don't worry about clumsy sentence structure or odd spelling. Your goal is to create a big, rough draft.

- *Try invisible writing.* Turn the contrast knob to darken the screen. This technique encourages you to look at your hardcopy outline or the keyboard. You don't stop typing so often because you are less tempted to revise what you have just written. If you get nervous because you don't see the words appearing on the screen, turn the contrast knob enough so that you can see the text faintly; in this way you will see that words are appearing on the screen but you won't be tempted to stop and revise.

- *Use abbreviations.* To speed up the drafting, use the *search-and-replace* function, which lets you find any phrase, word, or series of characters and replace it with any other writing. For example, you could type *w/* instead of the word *with* throughout the draft. Or if you will need to use the word *potentiometer* a number of times in your document, you can type **po* each time. Then, during the revising stage, you can instruct the computer to change every *w/* to *with* or every **po* to *potentiometer.* The search-and-replace function also can help you reduce the number of misspellings in your document, for you have to spell the word correctly only once. One word of caution: if you don't use an asterisk or some other unusual character to mark your abbreviation, you might accidentally turn every *po* into *potentiometer.* The word *potential* will become *potentiometertential.* After you make this mistake once in a long document, you will understand why it is called *search-and-destroy.*

- *When your time is up, stop in the middle of a section.* When it is time to stop drafting, stop in the middle of a paragraph, or even in the middle of a sentence. When you start to draft again, you will find it easy to conclude the idea you were working on. This technique will help you avoid writer's block—the mental paralysis that can set in when you stare at a blank page—to which you might be more vulnerable if you had to start a new major section.

Understanding How to Develop an Argument

In technical communication, as in other forms of nonfiction writing, you use different kinds of evidence to develop your argument as you draft and revise your document. An *argument* is simply your arrangement of facts and judgments about some aspect of the world. When you create an argument, you are asking your readers to accept your set of facts and judgments and, sometimes, to act according to your recommendations. An argument can be as short as a paragraph or as long as a multivolume report; it can take many forms, and it can discuss almost any kind of issue. Here are some examples:

FROM A DESCRIPTION OF A SITE:
Features A, B, and C characterize the site.

FROM A STUDY OF WHY A COMPETITOR IS OUTSELLING
YOUR OWN COMPANY:
Its dominance can be attributed to the following four major factors: A, B, C, and D.

FROM A FEASIBILITY STUDY CONSIDERING FOUR ALTERNATIVE
COURSES OF ACTION:
Alternative A is better than alternatives B, C, and D.

FROM A SET OF INSTRUCTIONS FOR PERFORMING A TASK:
The best and safest way to perform the task is to complete task 1, then task 2, and so on.

To develop an effective argument, you must

- understand your audience's goals
- understand the nature of persuasive evidence
- present yourself effectively
- understand political realities

Understanding Your Audience's Goals

Chapter 4 discussed how to use audience profile sheets to help you think about how individual readers will receive your message and use the document you are preparing. In addition to seeking information about particular readers, however, you should think about your audience's broader goals. Certainly, most people want their company to prosper. Yet most people are concerned primarily about their own welfare and interests within the company. Most people value two qualities: *security* and *recognition.*

People's need for security leads them to prefer safe actions to controversial ones, especially if the controversial ones have a significant chance of jeopardizing their status or power within the organization or threaten their livelihood. Therefore, people will oppose an action if they perceive that it might hurt them personally, even if it seems certain to help the organization. Another aspect of security to consider is workload; most people will resist an argument that calls for them to do more work.

People's need for recognition leads them to defend their own actions, even if they would oppose the same actions by someone else. For this reason, you can expect opposition from the person or people associated with any action or decision your argument criticizes. It is wise, therefore, to present your argument so that you are not openly criticizing anyone. Rather, present it so that you are responding to the company's present and future needs. Look forward, not back, and be diplomatic.

Understanding the Nature of Persuasive Evidence

Every argument calls for a different kind of evidence, but people in general react most favorably to four kinds of evidence:

- *"Commonsense" arguments.* As used in this phrase, the word *commonsense* has a special meaning: "most people would think that . . ." For example, the following sentence presents a commonsense argument that binge drinking is harmful:

  Binge drinking cannot be good for you; it makes you sick while you are doing it and for quite a few hours afterward.

  A commonsense argument, therefore, appeals to a person's understanding and experience of the way the world works. It says, "I don't have any hard evidence to support my conclusion, but it stands to reason that . . ." In this case, the argument is that behavior that the body rejects is probably not healthy. If your audience's sense of what is commonsense matches yours, your argument is likely to be persuasive.

- *Numerical data.* Numerical data—statistics—are generally more persuasive than commonsense arguments.

  Compared with students who do not binge-drink, binge drinkers are three times as likely to have unprotected sex; female students who binge-drink are twice as likely as their non-binge-drinking peers to be the victims of sexual abuse and rape.

- *Illustrations.* An example or illustration makes the abstract more concrete and therefore more vivid and more memorable.

  John M. began to binge-drink when he was a junior in high school. Three or four nights a week he and his friends would get together and each drink 6–12 beers. Because they were underage and wanted to hide their activities from their parents, they did their drinking away from home—in alleys, in parking lots, at the school athletic fields. One night, as they were walking back from the athletic fields . . .

  Illustrations are often used in conjunction with numerical data. The illustration provides a memorable incident, and the numerical data show that the illustration is part of a pattern, not a fluke or some sort of coincidence. For example, the writer using the illustration of a teenage boy hit by a car when he is binge drinking needs some numerical data to prove that a significant number of binge drinkers suffer similar fates; otherwise, the illustration, although powerful, doesn't support the argument.

- *Expert testimony.* A message delivered by an expert is more persuasive than the same message delivered by someone without credentials. A well-researched article on binge drinking written by a respected public-health official and published in a reputable medical journal is likely to be persuasive. When you make arguments you will often cite expert testimony from published sources or from interviews you have conducted.

Presenting Yourself Effectively

Just as testimony from an expert is more persuasive than the same testimony from a nonexpert, an argument from a person the audience likes is more persuasive than the same argument from a person the audience dislikes. As you create your argument, try to project the following characteristics:

- *Cooperativeness.* As discussed in Chapter 3, you will work with groups much of the time in your professional career. Show a cooperative spirit by focusing on the goals and values of the organization, not on your own interests.

- *Moderation.* Be moderate in your judgments. The problem you are describing will not likely spell doom for your organization, and the solution you propose will not solve all the company's problems.

- *Fair-mindedness.* Acknowledge opposite points of view, even as you offer counterarguments. Don't let your readers think that you dismiss everyone else's opinions.

- *Modesty.* If you fail to acknowledge that you don't know everything, someone else will be sure to volunteer that insight.

The following paragraph shows how a writer can demonstrate the qualities of cooperativeness, moderation, fair-mindedness, and modesty:

In the first three sentences, the writer acknowledges the problems with his recommendation.

The use of "I think" adds an attractive modesty: the recommendation might turn out to be unwise.

This plan is certainly not perfect. For one thing, it calls for a greater up-front investment than we had anticipated. And the return-on-investment through the first three quarters is likely to fall short of our initial goals. However, I think this plan is the best of the three alternatives for the following reasons. . . . Therefore, I recommend that we begin planning immediately to implement the plan. I am confident that this plan will enable us to enter the flat-screen market successfully, building on our fine reputation for high-quality advanced electronics.

The recommendation itself is moderate; the writer does not assert that the plan will save the world.

In the last two sentences, the writer shows a spirit of cooperativeness by moving the focus from his own analysis of the alternatives to the goals of the company.

Understanding Political Realities

Choose your battles. There are some goals that you can reasonably hope to achieve, and others that you can't; don't use up all your energy and sacrifice all your credibility on a losing cause.

The same advice applies in the workplace. If you know that a particular course of action is clearly in the best interests of the company, but you also know that upper management disagrees or that the company can't afford to do it, don't be a martyr to the cause. Instead, try to figure out what aspects of the course of action might be achieved through some other means, or scale back the idea so that it is affordable.

Two big exceptions to this rule are: matters of ethics or safety. As discussed in Chapter 2, under certain circumstances compromise is unacceptable.

Understanding Basic Patterns of Organizing Arguments

To create a successful argument, you need to understand your audience's goals and the nature of persuasive evidence, present yourself effectively, and understand the political realities. But how do you actually organize the information that makes up your argument? Every argument is different and calls for a unique organizational pattern. Long, complex arguments often contain a number of different organizational patterns at once. For instance, one of your arguments might be a comparison and contrast of two pieces of equipment, but within that argument you might organize the features of the two pieces of equipment from more important to less important.

It is useful to think about basic organizational patterns for two reasons:

- *The basic patterns can serve as templates for you to modify.* There is no reason to reinvent the wheel. If a standard pattern will work well with the information you want to convey, start with it, then modify it as necessary.

- *Your readers will understand the basic patterns.* Because they are accustomed to the patterns, they will be able to concentrate better on your message.

This section discusses the eight organizational patterns shown in Figure 8.6.

Chronological   The chronological pattern—the time pattern—commonly describes events. Here are some examples of documents that use chronology as an organizing principle in some sections:

- In an *accident report,* you describe the events in the order they occurred.
- In a *trip report,* you describe what you did on the trip.
- In the background section of a *report,* you describe the events that led to the present situation.

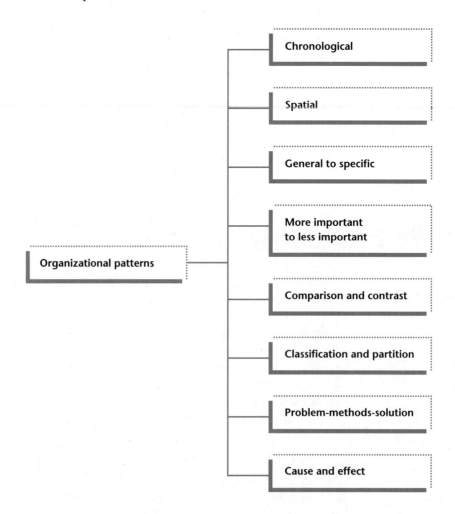

FIGURE 8.6

Organizational Patterns

- In a *reference manual,* you explain how to carry out a task by describing the steps in order.
- In the methods section of a *journal article,* you describe the actions you took in carrying out the experiment.

When writing a chronologically organized argument, follow three guidelines:

- *Provide signposts.* If the passage is more than a few hundred words long, use headings. Use words such as *step, phase, stage,* and *part,* and consider numbering them. Add substantive phrases to increase clarity:

  Phase One: Determining Our Objectives
  Step 3: Installing the Lateral Supports

  You can also use signposts at the paragraph and sentence levels. Transitional words such as *then, next, first,* and *finally,* used in topic sentences and in support sentences, help your reader follow your discussion. For more information on transitional words and phrases, see Chapter 14.
- *Consider using graphics to complement the text.* As Chapter 12 suggests, graphics can clarify and emphasize chronological passages. Flowcharts are used commonly in writing for all kinds of readers, from the most expert to the general reader.
- *Analyze events where appropriate.* Chronology is perhaps the easiest pattern to use because the narrative provides its own momentum; you just tell your story. Remember, however, that chronology is limited in that it tells the reader what happened but doesn't explain why or how it happened, or what it means. For instance, in an accident report, the largest section is usually devoted to the chronological discussion, but the report is of little value if it lacks sections that explain what caused the accident, who bears liability for the accident, and how such accidents can be prevented.

Figure 8.7 on pages 160–162 (based on Bell, 1993, pp. 32–35) shows a chronologically organized passage.

Spatial   The spatial pattern is used commonly to describe objects and physical sites. Here are some examples of documents that might use spatial organization as an organizing principle in some sections:

- In an *accident report,* you describe the physical scene where the accident occurred.
- In a *feasibility study* about building a facility, you describe the property on which it would be built.
- In a *proposal* to design a new microchip, you describe the new chip.

The writer introduces the topic.

Introduction

The following discussion explains the process by which Panasonic builds semicustom bicycles. What is a semicustom bicycle, and how does it differ from a custom bicycle? A custom bicycle is one whose frame is built to the buyer's individual measurements (torso length, leg length, and arm length) and style of riding (racing, touring, or off road). The frame is then fitted with the buyer's choice of components. A custom bicycle costs $2,500 or more.

Panasonic's semicustom bicycles are halfway between a custom bike and a mass-produced one. For a road bike, say, customers choose one of 10 styles of frame from 15 sizes of chrome-molybdenum or chromoly steel (from 46 cm to 60 cm in 1-cm increments) or 13 sizes of aluminum or carbon fiber. They specify one of three head angles for the front fork, six variations in the stem length for the handlebars (to fine tune for torso length), four in the handlebar width (to adjust for shoulder width), and two in gearing. Finally, customers choose among 191 color schemes of up to three colors each, and decide which of two places to put a decal bearing the owner's name in five different styles of type. Figure 1 shows that Panasonic offers more than 11 million variations on its semicustom bikes.

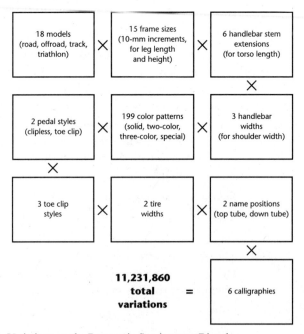

Figure 1. Variations on the Panasonic Semicustom Bicycle

FIGURE 8.7

Argument Organized
Chronologically

(Figure 8.7 continued)

Assembling a semicustom bike entails the six steps shown in Figure 2:

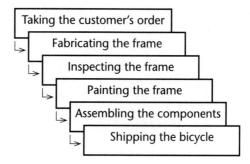

Figure 2. Steps in Assembling a Semicustom Bike

The steps of the process are presented in sequence.

Taking the Customer's Order

The customer places an order at one of 1,500 bike shops in Japan. The customer's body measurements are taken and entered, along with his or her preferences, on a detailed order form. That form is then faxed directly to the factory, where operators enter the information into a computer.

Fabricating the Frame

The computer generates a CAD/CAM mechanical drawing of the bicycle and assigns a bar code number to each component. A veteran craftsman is assigned to oversee the fabrication of the bicycle. The three main tubes are automatically cut to length with a machine tool; the computer calculates the exact tube lengths and miter angles and sets the cutters at the exact position.

Inspecting the Frame

Although the details of fabricating the frames differ with the material chosen by the customer, each frame undergoes an automatic three-dimensional measuring-system check. All dimensions and data are processed by a computer to yield a good/no-good decision on a video display.

Painting the Frame

If the frame passes inspection, robots prime and paint the frame, according to the customer's specifications, with six thin coats of paint. For two- and three-color jobs, the craftsman assists the robots. Counting the curing time for the paint, the process of fabricating and painting the frame takes five days.

Assembling and Attaching the Components

The craftsman then assembles and attaches the components—the gears, brakes, pedals, and so on. The different components are stored at parts stations

(Figure 8.7 continued)

on shelves. Lights at each station are wired into the computer system. As each order is processed, the proper station lights up, enabling the craftsman to find the correct components quickly and accurately. The craftsman then inspects the completed bicycle.

Shipping the Bicycle

The completed bicycle is partially disassembled for compactness, then shipped to the shop where the order was taken. The shop owner assembles the bike and notifies the customer, who also receives a printout of the CAD mechanical drawing of the bike's specifications and a personal thank-you letter from the company.

Conclusion

Panasonic produces 50-60 semicustom bikes per day. The production time is 8-10 working days, about twice that of a mass-produced bicycle. At a price of about $1,300, the semicustom bike is about four times the price of a mass-produced bike, but only half the price of a custom bike.

- In the results section of a *journal article* about the effectiveness of an antitoxin, you describe the tissue sample on which you tested the antitoxin.

When you write a spatially organized argument, follow three guidelines:

- *Provide signposts.* Just as with chronologically organized passages, help your readers follow the argument by using words and phrases indicating physical location (*to the left, above, in the center*) in headings, topic sentences, and support sentences.
- *Consider using graphics to complement the text.* Diagrams, drawings, photographs, and maps are often used to add a visual dimension to the text.
- *Analyze events where appropriate.* As is the case with chronology, a spatial organization doesn't explain itself; you have to do the analysis. A diagram of a floor plan cannot explain why the floor plan is effective or ineffective.

Figure 8.8 is an excerpt from a process description in which the argument is organized spatially.

Description of a Tropical-Fish Aquarium

Tropical fish require an artificial environment that simulates their natural habitat. Although aquariums vary in size, shape, and material, they all contain certain features. This paper describes a suitable microcosm for most tropical fish (usually bright-colored, fresh- and warm-water fish, originally from the tropics). The three critical aspects of the environment are the following:

- aquascape (gravel, plants, and decorative items)
- water
- light

The writer states and justifies his pattern of organization.

These aspects of the environment will be described from the bottom of the tank to the lid. This order reflects the process you would follow in setting up the tank: you would establish the aquascaping, add the water, and set up the light.

Aquascape

Aquascaping is the process of beautifying and enhancing an aquarium in order to simulate a natural habitat. Three items are important in aquascaping a freshwater aquarium: gravel, plants, and decorative objects. The plants and decorative objects promote breeding by providing a hiding place for fish eggs. If the eggs are not hidden and protected, the parents and other fish in the aquarium will eat them. Plants and decorative items also provide the fish with security. Less active fish need quiet, hidden areas; agile, good swimmers need open, turbulent areas.

Gravel

Gravel is not only decoration; it is also important for anchoring plants to the aquarium floor. Gravel 2–3 millimeters in diameter is best because it is small enough to anchor plants, yet not large enough to trap uneaten food particles. The gravel should make a bed about two inches thick. A light-colored gravel is recommended because gravel color will be an indicator of overfeeding. Excess food will decay and turn black very rapidly.

Plants

Plants, whether live or artificial, are not essential for the well-being of fish, but they . . .

Decorative Items

Decorative items should be thoroughly cleaned . . .

Water

Water conditioning, pH, and temperature are important to the survival of the fish. Many people believe that fish acquire their oxygen from separating the hydrogen and oxygen that bond to create the water, but they don't. They use the

FIGURE 8.8

Argument Organized Spatially

(Figure 8.8 continued)

oxygen that is dissolved in the water from the atmosphere. Distilled water (water without dissolved solids and oxygen) is unsuitable for aquarium use.

Conditioning

If the aquarium water is not properly conditioned, the fish . . .

pH

Water pH is the measure of alkalinity and acidity . . .

Temperature

Most tropical fish thrive between 75° and 80° F. . . .

Light

If the proper amount of light is not provided for the fish and the plants, both will die. Fish do see colors that help them to identify food, as well as fish of the opposite gender. Fish are diurnal (awake during the day), and most species require twelve hours of light. However, direct sunlight is harmful. The ideal area for an aquarium using sunlight would be a northern exposure. Artificial light is just as good as sunlight. Artificial light can be turned on when the room is dark. This artificial light illuminates the tank and provides beauty. Too much light allows green algae to grow.

Conclusion

Tropical fish are, in general, more fragile than nontropical fish, and therefore they need greater care. An aquarium consisting of aquascaping to simulate the natural environment, properly conditioned water of the right temperature and alkalinity, and adequate light will enable the tropical fish to prosper.

General to Specific

The general-to-specific pattern is fundamental to technical communication. Based on the idea that readers need a general understanding of a subject before they can understand and remember the details, the general-to-specific pattern is used in many kinds of technical documents:

- In a *process description,* you describe the overall process before you describe each step in detail. (In the detailed descriptions of each step, you will probably use other organizational patterns.)
- In a *report,* you provide an executive summary—an overview for managers—before the body of the report.
- In a set of *instructions,* you provide general information about the tools and materials needed and about safety measures before providing the step-by-step instructions.

- In a *memo,* you describe the background before you provide the details of your message.

When you write an argument organized from general to specific, follow these two guidelines:

- *Provide signposts.* In the introduction, explain that you are using the general-to-specific pattern. If appropriate, incorporate the words *general* and *specific* or other appropriate terms in the major headings or at the start of the text for each item you are describing.
- *Consider using graphics to complement the text.* Diagrams, drawings, photographs, and maps are often used to add a visual dimension to the text.

Figure 8.9 is an example of an argument organized according to the general-to-specific pattern. The document is a description of types of optical telescopes used in astronomy.

More Important to Less Important

The more-important-to-less-important pattern is useful when you want to direct your readers' attention to the most critical information first. This organizational pattern recognizes that readers of technical communication often want the bottom line—the most critical information—on top.

The writer provides general information about the history, size, and operating principle of telescopes used in astronomy.

Introduction

The first telescope was invented in 1608 by the Dutch optician Hans Lippershey, who mounted two lenses in a tube. One year later, Galileo refined this design and began to study space. He soon discovered, for instance, that Jupiter has several moons. In 1668, Isaac Newton invented a telescope that used a mirror. His design is the basis of modern astronomical optical telescopes.

Optical telescopes vary tremendously in size and in power. A small telescope has a 1" diameter lens; a large one, 20" or even 30". But both instruments work essentially the same way: they use a lens to collect and focus light waves. In simple astronomical optical telescopes, photographic film is used to record the image created by the lens. In more sophisticated ones, computers are used to enhance the process of amplifying, interpreting, and recording the image.

Three basic types of optical telescopes are used in astronomy: refracting telescopes, reflecting telescopes, and refracting-reflecting telescopes.

The Refracting Telescope
The refracting telescope . . .

The writer begins his detailed discussion of the types of optical telescopes.

FIGURE 8.9

Argument Organized from General to Specific

Here are examples of documents that use the more-important-to-less-important pattern as an organizing principle in some sections:

- In an *accident report,* you describe the three most important factors that led to the accident.
- In a *feasibility study* about building a facility, you describe the major reasons that the site is appropriate.
- In a *proposal* to design a new microchip, you describe the major applications for the new chip.
- In the conclusion section of a *journal article* about the effectiveness of an antitoxin, you describe the major conclusions of the experiment.

When writing an argument organized from more important points to less important ones, follow three guidelines:

- *Provide signposts.* Tell your readers how you are organizing the passage. For instance, in the introductory text of a proposal to design a new microchip, write something like the following: "These three applications for the new chip are arranged from most important to least important."

  Be straightforward with your readers. If you have two very important points and three less important points, present them that way: group the two big points together and label them, as in "Major Reasons to Retain Our Current Management Structure." Then present the less important factors as "Other Reasons to Retain Our Current Management Structure." Being straightforward is effective for two reasons: the material is easier to follow because it is clearly organized, and your credibility is increased.
- *Make clear in your discussion why one point is more important than another.* It is not enough to say that you will be arranging the items from more important to less important; you have to explain why the more important point is in fact more important.
- *Consider using graphics to complement the text.* Diagrams and lists are often used to suggest levels of importance.

Figure 8.10, from a memo written by an executive at a company that sells equipment to manufacture semiconductors, shows the more-important-to-less-important organizational structure.

Comparison and Contrast

Typically, the comparison-and-contrast organizational pattern is used when you want to evaluate two or more options. This pattern lies at the heart of the feasibility study (see Chapter 23 for more information). Here are examples of documents that use the comparison-and-contrast pattern as an organizing principle in some sections:

A Three-Point Program to Improve Service

What we have learned from the recent conference of semiconductor purchasers and from our own focus groups is that customers expect and demand better service than the industry currently provides. By this I don't mean returning phone calls, although that sort of common courtesy remains important. I mean something much more ambitious and difficult to attain: helping our customers do their job by anticipating and addressing their total needs. For this reason, I have formed a Customer Satisfaction Panel, chaired by Maureen Bedrich, whose job will be to develop policies that will enable us to improve the quality of the service we offer our customers.

I have asked the panel to consider three major areas:

- improving the ease of use of our equipment
- improving preventive and corrective maintenance
- improving our compatibility with other vendors' products

Improving the Ease of Use of Our Equipment

The most important area to improve is user friendliness. When we deliver a new product, we have to sit down with the customer and explain how to integrate the new product into their own manufacturing process. This process is time consuming and costly not only for us but also for them. For this reason, we must explore the option of automating this integration process. Automated process control, which is already common in such industries as machine-vision, would allow our customers to integrate our semiconductors much more efficiently. In addition, it would help them determine, through process control, whether our equipment is functioning according to specification.

Improving Preventive and Corrective Maintenance

The second most important area for study is improving preventive and corrective maintenance. Our customers will no longer tolerate down times approaching 10 percent; they will accept no more than 2 to 3 percent. In the semiconductor equipment field, preventive maintenance is critical because gases used in vapor-deposition systems periodically have to be removed from the inside of the equipment. Customers want to be able to plan for these stoppages to reduce costs. Currently, we have no means of helping them plan.

We also have to assist our customers with unscheduled maintenance. Basically, this means modularizing our equipment so that the faulty module can be pulled out of the equipment and replaced on the spot. This redesign will take many months, but I'm sure it will become a major selling point.

Improving Our Compatibility with Other Vendors' Products

Finally, we have to accept the fact that no one in our industry is likely to control the market, and so we have to make our products more compatible with other manufacturers'. Making our products more compatible means that we must be willing to put our people on-site to see what the customer's setup is and help them determine how to modify our product to fit in efficiently. We can no longer offer a "take-it-or-leave-it" product.

I hope you will extend every effort to work constructively with Maureen and her committee over the coming months to ensure that we improve the overall service we offer our customers.

The writer suggests his organizational pattern in the topic sentence.

The writer again indicates his organizational pattern.

FIGURE 8.10

Argument Organized from More Important to Less Important

- In a *feasibility study* about building a facility, you compare and contrast two sites.

- In a *memo,* you compare and contrast the three finalists for a position in your department.

- In a *proposal* to design a new microchip, you compare and contrast two different strategies for designing the chip.

- In an internal *report* describing a legal challenge that your company faces, you compare and contrast several options for responding.

Writing a comparison and contrast argument requires two basic steps:

- *Establishing criteria for the comparison and contrast.* To compare and contrast two or more items, you must determine the criteria: the standards or needs against which you will study the items. For instance, if you need to choose an elective course to take next semester, your only criterion might be the time that it is offered: it must be offered at ten o'clock on Mondays, Wednesdays, and Fridays. For this criterion, MWF 10 is its *required characteristic.* For you to consider taking the course, it has to meet MWF 10. However, you probably have other criteria in mind: you would like the course to be interesting, to look good on your transcript, and so on. That the course be interesting is an example of a *desired characteristic.* If necessary, you might take a course that holds little interest for you—provided it meets your required characteristic.

  Almost always, you will need to consider several criteria in carrying out a comparison and contrast. For a recommendation report on which kind of computer your company should buy, for example, you would probably consider a number of criteria, some of which have required characteristics and some of which have desired characteristics. For instance, a required characteristic might be that you be able to attach the computer to your company's network. If a particular computer cannot be attached, you will not consider it. But most of the criteria might call for desired characteristics, such as ease of operation, reliability, and accessibility of maintenance and service personnel. You will evaluate each option—each computer for sale—first by eliminating those that fail to meet the required characteristics and then by comparing and contrasting the remaining options according to the desired characteristics.

- *Organizing the discussion.* Two basic patterns are used for organizing the discussion: *whole by whole* and *part by part.* Figure 8.11 shows the difference between these two patterns. In this figure, two different printers—Model 5L and Model 6L—are being compared and contrasted according to three criteria: price, resolution, and print speed.

Whole by Whole	Part by Part
Model 5L	Price
• price	• Model 5L
• resolution	• Model 6L
• print speed	Resolution
Model 6L	• Model 5L
• price	• Model 6L
• resolution	Print Speed
• print speed	• Model 5L
	• Model 6L

Whole-by-Whole and Part-by-Part Patterns

The whole-by-whole structure is effective if you want to focus on each item as a complete unit rather than on individual aspects of different items. For example, suppose you are writing a feasibility report on purchasing an expensive piece of laboratory equipment. You have narrowed your choice to three different models, each of which meets your required characteristics. Each model has its advantages and disadvantages. Because your organization's decision on which model to buy will depend on an overall assessment rather than on any single aspect of the three models, you choose the whole-by-whole structure; it gives your readers a good overview of the different models.

In contrast, the part-by-part structure focuses on the individual aspects of the different items. Detailed comparisons and contrasts are more effective when they use the part-by-part structure. For example, suppose you are comparing and contrasting the three pieces of laboratory equipment. The one factor that distinguishes the three models is reliability: one model has a much better reliability record than the other two. The part-by-part structure lets you create a section on reliability to highlight this aspect. Comparing and contrasting the three models in one place in the document rather than in three places emphasizes your point. You sacrifice a coherent overview of the different items for a clear, forceful comparison and contrast of an aspect of the three models.

Of course, you can have it both ways; if you want to use a part-by-part pattern to emphasize particular aspects, you can begin the discussion with a general description of the different items. The general description provides the overview of a whole-by-whole discussion, whereas the part-by-part discussion emphasizes particular aspects of the different options you are comparing and contrasting.

Once you have chosen the overall pattern—whole-by-whole or part-by-part—you must decide how to organize the second-level items. That is, in a whole-by-whole passage, you have to sequence the "aspects"; in a part-by-part passage, you have to sequence the "options." For most documents, a more-important-to-less-important pattern will work well because your readers want to get to the bottom line as soon as possible.

For some documents, however, other patterns might work better. People who write for readers outside their own company often reverse the more-important-to-less-important pattern because they want to make sure their audience reads the whole discussion. This pattern is also popular with writers who are delivering bad news. If, for instance, you want to explain why you are recommending that your organization not go ahead with a popular plan, the reverse sequence enables you to show your readers the problems with the popular plan before you present the plan you recommend. Otherwise, you might surprise the readers, and they might start to formulate objections in their minds before you have had a chance to explain your position.

Figure 8.12 (Murphy, 1994), an excerpt from a report, shows the part-by-part form of the comparison-and-contrast organizational structure.

Classification and Partition

Classification is the process of assigning items to different categories. For instance, all the students at a university could be classified by sex (males and females), age (18 years old, 19, and so forth), major (nursing, forestry), and any number of other characteristics. You can also create categories within categories. For instance, within the category of students majoring in business at your college or university, you can create subcategories: male business majors and female business majors.

Here are examples of documents that use classification as an organizing principle:

- In a *feasibility study* about building a facility, you classify different sites into two categories: domestic and foreign.
- In a *journal article* about ways to treat a medical condition, you classify the treatments as surgical and nonsurgical.
- In a description of a major in a *college catalog,* you classify courses as required or elective.

Partition is the process of breaking down a unit into its components. A stereo system could be partitioned into the following components: cassette deck, CD player, tuner, amplifier, and speakers. Each component is separate, but together they form a stereo system. Each component can of course be partitioned further.

Partition is used in descriptions of objects, mechanisms, and processes, as explained in Chapter 10. Here are examples of documents that use partition as an organizing principle in some sections:

- In an *equipment catalog,* you use partition to describe one of your products.

The writer introduces her topic by defining the two items she will compare and contrast.

On-line and Off-line Debit Cards:
First-Year Costs of Implementation at John's Market

The debit card, a plastic card that transfers funds directly from a consumer's bank account to a retailer's bank account, is becoming very popular in our industry. At your request, I have investigated the two basic forms of debit cards: on-line and off-line.

Debit-card transactions are either on-line (instantaneous like automatic-teller-machine transactions) or off-line (in which the transaction takes about two days, like a check transaction). This paper focuses on the costs and benefits of implementing on-line or off-line debit card programs throughout John's Market's 35 stores.

The writer presents her main recommendation and a cross-reference to the sections where she discusses it in more detail.

I recommend that John's Market implement the on-line debit-card system as soon as possible. See the recommendations section for more details.

The writer explains the criteria by which she will compare and contrast the two kinds of debit cards.

Costs to the Company

Costs of the debit-card program are evaluated on the basis of four criteria:

- new equipment
- fees
- bill payment and delinquency
- handling the transaction

The writer is using the part-by-part pattern. For each criterion, she compares and contrasts the two cards.

New Equipment

The on-line debit card requires the purchase of a point-of-sale (POS) terminal for each cash register. These POS terminals, configured like an ATM keypad, cost about $400 each. The 35 stores have an average of 10 cash registers each, meaning that 350 POS terminals would need to be purchased. The start-up cost for equipment would total about $140,000 if the on-line program were adopted throughout John's chain.

No new equipment purchases would be necessary to implement the off-line debit card because it operates on John's existing credit-card system. The debit card is run through the credit-card scanners already connected to each register. The transaction is stored and processed at a later time.

Fees

A one-time, $5,000 hookup fee to access telecommunications and the ATM network would be charged for each store that implements the on-line program. This communication fee would total $175,000 if the on-line program were implemented at all 35 stores. Banks would thereafter assume the responsibility for monthly transaction fees and pass that cost to their members in the form of monthly ATM transaction fees.

The off-line program requires John's Market to pay a 5 percent service charge on each purchase to the banks that guarantee the debit cards. This is

FIGURE 8.12

Argument Organized by Comparison and Contrast (Part by Part)

(Figure 8.12 continued)

The writer explains the procedures for bill payment and delinquency for the two kinds of cards.

much like a bank/merchant relationship with credit-card transactions. Customer response to off-line debit cards has been high in test areas. An average of 20 percent of customers adopted the off-line payment system when offered. This means John's could expect to pay about $750,000 a year in service charges on its $75 million gross sales.

Bill Payment and Delinquency

John's Market would receive instant payment. . . .

The writer explains the procedures for handling the transaction with the two kinds of cards.

Handling

The costs associated with handling . . .

Conclusions

The on-line debit program would require initial investments of $140,000 for equipment and $175,000 for hookup fees. This card could possibly pay for part of its own setup costs in staff reductions in the accounting department. A $525,000 savings could be expected in staff reductions. The card would not only improve cash flow in the stores but also yield an estimated savings of $210,000 in the first year.

The off-line debit card program would not require setup costs, but banks would assess service charges for guaranteeing their cards. Service charges could be expected to total $750,000 the first year. Cash would flow one day faster with the on-line program, and staffing costs may increase with increased use of the off-line program. The off-line debit card would cost an estimated $750,000 in the first year.

Cost in Year One

Type of Card	Equipment	Fees	Payment/ Delinquency	Handling	Total Cost
On-line Card	+140,000	+175,000	N/C	-525,000	-210,000
Off-line Card	N/C	+750,000	N/C	N/C	+750,000

N/C = No Change

- In a *proposal,* you use partition to describe an instrument you propose developing.
- In a *brochure* describing how to use a product, you use partition to describe it.

When writing an argument organized by classification or partition, follow these six guidelines:

- *Choose a basis of classification or partition that is consistent with your audience and purpose.* Never lose sight of your audience—their backgrounds, skills, and reasons for reading—and your purpose. If you are writing a brochure for hikers in a particular state park, you will probably classify the local snakes according to whether they are poisonous or non-poisonous, not according to size, color, or any other basis.
- *Use only one basis of classification or partition at a time.* If you are classifying graphics programs according to their technology—paint programs and draw programs—do not include another basis of classification, such as cost.
- *Avoid overlap.* In classifying, make sure that no single item could logically be placed in more than one category; in partitioning, make sure that no listed component includes another listed component. Overlapping generally occurs when you change the basis of classification or the level at which you are partitioning a unit. In the following classification of bicycles, for instance, the writer introduces a new basis of classification that results in overlapping categories:

– mountain bikes
– racing bikes
– touring bikes
– ten-speed bikes

The first three items share a basis of classification: the type of bicycle. The fourth item has a different basis of classification: number of speeds. Adding the fourth item is illogical because a particular ten-speed bike could be a mountain bike, a touring bike, or a racing bike.

- *Be inclusive.* When you classify or partition, be sure to include all the categories. For example, a classification of music according to type would be incomplete if it included popular and classical music but not jazz. A partition of an automobile by major systems would be incomplete if it included the electrical, fuel, and drive systems but not the cooling system.

   If your purpose or audience requires that you omit a category, tell your readers what you are doing. If, for instance, you are writing in a classification passage about recent sales statistics of General Motors, Ford, and Chrysler, don't use "American car manufacturers" as a classifying tag for the three companies. Although all three belong to the larger class "American car manufacturers," there are other American car manufacturers. Rather, classify General Motors, Ford, and Chrysler as "the big three" or refer to them by name.

- *Arrange the categories in a logical sequence.* After establishing your categories and subcategories of classification or the components in a partition, arrange them according to some reasonable plan: chronology (first to last), spatial development (top to bottom), importance (most to least), and so on.

- *Consider using graphics to complement the text.* Block diagrams are commonly used in classification arguments; drawings and diagrams are often used in partitions arguments.

In Figure 8.13, a classification of nondestructive testing techniques, the writer uses classification and subclassification effectively in introducing nondestructive testing to a technical audience.

The writer classifies non-destructive testing into four categories. Notice that she justifies the sequence she has chosen.

Types of Nondestructive Testing

Nondestructive testing of structures permits early detection of stresses that can cause fatigue and ultimately structural damage. The least sensitive tests isolate macrocracks. More sensitive tests identify microcracks. The most sensitive tests identify slight stresses. All sensitivities of testing are useful because some structures can tolerate large amounts of stress—or even cracks—before their structural integrity is threatened.

Currently there are four techniques for nondestructive testing, as shown in Figure 1. These techniques are presented from least sensitive to most sensitive.

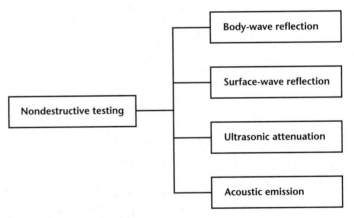

Figure 1. Types of Nondestructive Testing

Body-Wave Reflection

In this technique, a transducer sends an ultrasonic pulse through the test material. When the pulse strikes a crack, part of the pulse's energy is reflected back to the transducer. Body-wave reflection cannot isolate stresses: the pulse is sensitive only to relatively large cracks.

FIGURE 8.13

Argument Organized by Classification and Sub-classification

(Figure 8.13 continued)

Surface-Wave Reflection

The transducer generates an ultrasonic pulse that travels along the surface of the test material. Cracks reflect a portion of the pulse's energy back to the transducer. Like body-wave reflection, surface-wave reflection picks up only macrocracks. Because cracks often begin on interior surfaces of materials, surface-wave reflection is a poor predictor of serious failures.

Ultrasonic Attenuation

The transducer generates an ultrasonic pulse either through or along the surface of the test material. When the pulse strikes cracks or the slight plastic deformations associated with stress, part of the pulse's energy is scattered. Thus, the amount of the pulse's energy decreases. Ultrasonic attenuation is a highly sensitive method of nondestructive acoustic testing.

There are two methods of ultrasonic attenuation. One technique reflects the pulse back to the transducer. The other uses a second transducer to receive the pulses sent through or along the surface of the material.

Acoustic Emission

When a test specimen is subjected to a great amount of stress, it begins to emit waves; some are in the ultrasonic range. A transducer attached to the surface of the test specimen records these waves. Current technologies make it possible to interpret these waves accurately for impending fatigue and cracks.

> *The writer introduces a second level of classification.*

Notice that the writer could have used another basis for classification: sensitivity. The four techniques range from very sensitive to less sensitive.

Figure 8.14 (Larson, 1990, p. 674) is an example of partition.

The Ear

The two main responsibilities of the ear are hearing and balance. The ear has three parts: outer ear, middle ear, and inner ear. The outer ear, the part that we see, is composed of the pinna or auricle (the folds of skin and cartilage that we usually refer to as the ear) and the outer ear canal, which delivers sound to the middle ear. Within the outer ear canal are wax-producing glands and hairs that protect the middle ear.

> *The writer partitions the ear into three components and arranges them spatially.*

FIGURE 8.14

Argument Organized by Partition

(Figure 8.14 continued)

The writer partitions the middle ear into three components.

The function of the middle ear is to deliver sound to the inner ear, where it is processed into a signal that our brain recognizes. The middle ear is a small cavity with the eardrum (tympanic membrane) on one side and the entrance to the inner ear on the other. Within the middle ear are three small bones known as the hammer (malleus), anvil (incus), and stirrup (stapes) because of their shapes. These bones act like a system of angular levers to conduct sound vibrations into the inner ear. The hammer is attached to the lining of the eardrum, the anvil is attached to the hammer, and the stirrup links the anvil to the oval window, the opening to the inner ear.

As the drawing makes clear, the writer is partitioning and subpartitioning the ear.

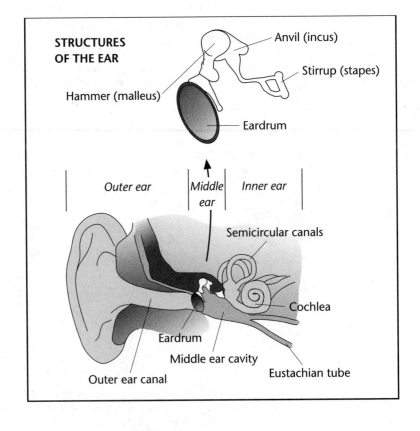

For more examples of partition, see Chapter 10, which covers descriptions of objects, mechanisms, and processes.

Problem-Methods-Solution

The problem-methods-solution pattern is an excellent way to organize an argument about most kinds of problem-solving projects. It is easy to write—and easy to read—because it reflects the logic used in carrying out a project.

The three components of this pattern are simple to identify:

- *Problem.* The problem is a description of what was not working (or not working as effectively as it should), or what opportunity exists for improving current processes.
- *Method.* The method is the procedure you performed either to confirm your analysis of the problem or to solve it, or to exploit the opportunity that exists.
- *Solution.* The solution is a statement of whether your analysis of the problem was correct, or of what you discovered or devised to solve the problem or capitalize on the opportunity.

Here are some examples of documents that use problem-methods-solution as an organizing principle:

- In a *proposal,* you describe a problem in your business, how you plan to carry out your research, and how your end product (an item or a report) can be used to solve the problem.
- In a *completion report* about a project to improve a manufacturing process, you describe the problem that motivated the project, the methods you used to carry out the project, and the findings: the results, conclusions, and recommendations.
- In a *journal article* about ways to treat a medical condition, you describe the problem (the disadvantages of the current techniques), your methods, and your results and conclusions.

The problem-methods-solution pattern is discussed in more detail in Chapters 21–23.

When you write a problem-methods-solution argument, follow these four guidelines:

- *In describing the problem, be clear and specific.* Don't say that our energy expenditures are getting out of hand. Instead, say that the energy usage has increased 7 percent in the last year and that the utility costs have risen 11 percent. Then calculate the total increase in energy costs.
- *In describing your methods, make sure your readers understand not only exactly what you did but also why you did it that way.* Most technical problems can be approached using several different methods; therefore, you might have to justify your choices. Why, for example, did you use a *t*-test in calculating the statistics in an experiment? If you can't defend your choice, you lose your credibility.

- *In describing the solution, don't overstate.* Be careful about claiming that "this project will increase our market share from 7 percent to 10 percent within twelve months." The world is too complicated for that sort of self-assurance. Instead, be cautious: "This project promises to increase our market share significantly: an increase from 7 percent to 10 percent or even 11 percent is quite possible." This way, your document won't come back to haunt you if things don't turn out as well as you had anticipated.

- *Choose a logical sequence.* The most common sequence, naturally, is to start with the problem and conclude with the solution. Sometimes, however, you might want to present the problem first and then go directly to the solution, leaving the methods for last. This sequence deemphasizes the methods, an appropriate strategy if your readers either already know them well or have little need to understand them. When readers want to focus on the solution, you can begin with the solution and then discuss the problem and methods.

    Different sequences work equally well so long as you provide some kind of preliminary summary to give your readers an overview and provide headings or some other design elements (see Chapter 13) to help your readers find the information they want.

The problem-methods-solution argument in Figure 8.15 is an excerpt from a newsletter article written by the president of a company that manufactures personal computers.

Cause and Effect   Technical communication often involves cause-and-effect arguments. Sometimes you will reason forward: cause to effect. If we raise the price of a particular product we manufacture (cause), what will happen to our sales (effect)? The government has a regulation that we may not use a particular chemical in our production process (cause); what will we have to do to keep the production process running smoothly (effect)? Sometimes you will reason backward: from effect to cause. Productivity went down by 6 percent in the last quarter (effect); what factors led to this decrease (causes)? The federal government has decided that used-car dealers are not required to tell potential customers about the cars' defects (effect); why did the federal government reach this decision (causes)?

Cause-and-effect reasoning, therefore, provides a way to answer the following two questions:

- What will be the effect(s) of X?
- What caused X?

Here are examples of documents that use cause and effect as an organizing principle:

The Problem

Earlier this year, we were proud to offer the industry's largest array of add-on multimedia products for both our own computers and those of other manufacturers. Our offerings in cards, CD-ROM drives, speakers, and other peripherals were unrivaled in both quantity and quality. And the response was terrific: in our first three months we sold more than 12,000 multimedia kits and 58,000 other peripheral units.

But growing pains soon became apparent: we logged more than 9,000 multimedia-related customer-support calls in that same period. What was the cause of this unprecedented customer-support problem? After considerable analysis of our customer-support data, we concluded that two factors were at work:

- Add-on multimedia kits, even those meant for our own computers, were not necessarily compatible with the hardware or the software our customers were using. We heard too many horror stories about how the kits were installed properly, but when the customer tried to reboot, the operating system was gone.
- Some 70 percent of the customers were novices, as opposed to less than 40 percent for our other product lines, and our documentation was simply inadequate to the task.

Meeting the Challenge

We recognized that being a pioneer in the industry had its costs: we were the first to encounter the problems that are now pervasive in the industry and well publicized in the literature. And because we were first, we took our lumps from the trade journals for the resulting problems with customer satisfaction.

We instituted a four-point plan to meet the challenge:

- We instituted a new quality-control program. Now every product is treated just the way a customer treats it. It is taken out of the box, plugged in, and turned on. We make sure that the printer setup is accurate and that the hardware and the bundled software are compatible. At our weekly audit meetings, we review that week's quality-control data; each team leader is now empowered to stop production to investigate a recurrent or unexplained problem.
- We expanded our use of novices in our preproduction focus groups and in the quality-control program. We are concentrating on learning how the novice uses our products, for in our expansion into the family market we will find that an increasing percentage of our customers are first-time computer owners.
- We instituted a Process-Improvement Team, a group of 12 veteran employees committed to improving customer support and customer satisfaction. Among the first innovations of the Process-Improvement Team was the creation of more than 200 documents to assist users with com-

The writer describes the problem in detail and analyzes its causes.

The writer speculates on the causes of the problem.

The writer describes the methods—the steps the company took to solve the problem.

FIGURE 8.15

Argument Organized by the Problem-Methods-Solution Pattern

(Figure 8.15 continued)

mon problems encountered when installing our kits and using common software. These documents are faxed to customers at no charge when they call a special toll-free number.
- We instituted a Quality Team of 15 employees charged with the responsibility of seeking Manufacturing's ideas about quality and efficiency standards.

The Results

These measures have been in place for only two months, and it is too early to declare total victory, but the preliminary data are encouraging. Customer-support calls on our multimedia kits are down over 15 percent the last two months. As reported by Customer Support, the incidence of catastrophic problems—such as destruction of the operating system—are down over 30 percent. The increased use of novices in design and in focus groups has led to three interface improvements that were noted in a *PC Week* article earlier this month. The work of the Quality Team has resulted in a 7 percent decrease in rejection rates of our multimedia kits.

In short, I think we are on the right track. But quality improvement is a frame of mind and a commitment, not a goal that can ever be reached. I pledge to you that we shall continue to strive to make RST the best place to buy PCs and PC-related products.

> The writer describes the solution—the data on the reduction in the problem.

- In an *environmental impact statement,* you argue that the proposed construction would have three important effects on the ecosystem.
- In the recommendation section of a *report,* you argue that your recommended solution would improve operations in two major ways.
- In the introduction to a *journal article,* you argue that your topic is worthy of study because it helps readers understand an important but overlooked aspect of the subject you are addressing.

The following discussion of creating causal links is based on the theory of philosopher Stephen Toulmin (1980).

The Three Elements of a Causal Argument

A persuasive causal argument has three elements, as shown in Figure 8.16.

The *assertion* is the claim that you wish to prove or make compelling. Here are examples of assertions:

- Our company should institute flextime.
- Our company should not institute flextime.

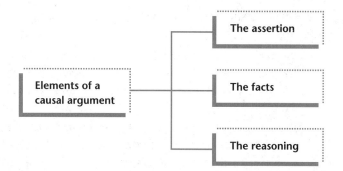

FIGURE 8.16

Elements of a Causal
Argument

- The search for alternative fuels has been hindered by myopic government policy.
- By the end of next year, we will have the new rolling machine in full production.
- The key to success in our industry is commitment to the customer.

The assertion is the conclusion you want your readers to accept—and, if appropriate, act on—after they have finished reading your discussion.

The *facts* are the information you want your audience to consider as they read your discussion. For a discussion of the merits of flextime, for example, the facts might include the following:

- The turnover rate of our female employees is double that of our male employees.
- Replacing a staff-level employee costs us about one-half the employee's annual salary; replacing a professional-level employee, a whole year's salary.
- At exit interviews, over 40 percent of our female employees under the age of 38 state that they quit because they cannot work and also be home for their school-age children.
- Other companies have found that flextime significantly decreases the turnover rate among female employees under the age of 38.
- Other companies have found that flextime has additional benefits and introduces no significant problems.

The *reasoning* is the way you derived the assertion from the facts. In the discussion of flextime, the reasoning involves three links:

- Flextime appears to have reduced the problem of high turnover among younger female employees at other companies.
- Our company is similar to other companies.
- If flextime has proven helpful at other companies, it is likely to prove helpful at our company.

For your readers to be convinced by your argument, you have to achieve four goals.

- *Readers must understand your facts and your reasoning.* You have to communicate your facts clearly, using whatever verbal and visual resources you have. And you have to express your reasoning clearly as well.

- *Readers must accept that your facts are accurate and relevant to the subject.* If you are not a well-known authority writing to a general reader, you have to cite your sources or use other appropriate techniques to show your readers that you are credible. And you have to show your readers that these are the appropriate facts for this discussion.

- *Readers must believe that you have not left out any relevant facts.* This is a challenge, because research standards and practices vary from field to field. In general, however, common sense is the best guide. For instance, if one of your readers believes, from talking to his neighbor, that flextime causes new problems—decreased car-pooling opportunities and increased utility bills, for example—but you haven't addressed these potential problems, that reader might doubt your assertion. You have to meet the skeptical or hostile reader's possible objections to your case by qualifying your assertion or conceding its drawbacks.

- *Readers must accept that your reasoning is logical.* How do you do that? If there were an easy answer, the world would be much more peaceful than it is. But again, use your common sense. If your argument depends on the idea that your company is like the other companies that have found flextime to be successful, you have to decide whether to make that point explicitly. If all the examples you can cite are from the same industry as yours, from companies of about the same size, with employees of approximately the same background, it might be necessary only to list some of those companies. But if the companies you can cite are not apparently similar to yours, you might want to explain why you think that a plan that worked well for them is likely to work well for you, despite the differences.

Logical Fallacies

As you create causal arguments, be careful to avoid three basic logical errors:

- *Inadequate sampling.* Inadequate sampling is the fallacy of drawing conclusions on the basis of an insufficient sample of cases. It would be illogical to draw the following conclusion:

  The new Gull is an unreliable car. Two of my friends own Gulls, and both have had reliability problems.

You would have to study a much larger sample and compare your findings with those for other cars in the Gull's class before reaching any valid conclusions.

Sometimes, however, you have only incomplete data. In these cases, qualify your conclusions:

The Martin Company's Collision Avoidance System has performed according to specification in extensive laboratory tests. However, the system cannot be considered effective until it has performed satisfactorily in the field. At this time, Martin's system can only be considered very promising.

- *Post-hoc reasoning*. The complete phrase—*post hoc, ergo propter hoc*—means "After this, therefore because of this." The fact that A precedes B does not mean that A caused B. The following statement is illogical:

There must be something wrong with the new circuit breaker in the office. Ever since we had it installed, the air conditioners haven't worked right.

The air conditioners' malfunctioning *might* be caused by a problem in the circuit breaker; but malfunctioning might be caused by inadequate wiring or by problems in the air conditioners themselves.

- *Oversimplifying*. The logical fallacy of oversimplifying occurs when a writer omits important information in establishing a causal link.

The way to solve the balance-of-trade problem is to improve the quality of the products we produce.

Although improving quality is important, international trade balances are determined by many factors, including tariffs and currency rates, and therefore cannot be explained by simple cause-and-effect reasoning.

Figure 8.17 on page 184, from a discussion of why mammals survived but dinosaurs did not (Jastrow, 1977, pp. 104–105), illustrates a very effective cause-and-effect argument.

Introducing and Concluding the Body

Drafting the arguments in the body of a document involves using and modifying the basic organizational patterns described in this chapter. Two more elements of the body—the introduction and conclusion—are also fundamental to the success of your argument.

Introducing the Body

An introduction has one main goal: to help your readers understand your argument by explaining *what* information you are going to present, *how* you are going to present it, and *why* you choose to present it that way. If you have clearly communicated this information, your readers will be more will-

The writer asserts here that one reason ancestral mammals were more intelligent than the dinosaurs is that their environment demanded great intelligence. The facts are that the mammals were outnumbered by their predators and had to search for food at night, when especially sharp sensory perception is required. The reasoning, which remains implicit, is the general theory of natural selection: that only those mammals with great intelligence survived and reproduced, and that this fact improved the intelligence of these mammals.

FIGURE 8.17

Argument Organized by the Cause-and-Effect Pattern

Why were the ancestral mammals brainier than the dinosaurs? Probably because they were the underdogs during the rule of the reptiles, and the pressures under which they lived put a high value on intelligence in the struggle for survival. These little animals must have lived in a state of constant anxiety—keeping out of sight during the day, searching for food at night under difficult conditions, and always outnumbered by their enemies. They were Lilliputians in the land of Brobdingnag. Small and physically vulnerable, they had to live by their wits.

The nocturnal habits of the early mammal may have contributed to its relatively large brain size in another way. The ruling reptiles, active during the day, depended mainly on a keen sense of sight; but the mammal, who moved about in the dark much of the time, must have depended as much on the sense of smell, and on hearing as well. Probably the noses and ears of the early mammals were very sensitive, as they are in modern mammals such as the dog. Dogs live in a world of smells and sounds, and, accordingly, a dog's brain has large brain centers devoted solely to the interpretation of these signals. In the early mammals, the parts of the brain concerned with the interpretation of strange smells and sounds also must have been quite large in comparison with their size in the brain of the dinosaur.

Intelligence is a more complex trait than muscular strength, or speed, or other purely physical qualities. How does a trait as subtle as this evolve in a group of animals? Probably the increased intelligence of the early mammals evolved in the same way as their coats of fur and other bodily changes. In each generation, the mammals slightly more intelligent than the rest were more likely to survive, while those less intelligent were likely to become victims of the rapacious dinosaurs. From generation to generation, these circumstances increased the number of the more intelligent and decreased the number of the less intelligent, so that the average intelligence of the entire population steadily improved, and their brains grew in size. Again the changes were imperceptible from one generation to the next, but over the course of many millions of generations the pressures of a hostile world created an alert and relatively large brained animal.

ing to read the document, better able to understand it, and better able to remember it.

Your document can have one introduction (at the beginning) or several introductions (one at the beginning and one at the start of each major section).

Every document calls for a different kind of introduction. A brochure on backyard pool safety obviously needs little introduction; a brief review of the statistics on injuries and deaths is all that you need. But the introduction of a scholarly article, which needs to conform to the practices in a particular academic field, might require many elements. To figure out how to draft the introduction to the body of your document, use your common sense and study similar documents to understand the conventions in your field and the expectations of your readers.

Figure 8.18 on page 186 shows the seven critical questions you should consider answering for your readers in an introduction.

- *What are the key terms that will be used in the argument?* Define any key terms that you will discuss in your document.

  Phantom risk refers to alleged but uncertain risks associated with a scientific or technological practice. For instance, weak magnetic fields represent a phantom risk in that the scientific community is uncertain whether they represent any health risk at all, or, if they do, what the threshold for such risks might be.

  If the reader needs to know a key term to understand the introduction, the definition should be presented early in the introduction. See Chapter 9 for more information on definitions.

- *What is the subject?* Answer this question directly, even though your readers might already know.

  This report describes the relationship between the courts and the scientific community on the subject of phantom risk. We live in an age of tremendous technological advances, and some new products and practices might pose health risks. The conflict explored in this report is that whereas the courts require certainty—they need to know whether a plaintiff was or was not injured by a technology—the scientific community often cannot supply a definitive answer.

- *What is the purpose of the argument?* Explain what you hope to achieve in the discussion.

  This report has two main purposes: to summarize the conflict between the legal and scientific communities and to propose three principles to guide the future debate on the legal ramifications of phantom risk.

- *What is the background of the subject?* Explain the background: the information that readers need to understand the discussion. Many of your readers might not be up-to-date on the subject.

  It is inevitable that some people will be injured by technological advances, and one of the foundations of our legal system is that victims be empowered to seek redress of their injuries through the legal system. However, science does not offer the kind of certainty that the legal system requires. Are breast implants dangerous? Cautious scientists would have to admit that they don't know, but different constituencies answer "yes" and "no" with self-assurance and conviction. When courts render false positive judgments—by finding that harmless products and services have caused injuries—they put companies out of business and deprive people of their products and services. When courts render false negatives—by finding that harmful products and services have not injured anyone—they endanger the public. Currently, there is

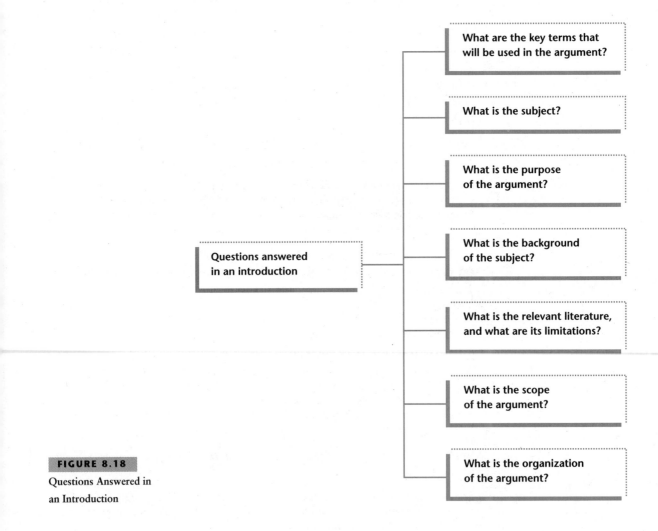

FIGURE 8.18

Questions Answered in
an Introduction

no way to reconcile the different cultures and practices of the legal and scientific
communities.

- *What is the relevant literature, and what are its limitations?* For books,
  journal articles, proposals, and reports, you want to show your readers
  that you have done the necessary research on your topic.

  A number of researchers have studied the topic of phantom risk, but most of the
  studies have examined individual cases from either the legal or scientific perspective.
  For instance, the asbestos issue has been discussed from the scientific perspective by
  Wilkins (1992), Thomas (1993), and Rivera (1993), and the legal perspective by Hal-
  loran (1994), Bradford (1993), and De Moss (1992). The edited collection by Foster
  et al. (1993) is the first to attempt a detailed look at the conflict; however, it offers
  piecemeal opinions about the different phantom-risk issues but does not provide a
  theoretical framework for reconciling the two cultures.

- *What is the scope of the argument?* Describe the scope of your discussion: what you are including or excluding.

  This report focuses on four phantom-risk issues: electrical and magnetic field risks, spermicide cancer risks, asbestos risks, and secondhand smoke risks. It reviews the scientific evidence and the legal opinions for each issue. Where appropriate, it discusses the social policy resulting from the legal actions.

- *What is the organization of the argument?* Explain how you are organizing your argument so your readers can concentrate on the information without worrying about what will come next.

  In the sections that follow, I treat each phantom-risk issue separately, as a case study. After a background discussion, I fill in the scientific consensus, followed by the legal precedents and, where appropriate, public policy. I conclude the article with a recommendation that we discuss three principles for reconciling the legal and scientific approaches to phantom-risk issues.

Concluding the Body

In discussions of writing, the word *conclusion* has two different meanings. One meaning refers to the inferences drawn from technical data. If, for instance, you are doing a feasibility report on whether to switch from paper-based to on-line documentation, you might conclude that the change would be unwise at this time. This kind of conclusion is discussed in Chapter 23.

Conclusion also refers to the final part of a document or a section of a document. The following discussion concerns this second sense of the word.

Although some kinds of documents, such as parts catalogs, do not generally have a conclusion, most do. When drafting a conclusion, be sure you answer the appropriate questions from the four shown in Figure 8.19.

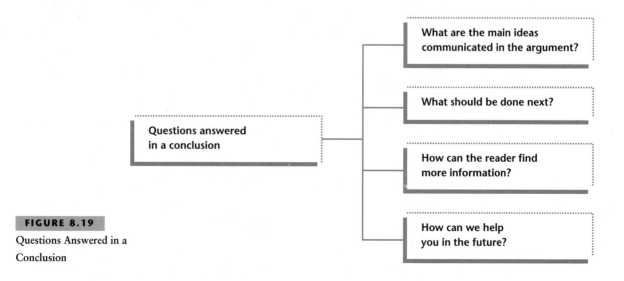

FIGURE 8.19

Questions Answered in a
Conclusion

- *What are the main ideas communicated in the argument?* Readers can forget material, especially material from the beginning of a long document. Therefore, it is a good idea to summarize important ideas in a paragraph. The following examples are from a report on reuse-adoption programs: companywide programs to increase the reuse of systems and software.

  Our analysis yielded two main conclusions. First, reuse adoption is good engineering and good business; it can reduce expenses and increase productivity and quality. Second, reuse adoption is a complex procedure that requires a substantial amount of planning and supervision. If the program is carried out casually or thoughtlessly, it can backfire and cause more problems than it solves.

- *What should be done next?* Offer recommendations on the course of future work.

  I recommend that we convene an ad hoc committee to study the feasibility of converting to a reuse-adoption plan. Specifically, we should carry out a Phase I analysis to determine the answers to the following four questions:

  1. Is demand in our market sufficient for the kinds of products that are appropriate for reuse adoption?
  2. Do we have sufficient assets—code, designs, templates, and so on—to justify reuse adoption?
  3. Are technological developments in our industry sufficiently predictable to justify reuse adoption?
  4. Are industry standards and government standards sufficiently predictable to justify reuse adoption?

- *How can the reader find more information?* Sometimes this is a sales message; sometimes it is not.

  I have asked Corporate Information to get the six articles listed in the references section. These articles will be gathered and routed next week. Please read them and be prepared to offer any questions or concerns at our meeting on August 15, at which time we will discuss the recommendations.

- *How can we help you in the future?* Offer to provide future services if you can.

  I think reuse adoption offers the promise to improve our business in a number of ways. Please feel free to get in touch with me, either before or after you read the articles, to discuss this initiative.

WRITER'S CHECKLIST

The following four questions relate to planning the argument:

1. Have you analyzed your audience's goals?
2. Is your evidence persuasive?
3. Have you presented yourself effectively?
4. Have you analyzed the political realities?

The following checklists cover the eight organizational patterns discussed in this chapter, as well as introductions and conclusions.

Chronological and Spatial

1. Have you provided signposts, such as headings or transitional words or phrases?
2. Have you considered using graphics to complement the text?
3. Have you analyzed events where appropriate?

General to Specific

1. Have you provided signposts, such as headings or transitional words or phrases?
2. Have you considered using graphics to complement the text?

More Important to Less Important

1. Have you explained clearly that you are using this organizational pattern?
2. Has your discussion made clear why the first point is the most important, the second is the second most important, and so forth?
3. Have you considered using graphics to complement the text?

Comparison and Contrast

1. Have you included the necessary criteria?
2. Have you chosen a structure—whole by whole or part by part—that is appropriate for your audience and purpose?
3. Have you chosen appropriate organizational patterns for your second-level items?

Classification and Partition

1. Have you chosen a basis consistent with the audience and purpose of the document?
2. Have you used only one basis at a time?
3. Have you avoided overlap?
4. Have you included all the appropriate categories?
5. Have you arranged the categories in a logical sequence?

Problem-Methods-Solution

1. Have you described the problem clearly and specifically?
2. If appropriate, have you justified your methods?
3. Have you avoided overstating your solution?
4. Have you sequenced the discussion in a way that is consistent with the audience and purpose of the document?

Cause and Effect

1. Have you clearly expressed your assertion, your facts, and your reasoning?
2. Have you sequenced your assertion, your facts, and your reasoning appropriately for your audience and purpose?
3. Have you avoided inadequate sampling?
4. Have you avoided post-hoc reasoning?
5. Have you avoided oversimplifying?

The following questions cover introductions and conclusions.

Introductions

1. Have you explained the subject of the document?
2. Have you explained the purpose of the discussion?
3. Have you explained the background of the subject?
4. Have you reviewed the relevant literature, explaining its limitations?
5. Have you explained the scope of the discussion?
6. Have you explained the organization of the discussion?
7. Have you defined any key terms that will be used in the discussion?

Conclusions

1. Have you summarized the main ideas communicated in the document?
2. Have you recommended what should be done next?
3. Have you explained how the reader can find out more information?
4. Have you described how you can help in the future?

EXERCISES

The first nine exercises call for you to write an argument using a different organizational pattern. For each exercise, attach a statement indicating your audience and purpose, as well as the type of document in which it might be included. For example, "This argument is addressed to first-semester college students. Its purpose is to help them understand the policies they are to follow in using the university's mainframe. The argument would be part of the *Student Handbook.*"

1. Using the chronological organizational pattern, write a passage of 500–1,000 words on one of the following topics or a topic of your choice:
   a. how to register for courses at your college or university
   b. how to learn to use a software package
   c. how to buy the right car for your needs
   d. how to prepare for a job interview
   e. how to determine job prospects in your field

2. Using the spatial organizational pattern, write a passage of 500–1,000 words on one of the following topics or a topic of your choice:
   a. your bicycle
   b. your car's dashboard
   c. the room in which you are sitting
   d. the space shuttle
   e. the remote-control device from a television or stereo set

3. Using the general-to-specific organization, write a passage of 500–1,000 words on one of the following topics or a topic of your choice:
   a. advances in manufacturing technology
   b. alternatives to incarceration for nonviolent criminals
   c. pay scales for general practitioners and medical specialists
   d. cooperative education and internships for college students
   e. energy efficiency in computers and printers

4. Using the more-important-to-less-important organizational pattern, write a passage of 500–1,000 words on one of the following topics or a topic of your choice:
   a. the reasons you chose your college or your major
   b. the effects of acid rain on a particular area
   c. the reasons you should (or should not) be required to study a foreign language (or learn a computer programming language)
   d. the three most important changes you would like to see at your school
   e. the reasons that recycling should (or should not) be mandatory

5. Using a comparison-and-contrast organizational pattern, write a passage of 500–1,000 words on one of the following topics or a topic of your choice:
   a. lecture classes and discussion classes
   b. the tutorials that come with two different software packages
   c. manual transmission and automatic transmission automobiles
   d. black-and-white and color photography
   e. two different word-processing programs

6. Using the problem-methods-solution pattern, write a passage of 500–1,000 words on one of the following topics or on a topic of your choice:
   a. how you solved a recent problem related to your education
   b. how you went about deciding on a recent major purchase, such as a car, a personal computer, or a bicycle
   c. how you would propose reducing the time required to register for classes or to change your schedule
   d. how you would propose increasing the ties between your college or university and local business and industry

7. Using classification, write a passage of 500–1,000 words on one of the following topics or a topic of your choice:
   a. foreign cars
   b. smoke alarms
   c. college courses
   d. personal computers
   e. cameras

8. Using partition, write a passage of 500–1,000 words about a piece of equipment or machinery you are familiar with, or about one of the following topics.
   a. a student organization on your campus
   b. an audio cassette tape
   c. a portable radio
   d. a bicycle
   e. a guitar

9. Using causal reasoning—either forward or backward—write a passage of 500-1,000 words on one of the following topics or a topic of your choice:
   a. women serving in combat in the military
   b. the price of gasoline
   c. the emphasis on achieving high grades in college

d. the prospects for employment in your field

10. In one paragraph each, describe the logical flaws in the following items:

 a. The election couldn't have been fair—I don't know anyone who voted for the winner.

 b. Increased restrictions on smoking in public are responsible for the decrease in smoking.

 c. Since the introduction of cola drinks at the start of this century, cancer has become the second greatest killer in the United States. Cola drinks should be outlawed.

 d. The PC industry is enjoying record profits; just look at the profits of Compaq and Micron.

 e. The value of the Americans with Disabilities Act has to be questioned, for since its passage, the number of discrimination suits based on disabilities has actually increased.

 f. The phenomenal growth of the Internet can be explained by the fiber-optic revolution of the 1980s.

11. Write a memo to your instructor evaluating the effectiveness of the introduction (page 192) from a student report about how audits are conducted. What works well in the introduction? How might the introduction be improved? See Chapter 16 for a discussion of memos.

12. The following conclusion is from the student report from which the introduction in exercise 11 is excerpted. Write a memo to your instructor evaluating the conclusion's effectiveness. What aspects of it are effective? How would you improve it? See Chapter 16 for more information on memos.

All major corporations and most minor corporations require an audit to be conducted every year. Audits are the primary tool used to ensure that financial irregularities such as embezzlement, conflict of interest, and inaccurate pricing are kept to a minimum.

REFERENCES

Bell, T. E. (1993). Bicycles on a personalized basis. *IEEE Spectrum, 30*(9), 32–35.

Bryzek, J., Petersen, K., & McCulley, W. (1994). Micromachines on the march. *IEEE Spectrum, 31*(5), 20–31.

Jastrow, R. (1977). *Until the sun dies.* New York: Norton.

Larson, D. E. (1990). *Mayo Clinic family health book.* New York: William Morrow.

Murphy, D. (1994). *On-line and off-line debit cards: First-year costs of implementation at John's Market.* Unpublished manuscript, Boise State University.

Toulmin, S. (1980). *The uses of argument.* London: Cambridge University Press.

Introduction

An audit is a formal examination and verification of financial accounts. According to the tenth edition of *Montgomery's Auditing* textbook, historians believe that audits were conducted as early as 4000 B.C. in Babylonia. Governments were concerned with establishing controls, including audits, to reduce errors and fraud in the tax collection system. Auditing has been seen in history ever since. The Bible (generally believed to cover the period of time between 1800 B.C. and A.D. 95) refers to internal controls and surprise audits. The earliest records in English-speaking countries are from England and Scotland. They have accounting records and references to audits dating back to A.D. 1130.

Auditing was fairly slow to progress in the United States. In the late 1800s the U.S. railroad companies employed auditors, and in 1887 the American Institute of Accountants, now the American Institute of Certified Public Accountants (AICPA), was established. In 1935 audits became more common for two reasons:

- The Securities Act of 1933 and the Securities Exchange Act of 1934, which required listed companies to file audited financial statements, were enacted
- The AICPA collaborated with the New York Stock Exchange to improve reporting standards

Today all major corporations conduct audits every year. Audits are used to analyze the financial condition of a company. Audits are usually conducted by certified public accountants who work for independent accounting firms such as Arthur Anderson. The purpose of performing an audit is to identify any problematic areas, such as embezzlement. An audit consists of six steps:

1. understanding the client's business
2. planning the audit
3. evaluating internal accounting controls
4. performing compliance tests
5. performing substantive tests
6. preparing the financial statement and issuing an opinion

CHAPTER 9

Drafting Definitions

The world of business and industry depends on clear definitions. Without written definitions, the working world would be chaotic. For example, suppose you learn at a job interview that the potential employer pays tuition and expenses for employees' job-related education. That's good news, of course, if you are planning to continue your education. But until you study the employee-benefits manual, you won't know with any certainty just what the company will pay for.

Who, for instance, is an *employee*? You would think an employee would be anyone who works for and is paid by the company. But you might find that for the purposes of the manual, an employee is someone who has worked for the company full time (40 hours per week) for at least six uninterrupted months.

Other terms also would have to be defined in the description of tuition benefits. What, for example, is *tuition*? Does the company's definition include incidental laboratory or student fees? What is *job-related education*? Does a course about time management qualify under the company's definition? What, in fact, constitutes *education*? All these terms and many others must be defined for the employees to understand their rights and responsibilities.

Definitions play a major role, therefore, in communicating policies and standards "for the record." When a company wants to purchase air-conditioning equipment, it might require that the supplier provide equipment certified by a professional organization of air-conditioner manufacturers. The organization's definitions of acceptable standards of safety, reliability, and efficiency will provide some assurance that the equipment is high quality.

Definitions, of course, have many uses outside legal or contractual contexts. Two such uses occur frequently.

- *Definitions clarify a description of a new development or a new technology in a technical field.* When zoologists discover a new animal species, for instance, they name and define it. When a scientist devises a new laboratory procedure, he or she defines and describes it in an article printed in a technical journal.
- *Definitions help specialists communicate with less knowledgeable readers.* A manual that explains how to tune up a car includes definitions of parts and tools. A researcher at a manufacturing company uses definitions in describing a new product to the sales staff.

Definitions, then, are crucial in many kinds of technical communication, from brief letters and memos to technical reports, manuals, and journal articles. All readers, from the general reader to the expert, need effective definitions to carry out their jobs.

Definitions, like every other technique in technical communication, require thought and planning. Before you can write a definition to include in a document, you must carry out three steps, as shown in Figure 9.1.

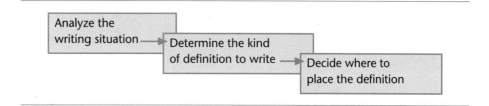

Analyzing the Writing Situation

The first step in writing effective definitions is to analyze the writing situation: the audience for and the purpose of your document.

- *Audience.* Unless you know whom you are addressing and how much that audience already knows about the subject, you cannot know which terms to define or the kind of definition to write. Physicists wouldn't need a definition of *entropy,* but a group of lawyers might. Builders know what a Molly bolt is, but some insurance agents don't. If you are aware of your audience's background and knowledge, you can easily devise effective informal definitions. For example, if you are describing a compact-disc player to a group of readers who understand automobiles, you can use a familiar analogy to define the function of the pause button: "The PAUSE button is the clutch pedal of the compact-disc player."

- *Purpose.* If you want to give your readers only a basic understanding of a concept—say, time-sharing vacation resorts—a brief, informal definition is usually sufficient. However, if you want your readers to understand an object, process, or concept thoroughly and be able to carry out tasks associated with it, a more formal and elaborate definition is required. For example, a definition of a "Class 2 Alert" written for operators at a nuclear power plant must be comprehensive, specific, and precise.

Determining the Kind of Definition to Write

As the preceding analysis of the writing situation has suggested, definitions can be short or long, informal or formal. Figure 9.2 on page 196 shows the three basic types of definitions.

Parenthetical Definition   A parenthetical definition is a brief clarification placed unobtrusively within a sentence. Sometimes a parenthetical definition is only a word or phrase:

The crane is located on the starboard (right) side of the ship.

Summit Books announced its intention to create a new colophon (emblem or trademark).

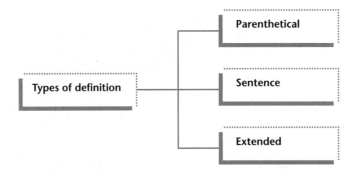

FIGURE 9.2

Types of Definition

> United Engineering is seeking to purchase the equity stock (common stock) of Minnesota Textiles.

A parenthetical definition can also take the form of a longer explanatory phrase or clause:

> Motorboating is permitted in the Jamesport Estuary, the portion of the bay that meets the mouth of the Jamesport River.

> Before the metal is plated, it is immersed in the pickle, an acid bath that removes scales and oxides from the surface.

Parenthetical definitions are not, of course, comprehensive. They serve mainly as quick and convenient ways of introducing new terms to the readers. Especially when you are using a parenthetical definition in writing addressed to general readers, make sure that the definition itself is clear. You have gained nothing if your readers don't understand your clarification:

> Next, check for blight on the epicotyl, the stem portion above the cotyledons.

If your readers are botanists, this parenthetical definition will be clear (although it might be unnecessary, for if they know the meaning of *cotyledons,* they are likely to know *epicotyl*). However, if you are addressing the general reader, this definition will merely frustrate them.

Sentence Definition

A sentence definition—a one-sentence clarification—is more formal than a parenthetical definition. Usually, a sentence definition follows a standard pattern: the item to be defined is placed in a category of similar items and then distinguished from the other items in that category:

Item	=	Category	+	Distinguishing Characteristics
A flip flop	is	a circuit		containing active elements that can assume either one of two stable states at any given time.

continued

Item	=	**Category**	+	**Distinguishing Characteristics**
An electrophorus	is	a laboratory instrument		used to generate static electricity.
Hypnoanalysis	is	a psychoanalytical technique		in which hypnosis is used to elicit information from a patient's unconscious mind.
A Bunsen burner	is	a small laboratory heating device		consisting of a vertical metal tube connected to a gas source.
An electron microscope	is	a microscope		that uses electrons rather than visible light to produce magnified images.

In many cases a sentence definition also includes a graphic. For example, the definitions of electrophorus, Bunsen burner, and electron microscope would probably be accompanied by photographs or drawings.

Sentence definitions are useful when your readers require a more formal or more informative clarification than parenthetical definition can provide. Writers often use sentence definitions to establish a working definition for a particular document: "In this report, the term *electron microscope* is used to refer to any microscope that uses electrons rather than visible light to produce magnified images." Such definitions are sometimes called *stipulative definitions,* for the writer is stating how the term will be used in the document.

In sentence definitions, follow four guidelines:

- *Be specific in writing the category and the distinguishing characteristics.* Remember, you are trying to distinguish the item from all other similar items. If you write "A Bunsen burner is a burner that consists of a vertical metal tube connected to a gas source," the imprecise category— "a burner"—defeats the purpose of your definition: many types of large-scale burners use vertical metal tubes connected to gas sources. If you write "Hypnoanalysis is a psychoanalytical technique used to elicit information from a patient's unconscious mind," the imprecise distinguishing characteristics— "used to elicit . . ."—ruins the definition: many psychoanalytical techniques are used to elicit information from a patient's unconscious mind. If your definition describes more than one item, you have to sharpen either the category or the distinguishing characteristics, or both.

- *Don't describe a specific item when you mean to define a general class of items.* If you wish to define *catamaran,* don't describe a particular catamaran. The catamaran you see on the beach in front of you might be made by Hobie and have a white hull and blue sails, but those characteristics are not essential to catamarans.

- *Avoid writing circular definitions, that is, definitions that merely repeat the key words of the item being defined in the category or the distinguishing characteristics.* In "A required course is a course that is required," what does "required" mean? Required of whom, by whom? The word is never defined. Similarly, "A balloon mortgage is a mortgage that balloons" is useless.

  However, you can use *some* kinds of words in the item as well as in the category and distinguishing characteristics; in the definition of *electron microscope* given earlier, for example, the word *microscope* is repeated. Here, *microscope* is not the difficult part of the item; readers know what a microscope is. The purpose of defining *electron microscope* is to clarify the *electron* part of the term.

- *Be sure the category contains a noun or a noun phrase rather than a phrase beginning with* when, what, *or* where.

  INCORRECT: A brazier is what is used to . . .

  CORRECT: A brazier is a metal pan used to . . .

  INCORRECT: An electron microscope is when a microscope . . .

  CORRECT: An electron microscope is a microscope that . . .

  INCORRECT: Hypnoanalysis is where hypnosis is used . . .

  CORRECT: Hypnoanalysis is a psychoanalytical technique in which . . .

Extended Definition   An extended definition is a long, detailed clarification—usually one or more paragraphs—of an object, process, or idea. Often an extended definition begins with a sentence definition, which is then elaborated. For instance, the sentence definition "An electrophorus is a laboratory instrument used to generate static electricity" tells you the basic function of the device, but it leaves many questions unanswered: How does an electrophorus work? What does it look like? What is it made of? An extended definition would answer these and other questions.

Extended definitions are useful, naturally, when you want to give your readers a reasonably complete understanding of the item. And the more complicated or more abstract the item being defined, the greater the need for an extended definition.

There is no one way to "extend" a definition. Your analysis of the audience and purpose of the communication will help you decide which method to use. In fact, an extended definition sometimes employs several of the nine techniques shown in Figure 9.3.

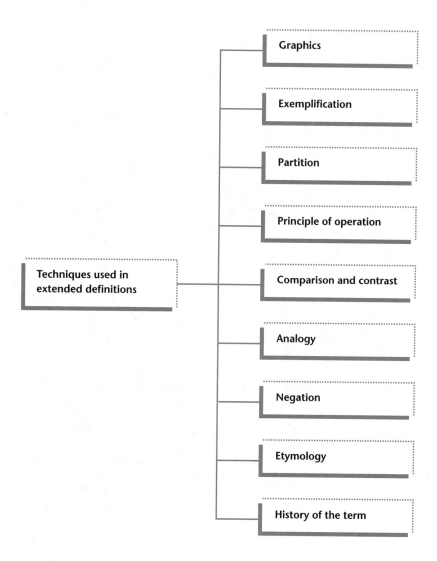

FIGURE 9.3

Techniques Used in
Extended Definitions

Graphics

Perhaps the most common way to create an extended definition in technical communication is to create some sort of graphic, such as a photograph, diagram, schematic, or flowchart, and then explain the graphic.

Graphics are particularly useful in defining physical objects. The simplest way to define a two-person saw, for instance, would be to draw a rough diagram. However, illustrations are also useful in defining concepts and ideas. A definition of the meteorological concept of temperature inversion, for instance, might include a diagram. So might definitions of lasers, compound fractures, and buying stocks on margin.

The following excerpt from an extended definition of *parallelogram* shows an effective combination of words and illustrations.

> A parallelogram is a four-sided plane whose opposite sides are parallel with each other and equal in length. In a parallelogram, the opposite angles are the same, as shown in the sketch below.

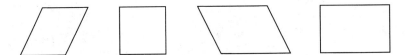

Exemplification

Exemplification—using an example (or examples) to clarify an object or idea—is particularly useful in making an abstract term easy to understand. The following paragraph is an extended definition of the psychological defense mechanism called *conversion* (Wilson, 1964, p. 84).

Notice that the first two sentences in this paragraph are essentially a sentence definition that might be paraphrased as follows: "Conversion is a mechanism of psychological defense by which the conflict is converted into the symptoms of a physical illness."

> A third mechanism of psychological defense, "conversion," is found in hysteria. Here the conflict is converted into the symptom of a physical illness. In a case of conversion made famous by Freud, a young woman went out for a long walk with her brother-in-law, with whom she had fallen in love. Later, on learning that her sister lay gravely ill, she hurried to her bedside. She arrived too late and her sister was dead. The young woman's grief was accompanied by sharp pain in her legs. The pain kept recurring without any apparent physical cause. Freud's explanation was that she felt guilty because she desired the husband for herself, and unconsciously converted her repressed feelings into an imaginary physical ailment. The pain struck her in the legs because she unconsciously connected her feelings for the husband with the walk they had taken together. The ailment symbolically represented both the unconscious wish and a penance for the feelings of guilt which it engendered.

This extended definition is effective because the writer has chosen a clear and interesting (and therefore memorable) example of the idea he is describing. No other examples are necessary in this case. If conversion were a more difficult concept to describe, an additional example might be useful.

Partition

Partition is the process of dividing a thing or idea into smaller parts so that the reader can more easily understand it. (See Chapter 8 for a detailed discussion of partition.)

The hitch is divided into its major components, each of which is then defined and described.

This extended definition of a load-distributing hitch would be accompanied by a diagram.

A *load-distributing hitch* is a trailer hitch that distributes the hitch load to all axles of the tow vehicle and the trailer. The crucial component of the load-distributing hitch is the set of spring bars that attaches to the trailer. For a complete understanding of the load-distributing hitch, however, the following other components should be considered first:

1. the shank
2. the ball mount
3. the sway control
4. the frame bracket

The shank is the metal bar that is attached to the frame of the tow vehicle . . .

Principle of Operation

The *principle of operation* is an effective way to develop an extended definition, especially that of an object or process. The following extended definition of a thermal jet engine is based on the mechanism's principle of operation.

This extended definition begins with a sentence definition.

A thermal jet engine is a jet-propulsion device that uses air, along with the combustion of a fuel, to produce the propulsion. In operation, the thermal jet engine draws in air, increases its pressure, heats it by combustion of the fuel, and finally ejects the heated air (and the combustion gases). The increased velocity of the ejected mixture determines the thrust: the greater the difference in velocity between air entering and leaving the unit, the greater the thrust.

Comparison and Contrast

Using *comparison and contrast,* the writer discusses similarities or differences between the item being defined and an item with which the readers are more familiar. The following definition of a bit brace begins by comparing and contrasting it to the more common power drill.

A *bit brace* is a manual tool used to drill holes. Cranked by hand, it can theoretically turn a bit to bore a hole in any material that a power drill can bore. Like a power drill, a bit brace can accept any number of different sizes and shapes of bits. The principal differences between a bit brace and a power drill are:

1. A bit brace drills much more slowly.
2. A bit brace is a manual tool, and so it can be used where no electricity is available.
3. A bit brace makes almost no noise in use.

The bit brace consists of the following parts. . . .

Analogy

An *analogy* is a specialized kind of comparison. In a traditional comparison, the writer compares the item being defined to a similar item. An electron microscope is compared to a common microscope, or a bit brace is compared to a power drill. In an analogy, however, the item being defined is compared to an item that is in some ways completely different but that shares some essential characteristic or characteristics.

For instance, the central-processing unit of a computer is often compared to a brain. Obviously, these two items are very different—except that the function of the central-processing unit in the computer is similar to that of the brain in the body. People often compare computer software to phonograph records, cassette tapes, or compact discs. Again, the differences are many, but the similarity is essential: the phonograph record contains the code that enables the stereo system to fulfill its function, just as the software contains the code that enables the computer to do its job.

The following example, from an extended definition of computer literacy, shows how an analogy can clarify an unfamiliar concept.

> *Computer literacy* is the ability to use computers effectively. If you can operate a personal computer to do word processing or create a database, you are computer literate. If you can operate a digital watch or program a VCR or use an automated teller machine at a bank, you also can be said to be computer literate. To use an analogy, computer literacy is like automotive literacy: if you know how to operate an automobile safely to get from one place to another, you possess automotive literacy. Just as you don't have to understand the principle of the internal-combustion engine to drive a car, you don't have to understand the concepts of RAM and ROM to use a computer.

Negation

A special kind of contrast is sometimes called *negation* or *negative statement*. Negation is the technique of clarifying a term by distinguishing it from a different term with which the reader might have confused it. The following example shows the use of negation.

> An *ambulatory patient* is *not* a patient who must be moved by ambulance. On the contrary, an ambulatory patient is one who can walk without assistance from another person.

Negation is rarely the only technique used in an extended definition. In fact, negation is used most often in a sentence or two at the start of a definition: for after you state what the item is not, you still have to state what it is.

Etymology

Etymology—the derivation of a word—is often a useful and interesting way to develop a definition.

> The word *mortgage* was originally a compound of *mort* (dead) and *gage* (pledge). The meaning of the word has not changed substantially since its origin in Old French. A mortgage is still a pledge that is "dead" upon either the payment of the loan or the forfeiture of the collateral and payment from the proceeds of its sale.

Etymology is a popular way to begin a definition of *acronyms,* which are abbreviations pronounced as words.

> *SCUBA* stands for self-contained underwater breathing apparatus.

COBOL, which is the Common Business-Oriented Language, was originally invented to . . .

Etymology, like negation, is rarely used alone in technical communication, but it is an effective way to introduce an extended definition.

History of the Term

A common way to define a term is to explain its history. Often the extended definition explains the original use of the term and then describes how the meaning changed over the years in response to historical events or technological advances. The following (Roblee & McKechnie, 1981, pp. 15–16) is an example of the use of the history of a term:

> In general, the common-law definition of *arson* was traditionally the willful burning of the house of another, including all outhouses or outbuildings adjoining thereto. The emphasis was on another's habitation, and his life and safety at the place where he or she resided. Then, many legal issues began to arise. Was a school a dwelling? a jail? a church? The common-law courts began to view the crime of arson as being against the habitation *or possessions* of another.
>
> Gradually, laws were enacted to plug the loopholes of the common-law definition of arson. The first laws brought all buildings or structures into the scope of arson, provided they had human occupancy of any kind on a regular basis. Later, the occupancy requirements were dropped. Today, *arson* is a term applied to the willful and intentional burning of all types of structures, vehicles, forests, fields, and so on.

Figure 9.4 on page 204 (based on Krol, 1992, p. 159) is an example of an extended definition addressed to a general audience.

Deciding Where to Place the Definition

In many cases, you do not need to decide where to place a definition. In writing your first draft, for instance, you may realize that most of your readers will not be familiar with a term you want to use. If you can easily provide a parenthetical definition that will satisfy your readers' needs, simply do so.

In assessing the writing situation before beginning the draft, however, you will often conclude that one or more terms will have to be introduced, and that your readers will need more detailed and comprehensive clarifications, perhaps sentence definitions and extended definitions. In these cases, you should plan, at least tentatively, where you are going to place the definitions.

Definitions can be placed in four different locations:

- *In the text.* The text itself can accommodate any of the three kinds of definition. Because they are brief and unobtrusive, parenthetical definitions are almost always included in the text; even if a reader already knows what the term means, the slight interruption will not be annoying.

Notice that the term being defined does not have to be introduced in the first sentence.

Nor does the term have to be defined in a traditional sentence definition.

In this paragraph, the author provides a basic definition of the term by analyzing it.

The writer introduces a comparison.

In describing his strategy for explaining ftp, the writer offers another definition—of anonymous ftp.

FIGURE 9.4

Extended Definition

   Often, you will find information on the Internet that you don't want to examine on a remote system: you want to have a copy for yourself. You've found, for example, the text of a recent Supreme Court opinion, and you want to include pieces of it in a brief you are writing. Or you found a recipe that looks good, and you want to print a copy to take to the kitchen. Or you found some free software that just might solve all your problems, and you want to try it. In each case, you need to move a copy of the file to your local system so you can manipulate it there.

   The tool for doing this is *ftp,* named after the application protocol it uses: the "File Transfer Protocol (FTP)." As the name implies, the protocol's job is to move files from one computer to another. It doesn't matter where the two computers are located, how they are connected, or even whether or not they use the same operating system. Provided that both computers can "talk" the FTP protocol and have access to the Internet, you can use the *ftp* command to transfer files. Some of the nuances of its use do change with each operating system, but the basic command structure is the same from machine to machine.

   Like *telnet, ftp* has spawned a broad range of databases and services. You can, indeed, find anything from legal opinions to recipes to free software (and many others) in any number of publicly available on-line databases, or archives, that can be accessed through *ftp.* For a sampling of the archives that you can access with *ftp,* look at the *Resource Catalog* in this book. If you're a serious researcher, you will find *ftp* invaluable; it is the common "language" for sharing data.

   Unfortunately, *ftp* is a complex program because there are many different ways to manipulate files and file structures. Different ways of storing files (such as binary or ASCII, compressed or uncompressed) introduce some complications, and may require some additional thought to get things right. First, we will look at how to transfer files between two computers on which you have an account (a log-in name and, if needed, a password). Next, we'll discuss *anonymous ftp,* which is a special service that lets you access public databases without obtaining an account. Most public archives provide anonymous ftp access, which means that you can get gigabytes of information for free—without even requiring that you have a log-in name. Finally, we'll discuss some common cases (accessing VMS, VM, DOS, or Macintosh systems), each of which requires some special handling.

Likewise, sentence definitions are often placed within the text. If you want all your readers to see your definition or you suspect that many of them need the clarification, the text is the appropriate location. Keep in mind, however, that unnecessary sentence definitions can bother your readers. Extended definitions are rarely placed within the text, because of their length. The obvious exception, of course, is the extended definition of a term that is central to the discussion; a discussion of recent changes in workers' compensation insurance will likely begin with an extended definition of that kind of insurance.

- *In footnotes.* Footnotes are a logical place for an occasional sentence definition or extended definition. The reader who needs the definition can find it easily at the bottom of the page; the reader who doesn't need it will ignore it. However, footnotes can make the page look choppy. If you will need more than one footnote for every two or three pages, consider creating a glossary.

- *In a glossary.* A glossary—an alphabetized list of definitions—can accommodate sentence definitions and extended definitions of fewer than three or four paragraphs. It is a convenient collection of definitions that otherwise might clutter up the document. A glossary can be placed at the start of a document (for example, after the executive summary in a report) or at the end, preceding the appendices. For more information on glossaries, see Chapter 11.

- *In an appendix.* An appendix is an appropriate place to put an extended definition that is one page or longer. A definition of this length would be cumbersome in a glossary or in a footnote and, unless it explains a crucial term, would be too distracting in the text. Because the definition is an appendix in the document, it will be listed in the table of contents. In addition, you can refer to it—with the appropriate page number—in the text.

WRITER'S CHECKLIST

This checklist covers parenthetical, sentence, and extended definitions.

1. Are all necessary terms defined?
2. Are the parenthetical definitions
   - ❏ appropriate for the audience?
   - ❏ clear?
3. Does each sentence definition
   - ❏ contain a sufficiently specific category and distinguishing characteristics?
   - ❏ avoid describing a particular item when a general class of items is intended?
   - ❏ avoid circular definition?
   - ❏ contain a noun or a noun phrase in the category?
4. Are the extended definitions developed logically and clearly?
5. Are the definitions placed in the most useful location for the readers?

EXERCISES

1. Add parenthetical definitions of the italicized terms in the following sentences:
   a. Reluctantly, he decided to *drop* the physics course.
   b. Last week the computer was *down*.
   c. The department is using *shareware* in its drafting course.
   d. The tire plant's managers hope they do not have to *lay off* any more employees.
   e. Please submit your assignments *electronically* and as *hard copy*.
2. Write a sentence definition for each of the following terms:
   a. catalyst
   b. compact-disc player
   c. job interview
   d. internal modem
   e. flextime

f. automatic teller machine

g. fax machine

h. local area network (LAN)

3. Write a 500- to 1,000-word extended definition of one of the following terms, or of a term used in your field of study. If you do secondary research, cite your sources clearly and accurately. If you photocopy graphics from a secondary source, cite them as well. In addition, check to be sure that the graphics are appropriate for your audience and purpose.

a. flextime

b. binding arbitration

c. robotics

d. an academic major (don't focus on any particular major; define what a major is)

e. quality control

f. bioengineering

g. fetal-tissue research

h. community policing

i. software

4. Revise any of the following sentence definitions that need revision:

a. Dropping a course is when you leave the class.

b. A thermometer measures temperature.

c. The spark plugs are the things that ignite the air-gas mixture in a cylinder.

d. Double parking is where you park next to another car.

e. A strike is when the employees stop working.

5. Identify the techniques used to create the following extended definition:

Holography, from the Greek holos (entire) and gram (message), is a method of photography that produces images that appear to be three-dimensional. A holographic image seems to change as the viewer moves in relation to it. For example, as the viewer moves, one object on the image appears to move in front of another object. In addition, the distances between objects on the image seem to change.

Holographs are produced by coherent light, that is, light of the same wavelength, with the waves in phase and of the same amplitude. This light is produced by laser. Stereoscopic images are created by incoherent light—random wavelengths and amplitudes, out of phase. The incoherent light, which is natural light, is focused by a lens and records the pattern of brightness and color differences of the object being imaged.

How are holographic images created? The laser-produced light is divided as it passes through a beam splitter. One portion of the light, called the reference beam, is directed to the emulsion—the "film." The other portion, the object beam, is directed to the subject and then reflected back to the emulsion. The reference beam is coherent light, whereas the object beam becomes incoherent because it is reflected off the irregular surface of the subject. The resulting dissonance between the reference beam and the object beam is encoded; it records not only the brightness of the different parts of the subject but also the different distances from the laser. This encoding creates the three-dimensional effect of holography.

6. Locate an extended definition in one of your textbooks or a journal article. In a memo to your instructor, describe the techniques the author uses to define the term. Then evaluate the effectiveness of these techniques. Submit a photocopy of the definition along with your assignment. For information on memos, see Chapter 16.

REFERENCES

Krol, E. (1992). *The whole Internet user's guide & catalog.* Sebastopol, CA: O'Reilly & Associates.

Roblee, C. L., & McKechnie, A. J. (1981). *The investigation of fires.* Englewood Cliffs, NJ: Prentice Hall.

Wilson, J. R. (1964). *The mind.* New York: Time, Inc.

CHAPTER 10 Drafting Descriptions

Technical communication is filled with descriptions—verbal and visual representations of objects, mechanisms, and processes.

- *Objects.* What is an object? The range is enormous: from physical sites such as volcanoes and other kinds of natural phenomena to synthetic artifacts such as hammers. A tomato plant is an object, as is an automobile tire or a book.

- *Mechanisms.* A mechanism is a synthetic object consisting of a number of moving, identifiable parts that work together as a system. A compact-disc player is a mechanism, as are a voltmeter, a lawnmower, a submarine, and a steel mill.

- *Processes.* A process is an activity that takes place over time: the earth was formed, steel is made, animals evolve, plants perform photosynthesis. *Descriptions of processes* differ from *instructions:* process descriptions explain how something happens; instructions tell how to do something. The readers of a process description want to understand the process, *but they don't want to perform the process.* A set of instructions however, is a step-by-step guide intended to enable the readers to perform the process. A process description answers the question "How is wine made?" A set of instructions answers the question "How do I go about making wine?" Instructions are discussed in Chapter 18.

Understanding the Role of Description

Descriptions of objects, mechanisms, and processes appear in virtually every kind of technical communication. Here are a few examples:

- A company studying the feasibility of renovating an old factory or building a new one creates a report that includes a description of the old factory. The old factory is a complex mechanism. The company's managers can make a rational decision only if they understand it thoroughly.

- A writer who wants to persuade his readers to authorize the purchase of some equipment includes a mechanism description in the proposal to buy the equipment.

- An engineer trying to describe to the research-and-development department the features she would like incorporated in a piece of equipment they are designing includes a mechanism description in her report.

- An engineer trying to describe to the sales staff how a product works, so that they can advertise it effectively, includes a mechanism description in the form of a product specification.

- A company manufacturing a consumer product includes a mechanism description as part of the operating instructions, to acquaint consumers with the product before they begin using it.

- A developer who wants to build a housing project includes in his environmental impact statement descriptions of the geographical area and of the process he will use in developing that area.

- An accountant who audits one of her company's branches describes the audit process in her audit report.

Notice that a description rarely appears as a separate document. Almost always, it is part of a larger document. For example, a maintenance manual for a boiler system might begin with a mechanism description to help the reader understand how the system operates. However, descriptions are so important in technical communication that they are discussed here in a separate chapter.

Analyzing the Writing Situation

Before you begin to write a description, consider carefully how the audience and purpose of the document will affect the way you write it.

- *Audience.* What does the audience already know about the general subject? If, for example, you are to describe an electron microscope, you first have to know whether your readers understand what a microscope is. If you want to describe how industrial robots will affect car manufacturing, you first have to know whether your readers already understand the current process. In addition, you need to know whether they understand robotics.

  The audience will determine not only your use of technical vocabulary but also your sentence and paragraph length. Still another audience-related factor is the use of graphics. Less knowledgeable readers need simple graphics; they might have trouble understanding sophisticated schematics or decision charts.

- *Purpose.* What are you trying to accomplish in the description? If you want your readers to understand how a personal computer works, you will write a *general description*: a description that applies to several different varieties of computer. If, however, you want your readers to understand a particular computer, you will write a *particular description*. A general description of personal computers might classify them by size, then go on to describe palmtops, laptops, and desktops. A particular description of a Micron 5200 will describe only that one model.

  An understanding of your purpose will determine every aspect of the description, including length, amount of detail, and number and type of graphics included.

Notice in the following introduction to a description of stress how the writer addresses his readers:

> All police officers are subjected to stress. For a police officer to perform his or her duties effectively—and for the police unit to perform effectively—stress must be recognized and managed.
>
> No officer should be forced to recognize and deal with the symptoms of stress in isolation. It is the responsibility of all police officers, but especially those in managerial positions, to recognize and identify the signs of stress and take an active role in reducing the stress. Otherwise, the stress can become debilitating for both the officer and the unit.
>
> The following description of stress . . .

Understanding the Structure of Object and Mechanism Descriptions

Object and mechanism descriptions have the same basic structure. In the following discussion, the word *item* refers to both objects and mechanisms.

Most descriptions of items have a four-part structure:

1. a title or section heading
2. a general introduction that tells the reader what the item is and what it does (definition)
3. a part-by-part description of the item (partition)
4. a conclusion that summarizes the description and tells how the parts work together

For more information on definitions, see Chapter 9. For more information on partition, see Chapter 8.

The Title or Section Heading
If the description of the item is to be a separate document, give it a title. If the description is to be part of a longer document, give it a section heading. (See Chapter 14 for detailed discussions of titles and headings.)

In either case, clearly state the subject and indicate whether the description is general or particular. For instance, the title of a general description might be "Description of a Minivan"; of a particular description, "Description of the 1995 Mazda MPV."

The General Introduction
The general introduction provides the basic information that your readers will need to understand the detailed description that follows. A general introduction usually answers the five questions shown in Figure 10.1.

Of course, in some cases the answers to one or more of the questions would be obvious and therefore should be omitted. For instance, everyone knows the function of a computer printer.

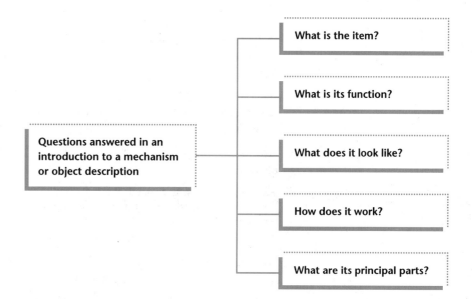

Questions Answered in
an Introduction to a
Mechanism or Object
Description

- *What is the item?* Generally, the best way to answer this question is to provide a sentence definition (see Chapter 9):

   An electron microscope is a microscope that uses electrons rather than visible light to produce magnified images.

   The Aleutian Islands are a chain of islands curving southwest from the tip of Alaska toward the tip of the Kormandorski Islands of the former Soviet Union.

   Then elaborate, if necessary, using any of the techniques described in Chapter 9 for writing an extended definition.

- *What is the function of the item?* State clearly what the item does:

   Electron microscopes magnify objects that are smaller than the wavelengths of visible light.

   Often, the function of a mechanism is implied in its sentence definition:

   A hydrometer is a sealed, graduated tube, weighted at one end, that sinks in a fluid to a depth that indicates the specific gravity of the fluid.

   Of course, some objects have no "function." The Aleutian Islands, as valuable as they may be, have no function in the sense that microscopes do.

- *What does the item look like?* Include a photograph or drawing if possible (see Chapter 12). If not, use an analogy or a comparison to a familiar item:

   The cassette that encloses the tape is a plastic shell, about the size of a deck of cards.

Mention the material, texture, color, and the like, if relevant. Sometimes an object is best pictured with both graphics and words. For example, a map showing the size and location of the Aleutian Islands would be useful. But descriptive, concrete language would be useful, too:

> The Aleutian Islands, a rugged string of numerous small islands, show signs of volcanic action. The islands are treeless and fog enshrouded. The United States military installations on the Aleutians are the main source of economic activity.

- *How does the item work?* In a few sentences, define the operating principle of the item:

> Essentially, a tire pressure gauge is a calibrated rod fitted snugly within an openended metal cylinder. When the cylinder end is placed on the tire nozzle, the pressure of the air escaping from the tire into the cylinder pushes against the rod. The greater the pressure, the farther the rod is pushed.

Sometimes objects do not "work"; they merely exist. For instance, a totem pole has no operating principle.

- *What are the principal parts of the item?* Few objects or mechanisms described in technical communication are so simple that you can discuss all the parts. And, in fact, rarely will you need to describe them all. Some parts will be too complicated or too unimportant to be mentioned; others will already be understood by your readers. A description of a bicycle would not mention the dozens of nuts and bolts that hold the mechanism together; it would focus on the chain, the pedals, the wheels, and the frame. Similarly, a description of the Aleutian Islands would mention the four main groups of islands but not all the individual islands, some of which are very small.

    At this point in the description, the principal parts should merely be mentioned; the detailed description comes in the next section. The parts can be named in a sentence or (if there are many parts) in a list. Of course, you will want to name the parts in the order in which you will describe them, to help your reader understand the description.

The introduction is an excellent place to put a graphic that complements your listing of major parts. Your goal is to enable the readers to follow the part-by-part description easily. For instance, if you partition the Aleutian Islands into four main island groups, your map should identify those four groups.

Figure 10.2 shows the introductory graphic accompanying a description of a VCR, as well as a graphic for one of the parts discussed in the detailed description.

The information provided in the introduction generally follows this pattern, focusing on the item's function, appearance, operating principle, and components. However, don't feel you have to answer each question in order, or in sepa-

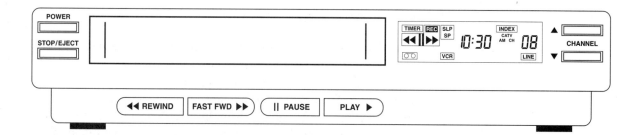

rate sentences. In fact, doing so often results in an unfocused and disjointed introduction. If your readers' needs or the nature of the item suggests a different sequence—or different questions—adjust the introduction accordingly.

The Part-by-Part Description

The body of the description—the part-by-part description—is essentially like the introduction in that it treats each major part as a separate item. That is, in writing the body you define what each part is, and then, if applicable, you describe its function, operating principle, and appearance. The discussion of the appearance should include shape, dimension, material, and physical details such as texture and color (if essential). For some descriptions, other qualities, such as weight or hardness, might also be appropriate. If a part has any important subparts, describe them the same way.

A description of an item therefore resembles a map with a series of detailed insets. A description of a computer system would include a keyboard as one of its parts. The description of the keyboard, in turn, would include the numeric keypad as one of its parts. And the description of the numeric keypad would include the arrow keys as one of its parts. This process of ever-increasing specificity continues as required by the complexity of the item and the needs of the readers.

In drafting the part-by-part description, follow these two guidelines:

- *Choose an appropriate organizing principle.* Three organizational principles are common:

  – *Functional.* The most common structure reflects the way the item works or is used. In a radio, for instance, the sound begins at the receiver, travels into the amplifier, and then flows out through the speakers.

– *Spatial.* Another common sequence is based on the physical structure of the item: from top to bottom, east to west, outside to inside, and so forth. Descriptions of places are often organized spatially.

– *Chronological.* The descriptions of some items are organized according to the chronology of assembling them. In describing a house, for example, you could describe the foundation, the floor joists, the subfloor, the exterior walls, the rafters, and the roof.

Most descriptions can be organized in various ways. For instance, the description of a house could be organized functionally (the different electrical and mechanical systems), spatially (top to bottom, inside to outside, east to west, and so on), and chronologically. And a complex description can use a combination of different patterns at different levels in the description.

• *Use graphics.* In general, try to create a graphic for each major part. Use photographs to show external surfaces, drawings to emphasize particular items on the surface, and cutaways and exploded diagrams to show details beneath the surface. Other kinds of graphics, such as graphs and charts, are often useful supplements. If, for instance, you are describing the Aleutian Islands, you might create a table that shows the total land area and the habitable land area of the island groups.

The Conclusion   Descriptions generally do not require elaborate conclusions. A brief conclusion is necessary, however, if only to summarize the description and prevent readers from placing excessive emphasis on the part discussed last in the part-by-part description.

A common technique for concluding descriptions of mechanisms and of some objects is to describe briefly how the parts function together. In a description of a telephone, for example, the conclusion might include a paragraph such as the following:

> When the phone is taken off the hook, a current flows through the carbon granules. The intensity of the speaker's voice causes a greater or lesser movement of the diaphragm and thus a greater or lesser intensity in the current flowing through the carbon granules. The phone receiving the call converts the electrical waves back to sound waves by means of an electromagnet and a diaphragm. The varying intensity of the current transmitted by the phone line alters the strength of the current in the electromagnet, which in turn changes the position of the diaphragm. The movement of the diaphragm reproduces the speaker's sound waves.

Figure 10.3 is a description of a modern long-distance running shoe (based on Kyle, 1986).

**General Description
of a Long-Distance Running Shoe**

When track and field events became sanctioned sports in the modern world some hundred and fifty years ago, the running shoe was much like any other: a heavy, high-topped leather shoe with a leather or rubber sole. In the last two decades, however, advances in technology have combined with increased competition among manufacturers to create long-distance running shoes that fulfill the two goals of all runners: decreased injuries and increased speed.

Introduction

This paper is a generalized description of a modern, high-tech shoe for long-distance running.

The modern distance running shoe has five major components:

- the outsole
- the heel wedge
- the midsole
- the insole
- the shell

Figure 1 is an exploded diagram of the shoe, showing each of the components listed above.

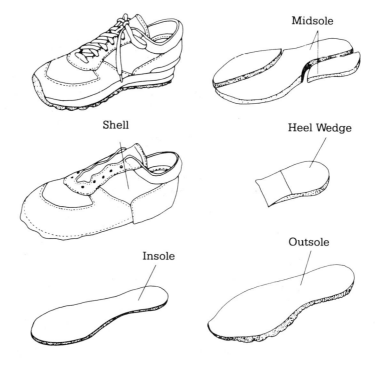

Figure 1. Exploded Diagram of a Long-Distance Running Shoe

(Figure 10.3 continued)

The writer explains the organization of the description.

The writer's approach is to provide a brief physical description of the component followed by its function.

The writer provides brief parenthetical definitions.

The Components

In the following sections, the five principal components of the shoe are discussed from bottom to top.

The Outsole

The outsole is made of a lightweight, rubberlike synthetic material. Its principal function is to absorb the runner's energy safely as the foot lands on the surface. As the runner's foot approaches the surface, it supinates—rolls outward. As the foot lands, it pronates—rolls inward. Through tread design and increased stiffness on the innerside, the outsole helps reduce inward rolling.

Inward rolling is a major cause of foot, knee, and tendon injuries because of the magnitude of the force generated during running. The force on the foot as it touches the running surface can be up to three times the runner's weight. And the acceleration transmitted to the leg can be 10 times the force of gravity.

The Heel Wedge

The heel wedge is a flexible platform that absorbs shock. Its purpose is to prevent injury to the Achilles tendon. Like the outsole, it is constructed of increasingly stiff materials on the inner side to reduce foot rolling.

The Midsole

The midsole is made of expanded foam. Like the outsole and the heel wedge, it reduces foot rolling. But it also is the most important component in absorbing shock. From the runner's point of view, running efficiency and shock absorption are at odds. The safest shoe would have a midsole of thick padding that would crush uniformly as the foot hit the running surface. A constant rate of deceleration would ensure the best shock absorption.

However, absorbing all the shock would mean absorbing all the energy. As a result, the runner's next stride would require more energy. The most efficient shoe would have a foam insole that is perfectly elastic. It would return all the energy back to the foot, so that the next stride required less energy. Currently, distance shoes have midsoles designed to return 40 percent of the runner's energy back to the foot.

The Insole

The insole, on which the runner's foot rests, is another layer of shock-absorbing material. Its principal function, however, is to provide an arch support, a relatively new feature in running shoes.

The Shell

The shell is made of leather and synthetic materials such as nylon. It holds the soles on the runner's foot and provides ventilation. The shell accounts for about one-third of the nine ounces a modern shoe weighs.

Conclusion

Today, scientific research on the way people run has led to great improvements in the design and manufacture of different kinds of running shoes. The results are a lightweight, shock-absorbing running shoe that balances the needs of safety and increased speed.

The conclusion summarizes the major points of the description.

Understanding the Structure of the Process Description

The structure of the process description is essentially similar to that of the object or mechanism description. The only important difference is that the process is partitioned into a reasonable (usually chronological) sequence of steps rather than parts. Most process descriptions have a four-part structure:

- a title or section heading
- a general introduction that tells the reader what the process is and what it is used for
- a step-by-step description of the process
- a conclusion that summarizes the description and tells how the steps work together

The Title or Section Heading

If the process description is to be a separate document, add a title. If the description is to be part of a longer document, add a section heading. (See Chapter 14 for detailed discussions of titles and headings.)

Whether you are writing a title or a section heading, clearly indicate the subject and whether the description is general or particular. For instance, the title of a general description might be "Description of the Process of Designing a New Production Car"; the title of a particular description might be, "Description of the Process of Designing the General Motors Saturn."

The General Introduction

The general introduction gives your readers the basic information they need to understand the detailed description that follows. In writing the introduction, answer six questions about the process, as shown in Figure 10.4 on page 218. Naturally, sometimes the answer to one or more of the questions is obvious and therefore should be omitted. For example, in a description of recording a football game for broadcast, there is no need to explain the function of the process.

- *What is the process?* Usually, the best way to answer this question is to provide a sentence definition (see Chapter 9):

  Debugging is the process of identifying and eliminating errors within a computer program.

  Then elaborate, if necessary.

- *What is the function of the process?* State clearly the function of the process:

  The central purpose of performing a census is to obtain up-to-date population figures, which legislators and government agencies use to revise legislative districts and determine revenue-sharing.

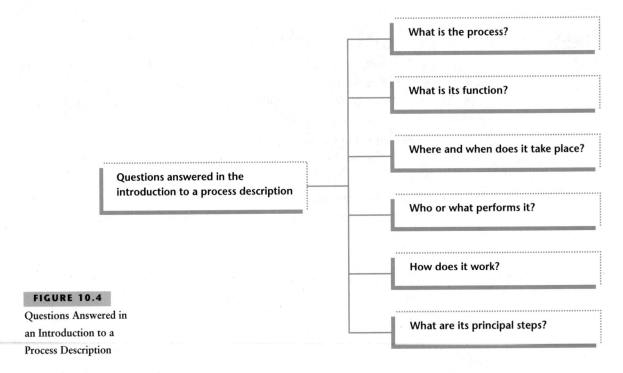

FIGURE 10.4

Questions Answered in
an Introduction to a
Process Description

Make sure the function is clear to your readers; if you are unsure whether
they already know the function, state it. Few things are more frustrating
than not knowing why a process is being performed.

- *Where and when does the process take place?* State clearly the location
  and time or occasion for the process:

Each year the stream is stocked with hatchery fish in the first week of March.

You can generally add these details simply and easily. Again, omit these
facts only if you are certain your readers already know them.

- *Who or what performs the process?* Most processes are performed by
  people, by natural forces, by machinery, or by some combination of the
  three. In most cases, you do not need to state explicitly that, for example,
  the young trout are released into the stream by a person; the context
  makes that clear.

  Do not assume, however, that your readers already know what agent
  performs the process, or even that they understand you when you have
  identified the agent. Someone who is not knowledgeable about computers, for instance, might not know whether a compiler is a person or a
  thing. The term *word processor* often refers ambiguously both to a piece
  of equipment and to the person who operates it. Confusion at this early
  stage of the process description can ruin the document.

- *How does the process work?* In a few sentences, define the principle or theory of operation of the process:

  The four-treatment lawn-spray plan is based on the theory that the most effective way to promote a healthy lawn is to apply different treatments at crucial times during the growing season. The first two treatments—in spring and early summer—consist of quick-acting organic fertilizers to promote rapid growth, and chemicals to control weeds and insects. The late summer treatment contains postemergence chemicals that control weeds and insects. The last treatment—in the fall—uses long-feeding fertilizers to encourage root growth over the winter.

- *What are the principal steps of the process?* Name the principal steps of the process, in one or several sentences or in a list. Name the steps in the order in which you will later describe them. The principal steps in changing an automobile tire, for instance, include jacking up the car, replacing the old tire with the new one, and lowering the car back to the ground. Changing a tire also includes secondary steps, such as placing blocks against the tires to prevent the car from moving once it is jacked up, and assembling the jack. Explain or refer to these secondary steps at the appropriate points in the description.

The information given in the introduction to a process description generally follows this pattern. However, don't feel you must answer each question in order, or in separate sentences. If your readers' needs or the nature of the process suggests a different sequence—or different questions—adjust the introduction.

As is the case with object and mechanism descriptions, process descriptions benefit from clear graphics in the introduction. Flowcharts that identify the major steps of the process are particularly useful.

The Step-by-Step Description

The body of the process description—the step-by-step description—is essentially like the introduction, in that it treats each major step as if it were a separate process. Of course, do not repeat your answer to the question about who or what performs the action unless a new agent performs a particular step. But do answer the other principal questions—what the step is, what its function is, and when, where, and how it occurs. In addition, if the step has any important substeps that the reader must know to understand the process, explain them clearly.

Follow these four guidelines in drafting the step-by-step description:

- *Structure the step-by-step description chronologically.* Discuss the initial step first and then discuss each succeeding step in the order in which it occurs in the process. If the process is a closed system and hence has no first step—such as the cycle of evaporation and condensation—explain to your readers that the process is cyclical, then simply begin with any principal step.

- *Explain causal relationships among steps.* Although the structure of the step-by-step description should be chronological, don't present the steps as if they have nothing to do with one another. In many cases, one step causes another. (See Chapter 8 for a discussion of cause and effect.) In the operation of a four-stroke gasoline engine, for instance, each step sets up the conditions under which the next step can occur. In the compression cycle, the piston travels upward in the cylinder, compressing the mixture of fuel and air. In the power cycle, a spark ignites this compressed mixture. Your readers will find it easier to understand and remember your description if you clearly explain the causality in addition to the chronology.

- *Use the present tense.* Discuss steps in the present tense, unless, of course, you are writing about a process that occurred in the historical past. For example, a description of how the earth was formed is written in the past tense: "The molten material condensed. . . ." However, a description of how steel is made is written in the present tense: "The molten material is then poured into. . . ." The present tense makes clear that the process is still being performed.

- *Use graphics.* Whenever possible, use graphics to clarify each point. Additional flowcharts are useful, but often other kinds of graphics, such as photographs, drawings, and graphs, are helpful as well. (See Chapter 12 for a discussion of graphics.) In the description of a four-stroke gasoline engine, you could use diagrams to illustrate the position of the valves and the activity occurring during each step. For example, you could show the explosion during the ignition step with arrows to indicate that the explosion is pushing the piston down within the cylinder.

The Conclusion   Process descriptions usually do not require long conclusions. If the description itself is brief—less than a few pages—a short paragraph summarizing the principal steps is all you need. Following is the conclusion from a description of the four-stroke gasoline engine in operation:

> In the intake stroke, the piston moves down, drawing the air-fuel mixture into the cylinder from the carburetor. As the piston moves up, it compresses this mixture in the compression stroke, creating the conditions necessary for combustion. In the power stroke, a spark from the spark plug ignites the mixture, which burns rapidly, forcing the piston down. In the exhaust stroke, the piston moves up, expelling the burned gases.

For longer descriptions, a discussion of the implications of the process might be appropriate. For instance, a description of the Big Bang, one theory of how

the universe began, might conclude with a discussion of how the theory is supported by recent astronomical discoveries and theories.

Following are two process descriptions. Figure 10.5 (Lewis, 1994, p. 3: 5)
shows the extent to which a process description can be based on graphics. The
topic is Clipper chips, which are used to code and decode electronic transmissions of information.

Figure 10.6 on pages 222–224 (U.S., 1991, pp. 45–47) is an excerpt from a
discussion of the process of diagnosing IAQ (indoor air quality) problems. The
audience consists of building owners and facility managers. After the excerpt
reprinted here, the authors provide detailed instructions for each of the steps,
starting with the initial walkthrough.

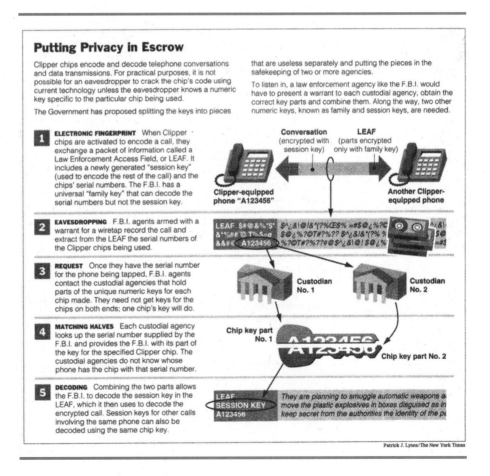

FIGURE 10.5

A Process Description
Based on a Graphic

Diagnosing IAQ Problems

6

This introduction begins by explaining the purpose of performing a diagnostic building investigation. The introduction explains the strategy for conducting the investigation and states who should perform it.

Notice the use of the flowchart to summarize the process.

The goal of the diagnostic building investigation is to identify and solve the indoor air quality complaint in a way that prevents it from recurring and that does not create other problems. This section describes a method for discovering the cause of the complaint and presents a "toolbox" of diagnostic activities to assist you in collecting information.

Just as a carpenter uses only the tools that are needed for any given job, an IAQ investigator should use only the investigative techniques that are needed. Many indoor air quality complaints can be resolved without using all of the diagnostic tools described in this chapter. For example, it may be easy to identify the source of cooking odors that are annoying nearby office workers and solve the problem by controlling pressure relationships (e.g., installing exhaust fans) in the food preparation area. Similarly, most mechanical or carpentry problems probably require only a few of the many tools you have available and are easily accomplished with in-house expertise.

The use of in-house personnel builds skills that will be helpful in minimizing and resolving future problems. On the other hand, some jobs may be best handled by contractors who have specialized knowledge and experience. In the same way, diagnosing some indoor air quality problems may require equipment and skills that are complex and unfamiliar. Your knowledge of your organization and building operations will help in selecting the right tools and deciding whether in-house personnel or outside professionals should be used in responding to the specific IAQ problem.

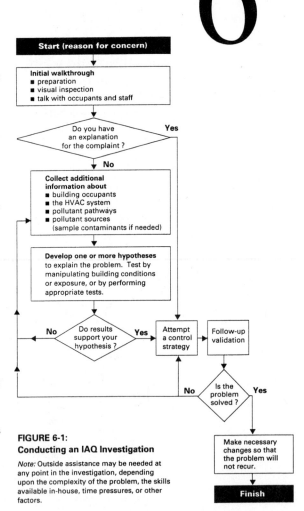

FIGURE 6-1:
Conducting an IAQ Investigation

Note: Outside assistance may be needed at any point in the investigation, depending upon the complexity of the problem, the skills available in-house, time pressures, or other factors.

FIGURE 10.6

A Process Description

continued

(Figure 10.6 continued)

The detailed section of the process description begins with this overview section that provides general information about the process and refers to the flowchart on the previous page.

The IAQ investigation is often a repetitive cycle of information-gathering, hypothesis formation, and hypothesis testing.

The overview contains a number of subsections, each of which has its own heading. Note that although the writer uses the second-person pronoun—"you"—the document is not a set of instructions but a process description. A set of instructions would include detailed information about tools and materials, as well as step-by-step instructions.

OVERVIEW: CONDUCTING AN IAQ INVESTIGATION

An IAQ investigation begins with one or more reasons for concern, such as occupant complaints. Some complaints can be resolved very simply (e.g., by asking a few common sense questions of occupants and facility staff during the walkthrough). At the other extreme, some problems could require detailed testing by an experienced IAQ professional. In this section "the investigator" refers to in-house staff responsible for conducting the IAQ investigation.

The flowchart on page 45 shows that the IAQ investigation is a cycle of information-gathering, hypothesis formation, and hypothesis testing. The goal of the investigation is to understand the IAQ problem well enough so that you can solve it. Many IAQ problems have more than one cause and may respond to (or require) several corrective actions.

Initial Walkthrough

An initial walkthrough of the problem area provides information about all four of the basic factors influencing indoor air quality (occupants, HVAC system, pollutant pathways, and contaminant sources). The initial walkthrough may provide enough information to resolve the problem. At the least, it will direct further investigation. For example, if the complaint concerns an odor from an easily identified source (e.g., cooking odors from a kitchen), you may want to study pollutant pathways as a next step, rather than interviewing occupants about their patterns of discomfort.

Developing and Testing Hypotheses

As you develop an understanding of how the building functions, where pollutant sources are located, and how pollutants move within the building, you may think of many "hypotheses," potential explana-

tions of the IAQ complaint. Building occupants and operating staff are often a good source of ideas about the causes of the problem. For example, they can describe changes in the building that may have occurred shortly before the IAQ problem was noticed (e.g., relocated partitions, new furniture or equipment).

Hypothesis development is a process of identifying and narrowing down possibilities by comparing them with your observations. **Whenever a hypothesis suggests itself, it is reasonable to pause and consider it.** Is the hypothesis consistent with the facts collected so far?

You may be able to test your hypothesis by modifying the HVAC system or attempting to control the potential source or pollutant pathway to see whether you can relieve the symptoms or other conditions in the building. If your hypothesis successfully predicts the results of your manipulations, then you may be ready to take corrective action. Sometimes it is difficult or impossible to manipulate the factors you think are causing the IAQ problem; in that case, you may be able to test the hypothesis by trying to predict how building conditions will change over time (e.g., in response to extreme outdoor temperatures).

Collecting Additional Information

If your hypothesis does not seem to be a good predictor of what is happening in the building, you probably need to collect more information about the occupants, HVAC system, pollutant pathways, or contaminant sources. Under some circumstances, detailed or sophisticated measurements of pollutant concentrations or ventilation quantities may be required. Outside assistance may be needed if repeated efforts fail to produce a successful hypothesis or if the information required calls for instruments and procedures that are not available in-house.

(Figure 10.6 continued)

Results of the Investigation

Analysis of the information collected during your IAQ investigation could produce any of the following results:

The apparent cause(s) of the complaint(s) is (are) identified.

Remedial action and follow-up evaluation will confirm whether the hypothesis is correct.

Other IAQ problems are identified that are not related to the original complaints.

These problems (e.g., HVAC malfunctions, strong pollutant sources) should be corrected when appropriate.

A better understanding of potential IAQ problems is needed in order to develop a plan for corrective action.

It may be necessary to collect more detailed information and/or to expand the scope of the investigation to include building areas that were previously overlooked. Outside assistance may be needed.

The cause of the original complaint cannot be identified.

A thorough investigation has found no deficiencies in HVAC design or operation or in the control of pollutant sources, and there have been no further complaints. In the absence of new complaints, the original complaint may have been due to a single, unrepeated event or to causes not directly related to IAQ.

Using Outside Assistance

Some indoor air quality problems may be difficult or impossible for in-house investigators to resolve. Special skills or instruments may be needed. Other factors can also be important, such as the benefit of having an impartial outside opinion or the need to reduce potential liability from a serious IAQ problem. You are best able to make the judgment of when to bring in an outside consultant. See *Section 8* for a discussion of hiring professional assistance to solve an IAQ problem.

WRITER'S CHECKLIST

The following questions cover descriptions of objects and mechanisms:

1. Does the title or section heading identify the subject and indicate whether the description is general or particular?

2. Does the introduction to the object or mechanism description
   - define the item?
   - identify its function (where appropriate)?
   - describe its appearance?
   - describe its principle of operation (where appropriate)?
   - list its principal parts?
   - include a graphic identifying all the principal parts?

3. Does the part-by-part description
   - answer, for each of the major parts, the questions listed in item 2?
   - describe each part in the sequence in which it was listed in the introduction?
   - include a graphic for each of the major parts of the mechanism?

4. Does the conclusion
   - summarize the major points made in the part-by-part description?
   - include (where appropriate) a description of the item performing its function?

The following questions cover process descriptions:

1. Does the title or section heading identify the subject and indicate whether the description is general or particular?

2. Does the introduction to the process description
   - define the process?
   - identify its function (where appropriate)?
   - identify where and when the process takes place?
   - identify who or what performs it?
   - describe how the process works?
   - list its principal steps?
   - include a graphic identifying all the principal steps?

3. Does the step-by-step description
   - answer, for each of the major steps, the questions listed in item 2?
   - discuss the steps in chronological order or in some other logical sequence?

☐ make clear the causal relationships among the steps?

☐ include graphic aids for each of the principal steps?

4. Does the conclusion

☐ summarize the major points made in the step-by-step description?

☐ discuss, if appropriate, the importance or implications of the process?

1. Write a 500- to 1,000-word description of one of the following items or of a piece of equipment used in your field. Include appropriate graphics. In a note preceding the description, specify your audience and indicate the type of description (general or particular) you are writing.

a. carburetor
b. locking bicycle rack
c. deadbolt lock
d. folding card table
e. lawn mower
f. photocopy machine
g. cooling tower at a nuclear power plant
h. jet engine
i. telescope
j. ammeter
k. television set
l. automobile jack
m. stereo speaker
n. refrigerator
o. personal computer

2. Write a memo to your instructor evaluating the effectiveness of the description that appears below and on pages 226–228. (See Chapter 16 for a discussion of memos.) Which aspects of the description

---

The Personal Computer

Introduction

This paper will briefly describe what a personal computer system is and the major components of the system. It is intended for a general reader who has a little knowledge about computers. Most readers will be familiar with the small amount of jargon contained herein since many of these words have crossed over into other professions.

A personal computer system is made up of input devices, a central processing unit, and output devices (see Figure 1). Information is put into the computer via input devices, processed in the central processing unit, then viewed via output devices.

This particular type of computer system is called a personal computer system because it has been scaled down small enough that most people have enough space for one in their home or business. The days of warehouse-sized computers are over—personal computers have become dominant in the computer industry.

Central Processing Unit

The central processing unit (CPU) is the "brain" of the computer (see Figure 1). Input is sent to the CPU via the input devices, processed in the CPU, then sent to output devices. The CPU contains components such as memory, a microprocessor, floppy disk drives, hard disk drives, graphics adapters, a power supply, modems, and so on. All of these components work together to process information. This processing can be in the form of documents such as letters, grants, and reports; number analysis such as statistics, spreadsheets, and mathematical calculations; or graphics such as slides.

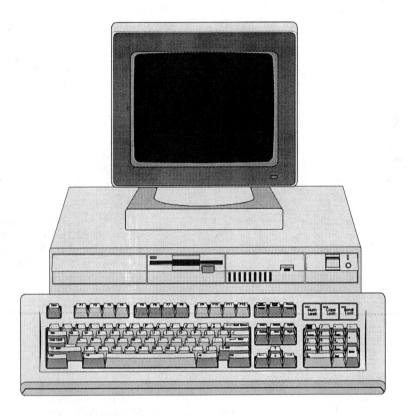

Figure 1

Input Devices

Input devices put data and commands into the central processing unit. There are many input devices such as bar code readers, scanners, touch screens, and voice-activated devices, but the two most common input devices are the keyboard and the mouse.

Keyboard

A keyboard is an input device that roughly resembles a typewriter keyboard in both size and arrangement (see Figure 2). However, there are some important differences. One difference is that the computer keyboard has a numeric keypad that resembles an adding machine keypad in addition to the usual letter, number, and character keys. Also, the computer keyboard has directional keys and a row of function keys (programmable keys used to perform routine tasks). The computer user can use the keyboard to type in commands to an operating system (a program that converts English words such as print, copy, move, and so on into series of electrical pulses that the machine can understand) or a software package, or type letters and numbers into software programs such as word processors and spreadsheets.

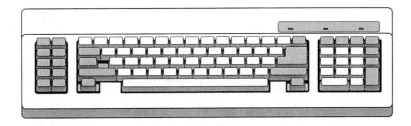

Figure 2

Mouse

A mouse is an input device about the size of a deck of cards (see Figure 3). The mouse has two or three buttons on the front edge that are depressed in order to choose commands from the computer screen. The mouse is connected by a wire to the back of the CPU. The purpose of the mouse is to bypass the directional keys on the keyboard and directly point to a command or graphical icon on the computer monitor's screen. To move the cursor on the screen, the user slides the mouse around on a rubber pad or even on a table or desktop.

Figure 3

Output Devices

Output devices display information that is contained inside the computer. There are many output devices such as plotters, slide makers, and projection screens, but the two most common output devices are the monitor and the printer.

Monitor

A monitor is an output device that resembles a small television in both size and appearance (see Figure 1). Monitors, like television sets, come in many screen sizes and in both black-and-white and color. The most common computer monitor uses a 14-inch diagonal screen with color video graphics. The monitor displays (much like a television) the characters and commands that the user types into the computer. The monitor also displays the results after a software package has processed data that was input by the user.

Printer

A printer is an output device that prints output onto paper. Printers vary in size from the size of a notebook to the size of a small desk. However, most printers are about 2 feet wide by 1 foot high by 2 feet deep (about the same size as the CPU). Documents, letters, spreadsheets, and so on are printed onto paper with the same quality as a typewriter. The printed copies are called "hard copies."

Conclusion

All of the components of the personal computer system are important. If one component was taken away, the user's ability to process information would be impaired. To remember the components, the reader should remember input devices, then the central processing unit, then the output devices.

are successful? Which parts of the description would you change? Why?

3. Rewrite the description in exercise 2.

4. Write a 500- to 1,000-word description of one of the following processes or a similar process with which you are familiar. Include appropriate graphics. In a note preceding the description, specify your audience and indicate the type of description (general or particular) you are writing. If you use secondary sources, cite them properly (see Chapter 11 for documentation systems).
   a. how steel is made
   b. how a nuclear power plant works
   c. how a food co-op works
   d. how a suspension bridge is constructed
   e. how we hear
   f. how a dry battery operates
   g. how a baseball player becomes a free agent
   h. how we see

5. The following topics are appropriate for both process descriptions and instructions. Write a 500- to

1,000 word process description. Later in the semester, when you are studying Chapter 18, write a set of instructions on the same topic.
   a. how to clean a fish
   b. how a fax machine operates
   c. how to study for a test
   d. how an audit is conducted
   e. how to winterize a recreational vehicle

REFERENCES

Kyle, C. R. (1986). Athletic clothing. *Scientific American, 254*(3), 104–110.

Lewis, P. H. (1994, April 24). Of privacy and security: The Clipper chip debate. *New York Times*, p. 3: 5.

U.S. Environmental Protection Agency. (1991). *Building air quality: A guide for building owners and facility managers*. Washington, DC: U.S. Environmental Protection Agency.

CHAPTER 11 Drafting the Front and Back Matter

Front matter consists of the elements that precede the body of a substantial document, such as title pages and tables of contents. *Back matter* consists of the elements that follow the body, such as glossaries and indexes. Front and back matter appear in various kinds of documents, including proposals, reports, and manuals. Some of the elements discussed in this chapter, such as glossaries, can appear in almost any kind of document, including brochures and flyers.

Different elements in the front and back matter play various roles:

- *To help readers find the information they seek.* The table of contents and the index exist so that readers can find what they want to read.
- *To help readers decide if they want to read the document.* The abstract—one of the summaries—enables readers to decide whether to take the time to read the whole document.
- *To substitute for the whole document.* The executive summary is written especially for managers, who often do not read anything else in the document.
- *To help readers understand the document.* The glossary—a list of definitions—helps readers who don't know the subject thoroughly.
- *To protect the document.* The cover, for instance, might contain no information at all.

Few documents have all the elements of the front matter and back matter in the order in which they are covered in this chapter; most organizations have their own format preferences. Therefore, you should study the style guide used in your company. If there is no style guide, study examples from the files to see how other writers have assembled their documents.

Using the Word Processor

A word processor simplifies assembling the front and back matter in three ways:

- *You can use word-processing software to determine the length of different elements.* Knowing the length is useful because many of the elements, such as abstracts, have length guidelines or restrictions.
- *You can use the copy function in drafting the different summaries and the transmittal letter.* You can make a copy of the body of the document and then revise it to make the elements you need. Using the copy function rather than drafting from scratch is also more accurate because you don't introduce as many errors.
- *You can use special features available in some word processors to help you create tables of contents and indexes.*

Writing the Front Matter

The front matter consists of the seven elements shown in Figure 11.1.

The Letter of Transmittal   The letter of transmittal introduces the purpose and content of the document to the principal reader. The letter is attached to the document, bound in with it, or simply placed on top of it. Even though the letter might contain no information that is not included elsewhere in the document, it is important, because it is the first thing the reader sees. It establishes a courteous and graceful tone for the document. Letters of transmittal are customary even when the writer and the reader both work for the same organization and ordinarily communicate by memo.

Transmittal letters generally contain the following seven elements:

- a statement of the title and the purpose of the document
- a statement of who authorized or commissioned the project and when

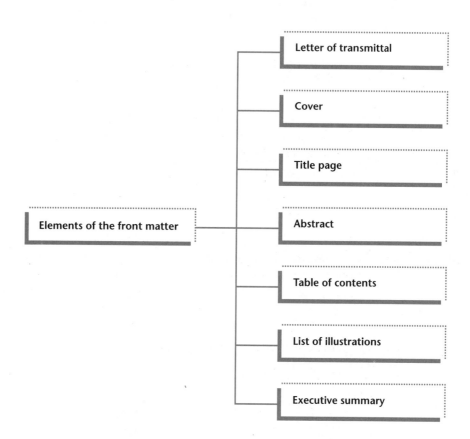

FIGURE 11.1

Elements of the Front Matter

- a statement of the methods used in the project (if they are noteworthy) or of the principal results, conclusions, and recommendations
- a statement urging that the organization carry out an additional project as a result of the findings of the current project
- an acknowledgment of any assistance the writers received in preparing the materials
- a gracious offer to assist in interpreting the materials or in carrying out further projects
- a reference to any errors or omissions (for example, the writers might want to include information that was gathered after the document was typed or printed)

Don't use the transmittal letter to apologize to the reader or ask for the reader's patience or compassion. Present your information professionally. Don't write sentences such as the following:

I didn't know a lot about hot air ballooning before I began this report, so please excuse any errors you might see.

I realize the figures aren't as neat as they should be, but I ran out of time and didn't have the chance to fix them.

Figure 11.2 shows a transmittal letter. (For a discussion of letter format, see Chapter 16.)

The Cover   The cover protects the document, either from normal wear and tear or from harsher environmental conditions such as salt water or grease. The cover usually contains the following information:

- the title of the document
- the name and position of the writer
- the date of submission
- the name or logo of the writer's company

Sometimes the cover also includes a security notice (if it is a military document) or a statement of proprietary information.

See Chapter 13 for information about different kinds of materials used in covers and different kinds of bindings.

The Title Page   The title page includes at least the following elements:

- the title
- the date of submission
- the names and positions of the writer and the principal reader
- the name or logo of the writer's company

ALTERNATIVE ENERGY, INC.
Bar Harbor, ME 00314

April 3, 19XX

Rivers Power Company
15740 Green Tree Road
Gaithersburg, MD 20760

Attention: Mr. J. R. Hanson, Project Engineering Manager

Subject: Project #619-103-823

Gentlemen:

We are pleased to submit "A Proposal for the Riverfront Energy Project" in response to your request of February 6, 19XX.

The windmill design described in the attached proposal uses the most advanced design and materials. Of particular note is the state-of-the-art storage facility described on pp. 14–17. As you know, storage limitations are a crucial factor in the performance of a generator such as this.

In preparing this proposal, we inadvertently omitted one paragraph on p. 26 of the bound proposal. That paragraph is now on the page labeled 26A. We regret this inconvenience.

If you have any questions, please do not hesitate to call us.

Yours very truly,

Ruth Jeffries

Ruth Jeffries
Project Manager

Enclosures 2

FIGURE 11.2

Letter of Transmittal

A good title answers two basic questions:

- What is the subject of the document?
- What is the purpose of the document?

For example:

> Choosing a Microcomputer: A Recommendation
>
> An Analysis of the Kelly 1013 Packager
>
> Open Sea Pollution-Control Devices: A Summary

Note that a convenient way to define the type of document is to use a generic term—such as *analysis, recommendation, summary, review, guide,* or *instructions*—in a phrase following a colon. For more information on titles, see Chapter 14.

Figure 11.3 shows a simple title page. A more complex title page might include other elements that identify the document, the different personnel associated with the project, and the distribution list.

The Abstract

An abstract is a brief technical summary—usually no more than 200 words—of the document. An abstract addresses readers who are familiar with the technical subject and who need to decide whether to read the full document. Therefore, in writing an abstract you can use technical terminology freely and refer to advanced concepts in your field.

You will see abstracts in three different places:

- *In the document itself.*
- *In different departments in the organization.* Because abstracts can be useful before and after a document is read—and can even be read in place of the document—often they are duplicated and kept on file in several locations within an organization: the division in which the project originated, as well as many or all of the higher-level departments. To facilitate this wide distribution, some organizations type abstracts on special forms.
- *In abstracting services.* As discussed in Chapter 6, abstract services print abstracts of journal articles and some government publications.

Abstracts are classified into two types: descriptive and informative. You will see the *descriptive abstract* most often where space is at a premium. Some government proposals, for example, call for descriptive abstracts to be placed at the bottom of the title page. You will see the *informative abstract* most often when more comprehensive information is required. If you are ever called on to write "an abstract" and don't know which kind the reader wants, write an informative one, not a descriptive one.

Abstracts often contain a list of about a half-dozen *key words*—words or phrases that are entered into electronic databases. Your job is to think of the different key words that people might use to access your document.

PETROLEUM PRICES AT THE END OF THE CENTURY:
A FORECAST

Prepared for: Harold Breen, President
Reliance Trucking Co.

by: Adelle Byner, Manager
Purchasing Department
Reliance Trucking Co.

April 19, 19XX

RELIANCE TRUCKING CO.
moving with the times since 1942

FIGURE 11.3

Simple Title Page

The Descriptive Abstract

The descriptive abstract, sometimes called the *topical* or *indicative* or *table-of-contents* abstract, does only what its name implies: it describes the kinds of information in the document. It does not provide the information itself: the important results, conclusions, or recommendations. It simply lists the topics covered, giving equal emphasis to each. Thus the descriptive abstract essentially duplicates the major headings in the table of contents. Figure 11.4 on page 236 is an example of a descriptive abstract.

ABSTRACT
"Design of a Radio-based System for Distribution Automation"
by Brian D. Crowe

At this time, power utilities' major techniques of monitoring their distribution systems are after-the-fact indicators such as interruption reports, meter readings, and trouble alarms. These techniques are inadequate in two ways. One, the information fails to provide the utility with an accurate picture of the dynamics of the distribution system. Two, after-the-fact indicators are expensive. Real-time load monitoring and load management would offer the utility both system reliability and long-range cost savings. This report describes the design criteria we used to design the radio-based system for a pilot program of distribution automation. It then describes the hardware and software of the system.

Key words: distribution automation, distribution systems, load, meters, radio-based systems, utilities

FIGURE 11.4

Descriptive Abstract

The Informative Abstract

The informative abstract presents the major information that the document conveys. Rather than merely list topics, it states the problem, the scope and methods (if appropriate), and the major results, conclusions, and recommendations.

The informative abstract includes three elements:

- *The identifying information.* Include the title of the document and your name. Sometimes, you will want to include your position, too.
- *The problem statement.* Write one or two sentences that define the problem or opportunity that led to the project.
- *The important findings.* Devote the final three or four sentences to the findings: the crucial information contained in the document. In reports, *findings* usually consist of results, conclusions, recommendations, and implications for further projects. Sometimes, however, the abstract presents other information. For instance, many technical projects focus on new or unusual methods for achieving results that have already been obtained through other means. In such a case, the abstract will focus on the methods, not on the results.

Figure 11.5 is an informative abstract.

ABSTRACT
"Design of a Radio-based System for Distribution Automation"
by Brian D. Crowe

At this time, power utilities' major techniques of monitoring their distribution systems are after-the-fact indicators such as interruption reports, meter readings, and trouble alarms. This system is inadequate in that it fails to provide the utility with an accurate picture of the dynamics of the distribution system, and it is expensive. This report describes a project to design a radio-based system for a pilot project. The basic system, which uses packet-switching technology, consists of a base unit (built around a personal computer), a radio link, and a remote unit. The radio-based distribution monitoring system described in this report is more accurate than the currently used after-the-fact indicators, it is small enough to replace the existing meters, and it is simple to use. We recommend installing the basic system on a trial basis.

Key words: distribution automation, distribution systems, load, meters, radio-based systems, utilities

FIGURE 11.5

Informative Abstract

The distinction between descriptive and informative abstracts is not absolute. Sometimes you will have to combine elements of both in a single abstract. For instance, suppose you are writing an informative abstract, but the report includes 15 recommendations, far too many to list. You might decide to identify the major results and conclusions, as you would in any informative abstract, but add that the report contains numerous recommendations, as you would in a descriptive abstract.

The Table of Contents

A table of contents has two main functions:

- *To help readers find the information they want.* Your table of contents is the most important guide readers have to help them navigate the document.

- *To help readers understand the scope and organization of the document.*

The table of contents uses the same headings that appear in the document itself. To create an effective table of contents, therefore, you first must make sure the document has effective headings—and that it has enough of them. If the table of contents shows no entry for five or six pages, the document could probably be divided into additional subsections. In fact, some tables of contents have an entry, or several entries, for every page in the document.

Insufficiently specific tables of contents generally result from the exclusive use of generic headings (those that describe entire classes of items) in the report. The following table of contents is relatively uninformative because it simply lists generic headings.

Table of Contents

This table of contents is ineffective because it does not provide enough detail. The methods section, for instance, runs from page 4 to page 18; the writer should add subentries to break up the text and to help readers find the information they seek.

To make the headings more informative, combine generic and specific items, as in the following examples:

Recommendations: Five Ways to Improve Information-Retrieval Materials Used in the Calcification Study

Results of the Commuting-Time Analysis

Then build more subheadings into the report. For instance, in the "Recommendations" example, make a separate subheading for each of the five recommendations.

Once you have created a clear system of headings within the document, transfer them to the contents page. Use the same text attributes—capitalization, boldface, italics, and outline style (traditional or decimal)—that you use in the body. (See the discussions of headings in Chapters 13 and 14.)

A word processor makes it simple to transfer headings from the body of the document to the table of contents. Some software programs do the job for you automatically. But even the simplest software helps you make the table of contents quickly and easily, without introducing errors. Simply make a copy of the document, so that you have two copies back to back in the same file. Then scroll through one copy, erasing the text. What you have left are the headings. If they are inconsistent, you will notice immediately and be able to fix the problems.

Figure 11.6 shows how you can combine generic and specific headings and use the resources of the word processor in a table of contents. The report is titled "Methods of Computing the Effects of Inflation in Corporate Financial Statements: A Recommendation."

Contents

FIGURE 11.6

Effective Table of Contents

The table of contents in Figure 11.6 works well for three reasons:

- Managers can find the executive summary quickly and easily.
- All other readers can find the information they are looking for because each substantive section is listed separately.

- The decimal numbering system used to identify the headings (such as 2.1 and 2.1.1) helps readers to understand the organization of the document. For more information on numbering systems, see Chapter 8.

Two notes about pagination:

- The table of contents page is not listed as an entry on the table of contents.
- Front matter is numbered with lowercase roman numerals (i, ii, and so forth). The title page of a document is not numbered, although it represents page i. The abstract is numbered page ii. The table of contents is usually not numbered, although it represents page iii. The roman numerals are often centered at the bottom of the page. The body of the document is numbered with arabic numerals (1, 2, and so on), usually in the upper outside corner of the page. See Chapter 13 for more information on pagination.

The List of Illustrations   A list of illustrations is a table of contents for the figures and tables of a document. (See Chapter 12 for a discussion of figures and tables.) If the document contains figures but not tables, the list is called a *list of figures*. If the document contains tables but not figures, the list is called a *list of tables*. If the document contains both figures and tables, figures are listed separately, before the list of tables, and the two lists together are called a *list of illustrations*.

Some writers begin the list of illustrations on the same page as the table of contents; others prefer a separate page. If the list of illustrations begins on a separate page, it is listed in the table of contents. Figure 11.7 shows a list of illustrations.

The Executive Summary   The *executive summary* (sometimes called the *epitome*, the *executive overview*, the *management summary*, or the *management overview*) is a one- or two-page condensation of the document. Its audience is managers, who rely on executive summaries to cope with the tremendous amount of paper crossing their desks every day. Managers do not need or want a detailed and deep understanding of the various projects undertaken in their organizations; this kind of understanding would in fact be impossible for them because of limitations in time and knowledge.

What managers *do* need is a broad understanding of the projects and how they fit together into a coherent whole. Consequently, a one-page (double-spaced) maximum for the executive summary has become almost an unwritten standard for documents of under 20 pages. For longer documents, the maximum length is often calculated as a percentage of the document, such as 5 percent.

List of Illustrations

Figures

Tables

FIGURE 11.7
List of Illustrations

The special needs of managers dictate a two-part structure for the executive summary:

- *Background.* Because managers are not necessarily technically competent in the writer's field, the background of the project is discussed clearly. The specific problem or opportunity is stated explicitly—what was not working, or not working effectively or efficiently; or what potential modification of a procedure or product had to be analyzed.
- *Major findings and implications.* Because managers are not interested in the details of the project, the methods rarely receive more than one or two sentences. The conclusions and recommendations, however, are discussed in a full paragraph.

For instance, if the research-and-development division at an automobile manufacturer has created a composite material that can replace steel in engine components, the technical details of the report might deal with the following kinds of questions:

- How was the composite devised?
- What are its chemical and mechanical structures?
- What are its properties?

The managerial implications, however, involve other kinds of questions:

- Why is this composite better than steel?

- How much do the raw materials cost? Are they readily available?
- How difficult is it to make the composite?
- Are there physical limitations to the amount we can make?
- Is the composite sufficiently different from similar materials to prevent any legal problems?
- Does the composite have other possible uses in cars?

The executives are less concerned about chemistry than about how this project can help them make a better automobile for less money.

In drafting an executive summary, follow these five guidelines:

- *Describe the background specifically.* In describing the problem or the opportunity, provide specific evidence. For most managers, the best evidence includes costs and savings. Instead of writing that the equipment we are now using to cut metal foil is ineffective, write that the equipment jams on the average of once every 72 hours, and that every time it jams we lose $400 in materials and $2,000 in productivity because the workers have to stop the production line. Then add up these figures for a monthly or annual total.

- *In describing opportunities you researched, use the same strategy.* Your company uses thermostats to control the heating and air conditioning. Research suggests that if you had a computerized energy-management system you could cut your energy costs by 20 to 25 percent. If your energy costs last year were $300,000, you could save $60,000 to $75,000. With these figures in mind, readers have a good understanding of what motivated the study.

- *Describe the methods briefly.* In most cases, your principal reader does not care how you did what you did. He or she assumes you did it competently and professionally. However, if you think your reader is interested, include a brief description—no more than a sentence or two.

- *Describe the findings in accordance with your readers' needs.* If your readers understand your subject sufficiently and want to know your principal results, provide them. Sometimes, however, your readers would not be able to understand the technical data or would not be interested. If so, go directly to the conclusions—the inferences you draw from the data. Most managers want your recommendations.

- *After drafting the executive summary, have someone review it.* Give it to someone who has had nothing to do with the project—preferably someone outside the field. That person should be able to read your summary and understand what the project means to the organization.

- *Decide how to integrate the executive summary with the body of the document.* The current practice in business and industry is to place it before the body. To highlight the executive summary further, writers commonly make it equal in importance to the entire detailed discussion. They signal this strategy in the table of contents, which shows the document divided into two units: the executive summary and the detailed discussion.

Figure 11.8 shows an effective executive summary.

Notice the differences between this executive summary and the informative abstract (Figure 11.5). The abstract focuses on the technical subject: whether the new radio-based system can effectively monitor the energy usage. The executive summary concentrates on whether the system can improve operations *at this one company*. The executive summary describes the symptoms of the problem in financial terms.

After a one-sentence paragraph describing the system design—the results of the study—the writer describes the findings in a final paragraph. Notice that this

---

Executive Summary

Presently, we monitor our distribution system using after-the-fact indicators such as interruption reports, meter readings, and trouble alarms. This system is inadequate in two respects.

- It fails to give us an accurate picture of the dynamics of the distribution system. To ensure enough energy for our customers, we must overproduce. Last year we overproduced by 7 percent, for a loss of $273,000.
- It is expensive. Escalating labor costs for meter readers and the increased number of difficult-to-access residences have led to higher costs. Last year we spent $960,000 reading the meters of 12,000 such residences.

This report describes a project to design a radio-based system for a pilot project on these 12,000 homes.

The basic system, which uses packet-switching technology, consists of a base unit (built around a personal computer), a radio link, and a remote unit.

The radio-based distribution monitoring system described in this report is feasible because it is small enough to replace the existing meters and because it is simple to use. It would provide a more accurate picture of our distribution system, and it would pay for itself in 3.9 years. We recommend installing the system on a trial basis. If the trial program proves successful, radio-based distribution-monitoring techniques will provide the best long-term solution to the current problems of inaccurate and expensive data collection.

FIGURE 11.8

Executive Summary

last paragraph clarifies how the pilot program relates to the overall problem described in the first paragraph.

Writing the Back Matter

The back matter usually consists of several or all of the four components shown in Figure 11.9.

The Glossary and List of Symbols

A *glossary* is an alphabetical list of definitions. It is particularly useful if your audience includes readers unfamiliar with the technical vocabulary in your document.

Instead of slowing down the detailed discussion by defining technical terms as they appear, you can use boldface or some other method of highlighting words to inform your readers that the term is defined in the glossary. A footnote at the bottom of the page on which the first boldfaced term appears serves to clarify this system for readers. For example, the first use of a term defined in the glossary might occur in the following sentence in the detailed discussion of the document: "Thus the **positron*** acts as the. . . ." At the bottom of the page, add a comment such as the following:

*This and all subsequent terms in boldface are defined in the Glossary, page 26.

Although the glossary is usually placed near the end of the document, right before the appendices, it can also be placed right after the table of contents. This placement is appropriate if the glossary is brief (less than a page) and defines terms that are essential for managers likely to read the body of the document. Figure 11.10, based on Ruiu (1994), provides an example of a glossary.

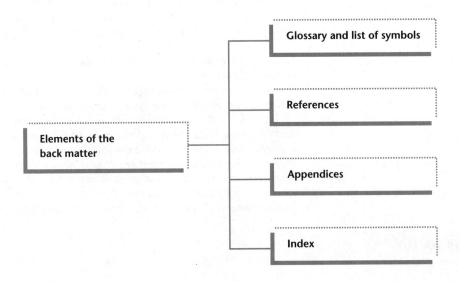

FIGURE 11.9

Elements of the Back Matter

Glossary

ISDN: integrated-services digital network. A completely digital public telecommunications network that can accommodate both voice and data traffic.

Sonet: synchronous optical network. A family of interfaces that defines the transportation and management of digital traffic on an optical-fiber network in North America. Outside North America, a similar system, called synchronous digital hierarchy (SDH), is used.

Switching fabric: the circuitry in a communications system that routes transmissions from their source to their destination.

Virtual channel: a defined connection through a network in which bandwidth is not allocated permanently, but only when a user's cells require transport.

FIGURE 11.10

Glossary

A *list of symbols* is structured like a glossary, but it defines not words and phrases, but the symbols and abbreviations used in the document. Don't hesitate to include a list of symbols if some of your readers might not know what your symbols and abbreviations mean or might misinterpret them. Like the glossary, the list of symbols may be placed before the appendices or after the table of contents. Figure 11.11 shows a list of symbols (in this case, including abbreviations).

List of Symbols

β	beta
CRT	cathode-ray tube
γ	gamma
Hz	hertz
rcvr	receiver
SNR	signal-to-noise ratio
uhf	ultra high frequency
vhf	very high frequency

FIGURE 11.11

List of Symbols

Documentation of Sources

Documentation of sources is the process of explicitly identifying the sources of the ideas and quotations used in your document. Documentation serves three basic functions:

- *To help you acknowledge your debt to your sources.* Complete and accurate documentation is primarily a professional obligation—a matter of ethics. Failure to document a source, whether intentionally or unintentionally, is plagiarism. At most universities and colleges, plagiarism means automatic failure of the course and, in some instances, suspension or dismissal. In many companies, it is grounds for immediate dismissal.
- *To help you establish credibility.* Effective documentation helps you place your document within the general context of continuing research and to define it as a responsible contribution to knowledge in the field. Knowing how to use existing research is a mark of a professional.
- *To help your readers find the source you have relied on in case they want to read more about a particular subject.*

Three kinds of material should be cited:

- *Any quotation from a printed source or an interview, even if it is only a few words.*
- *A paraphrased idea, concept, or opinion gathered from your reading.* There is one exception to this rule: an idea or concept so well known that it has become, in effect, general knowledge, such as Einstein's theory of relativity, needs no citation. If you are unsure whether an item is general knowledge, document it, just to be safe.
- *Any graphic from a printed or electronic source.*

Many organizations have their own documentation style; others use published style guides, such as the *U.S. Government Printing Office Style Manual*, the American Chemical Society's *Handbook for Authors*, or the *Chicago Manual of Style*. Find out what your organization's style is and abide by it.

Two documentation systems are shown here:

- *Publication Manual of the American Psychological Association*, 4th ed. (1994). This system, often referred to as *APA*, is used widely in the social sciences.
- *MLA Handbook for Writers of Research Papers*, 4th ed. (Gibaldi, 1995). This system, from the Modern Language Association, is used widely in the humanities.

Naturally, the discussion of these two systems is not comprehensive. Consult the APA or MLA manuals for detailed information on documenting the many different kinds of sources you use. And check with your instructor to see which documentation system to use in your documents for class.

APA Style

Like all documentation systems, the APA style consists of two elements: the citation in the text and the references at the end of the document.

Textual Citations   In the APA style, the citation is a parenthetical notation that includes the name of the source's author and the date of its publication immediately following the quoted or paraphrased material:

> This phenomenon was identified as early as 50 years ago (Wilkinson, 1948).

But if your sentence already includes the source's name, do not repeat it in the parenthetical notation:

> Wilkinson (1948) identified this phenomenon as early as 50 years ago.

If the author is an organization rather than a person, use the name of the organization:

> The causes of narcolepsy are discussed in a recent booklet (Association of Sleep Disorders, 1994).

You can refer to two sources in one citation. When you do, present the sources in alphabetical order:

> This phenomenon was identified as early as 50 years ago (Betts, 1949; Wilkinson, 1948).

Use first initials if two or more sources have the same last name:

> This phenomenon was identified as early as 50 years ago (B. Wilkinson, 1948; R. Wilkinson, 1949).

Sometimes, particularly if the reference is to a specific fact, idea, or quotation, the page (or pages) from the source is also listed:

> (Wilkinson, 1948, p. 36)

If two or more sources by the same author in the same year are listed in the references, add the letters a, b, c, and so on after the date of each to prevent confusion. (In your references list, these works are alphabetized first by year—starting with the earliest year—and then by title.)

> (Wilkinson, 1948a, pp. 36–37)

The second source would be identified similarly:

> (Wilkinson, 1948b, p. 11)

For a source written by two authors, cite both names. Use an ampersand in the citation itself, but use the word *and* in regular text:

> (Allman & Jones, 1987)

> As Allman and Jones (1987) suggested, . . .

For a source written by three, four, or five authors, cite all authors the first time you use the reference; after that, include only the last name of the first author followed by *et al.*

Bradley, Edmunds, and Soto (1995) argued . . .

Bradley et al. (1995) found . . .

For a source written by six or more authors, use the first author's name followed by *et al.*

(Smith et al., 1987)

The Reference List A reference list provides the information your readers will need to find each source you have used. Note that a reference list includes only those items that you actually used in researching and preparing your document. Do not include any sources used in your background reading. In addition, do not list any sources (such as personal interviews, phone calls, e-mail, and messages from Usenet newsgroups and bulletin boards) that, in the APA's phrase (p. 174), "do not provide recoverable data." Cite such sources in your text, as follows:

C. Connors (personal communication, October 6, 1996) argued that . . .

The individual entries in the references list are arranged alphabetically by author (and then by date if two or more works by the same author are listed). Works by an organization are alphabetized by the first significant word in the name of the organization; works signed "Anonymous" are listed alphabetically by the word "Anonymous." Where several works by the same author are included, they are arranged by date, beginning with the earliest.

APA style calls for the references to be typed double spaced. In addition, the first line of each reference is indented five to seven spaces. Paragraph indentations are helpful for journal editors who will be typesetting the article. However, APA style permits students to use a hanging indent, with second and subsequent lines indented five to seven spaces. Check with your instructor to see which format you should use.

Chapman, D. L. (1995, June 12). Detroit makes a big comeback. *Motorist's Metronome, 12,* 17–26.

Chapman, D. L. (1995, June 12). Detroit makes a big comeback. *Motorist's Metronome, 12,* pp. 17–26.

Following are standard reference forms used with various types of sources. Note that this is only a partial list; there are many more types of sources that you might need to cite.

The author's surname is followed by the first initial or initials only.

The date of publication, in parentheses followed by a period.

The title of the book, italicized, or underlined if italics are not available. Initial uppercase letters are used only for the first word of the title and of the subtitle, as well as any proper names (such as Three Mile Island.)

1. A Book

Cunningham, W. (1980). *Crisis at Three Mile Island: The aftermath of a near meltdown.* New York: Madison.— The location and name of the publisher come last. Note that publishers' names are often abbreviated. For instance, Prentice Hall, Inc., is often abbreviated as Prentice; consult your style guide. Include the name of the publisher's city and the state or country unless the city is well known (such as New York, Boston, or London).

2. A Book by Two or More Authors

Note the use of the comma after the first author's name.

Note the use of the ampersand (&) in place of the word *and.*

Bingham, C., & Withers, S. (1995). *Neural networks and fuzzy logic.* New York: IEEE.

3. A Book Issued by an Organization

The organization takes the position of the author. Although APA style calls for the names of organizations to be written out in full, the U.S. Department of Energy uses "U.S.," not "United States."

U.S. Department of Energy. (1996). *The energy situation in the next decade.* Technical Publication 11346-53. Washington, DC: U.S. Government Printing Office.

Note that the report number follows the title of the report.

4. A Book Compiled by an Editor or Issued under an Editor's Name

Notice that a space separates the two initials of the editor: D. E. Morgan.

Morgan, D. E. (Ed.). (1995). *Readings in alternative energies.* Boston: Smith-Howell.

The abbreviation for *editors* is (Eds.).

5. A Book in an Edition Other Than the First

Schonberg, N. (1991). *Solid state physics.* (3rd ed.). London: Paragon.

The name of the article is presented without any extra punctuation or highlighting devices such as italics.

The editor's initials and name—unlike those of the author or authors—are presented in the normal order.

6. An Article Included in a Book

May, B., & Deacon, J. (1995). Amplification systems. In A. Kooper (Ed.), *Advances in electronics* (pp. 101–114). Miami, FL: Valley Press.

7. A Journal Article

The title of the journal is italicized or underlined.

The first letters of important words are capitalized.

Hastings, W. (1990). The space shuttle debate. *The Modern Inquirer, 13*, 311–318.

The volume number of the journal is given in italics.

The page numbers of the article

8. An Article in a Proceedings

In the phrase, "45th International Technical Communication Conference," initial letters are uppercase because it is a proper noun.

Carlson, C. T. (1995). Advanced organizers in manuals. In K. Rainey (Ed.), *Proceedings of the 45th International Technical Communication Conference* (pp. RT56–58). Fairfax, VA: Society for Technical Communication.

9. An Unsigned Journal Article

If the author of a journal article is not indicated, alphabetize it by title. If the title begins with a grammatical article such as *the* or *a*, alphabetize it under the first word following the article (in this case, *state*).

The state of the art in microcomputers. (1995, Fall). *Newscene, 56*, 406–421.

If the journal issue is identified by a phrase such as "Fall," that phrase is included in the date.

10. A Newspaper Article

The specific date is added after the year.

The abbreviation *p.* or *pp.* is used for newspapers.

Eberstadt, A. (1995, July 31). Carpal tunnel syndrome. *Morristown Mirror and Telegraph,* p. B3.

The page number "B3" refers to page 3 of section B.

11. A Technical Report

Identifying numbers are included for technical reports.

Birnest, A. J., & Hill, G. (1996). *Early identification of children with ATD* (Report No. 43-8759). State College, PA: Pennsylvania State University College of Education. (ERIC Document Reproduction Service No. ED 186 389).

The name of the service used to locate the item: in this case, the ERIC system.

12. On-line Abstract

The title of the item

The kind of item—in this case, an on-line abstract

The author, whether a person or an organization, is identified.

Rutherford, C., & Sahni, G. (1995). Carbon-fiber rotors [On-line]. *IEEE Spectrum, 26*, 63–65. Abstract from: DIALOG File: ElecENGR Item: 32-34586

The vendor

The item's "address": its file name and item number

Periods are not used in the locating information, for they could be misinterpreted as part of the path name.

Writing references for electronic forms, such as the entry above and those that follow, is a real challenge. They are similar to printed sources in that they have titles and sources; but they are also very different, because electronic forms of information delivery change from month to month. Although the pace of technological change is likely to remain high or even accelerate, you will fulfill your professional obligation, as well as help your readers, if you include clear information on what the item is and how the reader can find it.

13. Abstract on CD-ROM

Rumbaugh, K. (1994). Client-server relationships: Just the tip of the iceberg [CD-ROM]. *Networking, 4*, 604–614. Abstract from: ERIC File: CompLIT Item: 60-14432

14. On-line Journal Article

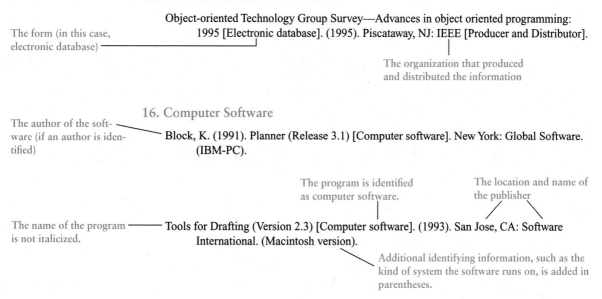

The length of the article (in paragraphs)

Nielsen, R. (1995, March). Radon risks [16 paragraphs]. *Carcinogens* [On-line serial], *4* (12). Available FTP: Hostname: princeton.edu. Directory: pub/carcinogens/1995. File: radon.95.3.12.radonrisks

The information needed to retrieve the article by FTP: hostname, directory, and file name. (See Chapter 5 for more information on FTP.)

15. Electronic Database or Data File

The form (in this case, electronic database)

Object-oriented Technology Group Survey—Advances in object oriented programming: 1995 [Electronic database]. (1995). Piscataway, NJ: IEEE [Producer and Distributor].

The organization that produced and distributed the information

16. Computer Software

The author of the software (if an author is identified)

Block, K. (1991). Planner (Release 3.1) [Computer software]. New York: Global Software. (IBM-PC).

The program is identified as computer software.

The location and name of the publisher

The name of the program is not italicized.

Tools for Drafting (Version 2.3) [Computer software]. (1993). San Jose, CA: Software International. (Macintosh version).

Additional identifying information, such as the kind of system the software runs on, is added in parentheses.

Figure 11.12 on page 252 is a sample reference list for use with the APA citation system:

References

Andress, K. (1995, July 12). More on the Scanlon Plan [36 paragraphs]. *Online Journal of Accounting Studies* [On-line serial]. Available: Doc. No. 113

Daly, P. H. (1993). Selecting and designing a group insurance plan. *Personnel Journal, 54,* 322–323.

Flanders, A. (1992). Measured daywork and collective bargaining. *British Journal of Industrial Relations, 9,* 368–392.

Goodman, R. K., Wakely, J., & Ruh, K. (1995). What employees think of the Scanlon Plan. *Personnel, 6,* 22–29.

Trencher, P., & Coughlin, C. (1988). *Recent trends in labor-management relations.* New York: Westly.

Zwicker, D. (1995, August 13). More on the Scanlon Plan: A response [Electronic datafile]. New York: American Institute of Insurance Underwriters [Producer and Distributor].

FIGURE 11.12

References in APA
Style

MLA Style

Like the APA documentation system, MLA style consists of two elements: the citation in the text and the references—called a *list of works cited*—at the end of the document.

Textual Citations   In MLA style, the citation is a parenthetical notation that includes the name of the source's author immediately following the quoted or paraphrased material. If you are referring to a specific fact, idea, or quotation, include the page (or pages) from the source:

This phenomenon was identified as early as 50 years ago (Wilkinson 134).

Here the citation shows that you are referring to page 134 of the Wilkinson source. If you are referring to the whole Wilkinson source, not to a particular page or pages, use only the name:

This phenomenon was identified as early as 50 years ago (Wilkinson).

If your sentence already includes the source's name, do not repeat it in a parenthetical notation:

Wilkinson identified this phenomenon as early as 50 years ago (134) .

If the author is an organization rather than a person, use the name of the organization:

The causes of narcolepsy are discussed in a recent booklet (Association of Sleep Disorders).

You can incorporate the name of the organization into the sentence:

The Association of Sleep Disorders discusses the causes of narcolepsy in a recent booklet.

You can refer to two or more sources in one citation:

This phenomenon was identified as early as 50 years ago (Betts 34; Wilkinson).

Use first initials if two or more sources have the same last name:

This phenomenon was identified as early as 50 years ago (B. Wilkinson; R. Wilkinson).

If you cite two or more sources by the same author in the list of works cited, add a brief form of the title after the author's name to prevent confusion:

(Wilkinson, *Particle Physics* 36–37)

(Wilkinson, "Ascent" 11)

For a source written by two or three authors, cite all names:

(Allman and Jones)

(Finn, Crenshaw, and Zander)

For a source written by four or more authors, you may list all the authors or you may list only the first author followed by *et al.*

Bradley, Edmunds, Stipe, and Soto argued . . .

Bradley et al. argued . . .

The List of Works Cited   A list of works cited provides the information your readers will need to find each source you have used. Note that the reference list includes only those items that you actually used in researching and preparing your document and that you cite in your document.

Alphabetize the entries in the list of works cited by author (and then by title if two or more works by the same author are listed). Works by an organization are alphabetized by the first significant word in the name of the organization.

Type the entries double spaced, with second and subsequent lines indented one-half inch. Leave one space between each item in the entry; for example, leave one space between the name of the author and the title of the work. Leave one space after any comma or colon.

Following are standard reference forms used with various types of sources. Note that this is only a partial list; there are many more types of sources that you might need to cite.

1. A Book

The author's full name is given, in reverse order.

The title of the book, italicized. All important words in the title are capitalized

Cunningham, Walter. *Crisis at Three Mile Island: The Aftermath of a Near Meltdown.* New York: Madison, 1980.

The location, name, and date of the publisher come last. Note that publishers' names are often abbreviated. For instance, Prentice Hall, Inc., is often abbreviated as Prentice; consult your style guide.

2. A Book by Two or More Authors

For a book by two or three authors, present the names in the sequence in which they appear on the title page.

Only the name of the first author is presented in reverse order.

Bingham, Christine, and Stephen Withers. *Neural Networks and Fuzzy Logic.* New York: IEEE, 1995.

A comma separates the names of the authors.

For a book by four or more authors, either present all the names or use *et al.*

Foster, Glenn, et al. *The American Renaissance.* Binghamton: Archive, 1995.

3. Two Books by the Same Author

For second and subsequent entries, use three hyphens followed by a period. Arrange the entries alphabetically by title of the work.

Smith, Louis. *International Standards.* New York: IEEE, 1995.

- - -. *Wave-Propagation Technologies.* Berkeley: Stallings, 1994.

4. A Book Issued by an Organization

The organization takes the position of the author.

National Commission on Corrections. *The Future of Incarceration.* Publication 11346-53. St. Louis, MO: Liberty, 1995.

5. A Book Compiled by an Editor or Issued under an Editor's Name

The abbreviation for editors is "eds."

Morgan, Donald E., ed. *Readings in Alternative Energies.* Boston: Smith-Howell, 1995.

6. A Book in an Edition Other Than the First

Schonberg, Nathan. *Solid State Physics.* 3rd ed. London: Paragon, 1991.

7. A Quotation of a Quotation

The title of Hanson's original article

The page numbers of Hanson's article

The source being quoted in the Ortiz book

Hanson, Alfred. "Multimedia for Tomorrow." *Technology Advances* 3 June 1995: 36–39, qtd. in Ortiz, Manuel. *Multimedia Systems.* Boston: Allied, 1996.

The book in which Ortiz quotes Hanson

In the parenthetical citation, give the pages in Ortiz's book on which Hanson is quoted.

According to Hanson (211), multimedia will be the dominant mode of instruction in colleges by 2005.

8. Magazine Article

Newman, Daniel. "Passive Restraint Systems." *Car Lore* 12 Dec. 1995: 41+.

Use the first three letters to abbreviate a month, but write out *May, June,* and *July.*

If the article has no author listed, alphabetize it by title:

"Passive Restraint Systems." *Car Lore* 12 Dec. 1995: 41+.

The plus sign signifies that the article begins on page 41 and then continues on subsequent pages interrupted by other articles or advertisements.

9. A Newspaper Article

Felder, Melissa. "Smokeless Tobacco: New Danger Signs." *New York Post* 4 May 1995, morning ed.: 13.

If the newspaper has more than one edition, cite the edition.

10. An Article Included in a Book

The name of the article is enclosed within quotation marks.

May, Bruce, and James Deacon. "Amplification Systems." *Advances in Electronics.* Ed. Alvin Kooper. Miami, FL: Valley Press, 1995.

11. An Article in a Journal with Continuous Pagination

The title of the journal is italicized.

Hastings, Wendy. "The Space Shuttle Debate." *The Modern Inquirer* 13 (1990): 311–18.

The pages on which the article appears

The volume of the journal in which the article appears

The year of publication

12. An Article in a Journal that Pages Each Issue Separately

Juneja, Gupta. "Asynchronous Transmission Techniques." *Video Quarterly* 6.2 (1994): 11–19.

The year of publication

The pages on which the article appears

The title of the journal is italicized.

The volume of the journal in which the article appears

The issue of the journal in which the article appears

13. An Article in a Proceedings

Carlson, Carl T. "Advanced Organizers in Manuals." *Proceedings of the 45th International Technical Communication Conference.* Ed. Ken Rainey. Fairfax, VA: Society for Technical Communication, 1995: RT56-58.

14. An Unsigned Journal Article

If the author of a journal article is not indicated, alphabetize it by title. If the title begins with an article (*the* or *a*), alphabetize it under the first word following the article (in this case, *state*).

"The State of the Art in Microcomputers." *Newscene* 56: 406–21.

15. An Interview That You Conducted

Gangloff, Richard. Personal interview. 24 Jan. 1995.

Abbreviate all months except May, June, and July.

16. A Lecture

Robbins, Bruce. "Trends in Secondary Education." Lecture in E382. Boise State University. 2 May 1994.

17. Material from a Periodically Published Database on CD-ROM

Cash, Rosalie. "Technology Management: The Last 40 Years." *New York Times* 21 Apr. 1995: 16. *Predicasts F and S Plus Text: United States.* CD-ROM. SilverPlatter. October 1995.

The date of issue of the CD-ROM

The name of the CD-ROM database

The name of the vendor of the CD-ROM

18. Material from a Nonperiodical on CD-ROM or Other Portable Medium

United States. Dept. of State. "Industrial Outlook for Petroleum and Natural Gas." 1994. *National Trade Data Bank.* CD-ROM. US Dept. of Commerce. Dec. 1995.

Indicate the medium, such as CD-ROM, diskette, or magnetic tape.

19. Material from an Online Database

Angier, Natalie. "Chemists Learn Why Vegetables Are Good for You." *New York Times* 13 Apr. 1993, late ed.: C1. *New York Times Online.* Online. 10 Feb. 1994.

If the material you are citing from the database also appeared in print, include the print publication information first. In this example, the article was first published in the print version of the *New York Times,* then in the online version. Note that dates are given for each separate publication.

20. Material from an Online Journal, Newsletter, or Conference Document

Moulthrop, Stuart. "You Say You Want a Revolution? Hypertext and the Laws of Media." *Postmodern Culture* 1.3 (1991); 53 pars. Online. BITNET. 10 Jan. 1993.

The name of the online journal

The length (in paragraphs) of the article. Sometimes the length of articles is expressed in pages, such as 12 pp.

The name of the computer network

The date of access

21. Electronic Mail

Selber, Stuart A. E-mail to the author. 17 Mar. 1995.

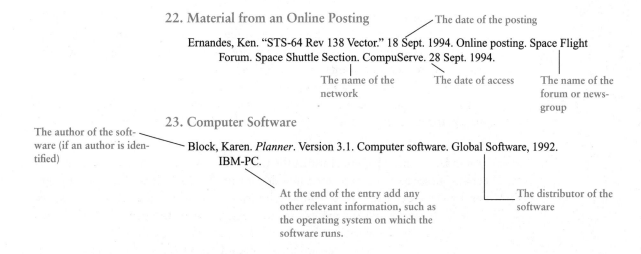

22. Material from an Online Posting

The date of the posting

Ernandes, Ken. "STS-64 Rev 138 Vector." 18 Sept. 1994. Online posting. Space Flight Forum. Space Shuttle Section. CompuServe. 28 Sept. 1994.

The name of the network

The date of access

The name of the forum or news-group

23. Computer Software

The author of the software (if an author is identified)

Block, Karen. *Planner*. Version 3.1. Computer software. Global Software, 1992. IBM-PC.

At the end of the entry add any other relevant information, such as the operating system on which the software runs.

The distributor of the software

Figure 11.13 is a sample list of works cited for use with the MLA citation system.

Works Cited

Andress, Klaus. "More on the Scanlon Plan." *Online Journal of Accounting Studies* July 1994: 432–36. DIALOG file 566, item 356764 948573.

Daly, Peter H. "Selecting and Designing a Group Insurance Plan." *Personnel Journal* 54 (Nov. 1995): 322–23.

Flanders, Andrew. "Measured Daywork and Collective Bargaining." *British Journal of Industrial Relations* 9.4 (1992): 368–92.

Goodman, Raymond K., Janice Wakely, and Kathleen Ruh. "What Employees Think of the Scanlon Plan." *Personnel* 6 (1995): 22–29.

Trencher, Patricia, and Christopher Coughlin. *Recent Trends in Labor-Management Relations*. New York: Westly, 1988.

FIGURE 11.13

References in MLA Style

The Appendices An appendix is any section that follows the body of the document (and the list of references or bibliography, glossary, or list of symbols). Appendices (or *appendixes*) conveniently convey information that is too bulky to be presented in the body or that will interest only a few readers. Appendices might include maps, large technical diagrams or charts, computations, computer printouts, test data, and texts of supporting documents.

Appendices, which are usually lettered rather than numbered (Appendix A, Appendix B, and so on), are listed in the table of contents and are referred to at the appropriate points in the body of the document. Therefore, they are accessible to any reader who wants to consult them.

Remember that an appendix including or consisting of a graphic is titled "Appendix," not "Figure" or "Table," even if it would have been so designated had it appeared in the body.

The Index   An index is an alphabetical list of the major items—topics, ideas, functions, and tasks—described and explained in a document. For long documents—100 pages or longer—the index is the most important accessing tool. Readers consult indexes to locate specific pieces of information; when they can't find what they are looking for, or when it takes them a long time to search for it, they become confused and frustrated. According to Bonura (1994, p. 4), the most common complaint about technical documents is that the index is poor or that the document lacks one altogether.

Who creates an index? In an organization with only a few technical communicators, the author of a document often creates its index. In larger technical-publications departments, there might be a specialist who does only indexes, or a technical editor might be responsible for indexing. Organizations can also turn to freelance indexers.

Each approach has its advantages and disadvantages. Writers who index their own documents have a deeper understanding of the subject and might be able to work faster than anyone else, but they also lack the objectivity of an outsider. For instance, an author/indexer might use the term *compact disc* throughout the document and the index, without realizing that many readers will be looking for the term *CD-ROM*. If this omission is not noticed and corrected, the index will be less useful than it should be.

Characteristics of an Effective Index

An effective index is accurate, comprehensive, precise, and easy to use.

- *Accuracy.* Because readers turn to the index to find where a topic is discussed, the page numbers listed must be accurate.
- *Comprehensiveness.* The index should include an entry for every term the reader is likely to seek. The average index might have three or four items listed for each page of text in the document, but that number is only a rough guide; the actual number might vary from one to more than ten.
- *Precision.* An index that includes entries such as "computers" followed by 30 page numbers is ineffective. Useful indexes have subentries and, when appropriate, sub-subentries. For instance, an index to a manual

about a word-processing program should have numerous subentries for
tables, including *creating, editing,* and *formatting,* and the subentry *formatting* should have sub-subentries such as *justifying, sizing,* and *shading.*

- *Ease of use.* The reader should not have to struggle to find an item. In
  creating the index, use two kinds of cross-references to help the reader
  navigate: *see* and *see also.*

  – *See* is used to direct readers to the proper entry, for example: "Drive,
    see disk drive." Use *see* only if the entry—*disk drive,* in this case—con-
    tains subentries. Otherwise, it is just as easy to list the page numbers
    for *drive* as well as for *disk drive.* Remember, your job is to help read-
    ers find what they are looking for; so if you can help them find it on the
    first try, do so, even if their choice of what to call the item is different
    from yours.
  – *See also* is used to direct readers to related entries. For example:
    "Search-and-replace, 17. *See also* Global searching; Editing text." This
    entry tells readers that two other entries—one on global searching and
    one on editing text—contain information related to the search-and-
    replace function.

Note that *see* and *see also* are usually italicized. An obvious point: readers
are not amused when the index does a double reverse: "floppy, *see* floppy disk,"
then "floppy disk, *see* disk drive." And documents actually go airborne when
there is an entry "floppy disk, *see* disk drive" but there is no entry for *disk drive.*

Sometimes it is useful to add an explanatory statement at the start of an
index. If, for example, you include names of people in a separate index, or you
choose to group all the function keys in an entry called *function keys,* rather
than alphabetizing them separately by the name of the individual function, alert
readers to your scheme in a brief headnote. When an index covers several documents, especially in software documentation, tell your readers in an explanatory
note how you will be abbreviating the names of the different documents.

Planning, Drafting, and Revising an Index

Creating an index is like creating any other part of a document. You have to
plan, draft, and revise.

- *Planning.* Think of your audience and purpose. How much do your readers know about the subject? If you are addressing novices, include many
  entries for basic items. And include numerous cross-references to help
  them get oriented and find related entries. Think about why they are
  reading. What kinds of information do they want? As with all kinds of
  technical communication, think in terms of tasks (what they want to

accomplish) in addition to functions (what the item does). If time permits, interview people who fit the profile of the reader.

In the planning stage, you might also want to create a list of items to index. For instance, if you are indexing a manual for a word-processing program, you might include items such as the following:

– commands
– control keys
– tasks (adding text, deleting text, printing)
– text attributes (boldface, italics, and so on)

A list such as this is particularly helpful when someone other than the writer is doing the indexing, for it helps the indexer understand what the writer thinks is important to include.

- *Drafting.* As with writing, people draft their indexes differently. The most popular method is the oldest: with index cards and a pencil. (Indexing software will be discussed later in this section.) First you read a section, such as a chapter or a part of one, circling the key terms. Don't forget to take advantage of the thinking you have already done: the key terms are likely to appear in chapter titles, section headings, and the topic sentences of the paragraphs.

  After finishing the section, go back and make a card for each entry. Some indexers like to create subentries at this point; others like to wait and then create all the subentries and sub-subentries later, after they have all the page references to the item. Many indexers like to check their page references as soon as they enter them on the card because checking will be much more difficult later. Some indexers like to alphabetize their cards as they go along; others wait because they don't want to break their concentration. Waiting can take longer, however, for you might end up making several different cards for the same entry, which you will have to consolidate later. At the end of this process, all the entries have to be word processed or typed.

  Some writers work directly on the computer, typing in their entries (either in the order in which they occur or alphabetically) during their first pass through the document.

- *Revising.* As you revise the index, check for six different items:

  – Have you entered all the items and are the entries easy to understand? For instance, in revising, you might want to change *location of files* to *files, location of,* to emphasize the key term.
  – Have you used terms consistently throughout the index? If you use *personal computer* in one place, don't use *microcomputer* in another.

– Do you need to create more subentries or sub-subentries? The *Chicago Manual of Style* (1993, p. 722) recommends creating subentries when an entry contains five or six page numbers.

– Have you used singular nouns in some places and plural nouns in others? Don't use *printer* in one location and *disk drives* in another; use all singular nouns (*printer, disk drive*) or all plural nouns.

– Do you need to change the way you indicated page numbers? For instance, in your first draft, you might have the entry *macros, 19–24*. In looking again at those pages, however, you see they should be listed as *19–22, 23, 24*, because the main explanation is really on pages 19–22, and macros are merely mentioned on 23 and 24.

– Do the page numbers make sense? For example, numbers might be inadvertently transposed (14–12 instead of 12–14) or the hyphen between numbers might be missing (1214 instead of 12–14).

In revising an index, you will find yourself going back and forth a lot; it simply is not possible to get it right the first time.

Computerizing the Indexing Process

If the index is part of a word-processed or typeset document, as it almost certainly will be, isn't there some way to take advantage of the power of the computer? Yes and no.

Indexing software automates the process of finding terms that you designate and then formatting them as entries, subentries, or sub-subentries. These programs let you change the format of the entries easily. Desktop-publishing software contains indexing functions as well, and even word-processing software lets you create an index by locating the items you designate. All indexing functions are basically a variation on the search function of standard word-processing programs.

But finding each reference to an item is not the hard part of indexing. The real challenge is to determine hierarchy (what is an entry and what is a subentry) and to create the *see* and *see also* cross-references. For this reason, indexing will always require an intelligent and informed person.

Guidelines for Indexing

Here are eight practical guidelines for indexing.

- Use either the run-in or indented format.
  Run-in format:

  Tables, 419–429
    converting text to, 423; converting to text, 424; copying, 426; creating, 419; importing spreadsheets to, 423; lines in, 428

Indented format:

Tables, 419–429
   converting text to, 423
   converting to text, 424
   copying, 426
   creating, 419
   importing spreadsheets to, 423
   lines in, 428

If a main entry takes up more than one line in the indented format, indent the second line more than the indent for the subentries:

Macros, 206–209, 233, 408. *See also* Keyboard map;
   Templates
   chaining of, 208
   definition of, 206

As you can see, the run-in format uses less space, but the indented form is easier to read.

- Format the index in two columns, in a type size two points smaller than the body text. See Chapter 13 for a discussion of columns and type sizes.
- Alphabetize either letter by letter or word by word. In the *letter-by-letter format,* ignore spaces between words:

box drawing
boxes
box shading

In the *word-by-word format,* alphabetize all entries that begin with the same word, then go on to the next word:

box drawing
box shading
boxes

- When you use acronyms and abbreviations, consider how most readers will search for the term. For instance, use "CD-ROM," not "Compact Disk-Read Only Memory." But you might want to include the spelled-out version in parentheses:

CD-ROM (Compact Disk-Read Only Memory), 18

You might also want to include a *see* entry under the spelled-out version.

- Alphabetize symbols and numbers as they are spoken. For instance, "486 microprocessor" is alphabetized under *f.*
- If the document has a glossary, make an entry for each term listed in the glossary. If the term also appears in the body of the document, list the glossary citation as a definition. For instance, if *curly braces* is defined in

the glossary on page 234 and mentioned in a discussion of math on page 135 of the body, here is how to make the index entry:

Curly braces
   definition of, 234
   in mathematical equations, 135

- Consider using boldface or italics to indicate the pages on which the main discussion of an item appears:

dot leaders, **171–172**, 211, 306

Be sure to add a note at the start of the index to explain your use of boldface.

- Find out the appropriate way to express inclusive numbers. The easiest and clearest way is to use all the digits: 289–293. Writing all the digits is important when you use section numbering (page 3-6 is the sixth page in section 3). To express inclusive numbers for a document with section numbering, use the following form: 3–6 to 3–9.

   Some organizations abbreviate the second number: 289–93. But abbreviating inclusive numbers can get complicated. For example, most style guides call for writing "100–104," rather than "100–04" or "100–4." Check your style guide to see how your organization expresses inclusive page numbers.

WRITER'S CHECKLIST

1. Does the transmittal letter
   - ❑ clearly state the title and, if necessary, the subject and purpose of the document?
   - ❑ state who authorized or commissioned the document?
   - ❑ briefly state the methods you used?
   - ❑ summarize your major results, conclusions, or recommendations?
   - ❑ acknowledge any assistance you received
   - ❑ courteously offer further assistance?

2. Does the cover include
   - ❑ the title of the document?
   - ❑ your name and position?
   - ❑ the date of submission?
   - ❑ the company name or logo?

3. Does the title page
   - ❑ include a title that clearly states the subject and purpose of the document?
   - ❑ list the names and positions of both you and your principal reader?
   - ❑ include the date of submission of the document and any other identifying information?

4. Does the abstract
   - ❑ list your name, the document title, and any other identifying information?
   - ❑ clearly define the problem or opportunity that led to the project?
   - ❑ briefly describe (if appropriate) the research methods?
   - ❑ summarize the major results, conclusions, or recommendations?

5. Does the table of contents
   - ❑ clearly identify the executive summary?
   - ❑ contain a sufficiently detailed breakdown of the major sections of the body of the document?
   - ❑ reproduce the headings as they appear in your document?
   - ❑ include page numbers?

6. Does the list of illustrations (or the list of tables or the list of figures) include all the graphics found in the body of the document?

7. Does the executive summary
   - ☐ clearly state the problem or opportunity that led to the project?
   - ☐ explain the major results, conclusions, recommendations, and managerial implications of your document?
   - ☐ avoid technical vocabulary and concepts that the managerial audience is not likely to know?

8. Does the glossary include definitions of all the technical terms your readers might not know?

9. Does the list of symbols include all the symbols and abbreviations your readers might not know?

10. Do the appendices include the supporting materials that are too bulky to present in the document body or that will interest only a small number of your readers?

11. Does the documentation system conform to the guidelines of the APA system, the MLA system, or another appropriate system?

12. Does the index
   - ☐ use either the run-in or indented format?
   - ☐ appear in a two-column format in a type size two points smaller than the body text?
   - ☐ use either letter-by-letter or word-by-word alphabetization?
   - ☐ alphabetize acronyms and abbreviations according to how most readers will search for the term?
   - ☐ alphabetize symbols and numbers as they are spoken?
   - ☐ include an entry for each item in the glossary?
   - ☐ express inclusive numbers appropriately?

EXERCISES

1. The following letter of transmittal is from a report written by an industrial engineer to his company president. Write a one-paragraph evaluation that focuses on the clarity, comprehensiveness, and tone of the letter.

Dear Mr. Smith:

   The enclosed report, "Robot and Machine Tools," discusses the relationship between robots and machine tools.
   Although loading and unloading machine tools was one of the first uses for industrial robots, this task has only recently become commonly feasible. Discussed in this report are concepts that are crucial to remember in using robots.
   If at any time you need help understanding this report, please let me know.

Sincerely yours,

2. The following informative abstract is from a report by an electrical engineer to her manager. Write a one-paragraph evaluation. How well does the abstract define the problem, methods, and important results, conclusions, and recommendations?

"Design of a New Computer Testing Device"
   The modular design of our new computer system warrants the development of a new type of testing device. The term *modular design* indicates that the overall computer system can be broken down into parts or modules, each of which performs a specific function. It would be both difficult and time consuming to test the complete system as a whole, for it consists of sixteen different modules. A more effective testing method would check out each module individually for design or construction errors prior to its installation into the system.

(continued on page 265)

This individual testing process can be accomplished by the use of our newly designed testing device.

The testing device can selectively call or "address" any of the logic modules. To test each module individually, the device can transmit data or command words to the module. Also, the device can display the status or condition of the module on a set of LED displays located on the front panel of the device. In addition, the device has been designed so that it can indicate when the module being tested has produced an error.

3. The following table of contents is from a report titled "An Analysis of Corporations v. Sole Proprietorships." Write a one-paragraph evaluation. How effective is the table of contents in highlighting the

CONTENTS

executive summary, defining the overall structure of the report, and providing a detailed guide to the location of particular items?

4. The following executive summary is from a report titled "Analysis of Large-Scale Municipal Sludge Composting as an Alternative to Ocean Sludge-Dumping." Write a one-paragraph evaluation. How well does the executive summary present concise and useful information to the managerial audience?

Coastal municipalities currently involved with ocean sludge-dumping face a complex and growing sludge management problem. Estimates suggest that treatment plants will have to handle 65 percent more sludge in 1995 than in 1985, or approximately seven thousand additional tons of sludge per day. As the volume of sludge is increasing, traditional disposal methods are encountering severe economic and environmental restrictions. The EPA has banned all ocean sludge-dumping as of next January 1. For these reasons, we are considering sludge composting as a cost-effective sludge management alternative.

Sludge composting is a 21-day biological process in which waste-water sludge is converted into organic fertilizer that is aesthetically acceptable, essentially pathogen-free, and easy to handle. Composted sludge can be used to improve soil structure, increase water retention, and provide nutrients for plant growth. At $150 per dry ton, composting is currently almost three times as expensive as ocean dumping, but effective marketing of the resulting fertilizer could dramatically reduce the difference.

5. Find a document, such as a textbook or a manual, that contains an index. Study a page of the index;

then write a memo to your instructor describing and evaluating the index. For instance, what form of alphabetization is used? Are the entries clear and easy to use? How effectively are *see* and *see also* used? Does the index appear comprehensive and accurate? See Chapter 16 for a discussion of memos.

6. Choose a journal article of some 8–15 pages and then write an index for it. Attach a note describing the audience and indicating what indexing style you are using. For instance, are you alphabetizing letter-by-letter or word-by-word? What style of pagination are you using? Attach a photocopy of the article.

REFERENCES

Bonura, L. S. (1994). *The art of indexing*. New York: Wiley.

Chicago manual of style (14th ed.). (1993). Chicago: University of Chicago Press.

Crowe, B. (1985). *Design of a radio-based system for distribution automation.* Unpublished manuscript, Drexel University.

Gibaldi, J. (1995). *MLA handbook for writers of research papers* (4th ed.). New York: Modern Language Association.

Publication manual of the American Psychological Association (4th ed.). (1994). Washington, DC: American Psychological Association.

Ruiu, D. (1994). Testing ATM systems. *IEEE Spectrum, 31*(6), 25.

Integrating Graphics

Graphics are the "pictures" used in technical communication: drawings, maps, photographs, diagrams, charts, graphs, and tables. In terms of appearance, graphics range from realistic, such as photographs, to highly abstract, such as organization charts. In terms of function, graphics range from the decorative, such as clip art that shows a group of people seated at a conference table, to highly informational, such as a table or a schematic diagram of an electrical device.

Graphics are important in technical communication because they can

- communicate some kinds of information that are difficult to communicate with words
- help you clarify and emphasize information
- interest the reader
- help nonnative speakers of English understand the document
- help communicate information to multiple audiences with different interests, aptitudes, and reading habits

This chapter begins by explaining the characteristics of effective graphics and describing the process of planning, creating, and evaluating them. Next it discusses the use of color and then covers the basic kinds of graphics used in technical communication. The chapter then explains how to show motion in graphics and how to create graphics for international readers. It concludes with a discussion of graphics software.

The most obvious reason to use graphics in your documents is that graphics motivate people to study the document. According to Gatlin (1988), 83 percent of what people learn derives from visual stimuli, whereas only 11 percent derives from our hearing. Because as a species we are attuned to acquiring information through sight, communication that includes a visual element beyond words on the page is more effective than communication that doesn't. People studying text with graphics learn about one-third more than people reading text without graphics (Levie & Lentz, 1982). And graphics help readers retain what they learn; people remember some 43 percent more when the document includes graphics (Morrison & Jimmerson, 1989). Readers want graphics. One survey indicates that readers of computer documentation consistently desire more graphics and fewer words (Brockmann, 1990, p. 203).

Graphics offer several benefits that words alone cannot provide.

- *Graphics are almost indispensable in demonstrating logical and numerical relationships.* For example, the organization chart used in most businesses is an effective way to represent the lines of authority in the organization. And graphics are useful in showing numerical relationships of all kinds. If you want to show the trends in the number of nuclear power plants completed each year over the last decade, a line graph would be much easier to understand than a paragraph full of numbers. Graphics can also show the relationships among several variables over time, such as the numbers of 4-cylinder, 6-cylinder, and 8-cylinder cars manufactured in the United States during each of the last five years.

- *Graphics can communicate spatial information more effectively than words alone.* Try to convey in words what a simple hammer looks like. It's not easy. But in 10 seconds you can draw a simple diagram that clearly shows the hammer. If you want to show your readers the location of the San Andreas fault, you can shade the fault area on a standard map of the United States. If you want to show the details of the derailleur mechanism on a bicycle, you can present a diagram of the bicycle with a close-up of the derailleur.

- *Graphics can communicate steps in a process more effectively than words alone.* A troubleshooter's guide, a common kind of table, explains how to understand what might be causing a problem in a process and what you might do to fix it. A diagram can explain clearly how acid rain forms.

- *Graphics can save space.* Consider the following paragraph:

  In the Wilmington area, some 90 percent of the population aged 18–24 watches movies or tapes on a VCR. They watch an average of 2.86 tapes a week. Among 35- to 49-year-olds, the percentage is 82, and the average number of movies or tapes is 2.19. Among the 50–64 age group, the percentage is 67, and the number of movies and tapes watched averages 2.5. Finally, among those people 65 years old or older, the percentage is 48, and the average number of movies and tapes watched weekly is 2.71.

  Presented as a paragraph, this information is uneconomical and hard to remember. Presented as a table, however, the information is more concise and more memorable.

Age	Percent Watching Tapes/Movies	Number of Tapes Watched per Week
18-24	90	2.86
35-49	82	2.19
50-64	67	2.50
65+	48	2.71

- *Graphics can reduce the cost of documents intended for international readers.* Translation costs can reach 60¢ per word; used effectively, graphics can reduce the number of words you have to translate (Horton, 1993).

As you draft your document, think of opportunities to use graphics to clarify, emphasize, summarize, and organize information. McGuire and Brighton

(1990) recommend that you be alert to certain words and phrases that often signal the opportunity to create a graphic:

categories	features	numbers	routines
components	fields	phases	sequence
composed of	functions	procedures	shares
configured	if and then	process	structured
consists of	layers	related to	summary of
defines			

If you find yourself writing a sentence such as "The three categories of input modules are . . . ," consider creating a diagram to reinforce your idea. If you write "The first step in the procedure is . . . ," think in terms of a flowchart or a logic box. If you write "This structure is related to . . . ," consider a diagram.

Characteristics of an Effective Graphic

An effective graphic must be clear and understandable, and it must be meaningfully related to the larger discussion. Six characteristics make a graphic clear and understandable:

- *A graphic should have a purpose.* Don't put a graphic in your document unless it will help your reader understand or remember the information. Beware of decorative, content-free clip art, the images that you can buy in hard copy or electronic form: drawings of businesspersons standing with clipboards, shaking hands, and so on. The novelty of meaningless clip art has worn thin.
- *A graphic should be honest.* Graphics can be dishonest, just as words can. You are responsible for making sure the graphic does not lie or mislead the reader. Following are some common ethical concerns to keep in mind as you create graphics:
  - If you did not create the graphic or generate the data, you must cite your source.
  - You must include all the relevant data in your graphic. For example, if you have a data point that you cannot explain, it is unethical to change the scale as a way to eliminate it.
  - You must show items as they really are. If the electrical appliance shown in your photograph requires an electrical cord, show it.
  - You should begin the axes in your graphs at zero—or mark them clearly—so that quantities are represented honestly.
  - You should not use a table to hide a data point that would be more obvious in a graph.
  - You should not use color or shading to misrepresent the importance of an item. A dark-shaded bar in a bar graph, for example, appears larger and nearer than a light-shaded bar of the same size.

Common problem areas are pointed out in the discussions of the different kinds of graphics throughout this chapter.

- *A graphic should present a manageable amount of information.* If you present too much information in a graphic, you risk confusing your readers. Although some kinds of graphics can accommodate much more information than others, the most important factors to consider are your audience and purpose: what kinds of graphics are your readers familiar with, how much do they already know about the subject, and what do you want the document to do? In general, readers learn best if you present information in small chunks. Therefore, it is better to create several simple graphics than to create one complicated one.

- *A graphic should meet the reader's format expectations.* Through experience, readers learn how to read different kinds of graphics. Follow the conventions for the different kinds of graphics—for instance, by using diamonds to represent decision points in a flowchart—unless you have a good reason not to.

- *A graphic should be simple and uncluttered.* Three-dimensional bar graphs are easy to make with modern software, but they are harder to understand than two-dimensional ones, as shown in Figure 12.1. Graphics scholar Edward Tufte (1983) uses the term *chartjunk* to describe the ornamentation that clutters up the graphic, distracting the reader from the message. Unnecessary 3-D is one example of chartjunk. In Figure 12.1 on page 272, the two-dimensional bar graph is clean and uncluttered, but the three-dimensional graph is more difficult to understand because the third dimension obscures the main data points. The number of uninsured emergency-room visits in February, for example, is very difficult to see in the three-dimensional graph.

- *A graphic should be clearly labeled.* Every graphic (except a brief, informal one) should have a unique, clear, and informative title. The columns of a table, and the axes of a graph, should be labeled fully, complete with the units of measurement. The lines on a line graph should also be labeled. Your readers should not have to guess whether you are using meters or yards, or whether you are including statistics from last year or just this year.

In addition, five characteristics of an effective graphic concern its integration with the surrounding text:

- *A graphic should be placed in an appropriate location.* If your readers need the information to understand the discussion, put the graphic directly after the pertinent point in the text—or as soon after that point as possible. If the information functions merely as support for a point that is already clear, or as elaboration of it, include it as an appendix.

TABLE 1
Insured and Uninsured Emergency Room Visits
Parkland Memorial Hospital
January–March 1994

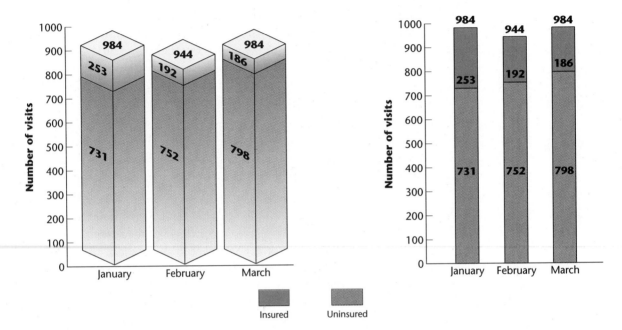

Chartjunk

- *A graphic should be introduced in the text.* Whenever possible, refer to a graphic before it appears. Ideally, place the graphic on the same page as the reference. Refer to the graphic by number (such as *Figure 7*). Do not refer to "the figure above" or "the figure below"; the graphic might be moved during the production process. If the graphic is an appendix item, tell your readers where to find it: "For the complete details of the operating characteristics, see Appendix B."

- *A graphic should be explained in the text.* Although a graphic is more visually interesting than text, you must state what you want your readers to learn from looking at it (Dean & Kulhavy, 1981). When introducing a graphic, ask yourself whether a mere paraphrase of its title will communicate its significance: "Figure 2 is a comparison of the costs of the three major types of coal gasification plants." Sometimes you need to explain why the graphic is important or how to interpret it. If you want the graphic to make a point, don't just hope your readers will understand it. Make the point explicitly:

As Figure 2 shows, a high-sulfur bituminous coal gasification plant is more expensive than either a low-sulfur bituminous or anthracite plant, but more than half of its cost is cleanup equipment. If these expenses could be eliminated, high-sulfur bituminous would be the least expensive of the three types of plants.

Often graphics are accompanied by captions—explanations ranging in length from a sentence to several paragraphs. Captions are useful for readers who are skimming the document; if the captioned material were included in the body text, they would not see it. Captioned text is usually smaller than the body text and presented in a different typeface.

- *A graphic should be clearly visible in the text.* Distinguish the graphic from the surrounding text by adding white space or rules—lines—or by enclosing it in a box. See Chapter 13 for a discussion of these design techniques.

- *A graphic should be accessible to your readers.* If the document is more than a few pages long and contains more than four or five graphics, consider including a list of illustrations—a table of contents for the graphics—so that readers can find them easily. See Chapter 11 for a discussion of lists of illustrations.

Planning, Creating, and Revising the Graphics

Using graphics involves planning, creating, and revising them. The following sections discuss these three steps.

Planning the Graphics

Some people, particularly those with a strong writing background, create documents by writing the words and then figuring out what kinds of graphics to add. Others, especially scientists, engineers, and technicians, think graphically first. They sketch in their graphics and then write the text. Some people combine the two approaches, and some use a text-first process for some kinds of documents and a graphics-first process for others.

Regardless of which process you use, however, at some point you must think about the role the graphics will play in the document. In planning your graphics, consider the following four aspects of the document, shown in Figure 12.2 on page 274.

- *Audience.* Some types of graphics, such as pie charts, are very easy for any reader to understand. Others, such as scatter graphs, are incomprehensible to the general reader. Do your readers understand certain kinds of graphics, such as schematics or computer flowcharts? Are they familiar with standard icons used in your field? Are they already motivated to read your document, or do you need to enliven the text with graphics to hold their attention or perhaps even use color for emphasis?

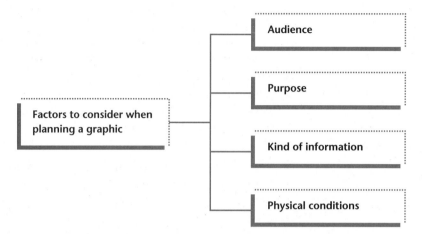

FIGURE 12.2

Factors to Consider
When Planning a Graphic

- *Purpose.* What point are you trying to make with the graphic? Figure 12.3 shows that even a few simple facts can yield a number of different points. Your responsibility is to determine what you want to show and then to figure out how best to show it. Don't rely on your software; it doesn't think.

  Each of these four graphs emphasizes a different point. Graph (a) focuses on the total number of cars disabled each month, classifying the problem by cause; graph (b) focuses on the three rail lines during one month; and so forth.

- *Kind of information.* The subject will help you determine what type of graphic to create. If you are writing about the number of languages spoken by citizens in your state, you will probably use tables for the mass of statistical data, maps to show the patterns of language use, and graphs to show trends in the statistics over time. If you are discussing how the sales of two products have changed over several years, a grouped bar chart might be the best type of graphic.

- *Physical conditions.* The physical conditions in which the document will be used—amount of lighting, amount of space available, exposure to water or grease, and so forth—might affect not only what kind of graphic to create but also the size and shape, the thickness of lines and size of type, and the color.

Creating the Graphics   You have thought about what kinds of graphics would help you communicate your information, and you might even have made some quick sketches. Now it is time to consider how you are going to create them. As you plan, consider four important factors:

- *Time.* Creating a complicated graphic can take a lot of time, so you need a schedule. See Chapter 5 for a discussion of planning.

TABLE 1
Number of Rail Cars Disabled by Electrical Problems November–January

Rail Line	November		December		January	
	Disabled by electrical problems	Total disabled	Disabled by electrical problems	Total disabled	Disabled by electrical problems	Total disabled
Bryn Mawr	19	27	17	28	20	26
Swarthmore	12	16	9	17	13	16
Manayunk	22	34	26	31	24	33

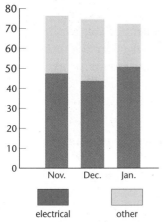

a. Number of rail cars disabled, November–January

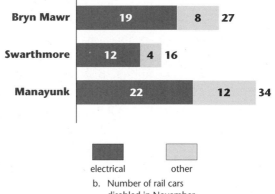

b. Number of rail cars disabled in November

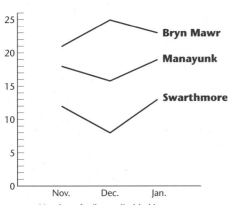

c. Number of rail cars disabled by electrical problems November–January

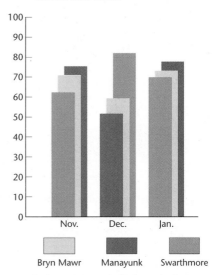

d. Range in percent of rail cars, by line, disabled by electrical problems, November–January

FIGURE 12.3

Different Graphics
Emphasizing Different
Points

- *Money.* A high-quality graphic can be expensive. How big is the budget for the project? How can you use that money effectively? See Chapter 5 for a discussion of budgeting.
- *Equipment.* Although some graphics can be made with a straight edge and a pencil, most require other tools, including word-processing software for tables, spreadsheets for tables and graphs, and graphics software for diagrams.
- *Expertise.* How much do you know about creating graphics? How much expertise do you have access to?

Time, money, equipment, and expertise are overlapping categories, of course, for if you have a large budget, you might be able to hire experts with the latest equipment and take the time to revise the graphics several times. Except for special projects, however, such as annual reports or extremely important proposals, people usually don't have all the resources they would like. Therefore, it is necessary to choose from among a range of different approaches to creating graphics, as shown in Figure 12.4.

- *Using existing graphics.* First, it is necessary to distinguish between two different kinds of documents: unpublished and published.

     For a student paper that will not be published, some instructors permit the use of photocopies of existing graphics; other instructors do not. If you are permitted to photocopy existing graphics, you are ethically obligated to cite them, just as you would any other information you are borrowing.

     For a published document, written by either a student or a nonstudent, the use of an existing graphic is permissible if the graphic is the property of the organization or if the organization has obtained permis-

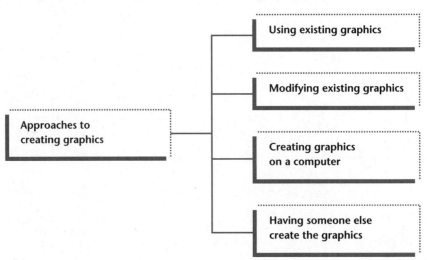

FIGURE 12.4

Approaches to Creating
Graphics

sion to use it. If the graphic was created under work-for-hire guidelines (see Chapter 2), the organization has an unlimited right to use it. If, however, the graphic was not created under work-for-hire guidelines, the organization must acquire the right to multiple use; that is, the contract under which the graphic was originally created for the organization must grant the organization multiple use, not merely one-time use.

Think carefully before using existing graphics, for four reasons:

- The style of the graphic might not match that of the other graphics you want to use.
- The graphic might lack some features you want or contain some you don't.
- The lines in a photocopied graphic often lack crispness.
- The shading in the original—the shades of gray or the colors—tends to "void out"; that is, it doesn't reproduce sharply, creating a blotchy appearance.

If you use an existing graphic, create your own number and title for it; the existing number will not match your numbering system, and the title is likely to be written in a different style.

- *Modifying existing graphics.* You can modify existing graphics in several ways. You can redraw an existing graphic, revising as you go, or you can use a scanner to digitize the graphic and then manipulate it electronically with graphics software. In either case, you are under the same ethical and legal obligations described in the previous section, just as if you were not modifying the graphic at all. Your source statement, however, should indicate that your graphic is "based on" or "adapted from" your source.

- *Creating graphics on a computer.* Although simple tables and diagrams can be done by hand, in most cases you will want to use a computer to create professional-quality graphics.

Computer graphics offer two main advantages over noncomputer graphics: variety and reusability. The variety and complexity of new computer graphics is considerable. With each passing month, new graphics software comes on the market with powerful capabilities. Three-dimensional effects are routine now, and animation is becoming common. Because they are digital, computer graphics can be stored, reused, and revised. After you create a graphic showing this month's budget, you can copy it and easily revise it next month. Remember, however, that simpler graphics are usually preferable to more complicated ones.

In addition to the general kinds of graphics software, there are specialized graphics packages for different applications. For instance, architecture students routinely create three-dimensional images using computer-assisted design (CAD) software. Computer scientists use flowcharting software to create flowcharts quickly. Proposal writers use project-management software to create network diagrams such as Gantt charts and PERT charts (see Chapter 21). Artists and designers use animation graphics that show movement over time. Physicians practice surgery using computer-generated graphics instead of cadavers, and aircraft designers use computer simulations rather than wind-tunnel experiments.

Computer graphics are not simple to create, however, and an unprofessional-looking graphic can hurt your credibility, leading the reader to question the professionalism of the whole document.

Visit the computer labs at your college or university to test some of the many excellent graphics software packages available. Also, consult the bibliography (Appendix D) for a list of books about computers and technical communication. But beware: any book about computer graphics is probably out of date as soon as it is published.

For more information on computer graphics, see the discussion later in this chapter.

- *Having someone else create the graphics.* If graphics software is so versatile, why would technical communicators choose not to make the graphics themselves on their own computers? One obvious reason: professional-level software can cost hundreds of dollars. Another less obvious reason: some people don't have the artistic ability to create professional-quality graphics. And even if the person does have the ability, proficiency with a sophisticated graphics package can require dozens or even hundreds of hours.

  In most cases, your choices for having someone else create your graphics are constrained by your budget and policies. Some companies have technical-publications departments with graphics experts. Others subcontract this work. Many print shops and service bureaus have graphics experts on staff or can direct you to them.

Revising the Graphics

As with any other aspect of technical communication, build in enough time and money to revise the graphics.

Create a checklist and evaluate each graphic to test its effectiveness. The Writer's Checklist at the end of this chapter is a good starting point.

Show the graphics to people with backgrounds like the intended readers' and ask them for suggestions. Revise the graphics and solicit more reactions.

Using Color Effectively

Color draws the reader's eyes to information you want to emphasize, helps you establish visual patterns to increase the reader's understanding, and adds interest. Color has been used for many decades in magazines and some books. But now that the price of color printers is dropping, color is becoming more and more common in technical communication. Color is a powerful tool that is easy to misuse. The following discussion is based on Jan V. White's excellent text *Color for the Electronic Age* (1990).

In using color in graphics and page design, keep in mind seven guidelines:

- *Don't overdo it.* Even if your computer monitor is capable of displaying 16 million colors, readers can understand only two or three at a time (White, 1990, p. 19). Use colors for small items, such as portions of graphics and important words. And don't use colors where black will work better. Use black for the main text; black is easier for the reader to interpret.

- *Use color to emphasize particular items.* Horton (1991) notes that we interpret color before we interpret shape, size, or placement on the page. Therefore, color effectively draws the reader's attention to a particular item or class of items on a page. Figure 12.5 shows how color emphasizes different kinds of items.

Color draws the reader's attention to the line.

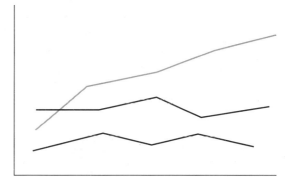

Color is useful for emphasizing short phrases. Used for longer passages, color loses its impact.

Pixel Driver Renders 600,000 Polygons Per Second

The colored frame focuses the reader's attention on the information in the graph.

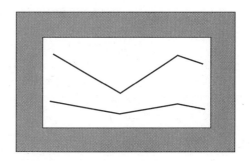

FIGURE 12.5

Color Used for Emphasis

(Figure 12.5 continued)

Table 2
Israeli Power Plant Sites

Site	Generating Units	Installed Capacity (MW)
Haifa	4	430
Hadera	4	1400
Tel Aviv	4	530
Ashdod	9	1210
Total	**21**	**3570**

Here the color emphasizes a row; it could also be used to emphasize a column or a single data cell.

Source: B. Golany, Y. Roll, & D. Rybak. (1994, August). Measuring efficiency of power plants in Israel by data envelopment analysis," *IEEE Transactions on Engineering Management 41*, 3, p. 292.

- *Use colors to create patterns.* We all recognize the pattern of black writing against a yellow background as a traffic sign. The same principle applies in graphics and document design. Choose a warm color such as red as the background or the text color for all your safety comments; choose a cool color such as green for your tips. Use different-colored paper for different sections of a manual. In creating patterns, consider using shapes too. For instance, use red for the safety comments, placing each one in an octagon resembling a stop sign. This way, you give your readers two visual characteristics to help them recognize the pattern. Figure 12.6 (Jain, 1994, p. 3) shows how the page of a technical journal uses color to establish patterns.

  In Figure 12.6, color is used to draw attention to three aspects of the page: the name of the department ("Editor-in-Chief's Message"); the title of the article ("What Is Multimedia, Anyway?"); and the marginal gloss (see Chapter 13 for a discussion of marginal glosses). Notice that the color in the department name and in the marginal gloss is *ramped;* that is, the color progresses from richer to paler. Ramping creates an impression of movement, drawing the reader's eye, in Figure 12.6, from left to right or top to bottom. Color can also be used effectively to emphasize other kinds of particular items, such as text boxes, rules, screens, and headers and footers. These design elements are discussed in Chapter 13.

- *Use contrast effectively.* The visibility of a color is a function of the background against which it appears. The strongest contrasts are between black and white and black and yellow. In Figure 12.7 on page 282, notice that a color washes out if the background color is too similar.

  The need to provide effective contrast applies not only in text but in graphics used in presentations (see Chapter 20). You want to provide a natural-looking background against which the text or graphics provide a vivid contrast. Transparencies are used with the room lights on; there-

Editor-in-Chief's Message

Ramesh Jain
University of California, San Diego

What Is Multimedia, Anyway?

Given its increasing popularity—and surrounding hype—many people are asking, "What *is* multimedia, anyway?" This question occurs not only to computer users but also to computer scientists. Almost every computer company claims leadership in multimedia; every software developer develops software for some aspect of multimedia; every application either targets multimedia or uses it for its "sexiness" as a marketing tool. Evidently everybody in computing and related fields is fascinated by multimedia. But what *is* it?

Contact editor-in-chief Jain at UC San Diego, Dept. of Electrical and Computer Engineering, MC 0407, 9500 Gilman Drive, La Jolla, CA 92093-0407, or by e-mail, jain@ece.ucsd.edu.

Most human activities, from mundane ones for basic survival to sophisticated intellectual and spiritual activities, involve gathering information about our environment that will help us perform particular tasks. This information comes to us from multiple disparate sources. We use our senses to acquire information directly from our surroundings, and we read books, listen to other people's experiences, look at pictures and maps, and watch movies. Our knowledge about our environment, including people and their feelings, is based on many dramatically varying information sources. We also get information from complementary sources that are very different at their representation level. We assimilate this disparate information to form a complete and unified model of the microcosm that concerns us. This model is vital to human activities. Through all our civilizations, the quest for efficient methods to enable people to form precise models of the world we live in has received maximum attention.

Not surprisingly, information storage, communication, and processing have been essential to the formation and advancement of civilization. Ancient cave paintings, artifacts, and stone documents exemplify the human desire to store information in perpetuity. Gutenberg's invention of movable type revolutionized civilization by allowing mass storage of textual information. This resulted in mass production of information sources and propagation of knowledge, which influenced the course of civilization. It also made text the primary means of propagating knowledge across generations, countries, and civilizations.

Technological advances have made storage and communication of image, audio, and video information possible in the last few years. Processing of all these forms of information using digital technology let us bring these disparate sources to one platform. People are excited about the merger of audio, video, images, text, and numeric modes of information. With technology, we can now store, communicate, and process these natural and complementary modes of expression. In this new, evolving environment, computers suddenly seem transparent. After working with alphanumeric computers for so long, we get a feeling of liberation.

Multimedia computing deals with storage, communication, and processing of information using disparate, complementary, and natural modes. The technology has finally arrived to let us deal with computers in a natural way. This is not an illusion or hype—it is a revolution in computing.

Obviously, multimedia computing involves most aspects of computing. The former dominance of alphanumeric information meant that all computing hardware and software was designed to deal primarily with it. Now, all aspects of computing, from networking to user interfaces, are designed to consider a multiplicity of representation and modes of information. The focus is now on information rather than the medium used to acquire, store, and communicate it.

Multimedia computing involves incorporating the modes of information naturally used by humans into computing. It aims to make computing natural to us. It attracts people from almost all fields, ranging from entertainment to rocket science, because it enhances our ability to deal with information and express our concepts and ideas. I am sure that in just a few more years, nobody will talk about "multimedia computing," just as nobody now talks about digital computers. Soon all computers will be adept in multimedia—along with their users. **MM**

FIGURE 12.6

Color Used to Establish Patterns

fore, the most popular background is the white provided by the screen, and the most popular color for the text or graphics is black or dark blue. For slides, however, the room lights are either dim or off. Therefore, the most popular background color for slides is dark blue, with yellow or white text and graphics. Figure 12.8 on page 282 shows the use of contrast in a transparency and a slide.

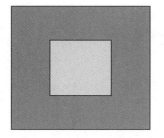

FIGURE 12.7

The Importance of
Background in Creating
Contrast

- *Take advantage of the meanings that colors already have.* In our culture, red signals danger, heat, or electricity. Yellow signals caution; orange signals warning. If you use any of these warm colors in ways that depart from these cultural meanings, you can confuse the reader. The cooler colors—blues and greens—are more conservative and subtle. Keep in mind, however, that different cultures interpret colors differently. See Chapter 4 for a discussion of these cultural differences. In the drawing of the electric car shown in Figure 12.9, the batteries are red. The warm red contrasts effectively with the cool green of the car body.
- *Realize that color swallows up text.* If you are printing against a colored background, you might need to make the text a little bigger, because the color will make text look smaller. Figure 12.10 shows that the text printed against a white background looks bigger than the same size text printed against a colored background. White type counteracts this effect, as shown in Figure 12.11.
- *Use light colors to make objects look bigger.* In Figure 12.12, the yellow box looks bigger than the blue box, even though they are the same size.

FIGURE 12.8

Contrast in a
Transparency (left) and a
Slide (right)

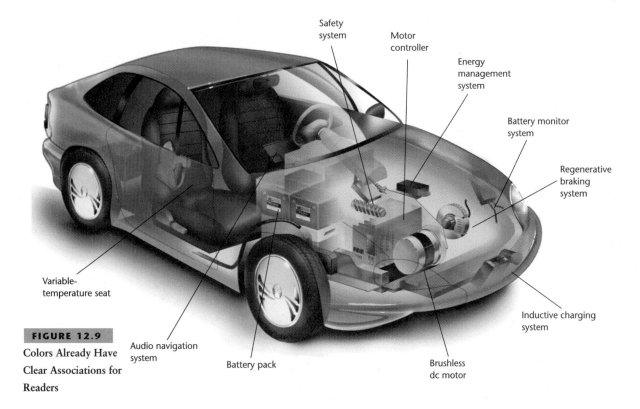

Safety system

Motor controller

Energy management system

Battery monitor system

Regenerative braking system

Inductive charging system

Brushless dc motor

Battery pack

Audio navigation system

Variable-temperature seat

Colors Already Have Clear Associations for Readers

A Colored Background Minimizes the Appearance of Text

This text looks bigger because of the white background.

This text looks smaller, even though it is the same size, because of the colored background.

White and Black Type against a Colored Background

This line of type appears to reach out to the reader.
This line of type appears to recede into the background.

Light-Colored Objects Appear Bigger than Dark-Colored Objects

Choosing the Appropriate Kind of Graphic

The graphics used in technical documents are classified into two basic categories: tables and figures. *Tables* are lists of data—usually numbers—arranged in columns. *Figures* are everything else: graphs, charts, diagrams, photographs, and the like. Generally, tables and figures have their own sets of numbers: the first table in a document is Table 1; the first figure is Figure 1. In documents of more than one chapter (as in this book), the graphics are usually numbered chapter by chapter. Figure 3.2, for example, would be the second figure in Chapter 3.

But this broad distinction between tables and figures offers little guidance about what kind of graphic to use in a particular situation. There is no simple system for choosing the type of graphic because for many kinds of situations several different types would do the job. In general, however, graphics can be categorized according to the kind of information they communicate. The discussion that follows is based on a classification system created by William Horton in his chapter "Pictures Please—Presenting Information Visually," in *Techniques for Technical Communicators* (Barnum & Carliner, 1993). I strongly recommend Horton's guide to graphics.

Technical information can be classified into four categories, as shown in Figure 12.13. Note that some kinds of graphics can convey several different kinds of information. For instance, tables can communicate both numerical values and procedures.

Figure 12.14 summarizes the following discussion.

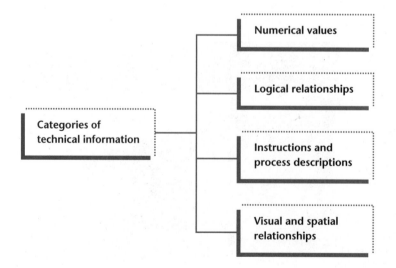

FIGURE 12.13

Four Categories of
Technical Information
Communicated by
Graphics

- Categories of technical information
  - Numerical values
  - Logical relationships
  - Instructions and process descriptions
  - Visual and spatial relationships

Purpose	Type of Graphic	What the Graphic Does Best
ILLUSTRATING NUMERICAL INFORMATION	Table	• Shows large amounts of numerical data, especially when there are several variables for a number of items.
	Bar graph	• Shows the relative values of two or more items.
	Pictograph	• Enlivens statistical information for the general reader.
	Line graph	• Shows how the quantity of an item changes over time. A line graph can accommodate much more data than a bar graph can.
	Pie chart	• Shows the relative size of the parts of a whole. Pie charts are instantly familiar to most readers.
ILLUSTRATING LOGICAL RELATIONSHIPS	Diagram	• Represents items or properties of items.
	Organization chart	• Shows the lines of authority and responsibility in an organization.
ILLUSTRATING INSTRUCTIONS AND PROCESS DESCRIPTIONS	Checklist	• Lists or shows what equipment or materials to gather, or describes an action.
	Table	• Shows numbers of items or indicates the state (on/off) of an item.
	Flowchart	• Shows the stages of a procedure or a process.
	Logic box	• Shows which of two or more paths to follow.
	Logic tree	• Shows which of two or more paths to follow.
ILLUSTRATING VISUAL AND SPATIAL CHARACTERISTICS	Drawing	• Shows simplified representations of objects.
	Map	• Shows geographical areas.
	Photograph	• Shows precisely the external surface of objects.
	Screen shot	• Shows what appears on a computer screen.

FIGURE 12.14

Choosing the Appropriate
Kind of Graphic (Based
on Horton [1993])

Illustrating Numerical Values

In technical communication, you will often have to display numerical information. Sometimes you will communicate precise numerical data; sometimes, relative values. Sometimes you will communicate numerical values at one point in time; sometimes, these values as they change over time. The basic kinds of graphics used to communicate numerical values are tables, bar graphs, pictographs, line graphs, and pie charts.

Tables

Tables easily convey large amounts of numerical data, and they often provide the only means of showing several variables for a number of items. For example, if you want to show the numbers of people employed in six industries in 10 states, a table would probably be the best graphic. Although tables lack the visual appeal of other kinds of graphics, they can handle much more information with complete precision.

Table 6. Test Results for Valves #1 and #2 ———— Title

Valve Readings	Maximum Bypass Cv	Minimum Recirculation Flow (GPM)	Pilot Threads Exposed	Main ΔP at Rated Flow (psid)
Valve #1				
Initial	43.1	955	+3	4.4
Final	43.1	955	+3	. . .
Valve #2				
Initial	48.1	930	+3	4.5
Final	48.2	950	+2	. . .

Number
Stub head
Row head
Column head

Source statement ——— Source: "Third Progress Report: Anderson Machine Tools Reconfiguration Project."

FIGURE 12.15

Parts of a Table

Keep in mind the needs and aptitudes of your audience. You can convey much more information in a table than your readers need or can understand easily. If you choose to create a large table—one with, say, 20 columns and 30 rows—consider placing it in an appendix and making a summary table for the body of the document.

Figure 12.15 shows the standard parts of a table. Tables are identified with both a number ("Table 1") and a substantive, informative title, one that encompasses the items being compared, as well as the basis (or bases) of comparison.

The basis of comparison is time: each year from 1990 to 1994.

Mallard Population in Rangeley, 1990-1994

The Growth of the Robotics Industry in Japan and the United States, 1995

The basis of comparison is the two nations: Japan and the United States.

Note that most tables are numbered and titled above the data. The number and title are centered horizontally or left justified.

Here are 10 guidelines for constructing tables:

• *Indicate the units in which you are presenting the data.* If all the data are expressed in the same unit, indicate that unit in the title:

Farm Size in the Midwestern States (in Hectares)

If the data in one column are expressed in a different unit from the data in another column, indicate the units in the column headings:

Population (in millions) | Per Capita Income (in thousands of U.S. dollars)

If all the data cells in a column use the same unit, indicate that unit in the column head, not in each data cell:

Speed (knots)
15
18
14

You can express data in both real numbers and percentages. A column heading and the first data cell under it might read as follows:

Number of Students (Percentage)
53 (83)

- *In the stub—the left-hand column—list the items being compared.* Arrange the items in the stub in some logical order: big to small, important to unimportant, alphabetical, chronological, geographical, and so forth. If the items fall into several categories, you can include the names of the categories in the stub:

Snow Belt States ....
      Connecticut...........................
      New York...............................
      Vermont ....

Sun Belt States ....
      Arizona ....
      California...............................
      New Mexico ....

If the items in the stub are not grouped in logical categories, skip a line every five rows to help the reader follow the rows across the table. Or use a shaded or colored background for every other set of five rows. Dot leaders, a row of dots that links the stub and the next column, are also useful.

- *In the columns, arrange the data clearly and logically.* Line up the numbers consistently by using the decimal tab function:

3,147.4
   365.7
46,803.5

In general, don't change units. If you use meters for one of your quantities, don't use feet for another, unless the quantities are so dissimilar that your readers would have a difficult time understanding them as expressed in the same units.

  3.4 hr
12.7 min
  4.3 sec

This list would probably be easier for most readers to understand than one in which all quantities were expressed in the same unit.

- *Perform the computations your readers will need.* If your readers will need to know the totals for the columns or the rows, provide them. If your readers will need to know percentage changes from one column to the next, present them:

Number of Students (Percentage Change from Previous Year)

1991	1992	1993
619	644 (+4.0)	614 (−4.7)

- *Use dot leaders if a column contains a "blank" spot: a place where there are no appropriate data:*

  3,147

  . . .

  46,803

  However, don't substitute dot leaders for a quantity of zero:

  3,147

  0

  46,803

- *Don't make the table wider than it needs to be.* The reader should be able to scan across a row easily. As Jan White (1984) points out, there is no reason to make the table as wide as the text column in the document. If a column heading is long—say, more than five or six words—stack the words rather than make the column wide:

  Computers Sold

  Without a CD-ROM

- *Minimize the use of lines.* Grimstead (1987) recommends using lines only when necessary: to separate the title and the headings, the headings and the body, and the body and the notes. When you use lines, make them thin rather than thick.

- *Try to keep the table on one page.* You can tell your software not to split a table between pages. If you have to split it, write *continued* at the bottom; on the next page, include the full title, followed by *continued*. Repeat the column headings.

- *Provide footnotes for any information that needs to be explained.* You should provide all the information your readers need to understand the graphic.

- *Indicate the source of your information if you did not generate it yourself.* A typical source statement is: "Source: Data from *IEEE Spectrum*, vol. 30, no. 8 (1994), p. 34."

  Tables can also convey textual information effectively, as shown in Figure 12.16 (U.S. Congress, *Making*, 1993: 41).

Bar Graphs

Like tables, bar graphs can communicate numerical values, but they are better at showing the relative values of two or more items. In considering whether to

Table 2.3 Types of Electronic Kiosks: Key Characteristics and Selected Applications

Type of kiosk	Key characteristics	Selected applications
Off-line: Stand-alone	For information that does not need updating: no telecommunications costs	GSA's Central Office Building directory
Off-line: Polled	Can update information, and retrieve queries and survey results over a telephone line and modem at night	USPS's "Postal Buddy"; "24-Hour City Hall"
On-line	Can process information immediately; can update rules and software in central computer; requires dedicated telephone line and central computer capacity.	Tulare County, CA's "Tulare Touch"; State of California's "InfoCalifornia"
On-line: Transactional	On-line, but can also collect money via credit or debit cards for bills and services	Long Beach, CA's "Auto Clerk"; State of California's "InfoCalifornia"

KEY: GSA = General Services Administration; USPS = U.S. Postal Service.
SOURCE: Office of Technology Assessment, 1993.

FIGURE 12.16
Text Table

make the bars vertical or horizontal, think about how people envision information. Vertical bars are best for showing quantities such as height, size, and amount. Horizontal bars are best for showing quantities such as speed and distance. However, these distinctions are not ironclad; as long as the axes are labeled carefully, your readers should have no trouble understanding the graph.

Figure 12.17 shows the structure of typical horizontal and vertical bar graphs.

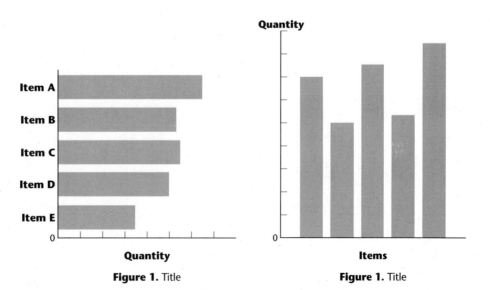

FIGURE 12.17
Structures of Horizontal and Vertical Bar Graphs

Figure 1. Title   **Figure 1.** Title

When you construct bar graphs, follow six basic guidelines.

- *Make the proportions fair.* For all bar graphs, number the axes at regular intervals. If you are drawing the graph (rather than using software), use a ruler or graph paper.

  For vertical bar graphs, choose intervals that make your vertical axis about three-quarters the length of the horizontal axis. If the vertical axis is much longer than that, the differences between the height of the bars will be exaggerated. If the horizontal axis is too long, the differences will be unfairly flattened. Make all bars equally wide, and make the amount of space between them about half the width of a bar. Here are two poorly proportioned graphs.

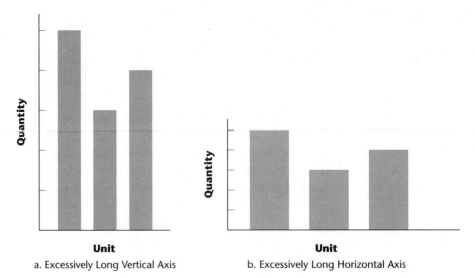

a. Excessively Long Vertical Axis   b. Excessively Long Horizontal Axis

- *If possible, begin the quantity scale at zero.* Doing so will ensure that the bars accurately represent the quantities. Notice how misleading a graph is if the scale doesn't begin at zero.

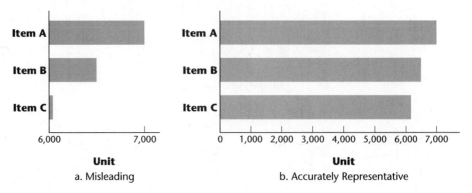

a. Misleading   b. Accurately Representative

Version (a) exaggerates the differences among items a, b, and c. If it is not practical to start the quantity scale at zero, break the quantity axis clearly.

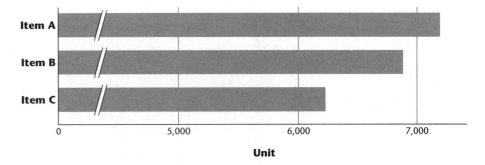

• *Use tick marks or grid lines to signal the amounts.* Ticks are the little marks drawn across the axis:

Grid lines are ticks extended through the bars:

In most cases, grid lines are necessary only if you have several bars (some of which would be too far away from tick marks for readers to gauge the quantity easily) and if the actual quantities are not indicated near the bars.

• *Arrange the bars in a logical sequence.* For a vertical bar graph, use chronology for the sequence when you can. For a horizontal bar graph, arrange the bars in order of descending size, beginning at the top of the graph, unless some other logical sequence seems more appropriate.

• *Place the title underneath the figure.* Unlike tables, which are usually read from top to bottom, figures are usually read from the bottom up.

• *Indicate the source of your information if you did not generate it yourself.*

Figure 12.18 on page 292 shows an effective bar graph.

Four variations on the basic bar graph can help you accommodate different communication needs.

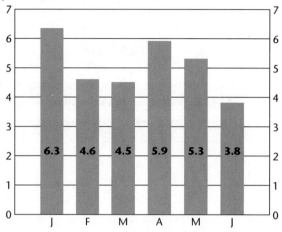

Inflation Rate (in percent)

FIGURE 12.18

Effective Bar Graph

Figure 1. Tri-County Inflation Rate This Year to Date

- The *grouped bar graph* lets you compare two or three quantities for each item you are representing. Grouped bar graphs are useful for showing information such as the numbers of full-time and part-time students at several universities. One kind of bar represents the full-time students; the other, the part-time. To distinguish the bars from each other, use hatching (striping), shading, or color, and either label one set of bars or provide a key.
- In the *subdivided bar graph,* Aspect I and Aspect II are stacked, just as wooden blocks are placed on top of one another. Although the totals are easy to compare in a subdivided bar graph, the individual quantities (except those that begin on the horizontal axis) are not.
- The *100-percent bar graph,* which shows the relative proportions of the elements that make up several items, is useful in portraying, for example, the proportion of full-scholarship, partial-scholarship, and no-scholarship students at a number of colleges.
- The *deviation bar graph* shows how various quantities deviate from a norm. Deviation bar graphs are often used when the information contains both positive and negative values, as with profits and losses. Bars on the positive side of the norm line represent profits; on the negative side, losses.

Figure 12.19 shows a grouped bar graph, a subdivided bar graph, a 100-percent bar graph, and a deviation bar graph.

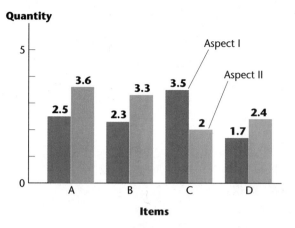

a. Grouped Bar Graph

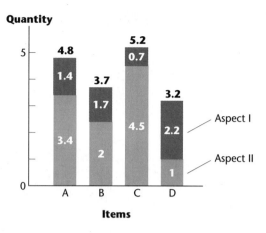

b. Subdivided Bar Graph

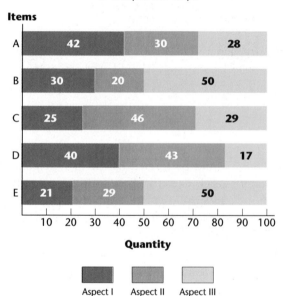

Aspect I Aspect II Aspect III

c. 100-Percent Bar Graph

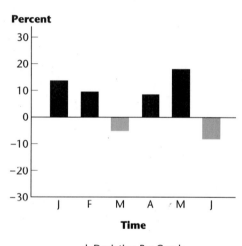

d. Deviation Bar Graph

FIGURE 12.19

Variations on the Bar Graph

Pictographs

Pictographs are bar graphs in which the bars are replaced by a series of symbols that represent the items, as shown in Figure 12.20 on page 294. Pictographs are used primarily to present statistical information for the general reader. The quantity scale is usually replaced by a statement that indicates the numerical value of each symbol. Thousands of clip-art symbols and pictures are available for creating pictographs.

Follow these two guidelines in creating pictographs:

- *In most cases, arrange pictographs horizontally rather than vertically.*
  Pictures of houses balanced on top of each other can look foolish.

Workers Using Data-Processing Equipment

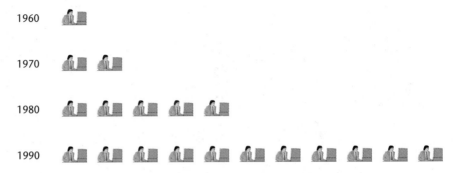

1960

1970

1980

1990

Each symbol represents one million workers

FIGURE 12.20

Pictograph

- *Represent quantities honestly.* Figure 12.21 shows that a picture drawn to scale can appear many times larger than it should. The reader sees the total area of the symbol rather than its height.

Line Graphs

In line graphs, the quantities are represented by points linked by a line. This line traces a pattern that in a bar graph would be formed by the highest point of each bar. Line graphs are used almost exclusively to show how the quantity of an item changes over time, such as the month-by-month production figures for a product. A line graph focuses the reader's attention on the change in quantity, whereas a bar graph emphasizes the quantities themselves.

An advantage of the line graph for demonstrating change is that it can convey much more data than a bar graph. Because you can plot three or four lines

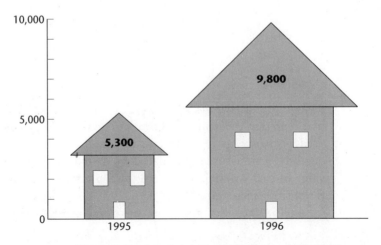

FIGURE 12.21

Misleading Pictograph

Figure 3. Housing starts in the Tri-State Area, 1995–1996

Million Barrels per Day

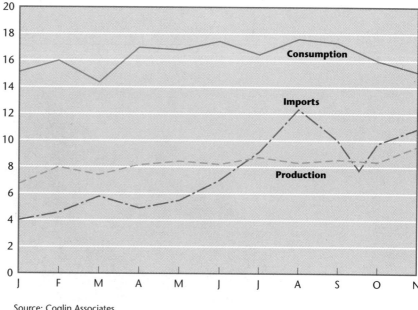

Source: Coglin Associates

FIGURE 12.22

Line Graph

Figure 1. U.S. Petroleum Consumption, Production, and Imports, January–November 1995

on the same graph, you can compare trends conveniently. If the lines intersect each other often, however, the graph will be unclear. If this is the case, draw separate graphs. Figure 12.22 shows a line graph. You can use a different color or a different pattern—or both—to distinguish the lines.

Use the following three guidelines in constructing line graphs:

- *If possible, begin the quantity scale at zero.* Just as in a bar graph, beginning at zero portrays the information most honestly. If you cannot begin at zero, clearly indicate a break in the axis.
- *Use reasonable proportions for the vertical and horizontal axes.* As with bar graphs, try to make the vertical axis about three-quarters the length of the horizontal axis.
- *Use grid lines—horizontal, vertical, or both—rather than tick marks when your readers need to read the quantities precisely.*

Two common variations on the line graph are the stratum graph and the ratio graph.

- *A stratum graph,* sometimes called an *area graph,* shows an overall change while partitioning that change into its parts. Figure 12.23 (U.S. Department of Commerce, 1994, p. 20-4) on page 296 is an example of a stratum graph. The figure shows that in 1990, DEC shipped some 30,000

Figure 26-10 U.S. RISC Workstations Shipments

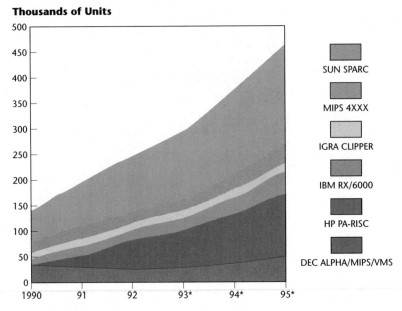

Thousands of Units

*Forecast
Note: RISC = reduced instruction set computing.
Source: InfoCorp.

FIGURE 12.23

Stratum Graph

Alpha/MIPS/VMS workstations; Sun shipped some 70,000 workstations. Together, the six companies represented in the stratum graph shipped 145,000 RISC workstations in 1990; by 1995, that figure is expected to rise to over 450,000.

- A *ratio graph* is a line graph used to emphasize percentages of change rather than the change in real numbers. A ratio graph charts data that could not be represented fairly on a standard line graph. For example, you might wish to compare the month-by-month sales of a large corporation with those of a small one. You would have great trouble making a vertical axis that would accommodate a company with monthly sales of $200,000 and one with monthly sales of $2,000,000. Even if you had a giant piece of paper, the graph could not reflect a true relation between the companies. If both companies increased their sales at the same rate (such as 1 percent per month), the small company's line would appear relatively flat, whereas the big company's line would shoot upward, just because of the large quantities involved.

To solve this problem, the ratio graph compresses the vertical axis more and more as the quantities increase. Figure 12.24 (Council, 1994, p. 5) shows how ratio graphs work. The bottom line—transfer pay-

Sources of Personal Income

Personal income increased $71.3 billion (annual rate) in February, following a decline of $18.8 billion in January. The changes were affected by a number of special factors, the most significant of which was uninsured losses to residential and business property from the California earthquake in January. Excluding the special factors, personal income increased $17.1 billion in February and $34.1 billion in January.

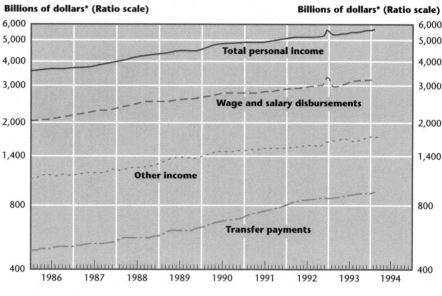

FIGURE 12.24

Ratio Graph

*Seasonally adjusted annual rates
Source: Department of Commerce

Council of Economic Advisors

ments—increases from about $520 billion in 1986 to about $920 billion in 1993. This increase is about 75 percent. During the same period, the top line, total personal income, increased from about $3.6 trillion to about $5.5 trillion, an increase of about 52 percent. Even though these quantities differ greatly, the ratio graph accurately represents how each one changes.

Pie Charts

The *pie chart* is a simple but limited design used for showing the relative size of the parts of a whole. Pie charts can be instantly recognized and understood by the untrained reader. Figure 12.25 on page 298 shows a typical pie chart.

Here are eight guidelines for creating pie charts:

- *Restrict the number of slices to six or seven.* Readers cannot easily handle more than seven slices because, as the slices get smaller, judging their sizes becomes more difficult.

- *Begin with the largest slice at the top of the pie and work clockwise in decreasing-size order, unless you have a good reason for arranging the slices in a different order.*

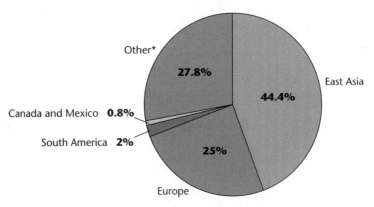

Where U.S. Tobacco Companies Export Cigarettes, 1993

*Other: Middle East, Eastern Europe, Central Asia, Central America, and the Pacific
Source: U.S. Department of Commerce

FIGURE 12.25

Pie Chart

- *Create a miscellaneous slice for very small quantities that would make the chart unclear.* Explain the contents of this slice in a footnote. This miscellaneous slice, sometimes called "other," appears after the other sections as you work in a clockwise direction.
- *Label the slices (horizontally, not radially) inside the slice, if space permits.* It is customary to include the percentage that each slice represents. Sometimes the absolute quantity is added.
- *To emphasize one of the slices, use a brightly contrasting color or separate the slice from the pie.* Do this, for example, when you want to introduce a discussion of the item represented by that slice.
- *Check to see that your software follows appropriate guidelines for pie charts.* Some software violates basic guidelines for creating pie charts. Many software programs violate the principle of simplicity; they add fancy visual effects to make them more exciting. These visual effects can actually hurt comprehension. For instance, many programs portray the pie in three dimensions, as in Figure 12.26 (Health, 1990, p. 7).

  The "shadows" misrepresent the slices to which they are attached by making them appear larger than they should be. The three-dimensional pie chart is misleading in another way: the slices at the bottom of the pie appear larger than they should because they seem to be closer to the reader. To communicate most clearly, make your pies two dimensional, not three.
- *Don't overdo fill patterns.* Fill patterns are the designs, shades, or colors that distinguish one slice from another. In general, use understated, sim-

Chart 2.

Population of Spanish origin by type of Spanish origin: United States, March 1987

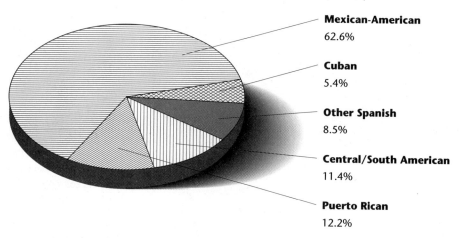

Mexican-American
62.6%

Cuban
5.4%

Other Spanish
8.5%

Central/South American
11.4%

Puerto Rican
12.2%

FIGURE 12.26

Three-Dimensional Pie Chart

Source: Bureau of the Census: Current Population Reports. Series P-20, No. 434, Page 12.

ple patterns, or use no patterns at all; instead, include information about the slice, such as raw numbers, percentages, or labels. This tactic reduces the need for legends.

- *Check to see that your percentages add up to 100.* If you are doing the calculations yourself, rather than using software, check your math.

Illustrating Logical Relationships

Graphics can help you communicate the logical relationships among different items. For instance, in describing a piece of hardware you might want to show its major components. In writing a proposal, you might want to show alternative courses of action and explain the advantages and disadvantages of each. The two kinds of graphics most useful for showing logical relationships are diagrams and organization charts.

Diagrams

A diagram is a visual metaphor, a drawing in which you use symbols to represent items or properties of items. Common kinds of diagrams in technical communication are blueprints, wiring diagrams, and schematics. Figure 12.27 on page 300 (U.S. Environmental, 1991, p. 124) is an example of a schematic diagram.

In this diagram, the artist has used visual symbols, such as the fan and the filter, to represent actual components and how they work together to perform a function. Note that diagrams are not representational. Although the silhouette of the person looks realistic, the HVAC system itself is drawn to show its func-

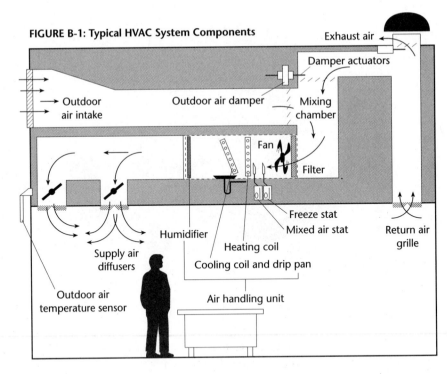

FIGURE B-1: Typical HVAC System Components

Exhaust air

Damper actuators

Outdoor air damper

Mixing chamber

Outdoor air intake

Fan

Filter

Freeze stat

Mixed air stat

Return air grille

Humidifier

Heating coil

Supply air diffusers

Cooling coil and drip pan

Outdoor air temperature sensor

Air handling unit

FIGURE 12.27

Diagram

tion, not its appearance. The clip art of the person helps the reader see the location of the HVAC equipment.

Diagrams can also clarify difficult concepts, as shown in Figure 12.28 (Labuz, 1984, p. 184). In Figure 12.28, the writer has effectively arranged a series of simple calculations to show how to convert binary numbers into decimal numbers.

A popular form of diagram is the *block diagram,* which uses simple geometric shapes—usually, rectangles—to suggest logical relationships. In Figure 12.29, notice how much clearer the block diagram is than the prose version of the same information.

FIGURE 12.28

Diagram Used to Clarify
a Difficult Concept

Converting a Binary Number into a Decimal Number

Binary Number:	1	1	1	0		
	8	4	2	1		
	x1	x1	x1	x0		
	8	4	2	0		
Total:	8 +	4 +	2 +	0	=	14 Decimal Number

Prose Version

DFC is composed of two modules: the
ADM and the COT. The ADM
microcode consists of three
components: the 120 protocol, the ADM
control, and the TCG module.

Block-Diagram Version

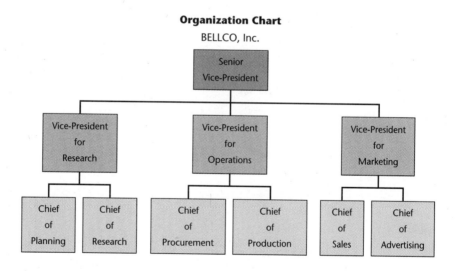

Block Diagram

Organization Charts

An organization chart is a form of block diagram that portrays the lines of
authority and responsibility in an organization. In most cases, the positions are
represented by rectangles. The more important positions can be emphasized
through the size of the boxes, the width of the lines that form the boxes, the
typeface, or color. If space permits, the boxes themselves can include brief
descriptions of the positions, duties, or responsibilities. Figure 12.30 is a typical
organization chart. Unlike most other figures, the title of an organization chart
generally appears above the chart because the chart is read from the top down.

Organization Chart

**Illustrating
Instructions and
Process
Descriptions** Graphics often accompany instructions (Chapter 18) and process descriptions
(Chapter 10). The following discussion covers some of the kinds of graphics
used in writing about actions: checklists, tables, flowcharts, logic boxes, and
logic trees.

Checklists

In explaining how to carry out a task, you often need to show the reader what equipment or materials to gather, or describe an action or series of actions to take. A *checklist* is simply a list of items, with check boxes before each one. If your reader might be unfamiliar with the items you are listing, it's a good idea to include drawings of the items, as shown in Figure 12.31.

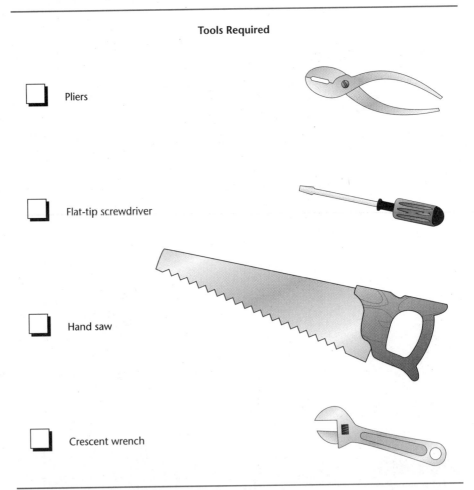

Tools Required

☐ Pliers

☐ Flat-tip screwdriver

☐ Hand saw

☐ Crescent wrench

FIGURE 12.31

Checklist

Often you need to indicate that your reader is to carry out certain tasks at certain intervals. A table is a useful graphic for showing this kind of information, as illustrated in Figure 12.32 (McComb, 1991, p. 133).

In Figure 12.32, the darkened check boxes indicate when an action is to be performed. For example, a VCR subjected to heavy use should have its exterior and interior cleaned and dusted at 6 months, 12 months, 18 months, 24 months, 30 months, and 36 months.

Light Use

PM (Months)	3	6	9	12	15	18	21	24	27	30	33	36
Clean/Dust Exterior/Interior												
Clean Heads												
Lubricate Transport												
Clean Rubber Parts												
Clean Remote Battery Terminals												
Clean Connectors												

Medium Use

PM (Months)	3	6	9	12	15	18	21	24	27	30	33	36
Clean/Dust Exterior/Interior												
Clean Heads												
Lubricate Transport												
Clean Rubber Parts												
Clean Remote Battery Terminals												
Clean Connectors												

Heavy Use

PM (Months)	3	6	9	12	15	18	21	24	27	30	33	36
Clean/Dust Exterior/Interior												
Clean Heads												
Lubricate Transport												
Clean Rubber Parts												
Clean Remote Battery Terminals												
Clean Connectors												

Your Machine

PM (Months)	3	6	9	12	15	18	21	24	27	30	33	36
Clean/Dust Exterior/Interior												
Clean Heads												
Lubricate Transport												
Clean Rubber Parts												
Clean Remote Battery Terminals												
Clean Connectors												

FIGURE 12.32

A Table Used to Communicate a Maintenance Schedule

Figure 5-1. Preventative maintenance schedule for VCRs subjected to light, medium, and heavy use.

Flowcharts

A flowchart, as its name suggests, shows the stages of a procedure or a process. A flowchart might be used, for example, to show the steps involved in transforming lumber into paper or in synthesizing an antibody. Flowcharts are useful, too, for summarizing instructions. The basic flowchart portrays stages with labeled rectangles or circles. If you are addressing general readers, consider using pictorial symbols instead of geometric shapes. If the process involves quantities (for example, paper manufacturing might "waste" 30 percent of the lumber), they can be listed or merely suggested by the thickness of the line used to connect the stages. Flowcharts can portray *open systems* (those that have a "start" and a "finish") or *closed systems* (those that end where they began).

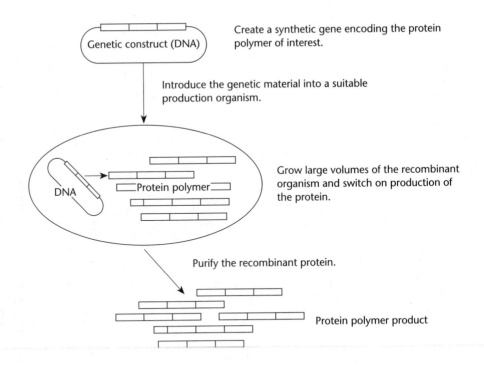

Create a synthetic gene encoding the protein polymer of interest.

Genetic construct (DNA)

Introduce the genetic material into a suitable production organism.

DNA   Protein polymer

Grow large volumes of the recombinant organism and switch on production of the protein.

Purify the recombinant protein.

Protein polymer product

FIGURE 12.33

Open-System Flowchart

SOURCE: Bioinformation Associates, Boston, MA.

Figure 2-3—Production of Recombinant Protein Polymers

Figure 12.33 (U.S. Congress, *Biopolymers*, 1993, p. 25) shows an open-system flowchart explaining the process of producing recombinant protein polymers. Figure 12.34 (McKay, McKay, & Duke, 1992, p. 236) shows a closed-system flowchart.

Figure 32
Regenerative Life Support System

In a controlled ecological life support system, as diagrammed here, biological and physico-chemical subsystems would produce plants for food and process wastes for reuse in the system. A bioprocessing unit, in which bacteria oxidize and catalyze the extraction of metals from their lunar or asteroidal ores, could be incorporated into this system. The bioprocessing unit would contribute to the gas and nutrient recycling, the biomass inventory, and the waste processing of the life support system.

FIGURE 12.34

Closed-System Flowchart

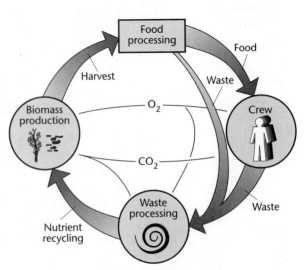

Logic Boxes

Logic boxes show the reader which of two or more paths to follow. Figure 12.35 shows a typical example.

Do you want to replace all occurrences of the first word with the second word?		
YES	NO	
Press Y. The computer replaces all occurrences of the first word with the second word.	Press N. The cursor moves to the next occurrence of the word.	
	Do you want to replace that occurrence of the word?	
	YES	NO
	Press SHIFT/F2.	Press F2.
	Press F4 to move the cursor to the next occurrence of the word. Then press SHIFT/F2 or F2, as you did above, to replace that occurrence of the word or to leave it as is.	
Do you want to use the search function again?		
YES	NO	
Press F8.	Press F10.	

FIGURE 12.35
Logic Boxes

Logic Trees

Logic trees are like logic boxes, but they use the metaphor of branches on a tree. Figure 12.36 shows a logic tree that helps students think through the process of registering for a course.

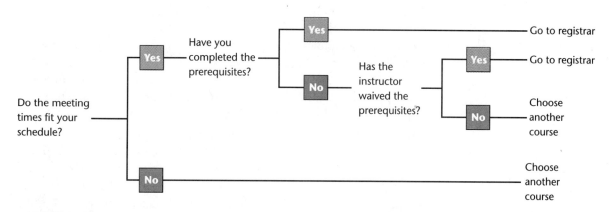

FIGURE 12.36
Logic Trees

Logic trees are probably somewhat easier to understand than logic boxes because the tree metaphor is visually more distinct than the box metaphor. However, logic boxes can accommodate more textual information than logic trees can.

Illustrating Visual and Spatial Characteristics

To show visual and spatial characteristics, use photographs, screen shots, line drawings, and maps.

Photographs

Photographs are unmatched for reproducing visual detail. If you want to show realistically the different kinds of tire-tread wear caused by various alignment problems, a photograph is best. If you want your readers to recognize a new product such as a new automobile model, you will probably use a photograph. And recent advances in specialized kinds of photography—especially in internal medicine and biology—are expanding the possibilities of the art.

Ironically, however, sometimes a photograph can provide too much information. In an advertising brochure for an automobile, a glossy photograph of the dashboard might be very effective. But if you are creating an owner's manual and you want to show how to find the trip odometer, a diagram will probably work better because it lets you focus on the one item you want to emphasize. Sometimes a photograph can provide too little information; the item you want to show can be inside the mechanism or obscured by some other component.

Taking pictures to be used in publications is best left to professionals, because including photographs in a document involves many complicated technical details. On occasion you will want to use existing photographs in a document; if so, follow these five basic guidelines:

- *Do not electronically manipulate the photograph.* Today it is routine to digitize a photograph, remove blemishes, and crop the photograph (cut it down to a desired size). There is nothing unethical about doing so. Figure 12.37 shows examples of cropped and uncropped photographs. However, manipulating the photograph—for example, enlarging the size of the monitor that comes with the computer system shown in the photograph—is unethical. As consumers we have all been angered and frustrated when we take a product out of the box and realize that it bears only a faint resemblance to the photograph on the box.

- *Help the reader understand the perspective.* If for some reason you are using an unexpected angle, tell your reader the orientation. For instance, state that you are using a top view. Figure 12.38 shows the top view of a computer printer. Most objects portrayed in magazines and journals are photographed from an angled view; this perspective provides more infor-

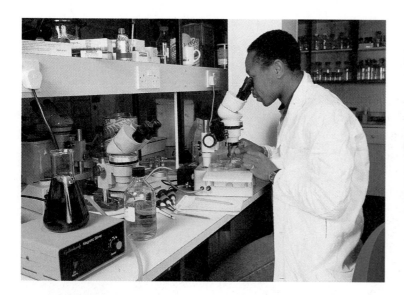

FIGURE 12.37

Cropping a Photograph
Emphasizes the Impor-
tant Information

mation because it shows the depth of the object as well as its height and
width.

- *If appropriate, include in the photograph some common object, such as a
  coin or a ruler, to give a sense of scale.*
- *Eliminate extraneous background clutter that can distract your reader.*
  Do this by cropping the photograph or by digitizing it and deleting
  unnecessary detail.
- *If appropriate, label components or important features.*

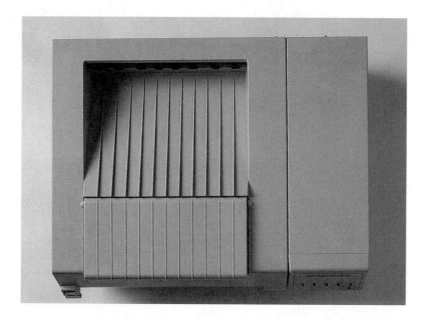

FIGURE 12.38

Photograph Taken from
an Unexpected Angle

Screen Shots

A screen shot is a computer graphic of what appears on the computer monitor. Screen shots are often used in software manuals to show the reader what the screen looks like at various points during the use of a program. The reader is reassured to see that the image on the screen looks like the screen shot in the documentation. Screen shots also help the reader locate information on the screen. When printing screen shots, reduce them to about 50 to 75 percent of their actual size. Figure 12.39 (Wu & Narasimhalu, 1994, p. 36) is an example of a screen shot.

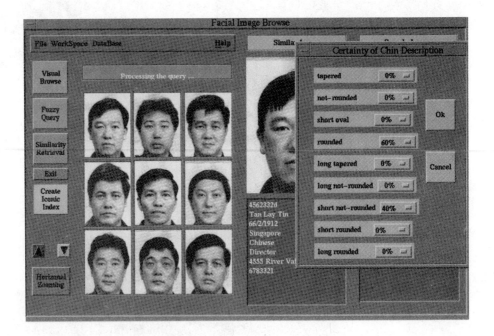

FIGURE 12.39

Screen Shot

Line Drawings

Line drawings are simplified visual representations of objects. Line drawings can have three advantages over photographs:

- Line drawings can focus the reader's attention on desired information better than a photograph can.
- Line drawings can highlight information that might be obscured by bad lighting or a bad angle in a photograph.
- Line drawings are sometimes easier for readers to understand than photographs are.

A line drawing such as the one in Figure 12.40 (Briggs & Stratton, 1989) shows readers enough of the lawnmower to orient them and highlights the important

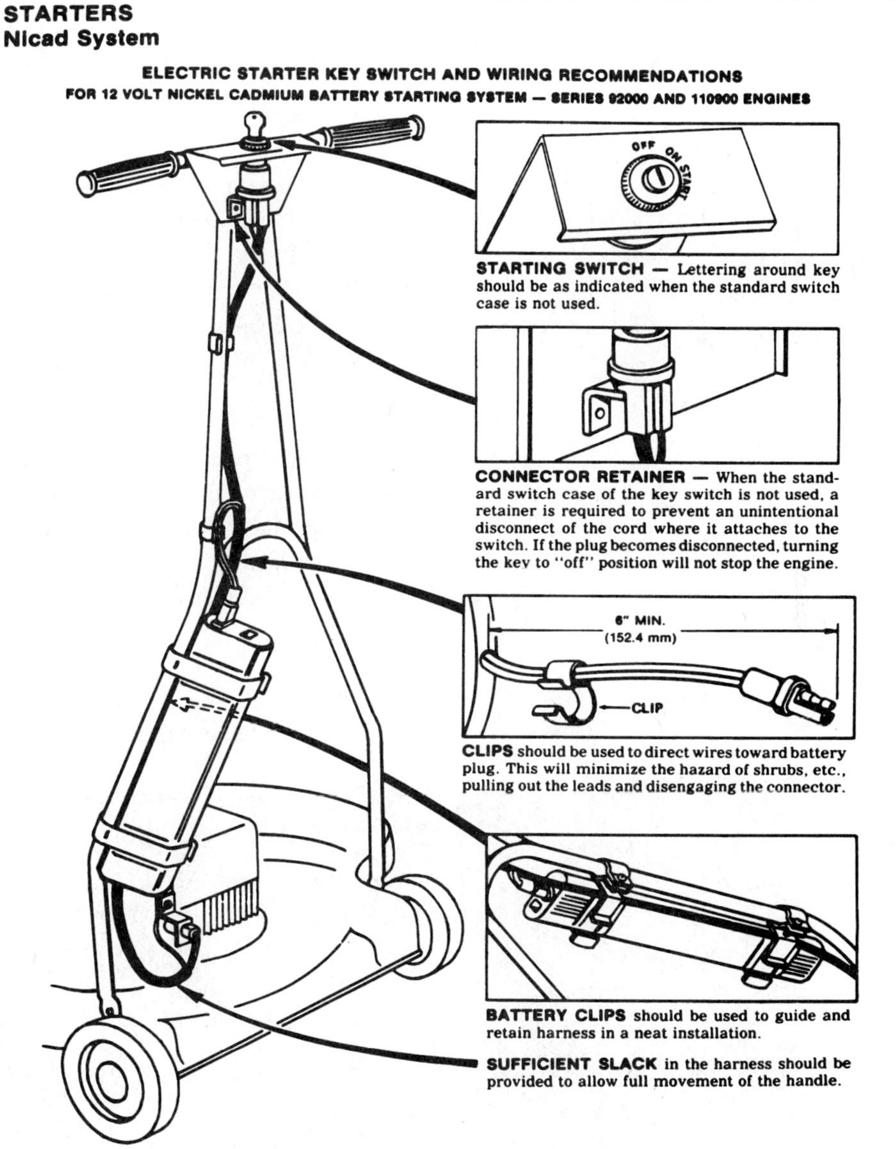

STARTERS
Nicad System

ELECTRIC STARTER KEY SWITCH AND WIRING RECOMMENDATIONS
FOR 12 VOLT NICKEL CADMIUM BATTERY STARTING SYSTEM — SERIES 92000 AND 110900 ENGINES

STARTING SWITCH — Lettering around key should be as indicated when the standard switch case is not used.

CONNECTOR RETAINER — When the standard switch case of the key switch is not used, a retainer is required to prevent an unintentional disconnect of the cord where it attaches to the switch. If the plug becomes disconnected, turning the key to "off" position will not stop the engine.

CLIPS should be used to direct wires toward battery plug. This will minimize the hazard of shrubs, etc., pulling out the leads and disengaging the connector.

BATTERY CLIPS should be used to guide and retain harness in a neat installation.

SUFFICIENT SLACK in the harness should be provided to allow full movement of the handle.

FIGURE 12.40

Line Drawing

information. Notice how close-ups effectively show details that could not be seen from an ordinary perspective.

You have probably seen three variations on the common drawing. *Phantom drawings* show parts that are hidden from view by outlining external items that obscure their view. *Cutaway drawings* "remove" a part of the surface to expose what is underneath. *Exploded drawings* separate components while maintaining their physical relationship. Figure 12.41 on page 310 shows phantom, cutaway, and exploded views.

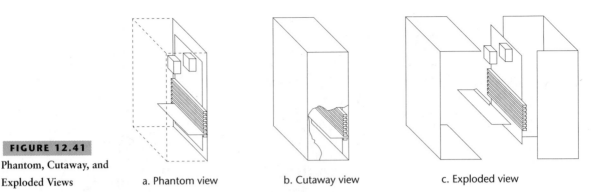

a. Phantom view   b. Cutaway view   c. Exploded view

Maps

Maps are readily available as clip art that can be modified with a graphics program. Include a scale and a legend if the map is not thoroughly familiar to your
readers. Also, use conventional colors, such as blue for water. Figure 12.42
shows a map made from clip art.

Showing Motion in Your Graphics

Often in technical communication you want to show motion. For instance, in an instruction manual for military helicopter technicians, you might want to show the process of removing an oil dipstick or tightening a bolt. Or you might want to show a warning light flashing. Although animation or video is often used today to show these sorts of processes, printed graphics are still needed to communicate this information. Following are two guidelines for showing motion:

- *If the reader is to perform the action, show the action from the reader's point of view.* In most cases, you need to show only the person's hands, not the whole body, as in Figure 12.43.

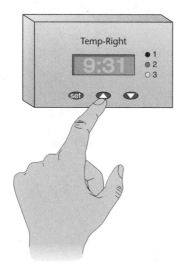

FIGURE 12.43

Showing the Action from the Reader's Perspective

- *Use arrows or other symbols to show actions.* Arrows show the direction in which something is moving or should be moved; shake lines suggest vibration; starburst lines suggest a blinking light. Another way to suggest motion is to show an object both before and after the motion. Figure 12.44 on page 312 shows these different symbols.

  Note that these symbols are conventional, not universal. If you are addressing international readers, consult a qualified person from your target culture to make sure the symbols you want to use are clear and inoffensive.

Creating Graphics for International Readers

Whether you are writing to people inside or outside your organization, consider the needs of readers who do not speak English as their primary language. If you

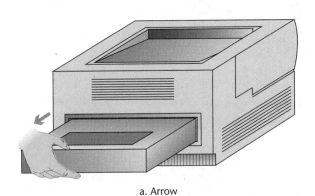

a. Arrow

c. Shake Lines

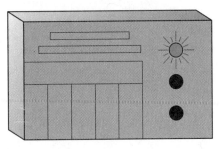

b. Starburst Lines

d. Before and After

FIGURE 12.44

Showing Motion

communicate with people in other countries, be aware that many do not speak English at all. In 1990, almost 32 million Americans spoke a language other than English at home (U.S. Bureau, 1993, p. 51). With the rate of immigration rising each year, that figure will surely increase.

Graphics are excellent tools for communicating with an international reader, but use them carefully. Like words, graphics have cultural meanings, and if you are unaware of them, you can communicate something very different from what you had intended. The following discussion, based on William Horton's article "The Almost Universal Language: Graphics for International Documents" (1993), presents six guidelines for using graphics when addressing an international audience:

- *Understand that there are different reading patterns.* In most countries, people read from left to right, but in some they read right to left or top to bottom. In some cultures, direction signifies value: the right-hand side is superior to the left, or the reverse. Therefore, you need to think about how to sequence graphics that show action, or where you put "before" and "after" graphics. If you want to show a direction, as in an informal flowchart, consider using arrows to indicate the direction in which the chart should be read.

- *Understand that there are different cultural attitudes toward giving instruction.* You might have noticed that instructions for products made in Japan are highly polite and deferential: "Please insert the batteries at this time." Some cultures favor spelling out general principles but leaving the reader to infer the details. In instructions written for people from these cultures, a detailed close-up of how to carry out a task might appear insulting. An instructional table with headings "When You See This . . ." and "Do This . . ." might be inappropriate.

- *Deemphasize trivial details.* Common objects such as plugs on the ends of power cords can take any number of shapes around the world; therefore, draw them so that they look generic rather than specific to one country.

- *Avoid culture-specific language, symbols, and references.* Don't use a picture of a mouse to symbolize a computer mouse, for the device is not known by that name everywhere. Avoid the casual use of national symbols (such as the maple leaf or national flags); you might make an error in a detail and unknowingly insult your readers. Use colors carefully: red means danger to most people from a Western culture, but it is a celebratory color to the Chinese. Don't portray an executive sitting in an office with a big window; the window is a sign of prestige in most cultures, but in Japan a "window watcher" is an employee who has disappointed the organization and is being paid to gaze out the window, not to do any productive work.

- *Portray people very carefully.* Every aspect of people's appearance, from their clothing to their hairstyles to their features, is culture-specific or race-specific. A photograph of a woman in casual Western attire seated at a workstation would not be effective in Saudi Arabia, for in Islamic cultures only the hands and eyes of a woman may be shown. Horton (1993) recommends using stick figures or silhouettes that do not reveal culture, race, or gender.

- *Be particularly careful in portraying hand gestures.* Many Western hand gestures, such as the "okay" sign, or the palm-forward gesture used by a traffic officer to stop cars, are obscene in other cultures. And long red fingernails would not be appropriate in many cultures. For this reason, use hands only when necessary, for example, when they are carrying out a task. Disguise the person's sex and race. When communicating with Arabic readers, try not to display the left hand, which they reserve for personal hygiene.

Cultural differences are many and subtle, and sometimes it seems as if everything you want to portray will offend someone. The best advice is to learn as much as possible about your readers and try to consider their habits and outlook. For more information on this subject, consult these three books:

Edward Hall. (1990). *Understanding cultural differences.* Yarmouth, ME: Intercultural Press.

Roger Axtel. (1991). *Gestures: The do's and taboos of body language around the world.* New York: John Wiley.

Scott Jones, Cynthia Kennelly, Claudia Mueller, et al. (1992). *Developing international user information.* Bedford, MA: Digital Press.

Also see Chapters 4 and 15 for additional discussions of writing to international readers.

Understanding Graphics Software

Two basic categories of graphics software are available for the personal computer:

- spreadsheet business graphics
- paint programs and draw programs

Spreadsheet Business Graphics

Spreadsheet programs—software created to help businesspeople calculate budgets and evaluate hypothetical business scenarios—offer business graphics capabilities. After you type in the numerical data you want to portray, such as the profits and losses of the four divisions of your company, you can ask the software to portray the data in various kinds of graphics and charts.

Spreadsheets make different kinds of tables, bar graphs, line graphs, and pie charts, in both two-dimensional and three-dimensional formats. You then add labels and a title and customize the graphic to suit your needs. Figure 12.45 shows how data entered on a spreadsheet can be displayed as a table and then as a bar chart. Most spreadsheets can display the same numerical data in dozens of different styles of graphics.

Some spreadsheets also let you produce slide shows; that is, the software presents the different graphics you have created, one after the other. You can choose the length of time that each graphic remains on the screen (such as 10 seconds) or you can advance to the next graphic by clicking the mouse.

Be careful in using spreadsheet business graphics, because the software doesn't provide advice on what kind of graphic to make or how to modify the basic presentation; it just draws whatever you ask it to. The software often

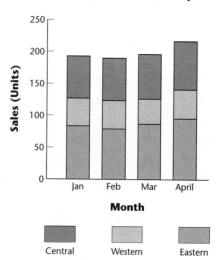

Regional Sales, Jan–April

	Eastern	Western	Central
Jan	83.5	43.7	65.6
Feb	78.9	44.6	66.3
Mar	86.7	39.5	69.5
April	95.3	45.7	75.4

FIGURE 12.45

Creating a Graphic from a Spreadsheet

a. Spreadsheet data

b. Grouped bar chart created from the spreadsheet data

makes poor choices: unnecessary 3-D, useless clip art, and lurid color combinations. To use the graphics capabilities effectively, you need to understand the basics of the different kinds of graphics.

Paint Programs and Draw Programs

Paint programs and draw programs let you create and then modify freehand drawings in a number of ways. For instance, you can

- modify the width of the lines used
- modify the size of the shapes you create
- copy, rotate, and flip images
- fill in shapes with different colors and textures
- add text with a simple text editor

With both kinds of programs, you can start in three different ways:

- By drawing on a blank screen.
- By importing an image from a clip-art library.
- By scanning an image. Scanning an image involves using a special piece of hardware—a scanner—and a special software program to translate a graphic on a piece of paper into a computerized image.

What is the difference between paint programs and draw programs?

Paint programs use pixel-based technology (also called bit-map technology). A pixel, short for *picture element,* is the smallest unit of space on the computer screen that can be treated as an individual element. In a paint program, each pixel is dark or lit (on a monochrome monitor) or dark or lit with a different shade of gray or color (on a gray scale monitor or color monitor, respectively).

One disadvantage of paint programs is that an enlarged image will look washed out and blurry, like a photograph that has been enlarged too much in a newspaper. In the larger image, you will notice the little dots that make up the picture. One way to avoid this problem is to create the image larger than you want it to appear and reduce it on a photocopy machine.

Draw programs use vector technology. In a draw program, you work with preexisting shapes, called graphics primitives, as shown here.

To create an image, manipulate the size and shape of the graphics primitives. The computer remembers the shapes you create by storing in memory the beginning and ending points of the lines that make up the shapes. These beginning and ending points are called vectors.

Figure 12.46 shows how paint programs and draw programs represent the letter *I*.

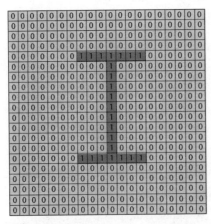

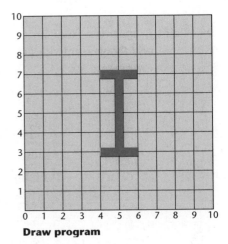

Paint program

Draw program

FIGURE 12.46

How an Image Is Represented in a Paint Program and a Draw Program

Draw programs are considered more sophisticated than paint programs for two reasons:

- *Vector information is independent of the display device.* You can output the image in any size or display it on any monitor without affecting its resolution.

- *Draw programs allow you to create more sophisticated three-dimensional effects.* If you draw a circle and a square, you can place the circle in front of the square. The computer recognizes and remembers the circle and the square as different shapes. With a paint program, the two shapes would merge; the pixels where the two original shapes overlap would assume whichever pattern is placed on them more recently, and you could not separate the two original shapes.

For these reasons, graphic artists and others use vector technology in sophisticated draw programs such as those used in CAD applications. As Figure 12.47 illustrates, you can create the same basic graphic with either a paint program or a draw program, but with a draw program you can manipulate the elements.

FIGURE 12.47

Images Created with a
Paint Program and a
Draw Program

Paint program   **Draw program**

This checklist focuses on the characteristics of an effective graphic.

1. Does the graphic have a purpose?
2. Is the graphic honest?
3. For an existing graphic, do you have the legal right to use it? If so, have you cited it appropriately?
4. Does the graphic present a manageable amount of information?
5. Does the graphic meet the reader's format expectations?
6. Is the graphic simple and uncluttered?
7. Is the graphic clearly labeled?
8. Is the graphic placed in an appropriate location in the document?
9. Is the graphic introduced clearly in the text?
10. Is the graphic explained in the text?
11. Is the graphic clearly visible in the text?
12. Is the graphic easily accessible to your readers?

EXERCISES

1. Find out from the admissions department of your college or university the numbers of students from the different states, or from the different counties in your state. Communicate this information using four different graphics:
   a. a map
   b. a table
   c. a bar graph
   d. a pie chart
   In three or four paragraphs, explain how the different graphics would be appropriate for different audiences and purposes, and how they emphasize different aspects of the information.

2. Create a flowchart for a process you are familiar with, such as registering for courses, applying for a summer job, studying for a test, preparing a paper, or performing some task at work. Your audience is someone who will be carrying out the process portrayed in the graphic.

3. Create an organizational chart for an organization you belong to or are familiar with: a department at work, a fraternity or sorority, a campus organization, or student government.

4. In two or three paragraphs, define an audience and purpose and then make a drawing of some object you are familiar with, such as a tennis racquet, a stereo speaker, or a weight bench. Be certain that the object and the drawing are appropriate to your audience and purpose.

5. In two or three paragraphs, define an audience and purpose and then draw a diagram of some concept, such as the effects of smoking or the advantages of belonging to the student chapter of a professional organization. Be certain that the concept and the diagram are appropriate to your audience and purpose.

6. Create a graphic for one of the following scenarios:
   a. In a user's manual for a computer, how to insert the floppy disk.
   b. In the owner's manual for a bicycle, how to locate its serial number.
   c. In a training manual for new waitpersons, a guide to explaining the menu to customers.
   d. In a set of instructions on how to change the oil in a car, a checklist for the equipment and materials you will need.
   e. In a student handbook at your college or university, the process of applying for graduation.

7. Create a set of graphics to complement a set of instructions on how to use an automated-teller machine. Your readers are nonnative speakers of English who are unfamiliar with automated-teller machines. For each graphic you create, write a one- or two-sentence explanation of what the text would be describing.

8. In the *Statistical Abstract of the United States* or an issue of *Forbes* or *Business Week,* find and photocopy a bar graph, a line graph, a pictograph, a pie chart, a flowchart, a photograph, and a diagram. For each graphic, write a brief discussion that responds to the following questions:
   a. Is the graphic necessary?
   b. Is it ethical, or does it misrepresent the information?
   c. Is it professional in appearance?
   d. Does it conform to the guidelines for that kind of graphic?
   e. Is it effectively integrated into the discussion?

9. The table on page 319 (U.S. Bureau, 1993, p. 134) provides statistics on AIDS in the United States. Study the table and then perform the following tasks.
   a. Create two different graphics, each of which communicates information about the incidence of AIDS among different age groups.
   b. Create two different graphics, each of which communicates information about the incidence of AIDS in women and in men.
   c. Create two different graphics, each of which communicates information about the incidence of AIDS among different racial/ethnic groups.

10. For each of the following graphics, write a paragraph evaluating its effectiveness and describing how you would revise it.
   a.

	1992	1993	1994
Civil Engineering	236	231	253
Chemical Engineering	126	134	142
Comparative Literature	97	86	74
Electrical Engineering	317	326	401
English	714	623	592
Fine Arts	112	96	72
Foreign Languages	608	584	566
Materials Engineering	213	227	241
Mechanical Engineering	196	203	201
Other	46	42	51
Philosophy	211	142	151
Religion	86	91	72

   b.

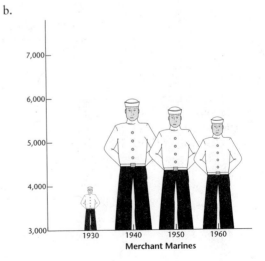

Merchant Marines

No. 203. AIDS Cases Reported, by Patient Characteristic: 1981 to 1992
[Provisional. For cases reported in the year shown. For data on AIDS deaths, see table 131. Data are subject to retrospective changes and may differ from those data in table 201]

Characteristic	Number of Cases										Percent Distribution	
	Total	1981–1984	1985	1986	1987	1988	1989	1990	1991	1992	1981–1984	1992
Total	**244,939**	**7,354**	**8,210**	**13,147**	**21,088**	**30,719**	**33,595**	**41,653**	**43,701**	**45,472**	**100.0**	**100.0**
Age:												
Under 5 years old	3,265	91	112	155	266	443	494	580	521	603	1.2	1.3
5 to 12 years old	765	5	17	28	56	128	102	143	146	140	0.1	0.3
13 to 29 years old	47,307	1,609	1,680	2,821	4,359	6,343	6,726	8,114	7,852	7,803	21.9	17.2
30 to 39 years old	112,084	3,446	3,856	6,100	9,617	14,118	15,524	18,944	19,954	20,525	46.9	45.1
40 to 49 years old	56,235	1,550	1,709	2,703	4,519	6,524	7,329	9.701	10,643	11,557	21.1	25.4
50 to 59 years old	17,835	536	631	964	1,562	2,146	2,413	2,934	3,246	3,403	7.3	7.5
Over 60 years old	7,448	117	205	376	709	1,017	1,007	1,237	1,339	1,441	1.6	3.2
Sex:												
Male	217,141	6,844	7,622	12,099	19,256	27,435	29.942	36,770	38,013	39,160	93.1	86.1
Female	27,798	510	588	1,048	1,832	3,284	3,653	4,883	5,688	6,312	6.9	13.9
Race/ethnic group:												
White[1]	132,573	4,331	4,968	7,837	12,962	17,063	18,569	22,325	22,193	22,325	58.9	49.1
Black[1]	75,929	1,951	2,081	3,388	5,379	9,118	10,293	13,220	14,609	15,890	26.5	34.9
Hispanic	33,827	1,039	1,099	1,803	2,543	4,260	4,358	5,658	6,405	6,662	14.1	14.7
Other/unknown	2,610	33	62	119	204	278	375	450	494	595	0.4	1.3

X Not applicable.   [1]Non-Hispanic.
Source: U.S. Centers for Disease Control, Atlanta, GA, unpublished data.

c.

d. Costs of the Components of a PC

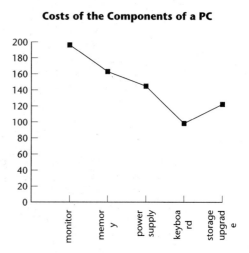

Expenses at Hillway Corporation

11. In each of the following exercises, translate the written information into at least two different kinds of graphics. For each exercise, write a two- or three-paragraph statement about which kinds work best. If one kind works well for one audience but not so well for another, explain.

a. Following are the profit and loss figures for Pauley, Inc., in early 1994: January, a profit of 6.3 percent; February, a profit of 4.7 percent; March, a loss of 0.3 percent; April, a loss of 2.3 percent; May, a profit of 0.6 percent.

b. The prime interest rate had a major effect on our sales. In January, the rate was 4.5 percent. It went up a full point in February, and another half point in March. In April, it leveled off, and it dropped one full point each in May and June. Our sales figures were as follows for the Crusader 1: January, 5,700; February, 4,900; March, 4,650; April, 4,720; May, 6,200; June, 8,425.

c. This year, our membership can be broken down as follows: 45 percent, electrical engineering; 15 percent, mechanical engineering; 20 percent, materials engineering; 10 percent, chemical engineering; and 10 percent, other.

d. In January of this year we sold 50,000 units of the BG-1, of which 20,000 were purchased by the army. In February, the army purchased 15,000 of our 60,000 units sold. In March, it purchased 12,000 of the 65,000 we sold.

e. The normal rainfall figures for this region are as follows: January, 1.5 in.; February, 1.7 in.; March, 1.9 in.; April, 2.1 in.; May, 1.8 in.; June, 1.2 in.; July, 0.9 in.; August, 0.7 in.; September, 1.3 in.; October, 1.1 in.; November, 1.0 in.; December, 1.2 in. The following rainfall was recorded in this region: January, 2.3 in.; February, 2.6 in.; March, 2.9 in.; April, 2.0 in.; May, 1.6 in.; June, 0.7 in.; July, 0.1 in.; August, 0.4 in.; September 1.3 in.; October, 1.2 in.; November, 1.4 in.; December, 1.8 in.

f. You can access the Rightfile from programs written in six different languages. Rightfile classifies these languages as two groups (A and B) and provides separate files for each group. The A group includes C, PL/1, and Fortran. The B group includes COBOL and RPGII. The object module for the A group is GRTP. The object module for the B group is GRAP.

12. Interview three different people: an engineer or scientist, a technical communicator, and a graphic artist or illustrator who works in technical communication. Find out about their different approaches to creating graphics for technical documents. Specifically, ask about their education in graphics, their use of tools or software, the way they integrate graphics into the process of writing a document, and their views of the importance of graphics in the work they do. Write your results in a memo to your instructor. See Chapter 16 for a discussion of memos.

13. Locate a graphic that you feel would be inappropriate for an international audience because it might be offensive or unclear to some cultures. Choose an intended audience, such as people from the Middle East, and write a brief statement explaining the potential problem. Finally, revise the graphic so that it would be appropriate for its intended audience.

14. At one of your campus's computer centers, study a popular piece of software, such as a spreadsheet or graphics package. Print out three of the basic kinds of graphics, such as graphs, tables, and pie charts. Write a memo to your instructor, describing how easy or difficult it was to learn how to create the graphics and evaluating the quality of the graphics produced. How well do the graphics conform to the basic guidelines presented in this chapter? Would you recommend the software to other members of your class? Why or why not? See Chapter 16 for a discussion of memos.

REFERENCES

Barnum, C. M., & Carliner, S. (1993). *Techniques for technical communicators.* New York: Macmillan.

Briggs & Stratton Corporation. (1989). *Briggs & Stratton service and repair instructions.* Part #270962.

Brockmann, R. J. (1990). *Writing better computer user documentation: From paper to hypertext.* New York: Wiley.

Council of Economic Advisers. (1994, March). *Economic indicators.* Washington, DC: U.S. Government Printing Office.

Dean, R. S., & Kulhavy, R. W. (1981). Influence of spatial organization in prose learning. *Journal of Educational Psychology, 73,* 57–64.

Gatlin, P. L. (1988). Visuals and prose in manuals: The effective combination. In *Proceedings of the 35th International Technical Communication Conference* (pp. RET 113–115). Arlington, VA: Society for Technical Communication.

Grimstead, D. (1987). Quality graphics: Writers draw the line. *Proceedings of the 34th International Technical Communication Conference* (pp. VC 66–69). Arlington, VA: Society for Technical Communication.

Health status of the disadvantaged chartbook. (1990). U.S. Department of Health and Human Services. DHHS Publication HRS-P-DV-90-1, p. 7.

Horton, W. (1993). The almost universal language: Graphics for international documents. *Technical Communication, 40,* 682–693.

Horton, W. (1991). *Illustrating computer documentation: The art of presenting information graphically on paper and online.* New York: Wiley.

Jain, R. (1994, Fall). What is multimedia, anyway? *IEEE MultiMedia, 1,* 3.

Labuz, R. (1984). *How to typeset from a word processor: An interfacing guide.* New York: R. R. Bowker.

Levie, W. H., & Lentz, R. (1982). Effects of text illustrations: A review of research. *Journal of Educational Psychology, 73,* 195–232.

McComb, G. (1991). *Troubleshooting and repairing VCRs* (2nd ed.). Blue Ridge Summit, PA: Tab Books.

McGuire, M., & Brighton, P. (1990). Translating text into graphics. Session at the 37th International Technical Communication Conference. Dallas, TX.

McKay, M. F., McKay, D. S., & Duke, M. B. (1992). *Space resources, vol. 3. Materials.* Washington, DC: National Aeronautics and Space Administration.

Morrison, C., & Jimmerson, W. (1989, July). Business presentations for the 1990s. *Video Manager, 4,* 18.

Tufte, E. R. (1983). *The visual display of quantitative information.* Cheshire, CT: Graphics Press.

U.S. Bureau of the Census. (1993). *Statistical abstract of the United States: 1993.* Washington, DC: U.S. Government Printing Office.

U.S. Congress, Office of Technology Assessment. (1993). *Biopolymers: Making materials nature's way—Background paper,* OTA-BP-E-102. Washington, DC: U.S. Government Printing Office.

U.S. Congress, Office of Technology Assessment. (1993). *Making government work: Electronic delivery of federal services,* OTA-TCT-578. Washington, DC: U.S. Government Printing Office.

U.S. Department of Commerce. (1994). *U.S. industrial outlook 1994: Forecasts for selected manufacturing and service industries.* Washington, DC: U.S. Government Printing Office.

U.S. Environmental Protection Agency. (1991). *Building air quality: A guide for building owners and facility managers.* Washington, DC: U.S. Government Printing Office.

White, J. V. (1990). *Color for the electronic age.* New York: Watson-Guptill.

White, J. V. (1984). *Using charts and graphs: 1000 ideas for visual persuasion.* New York: R. R. Bowker.

Wu, J. K., & Narasimhalu, A. D. (1994). Identifying faces using multiple retrievals. *IEEE MultiMedia, 1*(2), pp. 27–38.

CHAPTER 13 Integrating Design Elements

Most of us groan inwardly when we have to read a document made up of gray pages: all paragraphs, no graphics, and skinny margins. As readers, we fear that such a document is going to be both difficult and boring—a bad combination.

To a large extent, the effectiveness of a document depends on how well it is designed. Of course, the information it contains is critical, but before your readers actually read and understand a document, they *see* it. In less than a second, the document has started to make an impression on them, and that impression might determine how effectively they read it. (A first impression might even determine *whether* they read it at all.) Of course, your goal is not to entertain your readers with flashy colors and outrageous designs, like those in mass-market magazines. Still, an understanding of effective design principles can help you make a document that your readers will want to read—and that will help them understand and remember what you say.

Design principles are not new, but we are becoming more and more conscious of how powerful they are, largely because of the word processor. Modern word-processing programs are sophisticated design tools that let us use different typefaces and sizes, include graphics, and make multicolumn layouts. More powerful still are desktop-publishing programs, which provide extraordinarily sophisticated design options.

Where does this leave technical communicators and technical professionals? Today's technical communicators are expected to be familiar not only with word processing but also with desktop publishing, for they are likely to be responsible for producing a monthly newsletter, an internal procedures manual, or even a user's guide for customers. Technical professionals such as scientists are not expected (yet) to have desktop-publishing skills and experience, but the more they know about design, the more effective their documents will be.

This chapter covers the basics of designing documents and pages. The chapter first describes the goals of document and page design. It then discusses planning the design of the document and page. Next, the chapter explains how to design a document and how to design the individual pages.

Goals of Document Design and Page Design

In designing a document, you have five major goals, as shown in Figure 13.1 on page 324.

- *To make a good impression on readers.* All documents should look professional, if only because you want to show your readers that you take your work seriously. If you are writing to people outside your own organization, the document should reflect not only your own professional standards but also those of your organization.

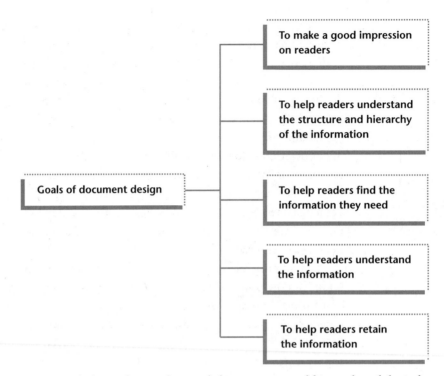

FIGURE 13.1

Goals of Document

Design

- *To help readers understand the structure and hierarchy of the information.* Regardless of what kind of document they are reading, from a one-page data sheet to a multivolume proposal, your readers should know where they are in the document. And they should be able to see the hierarchy of the information; that is, they need to know whether the portion they are reading is more important or less important than the portion that preceded it. Design helps communicate these relationships.

- *To help readers find the information they need.* One obvious characteristic of technical communication is that, most of the time, people don't read documents from cover to cover; they refer to documents to find answers to particular questions. Therefore, a well-designed document enables readers to find information quickly and easily. A table of contents and an index help in some kinds of documents, but design elements such as tabs, icons, and color can be extremely effective too. In addition, page design and choice of type are critical to a reader's ability to find information.

- *To help readers understand the information.* If you design the document and the page carefully, you help readers understand the information more quickly and easily. For instance, if you design a set of instructions so that readers know the text is right next to its accompanying graphic, you have made it easier for readers to understand the instructions.

- *To help readers retain the information.* An effective design enables readers to create a visual image of the information, enhancing retention. Design elements such as text boxes and pull quotes help readers remember important explanations and quotations.

Planning the Design of the Document and the Page

The first step in creating a design for any kind of technical document is to plan. Start by analyzing your audience's needs and expectations, and then consider your resources.

Analyzing Your Audience's Needs and Expectations

Use the techniques described in Chapter 4 for analyzing your audience. Consider such factors as their knowledge of the subject, their attitudes, their reasons for reading, the way they will be using the document, and the kinds of tasks they will be performing. For instance, if the document is a benefits manual for employees, you know that few people will read it like a book, but many people will refer to it often. Therefore, you will want to build in as many accessing tools as you can: a table of contents and index, of course, but perhaps tabs to separate the different sections or a different color paper or type for different sections. You might create a separate section containing the numerous forms the readers might have to fill out.

Think too about your audience's expectations. Readers expect to see certain kinds of information presented in certain ways, and unless you have a good reason to present information in other ways, you should fulfill their expectations. For instance, if you have worked with the documentation sets that come with word-processing programs, you have learned that different kinds of information—tutorial instructions, installation guides, quick-start guides, and reference material—come in different documents of different sizes and shapes. Tutorial information, for instance, is often presented in a small-format book, bound so that it lies flat on the table next to the keyboard.

Determining Your Resources

Once you have considered what your readers need and expect, think about your major resources: time, money, and equipment. Small, informal documents are generally made in-house, but for more ambitious projects, you need to decide how much of the job should be subcontracted to professionals. If your organization has a technical-publications department, you should consult the professionals there for information on scheduling and budgeting.

- *Time.* What is your schedule? A sophisticated design might require the expertise of professionals at service bureaus and print shops, and their services can require weeks. Working up even a relatively simple design for a newsletter can require many hours.

- *Money.* What is your budget for the project? Can you afford the services of professional designers and print shops? An in-house newsletter should look professional and attractive, but most managers would be unwilling to authorize thousands of dollars for a sophisticated design. They would authorize many thousands of dollars for the design of an annual report, however.

- *Equipment.* With a pencil and a straight edge you can create a terrific design, but to make that design come to life in the document itself, you need equipment. Although a word-processing program by itself can fulfill most of the routine design needs in most organizations, more complicated designs also require graphics and desktop-publishing programs. A good laser printer, less expensive than the average personal computer, will produce attractive documents in black and white. For high-resolution color, however, a more expensive printer is required. One common communication decision an organization faces is whether to invest more money in equipment and hire people with design and production skills, or to subcontract the design and production to professionals outside the company.

With an understanding of your readers' needs and expectations and of your own resources, you can begin to plan the design for the document and for the individual pages.

Designing the Document

Before you begin to design individual pages, think through the design of the whole document—how you want the different elements to work together to accomplish your objectives.

There are four major elements to consider in designing the whole document: size, paper, accessing tools, and bindings. Table 13.1 shows how these elements appear in typical documents produced in the United States. In other countries, conventions vary.

Size Size refers to two different aspects of document design: page size and page count.

- *Page size.* Your readers' needs are critical. Think about the best page size for communicating your information, and about how the document will be handled. For a procedures manual that will sit on a shelf most of the day, standard 8.5 × 11-inch paper, 3-hole-punched to fit in a ring binder, might be the obvious choice. But for a software tutorial, you will probably want to create relatively narrow columns in a document that fits easily on a desk while the reader works at the keyboard. Therefore, you might choose a 5.5 × 8-inch size. The physical dimensions of the

TABLE 13.1	*Document-Design Elements in Typical Documents*			
Typical Documents	**Document-Design Elements**			
	Size	Paper	Accessing Tools	Binding
MEMOS	8.5 × 11 inches, as well as smaller sizes.	Printer paper; bond.	Icons, headers and footers, page numbering.	Paper clip or staple.
LETTERS	8.5 × 11 inches.	Letterhead stationery for the first page; bond or printer paper for subsequent pages.	Icons, headers and footers, page numbering, color (sometimes).	Usually not bound.
INSTRUCTIONS	Varies, depending on the complexity of the subject and the size of the packaging.	Usually printer paper, but sometimes higher grade if environmental needs call for greater strength.	Icons, headers and footers, dividers and tabs, cross-reference tables, page numbering, color (sometimes).	Often, instructions are on a single sheet. Otherwise, saddle binding with staples, or (occasionally) ring binding.
MANUALS	Varies, up to 8.5 × 11 inches.	Varies, depending on environmental needs.	Icons, headers and footers, dividers and tabs, cross-reference tables, page numbering, color (sometimes).	Varies, but usually loose-leaf, ring, or saddle. Large manuals are often perfect bound.
PROPOSALS	Usually 8.5 × 11 inches.	Usually printer paper or bond.	Icons, headers and footers, dividers and tabs, cross-reference tables, page numbering, color (sometimes).	Varies, but usually loose-leaf, ring, or saddle. Often, proposals are bound like a book, with text on the left-hand page and graphics on the right-hand page.
REPORTS	Usually 8.5 × 11 inches.	Usually printer paper or bond.	Icons, headers and footers, dividers and tabs, cross-reference tables, page numbering, color (sometimes).	For brief, internal reports, usually ring binding. For longer or external reports, varies, including loose-leaf, ring, saddle, and perfect.
NEWSLETTERS	Usually 8.5 × 11 inches, but sometimes larger.	Usually newsprint or printer paper.	Icons, headers and footers, page numbering, color (sometimes).	Usually folded like a newspaper; sometimes saddle bound.
BROCHURES	Usually 8.5 × 11 inches folded into three panels of 8.5 × 3.67 inches. Sometimes larger.	Usually a heavier stock than printer paper, such as text paper.	Icons and colors.	None.

document can be important if it is to be used in a cramped area or has to fit in a standard-size compartment, such as in the cockpit of an airplane.

Paper comes precut in a number of standard sizes in addition to 8.5 × 11 inches, such as 4.5 × 6 inches and 6 × 9 inches. Of course, paper can be cut to whatever size you want, but costs increase substantially when you use a nonstandard size. Check with your technical-publications department or a print shop for current prices and availability of paper sizes.

- *Page count.* Although page count can be important because a document that is too thick might not fit where it must be stored, for the most part page count is a cost factor and a psychological factor.

  The cost factor is simple: paper is expensive and heavy, so you want to reduce the number of pages as much as you can, especially if you are printing large numbers of the document and mailing it. Professional journals usually have a very cramped design because mailing costs make it necessary to fit as much information as possible on every page. In the software industry, many companies now use the minimalist design, replacing 600-page manuals with 100-page versions—in part because of printing, warehousing, and mailing costs. (See Chapter 18 for a discussion of minimalist design.)

  The psychological factor is also easy to understand: everyone wants to spend as little time as possible reading technical communication. Therefore, if you can figure out a way to design the document so that it is 15 pages long rather than 30—but still easy to read—your readers will appreciate it.

Paper   Paper is made not only in different standard sizes but also in different weights and with different coatings.

The cheapest paper is newsprint. It is extremely porous, so it not only allows inks to bleed through to the other side but also picks up smudges and oil easily. In addition, it can turn yellow in as little as a few weeks. For these reasons, newsprint is generally used only for newspapers, informal newsletters, and other types of quick, inexpensive communication to large numbers of people.

The most common types of paper used in technical communication are bond (used for typing letters and memos), book paper (a higher grade that permits better print resolution), and text paper (an even higher grade used for more formal documents such as announcements and brochures). Heavier paper is also available for cover pages and for cards, such as quick-reference cards. But probably the most common kind of paper is the relatively inexpensive paper used in photocopy machines and laser printers; this paper is made to increase the resolution of the type.

Most types of paper are available uncoated or coated. The coating increases strength and durability, and it produces the best print resolution. However, some glossy coated papers produce an annoying glare. To deal with this problem, designers recommend choosing a paper with a slight tint; a bone white, for instance, produces less glare than a bright white.

The different types of papers are available in different grades. For small jobs—say, under 5,000 copies of a brief document—you probably want to use high-grade paper. For bigger jobs, the grade of paper will have a larger effect on the total cost.

Unless you have worked in the printing industry, you probably have not had a chance to learn much about paper. Work closely with printing professionals; they know, for example, about UV-coated paper, which greatly reduces the extent to which the sun fades the ink. And they know about the current costs of recycled paper, which is constantly improving in quality and becoming less expensive.

Bindings   Many technical documents are one sheet, or folded brochures, or held together with a paper clip or a staple. But you might soon be producing documents that require more sophisticated means of binding.

In technical communication, the four types of bindings shown in Figure 13.2 are used frequently.

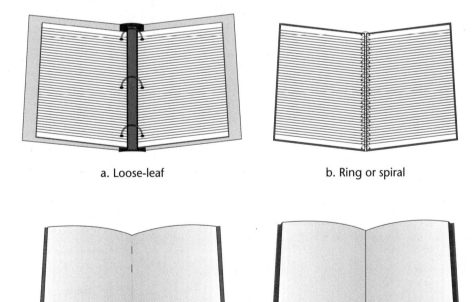

a. Loose-leaf   b. Ring or spiral

c. Saddle   d. Perfect

FIGURE 13.2

Common Types of
Bindings

- *Loose-leaf binders.* Loose-leaf binders are reusable, come in various sizes, and are convenient when pages must be added and deleted frequently. One drawback, however, is that most people don't keep their binders up-to-date, especially if there are frequent updates, so after a while few people in an organization have up-to-date binders. Another drawback is that a high-quality loose-leaf binder can cost several dollars.
- *Ring or spiral binders.* Similar to loose-leaf binders, ring binders use wire or plastic coils or combs to hold the pages together. Ring binders let you lay the document open on a desk or even fold it over, so that it takes up no more space than one page. This feature makes the document easy to hold in one hand. Neighborhood print shops can put plastic coils or combs on documents of almost any size for about a dollar each.
- *Saddle binding.* In saddle binding, a set of large staples is inserted from the outside of the document, when the document is opened to its middle pages. Saddle binding is not practical for large documents.
- *Perfect binding.* In perfect binding, the pages are glued together and a cover is attached. Perfect binding, which is used in book publishing, produces the most formal appearance, but it is relatively fragile, and the open document does not lie as flat as with other kinds of bindings.

Accessing Tools In a well-designed document, readers know how to move around easily and find the information they seek. Figure 13.3 shows six kinds of accessing tools that you can build into most documents.

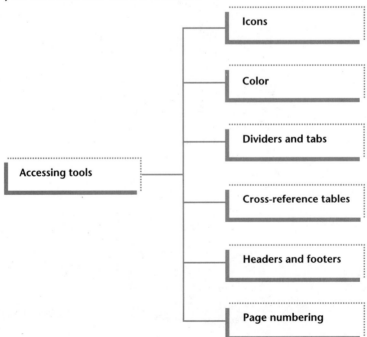

FIGURE 13.3
Accessing Tools

- *Icons.* Icons are pictures that symbolize actions. A garbage can on your computer screen symbolizes the task of erasing a file. An hourglass or a clock tells you to wait for the computer to perform a task. Perhaps the most common icon in technical communication is the stop sign, which often alerts the reader to a safety warning.

  Icons enable your readers to create a visual link between the icon and a particular kind of information. For instance, the icon of the pointing hand often introduces hints or suggestions in software manuals. Once readers learn that the icon has this function, they seek it out when they need help. For you as the writer, using the icon saves you the trouble of writing out repetitive comments such as "Following are some hints that might help you with this task."

  But beware of being too cute in thinking up icons. One computer manual uses a cocktail glass about to fall over to symbolize "tip." Not a good idea. And don't use too many different icons, or your readers will forget what they represent. Figure 13.4 presents common icons.

FIGURE 13.4

Icons

- *Color.* Color is perhaps the strongest visual attribute you can use (Keyes, 1993). If you have access to colors other than black and shades of gray, use them sparingly, for they can easily overpower everything else in your document.

  Just as people *see* documents before they read them, they see colors before they see shapes. Therefore, don't send mixed signals to your readers. Your third-level headings should not be in color, for example, if your first- and second-level headings are black; headings in color will stand out more, regardless of other visual elements you use, such as boldface or size variations. (See Chapter 14 for a discussion of heading levels.) Use colors to draw readers' attention to important features of the document, such as safety warnings, hints, major headings, and section tabs.

Colored paper is used frequently now as an accessing tool. Just as a telephone book has white pages, yellow pages, and blue pages, many technical documents now use colored paper to represent different sections. As always, however, use restraint; avoid using more than two or three colors, and avoid garish neons. See Chapter 12 for a detailed discussion of color.

- *Dividers and tabs.* You know dividers and tabs—the heavy-paper sheets with the paper or plastic extensions—from your loose-leaf notebooks. You may also have seen them in some textbooks. The tabs provide a place to add a label, and they enable readers to flip to a particular section. Often, a color scheme is used with dividers and tabs.

- *Cross-reference tables.* In a complicated document, you might want to refer readers to a related discussion. A cross-reference table, such as the one excerpted in Figure 13.5 (Fischer, 1992, p. 6), is an effective way to do so. As Figure 13.6 shows, a cross-reference table lets you direct readers with different backgrounds to various sections of the document.

- *Headers and footers.* A header is an accessing tool at the top of the page; a footer is an accessing tool at the bottom of the page. Headers and foot-

Table I.1

Where to Start

Topic	See Chapter/Appendix
Adding Commands or Command Names	20
Advanced 2D commands	8–13
Changing Command Names or Keystrokes	20
Changing or Customizing Menus	20
Changing the Coordinate System	1, 3, 7, 11
Controlling Drawing Accuracy	3

FIGURE 13.5

Cross-Reference Table

If you have never used the Internet:	
Read . . .	*To learn how to . . .*
Chapter 1	connect to the Internet
Chapter 2	use basic e-mail functions
etc.	
If you are familiar with the Internet, but not a user:	
Read . . .	*To learn how to . . .*
Chapter 5	join newsgroups
etc.	

FIGURE 13.6

Cross-Reference Table Addressed to Readers with Different Backgrounds

ers enable readers to see, at a glance, where they are in the document. In this book, for example, the headers on the left-hand pages indicate the chapter of the book you are in. Those on the right-hand pages indicate the most recent first-level heading. Note that, like page numbers, headers and footers are usually placed near the outside edge of the page, not near the spine of the document.

Sometimes writers build other identifying information into their header. For example, your instructor might ask you to identify your own assignments with the following information in a header: "Smith, Progress Report, English 302, page 6."

- *Page numbering.* Page numbering can be complicated. For memos, use arabic numbers in the upper right corner of the page. (Note that the first page of most documents is not numbered.) For documents printed on both sides of the page, such as this book, put the page numbers in the outside corners.

  Reports and other complex documents often use two sets of numbers: lowercase roman numerals (i, ii, and so on) for front matter and arabic numerals for the body. The title page is not numbered; the page after it is numbered ii.

  Appendices are often numbered and lettered. That is, Appendix A begins with page A-1, followed by A-2, and so on. Appendix B starts with page B-1.

  You should know about two other aspects of page numbering:

  – Sometimes documents that are photocopied and distributed in a meeting, through interoffice mail, or (especially) faxed include in the page numbering system the total number of pages in the document. The second page is numbered "2 of 17," and the third page is "3 of 17." This approach looks awkward, but at least readers are sure they have received the complete document.
  – Sometimes documents that will be updated use section numbering. Section 3 begins with page 3-1, followed by 3-2. Section 4 begins with 4-1. This way, a complete revision of one section does not affect the pagination of the subsequent sections.

Designing the Page

Not long ago, page design was a simple matter for technical communicators: text was double spaced, margins were one inch. Graphics pages followed the text pages. When readers needed to find the graphic that went with the text, they flipped forward through the document to find the table or chart. Now that sophisticated desktop-publishing software is available, we are studying the best ways to design the page so that it is attractive and functional.

This section begins with a discussion of page-design principles and of what learning theory can teach us about designing better pages. The section also discusses page grids, the visual plan for the page; typography; the design of titles and headings; and special design techniques (such as text boxes).

Principles of Effective Page Design

A page of technical communication is effectively designed if the reader recognizes a pattern—where to look for certain kinds of information, for instance—but isn't put to sleep by a monotonous, mechanical appearance. As shown in Figure 13.7, designers often refer to three qualities when they discuss page design: balance, consistency, and simplicity.

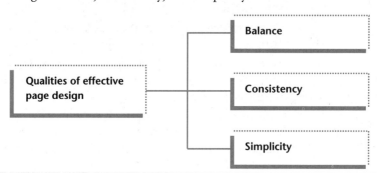

FIGURE 13.7

Qualities of Effective Page Design

- *Balance.* Balance refers to the visual stability of the page; it should not look as if it is tipping to one side or is top-heavy or bottom-heavy. In documents that open to a two-page spread, the concept of balance also applies to the two pages. On a balanced page, a graphic on one side can be offset by a similar size graphic on the other side, or one large graphic can be balanced by two smaller ones. Balance can be *symmetrical* or *asymmetrical.* Symmetrical balance is achieved when the two halves of the page are similar in appearance: a graphic balances a graphic, or white space balances white space. Asymmetrical balance is achieved when the two halves of the page are dissimilar but still stable. For example, an asymmetrically balanced page might have a black-and-white graphic in the upper left balanced by an icon in color in the lower right. Figure 13.8 shows some common designs for balanced pages.

    But don't be a slave to symmetry. Readers can be bored by a balance that is too regular; slight variations add visual interest to the look of the page, as seen in Figure 13.9.

- *Consistency.* A well-designed page uses consistent patterns, just as a well-designed document does. Margins, for example, should be consistent from page to page. Indentation should follow a regular pattern too: second-level text, for instance, might be indented five characters. Typefaces and sizes help you create a consistent pattern; first-level heads might be printed in 18-point Helvetica, for example, with body copy in 12-point Century Schoolbook. Other elements that can create consistent visual

On the left-hand page, symmetrical balance is achieved by centering key elements.

The page on the right is asymmetrically balanced because the graphic in the upper-left corner balances the two-column text on the right side of the page.

Symmetrical and Asymmetrical Balance

patterns are colors, icons, rules, and screens. See the discussion of page grids later in this chapter for more information on how to create and sustain consistent patterns.

- *Simplicity.* Keep in mind that the primary purpose of technical communication is to inform the reader. In general, a simple design is more effective than a complicated one. Don't be tempted to use all the design tricks loaded on the CD-ROM disk that came with your software; the philoso-

Each page of this two-page spread is asymmetrical, but viewed together the two pages are well balanced. Notice that white space can be an important element: the white space at the top of the right page balances the title at the top of the left page.

Variations on Balanced Pages

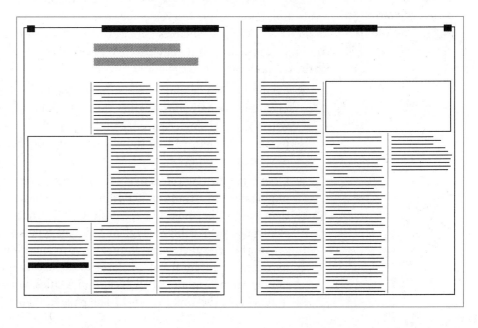

(Figure 13.9 continued)

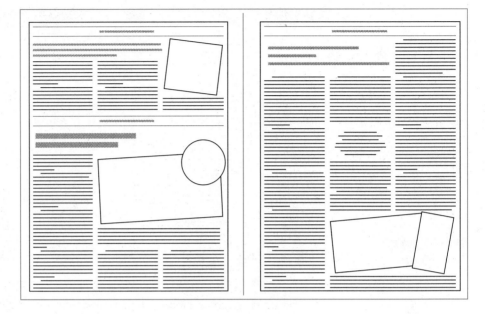

In this two-page spread, the angled placement of the art adds visual interest.

phy of some manufacturers is "If the feature fits on the disk, throw it in." For instance, typeface designers offer outlined and shadowed typefaces, but they are harder to read than regular typefaces. A page with four or five different fonts and type sizes is likely to confuse the reader.

Learning Theory and Its Relation to Page Design

Your job in designing the page is to create visual patterns so that readers can easily find, understand, and remember the information they need. A basic understanding of three principles of learning theory—chunking, queuing, and filtering—can help you design effective pages. (See Keyes [1993] for a summary of the research on learning theory.)

- *Chunking.* People understand information best if it is delivered to them in chunks—small units—rather than all at once. In a business letter, which is typed single spaced, chunking involves double spacing between paragraphs. Figure 13.10 shows how chunking helps readers see the units of related information.

- *Queuing.* Queuing refers to creating visual patterns that show levels of importance. In a traditional outline, the "I" heading is more important than the "1" heading. In page design, you want to use more emphatic elements—such as bigger type or boldface type—to suggest importance.

    Another visual element of queuing is indentation. Designers start the more important information closer to the left margin and indent the less important information. (An exception is titles, which are often centered in reports in the United States.) The page in Figure 13.11(a), for example,

FIGURE 13.10

Chunking

a. Without chunking

b. With chunking

shows queuing by size: the first-level head at the top is bigger than the second-level heads. The page in (b) shows queuing by size and by indentation: each of the three headings is a different size, and the headings and their accompanying text are aligned on a different left margin.

FIGURE 13.11

Queuing by Size and
by Indentation

a. Queuing by size

b. Queuing by size and indentation

a. Filtering using a text box   b. Filtering using a text box and an icon

FIGURE 13.12

Filtering

- *Filtering.* Filtering is the use of visual patterns to distinguish between different types of information so that readers can decide what they want to read. For instance, a stop sign often signals safety information in a set of instructions. Designers also use typography to create filters. Introductory material might be displayed in larger type, and notes might be displayed in italics or in a different typeface. Figure 13.12(a) shows a design that uses text boxes to set off notes. Figure 13.12(b) shows a text box with an icon used as a filtering device.

Page Grids, White Space, and Columns

Every page includes two kinds of space: white space and space for text and graphics. The best way to start designing a page is to make a grid—a picture of what it will look like. In making a grid, you will decide how to use white space and determine how many columns you want to have.

Table 13.2 shows how the page-design elements discussed in this section are used in the United States. In other countries, these conventions vary.

Page Grids

As the phrase suggests, a page grid is like a map on which you plan where the white space, the text, and the graphics will go. To create an effective grid, you must first understand your audience, their purpose in reading, and their reading behavior.

Many writers like to begin by drawing a thumbnail sketch. A *thumbnail sketch* is a rough drawing that helps you see how the different elements—the

TABLE 13.2	*Page-Design Elements in Typical Documents*				

	Page-Design Elements				
Typical Documents	Columns	Typeface	Type Sizes	Line Spacing	Justification
MEMOS	One.	Varies, but usually serif.	Usually 10–12.	Single spaced.	Left-justified or justified.
LETTERS	One.	Varies, but usually serif.	Usually 10–12.	Single spaced.	Left-justified or justified.
INSTRUCTIONS	Varies, but usually a multi-column design.	Varies.	Varies.	Varies.	Left-justified or justified.
MANUALS	Varies.	Varies, but usually serif.	Usually 10–12.	Varies, but usually single spaced.	Left-justified or justified.
PROPOSALS	Varies.	Usually serif.	Usually 10–12.	Varies.	Left-justified or justified.
REPORTS	Often one.	Usually serif.	Usually 10–12.	Often, double spaced.	Left-justified or justified.
NEWSLETTERS	Usually multicolumn.	Serif for body text; sans serif for display text.	Usually 10–12 point for body text.	Usually single.	Left-justified or justified.
BROCHURES	Usually one column per panel.	Varies.	Varies.	Varies.	Varies.

text and the graphics—will be arranged on the page. Figure 13.13 on page 340 shows two different thumbnail sketches for a page from the body of a manual.

Keep experimenting with thumbnail sketches of the different kinds of pages your document will have: body pages, front matter, and so on. When you are satisfied with your thumbnail sketches, make page grids.

You can use either a computer or a pencil and paper to create page grids. Or you can combine the two techniques. On a computer, you can use a word-processing program or a draw program or paint program (see Chapter 12) to create the grid. Many word-processing programs come with templates that you can use as is or modify. Desktop-publishing software offers the greatest variety of templates, many quite sophisticated.

You don't need sophisticated software to design a grid, however. With a pencil, you can draw the grid on a sheet of paper the size you will be using. Some writers like to use graph paper because the grid lines are convenient. You

FIGURE 13.13

Thumbnail Sketches

can use inches or the metric system for your measurements, or you can use the unit that printing professionals use—the pica—which equals one-sixth of an inch. Some writers like to experiment with pencil and paper, then transfer their grid to a computer. Figure 13.14 shows two simple grids: one using inches and one using picas.

Experiment with different grids until you think you have a page design that is attractive, able to meet the needs of your readers, and appropriate for the information you are conveying. Figure 13.15 on page 342 shows that the variety of grids you can create is limitless.

White Space

Sometimes called *negative space,* white space is the portion of the paper that does not have writing or graphics on it: the space between two lines of text, the space between text and graphics, and, most obviously, the margins. White space directs readers' eyes to a particular element, emphasizing it; in addition, white space helps readers see relationships among elements on the page.

Most people are surprised to learn that almost half the area of a typical page is devoted to margins. Why so much? Margins serve four main purposes:

- To cut down on the amount of information included on the page, making the page seem easier to read and use.
- To provide enough space that the document can be bound or that the reader can easily hold the page without covering up the text.
- To provide a neat frame. Pages on which the type extends all the way out to the edge of the paper look sloppy.

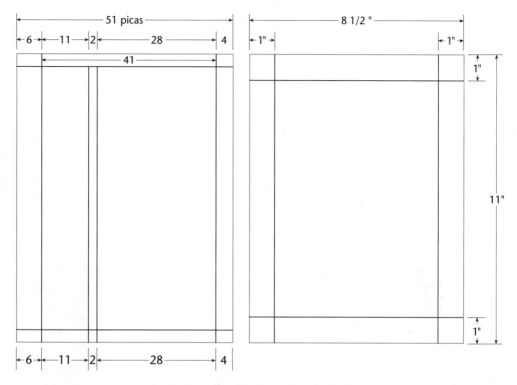

a. Grids using picas (left) and inches (right)

FIGURE 13.14

**Sample Grids Using Picas
and Inches**

b. Pages based on the grids

a. Double-column grid

b. Two-page grid, with narrow outside columns for notes

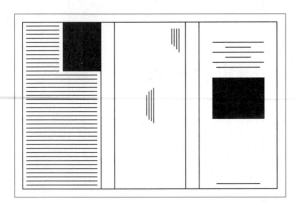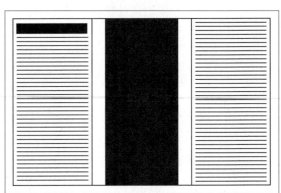

c. Three-panel brochure

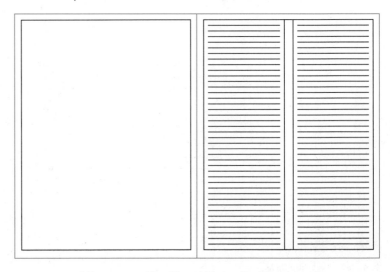

FIGURE 13.15

Popular Grids

d. Two-page grid, with graphics on the left page and
double-column text on the right page

- To provide space for marginal glosses. (Marginal glosses are discussed later in this chapter.)

Figure 13.16 shows the common margins used for left-hand and right-hand pages. The actual sizes of the margins are determined by several factors:

- *The size of the page.* In general, the smaller the page, the smaller the margins.
- *The amount of text that needs to fit on the page.* For an article in a journal that will be mailed, the margins tend to be small, to reduce the page count and, therefore, postage costs.
- *The difficulty of the text.* In general, the more technical the text, the larger the margins. This strategy makes the pages look less intimidating.
- *The background of the audience.* In general, the less knowledgeable the readers, the larger the margins.

White space other than margins also serves as a border that sets off and emphasizes an element on the page. For instance, white space around a graphic separates it from the text and draws the reader's eye to it. White space between columns helps the reader to read the text easily, without mistaking two adjacent columns for one. And white space between two sections of text helps the reader see that one section is ending and another is beginning.

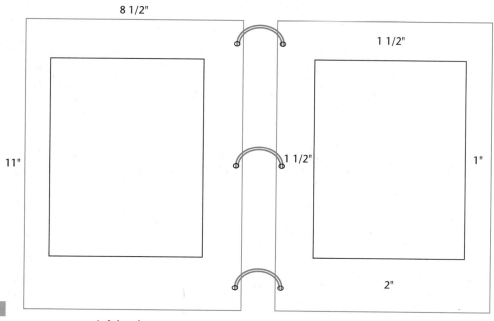

8 1/2"

1 1/2"

11"

1 1/2"

1"

2"

Left-hand page

Right-hand page

FIGURE 13.16

Typical Margins for a
Document that Is Bound
Like a Book

Columns

Most documents you write on the job will look much like the one-column documents you write as a student. But things become a lot more complicated when you go beyond the world of paper clips and staples. For example, you will have to leave bigger inside margins to accommodate the binding. But the greatest difference between the design of the pages you write as a student and that of the pages you will write on the job is in the use of multiple columns. A multicolumn design can have four major advantages for the technical communicator:

- Text in columns is generally easier to read because the lines are shorter.
- The use of columns lets you fit more information on the page because many of your graphics can fit in one column. Wider graphics can extend across two or more columns. In addition, you can fit more words on a page with a multicolumn design than with a single-column design.
- The use of columns allows you to create a visual pattern in which text is in one column and the accompanying graphic in an adjacent column.
- The use of columns gives you greater flexibility in sizing graphics.

Typography

Typography is the study of type and the way people read it. When designers and technical communicators study typography, they learn about typefaces, families, case, and sizes. They also consider the white space of typography: line spacing, line length, and justification.

Typefaces

A typeface is a set of letters, numbers, punctuation marks, and other symbols, all bearing a characteristic design. When the first movable, reusable metal type was invented in the Western world in the fifteenth century, the typefaces resembled ornate hand lettering. Today there are thousands of different typefaces, generally bearing the name of their designer or some historical name (such as Times Roman, Avant Garde, Helvetica, Garamond, or Zapf Chancery). More are designed every year. Figure 13.17 shows three different typefaces.

Typefaces are generally classified into two categories: *serif* and *sans serif*. A serif typeface, such as Times Roman, has short lines—serifs—at the ends of the major strokes in most of the letters. For instance, look at the serifs in the letter *N* in *New York Times* in the second paragraph in Figure 13.17. Sans serif typefaces, such as Univers, do not have these lines. Serif typefaces are generally considered easier to read because the short lines encourage the movement of the reader's eyes along the line. Sans serif typefaces are harder on our eyes than serif typefaces because the letters are less distinct from one another than they are in a serif typeface. For this reason, sans serif typefaces are used mostly for short documents and for headings. However, sans serif typefaces are easier to read on the screen and when printed on dot-matrix printers, because the letters are simpler; serif fonts can look smudged.

This paragraph, for example, was typed on a word processor using the Kaufman typeface. You are not likely to see this style of font in any technical communication because it is too ornate and too hard to read. It is better suited to wedding invitations and other formal announcements.

This paragraph was typed with the Times Roman typeface. It looks like the kind of type used by newspapers such as the *New York Times* in the nineteenth century. It is an effective typeface for body text in technical documents.

This paragraph was typed with the Univers typeface, which has a very modern, high-tech look. It is best suited for headings and titles in technical documents.

Typefaces

Cultural factors can be important, though. For instance, Europeans find sans serif typefaces easier to read than serif typefaces because sans serif is more common in Europe.

Most of the time, you will use a standard font, such as Times Roman, that your applications software includes and that your printer can reproduce. Just remember that different typefaces create different impressions—from highly formal and ceremonial to very modern and high tech—and that some typefaces cause less eye fatigue than others.

Type Families

Each typeface is actually a family of typefaces, consisting of a number of variations on the basic style. Figure 13.18, for example, is the Helvetica family.

As with typefaces, avoid the temptation to overdo the use of different members of a type family. Used sparingly and consistently, the different members of a type family help you direct the reader's attention to different kinds of text, such as warnings, notes, and characters that you want your readers to type (Felker, Pickering, Charrow, Holland, & Redish, 1981). You will need italics for book titles and other elements (what used to be underlined with a typewriter), and you might use bold type for emphasis and headings, but you can go a long time without needing condensed and expanded versions of typefaces. And you can live a full, rewarding life without ever using outlined or shadowed versions of typefaces.

Helvetica Light	***Helvetica Bold Italic***
Helvetica Light Italic	**Helvetica Heavy**
Helvetica Regular	***Helvetica Heavy Italic***
Helvetica Regular Italic	Helvetica Regular Condensed
Helvetica Bold	*Helvetica Regular Condensed Italic*

Helvetica Family of Type

Case

To make your technical communication easy to read, use uppercase and lowercase letters just as you would in any other kind of writing. Some writers mistakenly think that text printed in uppercase letters is easier to read or more emphatic than text printed in a mix of uppercase and lowercase letters. It isn't. Lowercase letters are actually easier to read because the individual variations from one letter to another are greater for lowercase than for uppercase, as shown in Figure 13.19.

Individual variations are greater in lowercase words

THAN THEY ARE IN UPPERCASE WORDS.

The average person requires 10 to 25 percent more time to read all-uppercase writing than writing using both cases (Haley, 1991). In addition, using lowercase and uppercase letters provides the visual cues that signal new sentences (Poulton, 1968). One more reason to use both cases: uppercase letters require up to 35 percent more space than do lowercase letters (Haley, 1991).

Type Sizes

Using different sizes of the same typeface can create other visual cues. Sizes are measured in a unit called a *point*. There are 12 points in a pica, or 72 points in an inch. Figure 13.20 shows the basic range of type sizes.

Software and printers with scalable typefaces can produce letters and other characters of almost any size, ranging from 0.25–999.75 points, in increments of 0.25 points. In most technical communication, body text is presented in 10-, 11-, or 12-point type.

This paragraph is printed in 10-point type. This size is easy to read, provided it is reproduced on a letter-quality impact printer or laser printer. On most impact printers, however, the resolution isn't high enough for type this small.

This paragraph is printed in 12-point type. If you are using a dot-matrix printer, 12-point type is the best size to use because it balances readability and economy.

This paragraph is printed in 14-point size. This size is appropriate for titles or headings.

Some other sizes that are used occasionally in technical communication:

footnotes	8- or 9-point type
indexes	2 points smaller than the body text
slides or transparencies	24- to 36-point type

6	ABCDEFGHIJKLMNOPQRSTUVWXYZABCDEFGHIJKLMNOPQRSTUVWXYZABC
8	ABCDEFGHIJKLMNOPQRSTUVWXYZABCDEFGHIJKLMNO
10	ABCDEFGHIJKLMNOPQRSTUVWXYZABCDEFG
12	ABCDEFGHIJKLMNOPQRSTUVWXYZAB
14	ABCDEFGHIJKLMNOPQRSTUVW
18	ABCDEFGHIJKLMNOPQR
24	ABCDEFGHIJKLMN
30	ABCDEFGHIJKL
42	ABCDEFG
48	ABCDEF
60	ABCDE

FIGURE 13.20

Helvetica Medium in 11 Sizes

In general, you want a 4 to 6 point difference between headings and body text, and between headings and titles.

Don't go overboard with size variations, or your document will look like a sweepstakes advertisement in the mail.

Line Length

Ironically, the line length that we use most often—about 80 characters on an 8.5 × 11 inch page—is somewhat difficult to read. In general, a short line—of perhaps 50 to 60 characters—is less tiring to read than a longer line, especially for long documents (Biggs, 1980). With a multicolumn format, you can choose a line length that is appropriate for the size of the type and the paper, as well as for the degree of difficulty of the text and the knowledge level of the readers.

Line Spacing

Sometimes called *leading* (pronounced *ledding*), *line spacing* refers to the white space between lines, or between one line of text and a graphic. If the line spacing is too big, the page looks diffuse, the text loses its coherence, and the reader tires quickly. If the line spacing is too small, the page looks crowded and is difficult to read; your eye sees the lines above and below the one you are reading, causing fatigue. Some research suggests that smaller type requires greater line spacing, that longer lines benefit from additional line spacing, and that sans serif typefaces need a little more than serif typefaces. Figure 13.21 shows the same text with three different amounts of line spacing.

Line spacing is usually determined by the kind of document you are writing. Memos are single spaced; other documents, such as reports and proposals, are double spaced or one-and-a-half spaced.

Line spacing can also separate one section of text from another. Breaks between single-spaced paragraphs are usually two spaces. Double-spaced and one-and-a-half-spaced text has no extra line spacing between paragraphs, but the first line of each paragraph is indented, as in handwritten documents.

For breaks between sections, the line spacing is usually greater than that between paragraphs. For instance, if you are typing a single-spaced letter, double

---

a.   Excessive line spacing

Aronomink Systems has been contracted by Cecil Electric Cooperative, Inc. (CECI)

to design a solid waste management system for the Cecil County plant, Units 1 and

2, to be built in Cranston, Maryland. The system will consist of two 600 MW pulver-

ized coal burning units fitted with high efficiency electrostatic precipitators and lime-

stone reagent FGD systems. The coal will contain an estimated 3% sulfur and 10%

ash. The station will output approximately 64 TPH (DWB) of FGD sludge and 24 TPH

fly ash at 100% load.

b.   Appropriate line spacing

Aronomink Systems has been contracted by Cecil Electric Cooperative, Inc. (CECI) to design a solid waste management system for the Cecil County plant, Units 1 and 2, to be built in Cranston, Maryland. The system will consist of two 600 MW pulverized coal burning units fitted with high efficiency electrostatic precipitators and limestone reagent FGD systems. The coal will contain an estimated 3% sulfur and 10% ash. The station will output approximately 64 TPH (DWB) of FGD sludge and 24 TPH fly ash at 100% load.

c.   Inadequate line spacing

Aronomink Systems has been contracted by Cecil Electric Cooperative, Inc. (CECI) to design a solid waste management system for the Cecil County plant, Units 1 and 2 to be built in Cranston, Maryland. The system will consist of two 600 MW pulverized coal burning units fitted with high efficiency electrostatic precipitators and limestone reagent FGD systems. The coal will contain an estimated 3% sulfur and 10% ash. The station will output approximately 64 TPH (DWB) of FGD sludge and 24 TPH fly ash at 100% load.

FIGURE 13.21

Line Spacing

space between paragraphs and triple space between sections. In other words, the line spacing *between* two sections should be greater than the line spacing *within* a section. Figure 13.22 shows the effective use of line spacing to distinguish one section from another and to separate text from graphics.

Line spacing is also used to set off block quotations. If the text is single spaced, set off the quotation above and below with double spacing; if the text is one-and-a-half or double spaced, use triple spacing.

FIGURE 13.22

Line Spacing Used to Distinguish One Section from Another

Justification

Justification refers to the alignment of words along the left and right margins. In technical communication, text is often *left-justified;* that is, except for paragraph indentations, the lines begin along a uniform left margin but end on an irregular right border. This irregular right border is often called *ragged right.* Ragged right is most common in typewritten and word-processed text (even though word-processing programs can justify the right margin).

In *justified text,* also called *full-justified text,* both the left and right margins are justified. Justified text is seen most often in typeset, formal documents. However, left-justified text offers an advantage even when you are typesetting: if you have to make minor revisions after the text is set, your changes won't affect as many lines because the irregular right margin gives you more room to make changes. For example, if you add a word to a short line, the change might not affect any subsequent lines.

The following passage (Berry, Mobley, & Turk, 1994, p. 647) is presented first in left-justified form and then in justified form.

Notice that the space
between words is uniform
in left-justified text.

Until recently, most software development has adhered to a linear development process, often referred to as the "waterfall" process. Rimbaugh (1992) describes this as several phases performed in sequence, one after another. Using this approach, a product is first designed, then developed, tested, and delivered. Nothing is assembled until the design is complete, and the design is not changed once assembly starts. The car being built moves straight down the assembly line and isn't complete until it reaches the end.

In justified text, the spacing between words is irregular, slowing down the reader. Because a big space suggests a break between sentences, not a break between words, the reader can become confused, frustrated, and fatigued.

Notice that the irregular
spacing not only slows
down reading but also
creates "rivers" of white
space. Readers are
tempted to concentrate
on the rivers running
south rather than on the
information itself.

Until recently, most software development has adhered to a linear development process, often referred to as the "waterfall" process. Rimbaugh (1992) describes this as several phases performed in sequence, one after another. Using this approach, a product is first designed, then developed, tested, and delivered. Nothing is assembled until the design is complete, and the design is not changed once assembly starts. The car being built moves straight down the assembly line and isn't complete until it reaches the end.

Should you justify your technical communication? If you are using a standard word-processing program rather than a sophisticated desktop-publishing system, probably not.

Justification can make the text harder to read in one more way. Some word-processing programs and typesetting systems automatically hyphenate words that do not fit on the line. Be careful about how your computer breaks words: it might split *therapist* into *the rapist,* even though most printed dictionaries would break the word differently. In addition, hyphenation slows down the reader and can be distracting. Left-justified text does not require hyphenation the way justified text sometimes does.

Designing Titles and Headings

Chapters 14 and 15 discuss the strategy of creating effective titles and headings. This section, however, explains how to integrate titles and headings into the page design. The principle is that you want titles and headings to stand out visually because they announce and communicate new information; your reader needs to see that you are beginning a new idea.

Designing Titles

Because a title is the most important heading in a document, it should be displayed clearly and prominently. If it is on a cover page or a title page, you might present it in boldface in a large size, such as 18 points. If it also appears at the top of the first page, you might make it slightly bigger than the rest of the text—perhaps 18-point type for a document printed in 12-point type—but not as big as on the cover page or title page. Titles are often centered on a page horizontally. For more information on titles, see Chapter 14.

Designing Headings

Your readers should be able to tell when you are beginning a new idea. The most effective way to distinguish one level of headings from another is to use size variations (Williams & Spyridakis, 1992). For example, a 20-percent difference in size between a first-level head and a second-level head will be clear for most readers. Boldface also sets off headings effectively. The least effective way to set off headings is by underlining them, perhaps because in many word-processing programs the underline obscures the descenders, the portions of letters such as *p* and *y* that extend below the body of the letters.

Indenting the headings can also help readers. In general, the more important the level, the closer to the left margin it appears. First-level headings, therefore, usually begin at the left margin. Second-level headings are often indented four or five characters; third-level headings, seven or eight. Another common design for third-level headings is to run the text in on the same line as the heading. When you indent a heading, consider indenting the text that follows it. This design accomplishes two goals:

- The indented heading remains clear and emphatic; it does not get swallowed up by the text that follows it.
- The indented text appears appropriately subordinate. In some documents, writers even decrease the type size for lower-level text. For instance, in a document that is printed in 12-point type, all third-level and fourth-level text is printed in 10-point type.

As mentioned in the discussion of margins, writers sometimes highlight their headings by using hanging indentation: extending the heading further out into the left margin than the accompanying text.

In designing headings, you also need to use line spacing well. A noticeable distance between a heading and the text increases the impact of the heading. Consider the three examples in Figure 13.23 on page 352.

Other Design Techniques

Technical communicators are adapting the design features of newspapers and magazines. Five design techniques that are appearing more frequently in technical communication are rules, boxes, screens, marginal glosses, and pull quotes.

- *Rules.* The designer's term for a straight line is *rule*. Using word-processing software, you can easily create horizontal or vertical rules for various

Summary

In this example, the writer has skipped a line between the heading and the text that follows it. The heading stands out clearly.

Summary
In this example, the writer has not skipped a line between the heading and the text that follows it. The heading stands out, but not as emphatically.

Summary. In this example, the writer has begun the text on the same line as the heading. This style makes the heading stand out the least.

FIGURE 13.23

The Effect of Different
Line Spacing on the
Visibility of Headings

purposes. Horizontal rules are often used to separate headers and footers from the body of the page, or to distinguish two sections. Vertical rules can separate columns in a multicolumn page or identify revised text in a manual (see Chapter 18).

- *Boxes.* Create boxes by adding rules on all four sides of an item. Boxes can enclose graphics or special sections of text; sometimes they are used as a border for the whole page. Boxed text is often placed so that it extends out into a margin, giving it increased emphasis.
- *Screens.* The shading that appears behind text or a graphic to give it emphasis is known as a *screen*. The degree of screen ranges from 1 percent to 100 percent. Usually, a 5 to 10 percent screen provides the desired emphasis, while leaving the boxed text legible. You can use screens with or without boxes. Figure 13.24 shows a box with a screen behind it.

3. Slide the old toner cartridge out.
4. Prepare the new toner cartridge.

Important: Do not remove the flexible plastic sheets. They need to remain in place.

FIGURE 13.24

Box with a Screen

- *Marginal glosses.* A marginal gloss is a brief text, usually a summary statement, placed in the margin of the document. Usually, marginal glosses are set in a different typeface—and sometimes a different color—from the main discussion, and often they are separated from the main discussion by a vertical rule. Marginal glosses can be helpful in providing an overview of the longer explanation; however, they can also compete with the main discussion for the reader's attention.

- *Pull quotes.* You might not have heard the term *pull quotes,* but you see them all the time, particularly in newspapers and magazines. A pull quote is simply a quotation—usually just a sentence—pulled from the text and displayed in a larger size and generally in a different typeface. Often a pull quote is boxed. The purpose of the pull quote is to attract readers' attention and make them want to read the document. Pull quotes are not appropriate for some kinds of documents, such as most reports, because they look too informal. But they are increasingly popular in newsletters. Figure 13.25 shows pull quotes.

FIGURE 13.25
Pull Quotes

WRITER'S CHECKLIST

1. Have you analyzed your audience—their knowledge of the subject, their attitudes, their reasons for reading, and the kinds of tasks they will be carrying out?

2. Have you thought about what your readers will expect to see when they pick up the document?

3. Have you determined your own resources: time, money, and equipment?

4. Have you considered the best size for the document?

5. Have you considered the best paper for the document?

6. Have you considered the best binding for the document?

7. Have you thought about the most appropriate accessing tools to include, such as icons, color, dividers, tabs, and cross-reference tables?

8. Have you devised a style for headers and footers?

9. Have you devised a style for page numbering?

10. Have you drawn thumbnail sketches and page grids that define columns and white space?

11. Have you chosen typefaces that are appropriate to your subject?

12. Have you used appropriate members of the type families?

13. Have you used type sizes that are appropriate for your subject and audience?

14. Have you decided on whether to use left-justified text or justified text?

15. Have you chosen a line length that is appropriate for your subject and audience?

16. Have you chosen line spacing that is appropriate for your line length, subject, and audience?

17. Have you designed your title for clarity and emphasis?

18. In designing your headings, have you worked out a logical, consistent style for each level?

19. Where appropriate, have you used rules, boxes, screens, marginal glosses, and pull quotes?

20. If you have access to color, have you used it appropriately to highlight particular items such as safety warnings?

EXERCISES

Several of the following exercises call for you to write memos. For a discussion of memos, see Chapter 16.

1. Photocopy a page from a book or a magazine article in your field. Write a 1,000-word memo to your instructor describing and evaluating its design. What aspects of design that are discussed in this chapter do you see in the page design? Does the design violate any of the guidelines offered in this chapter? Which aspects of the design work effectively, and which could be improved? Attach the photocopy to the memo. (If the original is in color, either attach the original to your memo or attach a color photocopy.)

2. Write a memo to your instructor describing and evaluating the design of the four-page brochure that accompanies the game Scrabble® (see opposite page). What aspects of design that are discussed in this chapter do you see in the brochure? Does the design violate any of the guidelines offered in this chapter? Which aspects of the design work effectively and which could be improved?

3. Write a memo to your instructor describing and evaluating the design on page 356 (Bonneville, 1993, p. 240) from an environmental impact statement. What aspects of design that are discussed in this chapter do you see in the page? Does the design violate any of the guidelines offered in this chapter? Which aspects of the design work effectively and which could be improved?

REFERENCES

Berry, R., Mobley, K., & Turk, K. (1994). Preparing to document an object-oriented project. *Technical Communication, 41,* 643–652.

Biggs, J. R. (1980). *Basic typography.* New York: Watson-Guptill.

Bonneville Power Administration. (1993). *Resource programs: Final environmental impact statement, volume 1: Environmental analysis.* Washington, DC: U.S. Department of Energy.

Felker, D. B., Pickering, F., Charrow, V. R., Holland, V. M., & Redish, J. C. (1981). *Guidelines for document designers.* Washington, DC: American Institutes for Research.

Fischer, J. M. (1992). *Inside DesignCAD.* Carmel, IN: New Riders Publishing.

Haley, A. (1991). All caps: A typographic oxymoron. *U&lc, 18*(3), 14–15.

Keyes, E. (1993). Typography, color, and information structure. *Technical Communication, 40,* 638–654.

Poulton, E. (1968). Rate of comprehension of an existing teleprinter output and of possible alternatives. *Journal of Applied Psychology, 52,* 16–21.

White, J. V. (1990). *Great pages: A common-sense approach to effective desktop design.* El Segundo, CA: Serif Publishing.

Williams, T., & Spyridakis, J. (1992). Visual discriminability of headings in text. *IEEE Transactions on Professional Communication, 35,* 64–70.

Note: These pages are reduced 50% from the original.

Back cover

Special Features

The Official SCRABBLE® Players Dictionary (Second Edition) offers lots of features to help you play faster, better, and more confidently:
• No other dictionary includes all the two-to-eight letter words you'll find here.
• Only words that are permissible in SCRABBLE® can be found here. Proper names, words requiring hyphens or apostrophes, words considered foreign, and abbreviations have been omitted.
• All main entries are given a part-of-speech label, followed by appropriate inflected forms.
• All acceptable variant forms are shown alphabetically.
• All noun plurals are shown, as well as comparatives and superlatives of adjectives and adverbs.

Where to Get It

The Official SCRABBLE® Players Dictionary (Second Edition) costs only $14.95* at your local bookstore.

How to Order It

If the dictionary is out of stock, you can order it directly from the publishers. Send a check or money order in U.S. funds for $15.95,* which includes postage and handling, to:

MERRIAM-WEBSTER INC.
47 Federal Street
P.O. Box 281
Springfield, MA 01102
Attn: Consumer Sales

*Prices subject to change.
©1991, 1992 Milton Bradley Company. All Rights Reserved.
The gameboard shown in the book jacket photograph is
©1948 by Milton Bradley Company.

SCRABBLE®
C R O S S W O R D   G A M E
by Milton Bradley

Front cover

JOIN THE CLUB!

♦

The National
SCRABBLE®
Association

TURN THE PAGE TO FIND OUT HOW

Also Inside:
• OFFICIAL SCRABBLE® PLAYERS DICTIONARY OFFER
• FREE TILE REPLACEMENT

Page 2

The National
SCRABBLE®
Association

As a member of the National SCRABBLE® Association, you'll receive a membership card; a roster of SCRABBLE® Players Clubs in the U.S. and Canada; 3 special word lists to help improve your play; and a 1-year subscription to the SCRABBLE® News. That's 8 exciting issues with...
• News about clubs and tournaments in your area
• Official National SCRABBLE® Association Tournament Rules
• Special word lists and helpful playing hints
• The latest in strategy
• Challenging quizzes & puzzles
Fill out the form below (print clearly) and send to: NATIONAL SCRABBLE® ASSOCIATION, P.O. Box 700, Greenport, NY 11944

YES! I'd like to join the National SCRABBLE® Association. I enclose a check or money order for $15.00 ($20.00 for Canadian membership; $25.00 outside the 50 states and Canada).
Only U.S. funds will be accepted.

☐ *I am interested in starting a club in my area.*

NAME

STREET

CITY

STATE   ZIP

Are any of your tiles missing or defective?

If so, we want to replace them free of charge. Please list them on the form below and send it, with any defective tiles, to: MILTON BRADLEY COMPANY, Attn. Customer Service Department, 443 Shaker Road, East Longmeadow, MA 01028

TILE REPLACEMENT FORM
Please print clearly:
NAME   STREET

CITY   STATE   ZIP

GAME:
☐ SCRABBLE   ☐ DELUXE SCRABBLE   ☐ TRAVEL SCRABBLE   ITEM # (from package):

MISSING TILES: DEFECTIVE TILES (please send with form):

Page 3

IT'S OFFICIAL!

Have you ever had a word challenged that you knew was acceptable—yet it couldn't be found in your dictionary? Now there's a dictionary that was created especially for SCRABBLE® players...

The Official SCRABBLE® Players Dictionary

Second Edition!
Over 5,000 new entries!

Any standard abridged dictionary can be used to settle a SCRABBLE® challenge. But this easy-to-use, 683-page hardcover volume includes well over 100,000 words—and they're *all* acceptable SCRABBLE® words! That's why it's the official dictionary of first reference for all SCRABBLE® game tournaments sponsored by the National SCRABBLE® Association.

2.2.5 Anadromous Fish
Columbia River Basin

The Pacific Northwest supports a large number of anadromous fish (species that migrate downriver to the ocean to mature, then return upstream to spawn). The principal anadromous fish runs in the Columbia Basin are chinook, coho, sockeye salmon, and steelhead trout.

These fish are an important resource to the Pacific Northwest, both for their economic value to the sport and commercial fisheries, and for their cultural and religious value to the region's Indian Tribes and others.

The development of dam and reservoir projects on the Columbia and Snake Rivers and tributaries has reshaped the natural flows of these rivers. The use of storage reservoirs to capture runoff for later release results in reduced flows during the spring and early summer, when juvenile salmon and steelhead are migrating downstream to the ocean. Water velocities have also been reduced as a result of the increased cross-sectional area of the river due to run-of-river projects. These changes have slowed juvenile fish migration, exposing juvenile salmon and steelhead to predation and disease and impairing their ability to adapt to salt water when they reach the ocean. Additional mortality occurs as fish attempt to pass each dam on their downstream migration to the ocean.

Flow: Flow plays an important part in moving juvenile fish downstream to the ocean. In 1982, the Northwest Power Planning Council established a specific volume of water, known as the water budget, to increase river flows during the April 15 through June 15 period. This coincides with the peak out-migration of spring fish (predominately yearling chinook, steelhead, and sockeye), which depend on adequate river flow, particularly velocity, for a successful migration. The Federal hydro system is operated to provide this water each year.

Not only is flow important for moving juvenile fish downstream past the dams, but flow is an important component of insuring successful spawning and emergence of fall chinook on Vernita Bar (a gravel bar used by spawning fall chinook, located in the Hanford Reach of the Columbia River, downstream from Priest Rapids Dam). In 1988, BPA and the mid-Columbia operators signed a long-term Vernita Bar Agreement, which specifies protection requirements for fall chinook spawning, incubation and emergence on Vernita Bar.

Spill: Until adequate bypass systems are installed at all the dams, spill remains a necessary means of moving juvenile fish past dams. Planned fish spill now includes the negotiated Spill Agreement, as well as a restricted operation at Bonneville Dam by the Corps. Planned spill also includes spill levels specified by the Federal Energy Regulatory Commission for non-Federal projects. Planned spill does not include overgeneration spill (water which is spilled because there is no market for the energy it would produce) and is not changed as a result of the resource additions. Planned fish spills are met under all water conditions.

CHAPTER 14 Revising for Coherence

Perhaps the most durable axiom about writing is only three words long: *Writing is revising*. This chapter deals with revising for coherence. The chapter begins by describing the basic techniques of revising and then discusses how to revise the whole document for coherence. Next, the chapter discusses revising titles, headings, lists, and paragraphs.

Chapter 15 concentrates on revising for sentence-level effectiveness.

Understanding the Given/New Strategy

The process of understanding text is much more complicated than we used to think. Once we believed that word length and sentence length were the major factors in understanding text, but now we think other factors are more important. Because understanding occurs within a person, not within a document, it is a process that involves the reader's own knowledge, attitude, and reading behavior. In reading, a person extracts new information and tries to associate it with existing information.

This simple idea suggests ways to make your writing easy to understand. It is the basis of coherent writing, that is, writing that hangs together, on the level of the whole document, long passages, paragraphs, and sentences.

In 1974, Susan Haviland and Herbert Clark presented their given/new strategy, which has since been recognized as a major breakthrough in understanding how people acquire new information. Their theory is simple and sensible.

Every sentence that we read or hear contains both given and new information. *Given information* is what the listener or reader is expected to know already. *New information,* obviously, is what the person is not expected to know already. When we listen or read, we try to understand the information by searching our memories for information that matches the given information in the sentence. Once we find a match, we can attach the new information to it, thereby changing our original information. The given/new strategy is shown in Figure 14.1.

For example, the sentence "The proposal we submitted last month won the contract" consists of two pieces of information: (1) we submitted a proposal last month and (2) that proposal won the contract. Readers will first try to understand the first part—we submitted a proposal last month—by searching for a memory of the company's having submitted that proposal. Then they will add the second part—that proposal won the contract—to the first part. Thereafter, their memories will contain the updated information.

However, if you wrote that same sentence, "The proposal we submitted last month won the contract," but your reader did not know that the company had submitted a proposal last month, communication would break down, at least momentarily, as the reader tried (unsuccessfully) to remember the proposal. The

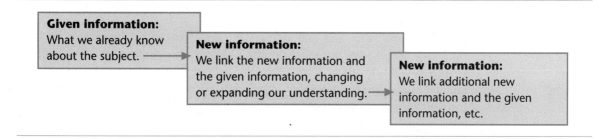

Given information:
What we already know about the subject.

New information:
We link the new information and the given information, changing or expanding our understanding.

New information:
We link additional new information and the given information, etc.

FIGURE 14.1

The Given/New Strategy

reader could continue to try to remember, give up in frustration, or create a meaning: "We must have submitted a proposal last month, and it apparently won the contract." If the created meaning is accurate, the only cost has been the reader's labor; if the created meaning is inaccurate, the price could be much higher.

Haviland and Clark tested their theory by studying the effects of different ways of presenting given and new information. They hypothesized that readers would have an easier time understanding the link if it were clear and explicit rather than less clear and implicit. Their research confirmed their hypothesis. Readers found links such as those in Set 1, below, easier to understand than those in Set 2.

SET 1:   We got some beer out of the trunk. The beer was warm.

SET 2:   We checked the picnic supplies. The beer was warm.

Again, these results are common sense. In each set, the first sentence provides the given information that readers rely on in reading the second sentence. Because the word *beer* is used in both sentences in Set 1, readers easily add the new information to the old. However, in Set 2, the writer has made comprehension a little more difficult for readers. When they get to the *beer* in the second sentence, they have to infer that the beer must have been part of the *picnic supplies* mentioned in the previous sentence. Obviously, this linkage—"beer = picnic supplies"—is less clear than "beer = beer."

What does the given/new strategy have to do with technical communication? It supports a simple, fundamental idea: you will be most successful at transmitting information to people if you deliver new information gradually, building on what they already know. This idea applies to all units of discourse, from large to small. The rest of this chapter will show how this idea applies to revising the different elements of a document.

Techniques Used in Revising

Revising is the process of making sure the document says what you want it to say—and that it says it professionally. As is the case with other phases of the writing process, everyone uses a different technique for revising. But the important point is that you must have a technique. You cannot hope to simply read through your document, waiting for the problems to leap off the page. Some of them might; most won't.

There are three basic ways to revise a document:

- Revise it by yourself.
- Obtain help from someone else.
- Test readers as they use the document.

Revising by Yourself

Before you give a document to someone else for help in revision, and before you test the document with real users, revise it yourself. Although this process won't help you find and fix all the problems, you'll make a lot of progress. Follow the five guidelines shown in Figure 14.2.

- *Let it sit.* You cannot revise your document effectively right after you finish writing it. Although you can immediately identify some of the smaller writing problems—such as errors in spelling or grammar—you cannot objectively assess the quality of what you have said: whether the information is clear, comprehensive, and coherent.

  To revise your draft effectively, set it aside for a time. Give yourself at least a night's rest. If possible, work on something else for a few days. This will give you time to "forget" the draft and approach it more as your readers will. In a few minutes, you will see problems that would have escaped your attention even if you had spent hours revising it right after completing the draft.

  You're probably thinking that you don't have time to set it aside for a night. There is no easy solution to the time problem. You just have to try

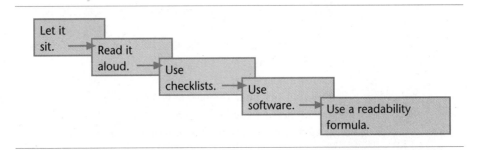

FIGURE 14.2

Revising by Yourself

to plan your schedule so that you are done drafting a few days before the document is due.

- *Read it aloud.* Reading a document aloud gives you a better sense of what it sounds like than reading it silently. Any teacher of writing has seen this effect a hundred times. A student submits a paper in which the first sentence is missing a necessary word. The teacher asks the student to look at the sentence and fix the problem, and she replies that she can't see any problem. When asked to read the sentence out loud, the student immediately "hears" the mistake. Big problems, too, such as illogical reasoning or missing evidence, also seem more obvious when you hear your document. Reading your draft out loud to identify possible areas for improvement will make you feel foolish. Welcome to the club.

- *Use checklists.* Before airplane pilots take off, they perform a series of safety checks written out in checklists; they can't count on remembering all the different safety checks they need to carry out. The same applies to writing.

  This text is full of checklists that summarize the major points you should look for in revising different kinds of documents. However, any checklist that someone else has written for you has a big drawback: it doesn't necessarily cover the things you need to check most. For example, you might often confuse the use of *its* and *it's* in your writing. Don't count on a generic checklist to include this issue. Instead, keep a list of the points your instructors make about your writing and then customize the checklists you use.

  Many writers construct checklists that help them remember to review the document three times: once for overall meaning and clarity, once for organization and development, and once for correctness. Of course, each of the three checklists includes numerous items.

- *Use software.* Many software programs can help you spot different kinds of style problems. Like a checklist, however, a program is a generalized tool that cannot understand your context—the purpose and audience of your communication—and therefore cannot be relied on to offer relevant and wise advice all the time. Revision software—spelling checkers, style programs, and thesaurus programs—is discussed in Chapter 15.

- *Use a readability formula.* A readability formula is a quasi-mathematical technique for assessing the relative difficulty of the writing. Chapter 15 describes these formulas and their limitations.

Obtaining Help from Someone Else

Once you have done what you can by yourself, enlist a colleague to help you revise. In general, you can turn to two kinds of people for help:

- *Subject-matter experts.* If, for instance, you have written an analysis of alternative fuels for automobiles, you could ask an automotive expert to review it. In the working world, important documents are routinely reviewed by technical experts, lawyers, and marketing specialists before being released to the public.

- *People who are like the eventual readers.* People who fit the profile of the eventual readers will help you see problems you didn't notice in revising by yourself. People more knowledgeable about the subject than your eventual readers will understand the document even if it isn't clear; therefore, they might not be sufficiently critical.

If possible, show the document to both kinds of people. But think twice before asking a personal friend or a spouse. You want objective advice. You don't want that advice to be affected by your relationship—and you don't want the relationship to be affected by the advice.

When you ask someone to help you, provide specific instructions. Tell the person as much as you can about your audience and purpose and about the tasks you still have to complete before you finish the document. Then tell the person—or, better yet, write out—the kinds of problems you want him or her to look for.

After reading the draft, your colleagues may be able to point out its strong points, unclear passages, and sections that need to be added, deleted, or revised.

Testing Readers as They Use the Document

Analyzing a document can provide only limited information; it cannot tell you how well the document will work when it gets into the hands of the people who will use it. To find out that information, you must test readers as they use the document.

Testing readers entails four commonsense steps, as shown in Figure 14.3.

- *Determining what you want to test.* The questions you want to answer are likely to vary from one kind of document to the next. For a sales brochure, you might want to determine whether it makes a positive impression on potential customers. For a set of instructions, you might want to see whether a general reader can carry out the task safely, easily, and effectively.

- *Choosing representative test participants.* A basic principle of testing is that the test participants must match the real users of the document as closely as possible. If the real users will be mostly men in their twenties, test men in their twenties. Try to match the test participants to the eventual users in terms of education, experience, and all the other factors involved in analyzing audience as discussed in Chapter 4.

- *Making the tests realistic.* After recruiting representative test participants, make the tests themselves realistic. If you are testing a user's manual for a

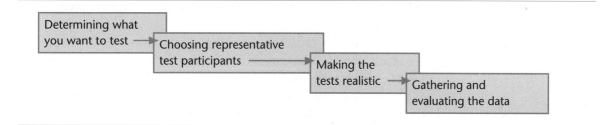

word-processing program, ask the test subject to use the manual to carry out representative word-processing tasks on the appropriate computer.

For documents that communicate information but don't require that the reader actually do anything, you have to be more inventive. You can ask participants to do the following:

– Tell you what they are thinking as they read the document.
– Answer questions after they read the document.
– Write a paraphrase of a particular piece of information after they read the document.
– Read the document, then read a second document and state whether the second document accurately reflects what was said in the first document.

- *Gathering and evaluating the data.* The data you collect from testing the documents must be evaluated as carefully as the data from any experiment. Basic principles of experimentation apply, such as having enough test subjects to ensure statistically valid results and making sure that the process of gathering data does not unintentionally influence the test subjects' behavior. If you interview the test subjects, ask each the same questions in the same way, and record the information accurately and precisely.

Testing is discussed in detail in Chapter 19.

Revising the Whole Document for Coherence

In looking for different kinds of problems to fix, most writers prefer a top-down pattern. That is, they look first for the largest and most important problems and proceed to the smaller, less important ones so that they don't waste time fixing awkward paragraphs they might eventually throw out. For this reason, many writers begin revising by considering the whole document.

In revising the whole document, answer the following five questions:

- *Have you left anything out in turning your outline into a draft?* In the process of drafting, it is easy to leave out a topic. Check the document against the outline to see that all the topics in the outline are presented in the document itself.

- *Have you included all the elements your readers expect to see?* In planning the document, you thought about what the readers would expect to see. For instance, readers of a report expect to see certain elements, such as a transmittal letter, a table of contents, and an abstract. If your document is missing any of the elements the readers expect, they might be distracted and be unable to understand the document easily.

- *Are all the elements presented consistently?* In drafting quickly, you probably were not paying close attention to consistency. Check to see that parallel items are presented consistently. For instance, if you are using 18-point boldface text for your first-level headings, make sure all your first-level headings are presented that way. And check for grammatical parallelism, particularly in lists.

- *Is the emphasis appropriate throughout the document?* If the document has a table of contents, study it carefully. If a relatively minor topic seems to be treated at great length, with numerous subheadings, check the text itself. Maybe you created more headings at that point than you did in treating some of the other topics. But if your treatment of the minor topic is in fact excessive, mark passages for possible condensing.

- *Are the cross-references accurate?* In a technical document you often refer to other elements in the document. For instance, you will cross-reference an appendix or refer to a table or a figure that appears on a different page. Check to see that items you refer to are where you say they are.

Revising Titles for Coherence

The title is a key element in a document because it becomes the readers' first piece of "given" information. If the document is to cohere, its various topics have to relate clearly to the title.

Readers look for the title to provide two important pieces of information: your *subject* and *purpose*. Don't revise your title until you complete the document. The reason is simple: until you finish the document, you cannot be sure of the extent to which your subject and purpose have changed since you began to plan the document.

Precision is the key to a good title. If you are writing a feasibility study on the subject of offering free cholesterol screening at your company, make sure the title contains the key terms *cholesterol screening* and *feasibility*. The following title would be effective:

Offering Free Cholesterol Screening at Thrall Associates: A Feasibility Study

If your document is an internal report discussing company business, you might not need to identify the organization. The following would be clear:

Offering Free Cholesterol Screening: A Feasibility Study

Of course, you could present the purpose before the subject:

A Feasibility Study of Offering Free Cholesterol Screening

Do not substitute general terms, such as *health screening* for *cholesterol screening*. Keep in mind that key terms from your title might be used in various kinds of indexes; the more precise your terms, the more useful your readers will find the title.

Your readers should be able to translate your title into a clear and meaningful sentence. For instance, the title "A Feasibility Study of Offering Free Cholesterol Screening" could be translated into the following sentence: "This document reports on whether it is feasible to offer free cholesterol screening for our employees." Notice, however, what happens when the title is incomplete: "Free Cholesterol Screening." All the reader knows is that the document has something to do with free cholesterol screening. But is the writer recommending that free cholesterol screening be instituted—or discontinued? Or is the writer reporting on how well an existing program is working out?

Following are more examples of effective titles:

Choosing a Laptop: A Recommendation

An Analysis of the Kelly 1013 Packager

Open Sea Pollution-Control Devices: A Summary

A Forecast of Smoking Habits in the United States in the Coming Decade

Revising Headings for Coherence

A heading is a lower-level title inside a document. A clear and informative heading is vital in technical communication because it announces the subject of the discussion that follows it. This announcement helps your readers understand what they will be reading or, in some cases, helps them decide whether to read it at all. For the writer, a heading eliminates the need to announce the subject in a sentence such as "Let us now turn to the advantages of the mandatory enrollment process."

In revising your headings, follow these six guidelines:

- *Avoid long noun strings.* They are hard to understand:

  Production Enhancement Proposal Analysis Techniques

Instead, put some of the prepositions back into the phrase to make the title clearer.

Techniques for Analyzing the Proposal for Enhancing Production

This version says, more clearly than the noun-string version, that the writer is going to describe the techniques. See Chapter 15 for a detailed discussion of noun strings.

- *Be informative.* You could add even more information to the preceding example: some indication of how many techniques will be described.

Three Techniques for Analyzing the Proposal for Enhancing Production

And you can go one step further, by indicating what you wish to say about the three techniques:

Advantages and Disadvantages of the Three Techniques for Analyzing the Proposal for Enhancing Production

Don't worry if the heading seems too long; clarity is more important than brevity.

- *Use the question form for less knowledgeable readers.*

What Are the Three Techniques for Analyzing the Proposal for Enhancing Production?

The question form is easier for less knowledgeable readers to understand (Benson 1985).

- *Use the "how to" form in instructional material, such as manuals.*

How to Analyze the Proposal for Enhancing Production

- *Use the verbal form (-ing) to suggest a process.*

Analyzing the Proposal for Enhancing Production

- *Avoid back-to-back headings.* Headings should be followed by text, not by lower-level headings. In other words, avoid the following:

3. Approaches to Neighborhood Policing

    3.1 Community Policing

Text should separate the two headings, as in this example:

3. Approaches to Neighborhood Policing
The past decade has seen unprecedented strains on the social institutions of society, as well as a further weakening of the family structure. One of the results of these twin problems is the increase in violent crime. Police departments throughout the country have struggled to keep up with the increase, especially the rise in violence among young people. In this section, we discuss two new approaches to neighborhood policing that appear to be working successfully in cities of different sizes and with different problems: community policing and police-spon-

sored activities for youth. For each of the two approaches, we will . . .

3.1 Community Policing

The text after the heading "3. Approaches to Neighborhood Policing," called an *advance organizer,* introduces the material in section 3. Like any other kind of introduction, the forecast indicates the background, purpose, scope, and organization of the discussion that follows it. In this way, the forecast improves coherence by giving readers an overview of the discussion before they encounter the details.

For information on how to format the headings, see Chapter 13.

Revising Lists for Coherence

Lists are fundamental to technical communication because they add a visual dimension to the text, making it easier for readers to understand the discussion. As you revise your text, look for opportunities to turn traditional paragraphs into lists.

Not every paragraph should be a list, of course. But lists work especially well in conveying any kind of information that can be enumerated or expressed in a sequence. For instance, the following sentences introduce material that would be well suited to a list:

Contractors typically use one of four methods to shore the walls of an excavation.

Solar-energy research stalled in the 1980s for three major reasons.

When you arrive at the site, please make the following arrangements.

The next chapter discusses the use of lists in individual sentences, but lists can also enhance coherence in longer passages. For readers, the chief advantage of a list format is that the information is easier to read and to remember. The reason is that the logic of the discussion is more clearly evident on the page. The key terms presented in the list are set off with bullets or numbers, enabling readers to see them at a glance. Thus the coherence of the discussion is enhanced: readers see the overall structure before reading the detailed discussion.

For you as the writer, turning paragraphs into lists has four advantages:

- *It forces you to look at the big picture.* As you draft, it is easy to lose sight of the information outside the paragraph you are working on. But when you revise, looking for opportunities to create lists, you force yourself to study the key idea in each paragraph. This practice increases your chances of noticing that an important item is missing or that an item needs to be clarified.

- *It forces you to examine the sequence of the items.* As you transform some of your paragraphs into bulleted or numbered lists with key phrases, you get a clearer look at the sequence of information. If you want to make a change, the list format makes it easy.

- *It forces you to create a clear lead-in.* In the lead-in, you can add a number:

  Auto imports declined last year because of four major factors:

  Of course, you can add the same number in a forecast leading into a traditional paragraph, but you are less likely to be thinking in those terms because you will not be focusing on the bulleted items.

- *It forces you to tighten and clarify the prose.* When you see an opportunity to make a list, look for a word, phrase, or sentence to identify each item. Your focus shifts from weaving sentences together in a paragraph to highlighting key ideas. And once you have formatted the list, look at it critically and revise it until it is as clear and concise as you can make it.

Figure 14.4 (based on Cohen & Grace, 1994, p. 15) shows the same passage displayed in a paragraph form and in a list form. The authors are discussing the idea that engineers have a special social responsibility.

Notice in Figure 14.4 that turning the paragraph into a list accomplishes two things:

- It forces the writer to create headings that provide a sharp focus for each bulleted entry.
- It enables the writer to delete the wordy topic sentences from the paragraph version. By deleting these sentences, the writer saves space so that the list version of the passage is not significantly longer than the paragraph version, despite the indentations and extra white space.

Revising Paragraphs for Coherence

The preceding discussion on the virtues of listing information is not meant to suggest that you could or should eliminate paragraphs. In fact, you will often write paragraphs within listed passages, as you saw in Figure 14.4. But regardless of whether you are writing traditional paragraphs or paragraphs within lists, try to enhance their coherence as you revise.

The two kinds of paragraphs are body paragraphs and transitional paragraphs.

A *body paragraph* is the basic structural unit for communicating the technical information. A body paragraph could be defined as a group of sentences (or sometimes a single sentence) that is complete and self-sufficient but that also contributes to a larger discussion. The challenge of creating an effective paragraph in technical communication is to make sure, first, that all the sentences

Currently, there are three conceptions of the relation between engineering as a profession and society as a whole.

The first conception is that there is no relation. Engineering's proper regard is properly instrumental, with no constraints at all. Its task is to provide purely technical solutions to problems.

The second conception is that engineering's role is to protect. It must be concerned, as a profession, with minimizing the risk to the public. The profession is to operate on projects as presented to it, as an instrument; but the profession is to operate in accordance with important safety constraints, which are integral to its performing as a profession.

The third conception is that engineering has a positive social responsibility to try to promote the public good, not merely to perform the tasks that are set for it, and not merely to perform those tasks such that risk is minimized or avoided in performing them. Rather, engineering's purpose as a profession is to promote the social good.

Currently, there are three conceptions of the relation between engineering as a profession and society as a whole.

- *There is no relation.* Engineering's proper regard is properly instrumental, with no constraints at all. Its task is to provide purely technical solutions to problems.

- *The engineer's role is to protect society.* Engineering is concerned, as a profession, with minimizing the risk to the public. The profession is to operate on projects as presented to it, as an instrument; but the profession is to operate in accordance with important safety constraints, which are integral to its performing as a profession.

- *The engineer's role is to promote social responsibility.* Engineering has a positive social responsibility to try to promote the public good, not merely to perform the tasks that are set for it, and not merely to perform those tasks such that risk is minimized or avoided in performing them. Rather, engineering's purpose as a profession is to promote the social good.

FIGURE 14.4

Paragraph Format and List Format

clearly and directly substantiate one main point, and second, that the whole paragraph follows logically from the material that precedes it.

Readers tend to pause between paragraphs (not between sentences) to digest the information given in one paragraph and link it with that given in the previous paragraphs. For this reason, the paragraph is the key unit of composition. Readers might forgive or at least overlook a slightly fuzzy sentence. But if they can't figure out what a paragraph says or why it appears where it does, communication is likely to break down.

A *transitional paragraph* helps readers to progress from one major point to another. In most cases, it summarizes the previous point and helps readers understand how it relates to the next one.

The following example of a transitional paragraph is taken from a manual explaining how to write television scripts. The writer has described six principles of writing for an episodic program, including introducing characters, pursuing the plot, and resolving the action at the end of the episode.

The six basic principles of writing for episodic television, then, are the following:

- Reintroduce the characters.
- Make the extra characters episode-specific.
- Present that week's plot swiftly.
- Make the characters react according to their personalities.

- Resolve the plot neatly.
- Provide a denouement that hints at further developments.

But how do you put these six principles into action? The following section provides specific how-to instructions.

Notice that the first sentence contains the word "then" to signal that it is introducing a summary. Note, too, that the final sentence of the transitional paragraph clearly indicates the relationship between what precedes it and what follows it.

Structure Paragraphs Clearly

If you draft quickly—as you should—your paragraphs will need a lot of revision before they are easy to read and understand. A hasty paragraph often starts off with a number of details: about who worked on the problem before and what equipment or procedure they used; about the ups and downs of the project, the successes and setbacks; about specifications, dimensions, and computations. The paragraph winds its way down the page until, finally, the writer concludes: "No problems were found."

This structure—moving from the particular details to the general statement—reflects the way we think: we accumulate the details and then draw conclusions from them. But the paragraph will be easier to read if the main point is presented first and then supported in the remainder of the paragraph.

The Topic Sentence

Help your readers. Put the point—the *topic sentence*—up front. Technical communication should be clear and easy to read, not full of suspense. If a paragraph describes a test you performed on a piece of equipment, include the result in your first sentence:

The point-to-point continuity test on Cabinet 3 revealed no problems.

Then go on to explain the details. If the paragraph describes a complicated idea, start with an overview:

Mitosis occurs in five stages: (1) interphase, (2) prophase, (3) metaphase, (4) anaphase, and (5) telophase.

Then describe each phase. In other words, put the "bottom line" on top.

Notice, for instance, how difficult the following paragraph is to read because the writer structured the discussion in the same order that she performed her calculations:

Our estimates are based on our generating power during eight months of the year and purchasing it the other four. Based on the 1995 purchased power rate of $0.034/KW (January through April cost data) inflating at 8 percent annually, and a constant coal cost of $45-$50, the projected 1996 savings resulting from a conversion to coal would be $225,000.

Putting the bottom line on top makes the paragraph much easier to read. Notice how the writer adds a numbered list after the topic sentence:

> The projected 1996 savings resulting from a conversion to coal are $225,000. This estimate is based on three assumptions: (1) that we will be generating power during eight months of the year and purchasing it the other four, (2) that power rates inflate at 8 percent from the 1995 figure of $0.034/KW (January through April cost data), and (3) that coal costs remain constant at $45-$50.

The topic sentence in technical communication functions just as it does in any other kind of writing: it summarizes or forecasts the main point of the paragraph.

The Support

The support follows the topic sentence. The purpose of the support is to make the topic sentence clear and convincing. Sometimes a few explanatory details can provide all the support needed. In the paragraph about estimated fuel savings presented earlier, for example, the writer simply fills in the assumptions used in making the calculation: the current energy rates, the inflation rate, and so forth. Sometimes, however, the support must carry a heavier load: it has to clarify a difficult thought or defend a controversial one.

Because every paragraph is unique, it is impossible to generalize about the exact function of the support. Usually, however, the support fulfills one of the following five roles:

- to define a key term or idea included in the topic sentence
- to provide examples or illustrations of the situation described in the topic sentence
- to identify causes: factors that led to the situation
- to define effects: implications of the situation
- to defend the assertion made in the topic sentence

The techniques that writers use in developing the support include those used in most nonfiction writing, including definition, comparison and contrast, classification and partition, and causal analysis. Another key to supporting a paragraph is the use of specific, concrete details. These techniques are described in detail in Part Three.

Keep Paragraphs to a Manageable Length

How long should a paragraph of technical communication be? In general, 75 to 125 words will provide enough length for a topic sentence and four or five supporting sentences. Long paragraphs are more difficult to read than short paragraphs simply because readers have to concentrate longer. Long, unbroken stretches of type can so intimidate some readers that they actually skip them.

But don't let an arbitrary minimum guideline about length take precedence over your analysis of the audience and purpose. Often you will need to write very brief paragraphs. You might need only one or two sentences—to introduce a graphic, for example. A transitional paragraph also is likely to be quite short. If a brief paragraph fulfills its function, let it be. Do not combine two ideas in one paragraph to achieve a minimum word count.

Although it is confusing to include more than one basic idea in a paragraph, often you will need to break up one idea into two or more paragraphs. An idea that would require 200 or 300 words to develop probably should not be squeezed into one paragraph.

The following example shows how a writer addressing a general audience divided one long paragraph into two:

> High-tech companies have been moving their operations to the suburbs for two main reasons: cheaper, more modern space and a better labor pool. A new office complex in the suburbs will charge anywhere from half to two-thirds of the rent charged for the same square footage in the city. And that money goes a lot further, too. The new office complexes are bright and airy, with picture windows looking out on lush landscaping. New office space is already wired for the computers; and exercise clubs, shopping centers, and even libraries are often on-site.
>
> The second major factor attracting high-tech companies to the suburbs is the availability of experienced labor. Office workers and middle managers are abundant; many suburbanites, especially women returning to the labor force after their children start school, are highly trained and willing to make the short trip to the office complex. In addition, the engineers and executives, who tend to live in the suburbs anyway, are happy to forgo the commuting, the city wage taxes, and the noise and stress of city life.

A strict approach to paragraphing would have required one paragraph, not two, because all the information presented supports the topic sentence that opens the first paragraph. Many readers, in fact, could easily understand a one-paragraph version. However, the writer found a logical place to create a second paragraph and thereby improved the communication.

Following is another approach: making each "reason for moving to the suburbs" a separate paragraph.

> High-tech companies have been moving their operations to the suburbs for two main reasons: cheaper, more modern space and a better labor pool.
>
> First, office space is a bargain in the suburbs. A new office complex will charge anywhere from half to two-thirds of the rent charged for the same square footage in the city. And that money goes a lot further, too. The new office complexes are bright and airy, with picture windows looking out on lush landscaping. New office space is already wired for the computers; and exercise clubs, shopping centers, and even libraries are often on-site.
>
> Second, experienced labor is plentiful. Office workers and middle managers are abundant; many suburbanites, especially women returning to the labor force after their children start school, are highly trained and willing to make the short trip to the office complex. In addition, the engineers and executives, who tend to live in the suburbs

anyway, are happy to forgo the commuting, the city wage taxes, and the noise and stress of city life.

The original topic sentence becomes a transitional paragraph that leads clearly and logically into the two body paragraphs.

Of course, the writer could also use a list format:

> High-tech companies have been moving their operations to the suburbs for two main reasons:
>
> - *Cheaper, more modern space.* Office space is a bargain in the suburbs. A new office complex will charge anywhere from half to two-thirds of the rent charged for the same square footage in the city. And that money goes a lot further, too. The new office complexes are bright and airy, with picture windows looking out on lush land-scaping. New office space is already wired for the computers; and exercise clubs, shopping centers, and even libraries are often on-site.
> - *A better labor pool.* Office workers and middle managers are abundant; many subur-banites, especially women returning to the labor force after their children start school, are highly trained and willing to make the short trip to the office complex. In addition, the engineers and executives, who tend to live in the suburbs anyway, are happy to forgo the commuting, the city wage taxes, and the noise and stress of city life.

Use Coherence Devices within and between Paragraphs

After you have revised the main structure of the paragraph—the topic sentence and the support—make sure that the paragraph is coherent. In a coherent paragraph, thoughts are linked together logically and clearly. Parallel ideas are expressed in parallel grammatical constructions. Even if the paragraph already moves smoothly from sentence to sentence, emphasize the coherence in three ways, as shown in Figure 14.5.

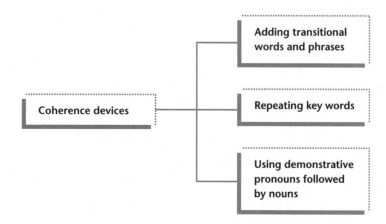

FIGURE 14.5

Coherence Devices

Transitional Words and Phrases

Transitional words and phrases help the reader understand a discussion by pointing out the direction the thoughts are following. Table 14.1 lists the most common logical relationships between two thoughts and some of the common transitions that express those relationships:

In the following examples, the first versions contain no transitional words and phrases. Notice how much clearer the second versions are.

WEAK: Neurons are not the only kind of cell in the brain. Blood cells supply oxygen and nutrients.

IMPROVED: Neurons are not the only kind of cell in the brain. *For instance,* blood cells supply oxygen and nutrients.

WEAK: The project was originally expected to cost $300,000. The final cost was $450,000.

IMPROVED: The project was originally expected to cost $300,000. *However,* the final cost was $450,000.

WEAK: The manatee population of Florida has been stricken by an unknown disease. Marine biologists from across the nation have come to Florida to assist in manatee disease research.

IMPROVED: The manatee population of Florida has been stricken by an unknown disease. *As a result,* marine biologists from across the nation have come to Florida to assist in manatee-disease research.

Use transitional words to maintain coherence *between* paragraphs just as you use them to maintain coherence *within* paragraphs. Place the transitions as close as possible to the beginning of the second element. For example, the link between two sentences within a paragraph should be near the start of the second sentence:

The new embossing machine was found to be defective. *However,* the warranty on the machine will cover replacement costs.

TABLE 14.1	*Transitional Words and Phrases*
Relationship	**Transitions**
addition	also, and, finally, first (second, etc.), furthermore, in addition, likewise, moreover, similarly
comparison	in the same way, likewise, similarly
contrast	although, but, however, in contrast, nevertheless, on the other hand, yet
illustration	for example, for instance, in other words, to illustrate
cause-effect	as a result, because, consequently, hence, so, therefore, thus
time or space	above, around, earlier, later, next, to the right (left, west, etc.), soon, then
summary or conclusion	at last, finally, in conclusion, to conclude, to summarize

The link between two paragraphs should be near the start of the second paragraph:

The complete system would be too expensive for us to purchase now _____
_____.

In addition, a more advanced system is expected on the market within six months
_____
_____.

Key Words

Repeating key words—usually nouns—helps readers follow the discussion. Notice in the following example how the first version can be confusing:

UNCLEAR: For months the project leaders carefully planned their research. The cost of the work was estimated to be over $200,000. (What is the work: the planning or the research?)

CLEAR: For months the project leaders carefully planned their *research*. The cost of the *research* was estimated to be over $200,000.

Out of a misguided desire to be interesting, some writers keep changing their important terms. *Plankton* becomes *miniature seaweed,* then *the ocean's fast food.* Leave this kind of word game to TV sportscasters; technical communication must be clear and precise.

Of course, too much repetition can be boring. You can vary nonessential terms, so long as you don't sacrifice clarity.

SLUGGISH: The purpose of the new plan is to *reduce the problems* we are seeing in our accounting operations. We hope to see a *reduction in the problems* by early next quarter.

BETTER: The purpose of the new plan is to *reduce the problems* we are seeing in our accounting operations. We hope to see an *improvement* by early next quarter.

Demonstrative Pronouns Followed by Nouns

In addition to using transitional words and phrases and repeating key phrases, using demonstrative pronouns—*this, that, these,* and *those*—can help you maintain the coherence of a discussion by linking ideas securely. In almost all cases, demonstrative pronouns should serve as adjectives rather than as pronouns; that is, they should be followed by nouns. In the following examples, notice that a demonstrative pronoun by itself can be confusing.

UNCLEAR: New screening techniques are being developed to combat viral infections. These are the subject of a new research effort in California. (What is being studied in California: new screening techniques or viral infections?)

CLEAR: New screening techniques are being developed to combat viral infections. *These techniques* are the subject of a new research effort in California.

UNCLEAR: The task force could not complete its study of the mine accident. This
was the subject of a scathing editorial in the union newsletter. (What
was the subject of the editorial: the mine accident or the task force's
inability to complete its study of the accident?)

CLEAR: The task force failed to complete its study of the mine accident. *This
failure* was the subject of a scathing editorial in the union newsletter.

Even when the context is clear, a demonstrative pronoun used without a noun
refers readers backward to an earlier idea and therefore interrupts readers'
progress.

INTERRUPTIVE: The law firm advised that the company initiate proceedings. This
caused the company to search for a second legal opinion.

FLUID: The law firm advised that the company initiate proceedings. *This
advice* caused the company to search for a second legal opinion.

Transitional words and phrases, the repetition of key words, and demon-
stratives cannot *give* your writing coherence: they can only help readers appreciate the coherence that already exists. Your job is, first, to make sure your
writing is coherent and, second, to highlight that coherence.

Turning a "Writer's Paragraph" into a "Reader's Paragraph"

To demonstrate the strategies that improve coherence, the following examples
show how these techniques can improve a weak paragraph. The paragraph is
from a status report written by a branch manager of a utility company. In it the
writer explains how he decided on a method to increase the company's business
within his particular branch. (The sentences are numbered to clarify the marginal comments.)

Sentence 1 focuses on the
two alternatives, not on
the final decision the
writer made.
Sentences 2 and 3
describe one alternative,
then the other.
Sentence 4 explains why
the first alternative was
undesirable.
Sentences 5 and 6 explain
why he chose the second
alternative.
Throughout the para-
graph, the focus is on the
process of the study, not
on the results.

(1) There were two principal alternatives considered for improving Montana Branch.
(2) The first alternative was to drill and equip additional sources of supply with sufficient capacity to provide for the present and projected system deficiencies. (3) The second alternative was to provide for said deficiencies through a combination of additional
sources of supply and a storage facility. (4) Unfortunately, groundwater studies which
were conducted in the Southeast Montana area by the consulting firm of Smith and
Jones indicated that although groundwater is available within this general area of our
system, it is limited as to quantity, and considerable separation between said sources.
(5) This being the case, it becomes necessary to utilize the sources that are available or
that can be developed in the most efficient manner, which means operating them in
conjunction with a storage facility. (6) In this way, the sources only have to be capable
of providing for the average demand on a maximum day, and the storage facility can
be utilized to provide for the peaking requirements plus fire protection. (7) Consequently, the second alternative as mentioned hereinabove was determined to be the
more desirable alternative.

First, let's be fair. The paragraph has been taken out of its context—a 17-page report—and was never meant to stand alone on a page. Also, it was written not for the general reader but for an executive of the water company—someone who, in this case, is technically knowledgeable in the writer's field. Still, an outsider's analysis of an essentially private communication can at least isolate the weaknesses.

Following is the writer's paragraph translated into a reader's paragraph. (The sentences are numbered to clarify the marginal comments.)

> *The topic sentence clearly states the main point of the paragraph. Sentence 2 justifies the assertion from sentence 1. Sentence 3 goes back to describe the other alternative they considered. Sentence 4 explains why that alternative was rejected.*

(1) We found that the best way to improve the Montana branch would be to add a storage facility to our existing supply sources. (2) Currently, we can handle the average demand on a maximum day; the storage facility will enable us to meet peaking requirements and fire-protection needs. (3) In conducting our investigation, we considered developing new supply sources with sufficient capacity to meet current and future needs. (4) This alternative was rejected, however, when our consultants (Smith and Jones) did groundwater studies that revealed that insufficient groundwater is available and that the new wells would have to be located too far apart if they were not to interfere with each other.

This revision is superior to the original in that it is shorter and more direct, and therefore easier to read and understand. Notice the use of transitional words ("currently" in sentence 2 and "however" in sentence 4); key words ("storage facility," "sources," and "needs"); and demonstrative pronouns with nouns ("this alternative").

The only possible objection to the streamlined version is that it is *too* clear, that it leaves the writer vulnerable in case his decision turns out to have been wrong. But if the decision doesn't work out, the writer will be responsible anyway, and poor writing will not endear him to his supervisor. Good writing is the best bet under any circumstances.

WRITER'S CHECKLIST

1. Have you revised the document yourself, by
   - ❑ letting it sit?
   - ❑ reading it aloud?
   - ❑ using checklists?
   - ❑ using revision software?
   - ❑ using a readability formula?

2. Have you revised the document by obtaining help from
   - ❑ appropriate subject-matter experts?
   - ❑ people similar to the eventual readers?

3. If possible and appropriate, have you tested readers as they use the document?

4. Have you revised the whole document by checking to make sure that
   - ❑ you haven't left anything out in turning your outline into a draft?
   - ❑ you have included all the elements your readers expect to see?
   - ❑ all the elements are presented consistently?
   - ❑ all the cross-references are accurate?
   - ❑ the emphasis is appropriate throughout the document?

5. Have you revised the title so that it
   - ❑ clearly states the audience and purpose of the document?
   - ❑ is sufficiently precise and informative?

6. Have you revised the headings?
   ❑ Have you avoided long noun strings?
   ❑ Is each heading sufficiently informative?
   ❑ Have you used the question form for less knowl-edgeable readers?
   ❑ Have you used the "how to" form in instruc-tional material, such as manuals?
   ❑ Have you used the verbal form (-ing) to suggest a process?
   ❑ Have you avoided back-to-back headings by including forecasts?
7. Have you looked for opportunities to turn tradi-tional paragraphs into lists?
8. Have you revised your paragraphs so that each one
   ❑ begins with a clear topic sentence?
   ❑ has adequate and appropriate support?
   ❑ is not too long for readers?
   ❑ uses coherence devices such as transitional words and phrases, repetition of key words, and demon-stratives followed by nouns?

EXERCISES

1. Write a one-paragraph evaluation for each of the following titles. How clearly does the title indicate the subject and purpose of the document? On the basis of your analysis, rewrite each title.
   a. Recommended Forecasting Techniques for Hal-dane Company
   b. Robotics in Japanese Manufacturing
   c. A Study of Disc Cameras
   d. Agriculture in the West: A 10-Year View
   e. Synfuels—Fact or Hoax?
2. Write a one-paragraph evaluation for each of the following headings. How clearly does the heading indicate the subject of the text that will follow it? On the basis of your analysis, rewrite each title to make it more clear and informative. Invent any nec-essary details.
   a. Multigroup Processing Technique Review Board Report Findings
   b. The Great Depression of 1929
   c. Low-level Radiation and Animals
   d. Minimize Down Time
   e. Intensive Care Nursing

3. Turn the following passage (based on Snyder, 1993) into a list format. The subject is bioremediation: the process of using microorganisms to restore natural environmental conditions.

Scientists are now working on several new research areas. One area involves using microorganisms to make some compounds less dangerous to the envi-ronment. Although coal may be our most plentiful fossil fuel, most of the nation's vast Eastern reserve cannot meet air-pollution standards because it emits too much sulfur when it is burned. The problem is that the aromatic compound dibenzothiothene (DBT) attaches itself to hydrocarbon molecules, producing sulfur dioxide. But the Chicago-based Institute of Gas Technology last year patented a bacterial strain that consumes the DBT (at least 90 percent, in recent lab trials) while leaving the hydrocarbon molecules intact.

A second research area is the genetic engineering of microbes in an attempt to reduce the need for toxic chemicals. In 1991, the EPA approved the first genetically engineered pesticide. Called Cellcap, it incorporates a gene from one microbe that produces a toxin deadly to potato beetles and corn borers into a thick-skinned microbe that is hardier. Even then, the engineered bacteria are dead when applied to the crops.

A third research area is the use of microorganisms to attack stubborn metals and radioactive waste. Microbes have been used for decades to concentrate copper and nickel in low-grade ores. Now researchers are exploiting the fact that if certain bacteria are given special foods, they excrete enzymes that break down metals and minerals. For example, researchers at the U.S. Geological Survey found that two types of bacteria turn uranium from its usual form—one that easily dissolves in water—into another—one that turns to a solid that can be easily removed from water. They are now working on doing the same for other radioactive waste.

4. Provide a topic sentence for each of the following paragraphs:
   a. ————————————————. All service cen-ters that provide gas and electric services in the tri-county area must register with the TUC, which is empowered to carry out unannounced inspections periodically. Additionally, all service centers must adhere to the TUC's Fair Deal Regulations, a set of standards that encom-passes every phase of the service-center opera-tions. The Fair Deal Regulations are meant to guarantee that all centers adhere to the same

standards of prompt, courteous, and safe work at a fair price.

b. _____. The reason for this difference is that a larger percentage of engineers working in small firms may be expected to hold high-level positions. In firms with fewer than 20 engineers, for example, the median income was $42,200. In firms of 20 to 200 engineers, the median income was $40,345. For the largest firms, the median was $38,600.

5. Develop the following topic sentences into full paragraphs:

   a. Job candidates should not automatically choose the company that offers the highest salary.

   b. Every college student should learn at least the fundamentals of computer science.

   c. The one college course I most regret not having taken is _____.

   d. Sometimes two instructors offer contradictory advice about how to solve the same kind of problem.

6. In the following paragraph, transitional words and phrases have been removed. Add an appropriate transition in each blank space. Where necessary, add punctuation.

   As you know, the current regulation requires the use of conduit for all cable extending more than 18″ from the cable tray to the piece of equipment. _____ conduit is becoming increasingly expensive: up 17% in the last year alone. _____ we would like to determine whether the NRC would grant us any flexibility in its conduit regulations. Could we _____ run cable without conduit for lengths up to 3′ in low-risk situations such as wall-mounted cable or low-traffic areas? We realize _____ that conduit will always remain necessary in high-risk situations. The cable specifications for the Unit Two report to the NRC are due in less than two months; _____

we would appreciate a quick reply to our request, as this matter will seriously affect our materials budget.

7. In each of the following exercises, the second sentence begins with a demonstrative pronoun. Add a noun after the demonstrative to enhance the coherence.

   a. The Zoning Commission has scheduled an open hearing for March 14. This _____ will enable concerned citizens to voice their opinions on the proposed construction.

   b. The university has increased the number of parking spaces, instituted a shuttle system, and increased parking fees. These _____ are expected to ease the parking problems.

   c. Congress's decision to withdraw support for the supercollider in 1994 was a shock to the U.S. particle-physics community. This _____ is seen as instrumental in the revival of the European research community.

REFERENCES

Benson, P. (1985). Writing visually: Design considerations in technical publications. *Technical Communication, 32*(4), pp. 35–39.

Cohen, S., & Grace, D. (1994). Engineers and social responsibility: An obligation to do good. *IEEE Technology and Society, 13*(3), pp. 12–19.

Haviland, S. E., & Clark. H. H. (1974). What's new? Acquiring new information as a process in comprehension. *Journal of Verbal Learning and Verbal Behavior, 13,* pp. 512–521.

Snyder, J. D. (1993). Off-the-shelf bugs hungrily gobble our nastiest pollutants. *Smithsonian, 24*(1), pp. 66+.

Weiss, E. H. (1982). *The writing system for engineers and scientists.* Englewood Cliffs, NJ: Prentice Hall.

Revising for Sentence Effectiveness

The previous chapter discussed techniques for improving the coherence of the whole document, of long passages, and of paragraphs. This chapter concentrates on sentences and the clauses, phrases, and words they contain.

Technical communication is workplace writing; it is meant to get a job done, not to show off the writer's skill. Therefore, the advice offered in this chapter is based on a simple idea: the sentences and words of technical communication should be clear, simple, concise, and easy to understand.

Your readers should not be aware of your presence. They should not notice that you have a wonderful vocabulary or that your sentences flow beautifully—even if those things are true. In the best technical communication, the writer fades into the background.

This is as it should be. Few people read technical communication for pleasure. Most readers either must read it as part of their work or want to keep abreast of new developments. People read technical communication to gather information, not to appreciate the writer's flair. For this reason, experienced writers do not try to be fancy. The old saying has never been more appropriate: *Write to express, not to impress.*

Using Revision Software Carefully

Revision software—spell checkers, style programs, and thesaurus programs—becomes more sophisticated and powerful all the time, and it can be of value. However, you need to know what it can and cannot do.

- A *spell checker* reviews the words you have typed, alerting you when it sees one word that isn't in its dictionary. Although that word might be misspelled, it also might be a correctly spelled word or a proper name that isn't in the dictionary. You can add the word to the dictionary so that the spell checker will know in the future that the word is not misspelled.

You are probably aware of the danger of relying on a spell checker: it cannot tell whether you have typed the correct word; it can tell you only whether the word is in its dictionary. Therefore, if you have typed "We need too dozen test tubes," the spell checker will not see a problem.

- A *thesaurus* lists synonyms or related words. A thesaurus program has the same strengths and weaknesses as a printed thesaurus: if you know the word you are looking for, but can't quite think of it, the thesaurus will help you remember it. But the listed terms might not be related to the key term closely enough to function as synonyms. Unless you are aware of the shades of difference, however, you might be tempted to substitute an inappropriate word. For example, the entries for the word *famous* in the thesaurus that is part of a well-known word processor include *infamous* and *notorious*. If you use either of them as a synonym for *famous*, you could embarrass yourself badly.

- A *style program* helps you identify and fix potential stylistic problems. For instance, a style program counts such factors as sentence length, number of passive voice constructions and expletives, and type of sentence. Many style programs identify abstract words and suggest more specific ones. Many point out sexist terms and provide nonsexist alternatives. Many point out fancy words, such as *recalcitrant*, and suggest substitutes, such as *stubborn*. Figure 15.1 (*Grammatik*, 1991) shows a screen from a popular style program.

The screen in Figure 15.1 shows some of the basic statistics for the document: readability scores (see the next section), as well as paragraph, sentence, and word statistics. Style programs provide numerous statistics, as well as line-by-line and word-by-word analyses of the document.

But remember that style programs do not understand who you are or what you are saying, and they do not know to whom you are writing. Although style programs can point out your use of the passive voice, they cannot tell you whether the passive voice is preferable to the active voice in the particular sentence. For instance, the style program in the word processor I am using has this to say about the first sentence in this paragraph: "It is preferable not to begin a sentence with *but*." Preferable according to whom? What about your audience, purpose, and subject? Do they have anything to do with the appropriateness of your word choice?

In a way, style programs make your job as a writer more challenging: by pointing out potential problems, they force you to decide about issues

```
┌──────────────────────────────────────────────────────────────────┐
│ [−]              Document Statistics for tour.txt              [▲] │
│ Problems marked/detected: 8/37                                     │
├────────────────────────────────────────────────────────────────── │
│                     READABILITY STATISTICS                         │
│                                                                    │
│ Flesch Reading Ease:    59          Flesch-Kincaid Grade Level:  9 │
│ Gunning's Fog Index:    13                                         │
├────────────────────────────────────────────────────────────────── │
│                     PARAGRAPH STATISTICS                           │
│                                                                    │
│ Number of paragraphs:  8            Average length:     4.2 sentences │
├────────────────────────────────────────────────────────────────── │
│                     SENTENCE STATISTICS                            │
│                                                                    │
│ Number of sentences:   34           Passive voice:        1        │
│ Average length:        16.6 words   Short  (< 14 words):  11       │
│ End with '?':          3            Long   (> 38 words):  1        │
│ End with '!':          2                                           │
├────────────────────────────────────────────────────────────────── │
│                     WORD STATISTICS                                │
│ Number of words:       567          Average length:     4.91 letters │
│ Prepositions:          49           Syllables per word:  1.55      │
└──────────────────────────────────────────────────────────────────┘
```

FIGURE 15.1

Screen from a Style
Program

that you might not have considered otherwise. But the payoff is that a wise use of the programs will give you a better document.

Revision software cannot replace a careful reading by you and by other people. Revision programs cannot identify unclear explanations, contradictions, inaccurate data, inappropriate tone, and so forth. Use the revision programs, revise your document yourself, and then get help from someone you trust.

Using Readability Formulas Cautiously

Readability formulas use mathematical techniques to help you determine how difficult a sample of writing is to read and understand. More than 100 readability formulas exist; government agencies and private businesses use them increasingly in an attempt to improve the writing their workers produce. And style programs often include readability formulas. For these reasons, you should become familiar with the concept behind them.

Most readability formulas are based on the idea that short words and sentences are easier to understand than long ones. One of the more popular formulas, Robert Gunning's "Fog Index," works like this:

1. Find the average number of words per sentence, using a 100-word passage.

2. Find the number of difficult words in that same passage. *Difficult words* are words of three or more syllables, except for proper names, combinations of

simple words (such as *horsepower*), and verbs whose third syllable is *-es* or *-ed* (such as *contracted*).

3. Add the average sentence length and the number of difficult words.
4. Multiply this sum by 0.4.

1. Average number of words per sentence: 13
2. Number of difficult words: <u>16</u>
3. Sum of 1 and 2:   29
4. $29 \times 0.4 = 12$

The result—12—represents an approximate grade level.

Readability formulas are easy to use, and it is appealing to think you can be objective in assessing the difficulty of your writing, but unfortunately they have not been proven to work. They do not reflect accurately how difficult it is to read a piece of writing (Battison & Goswami, 1981; Selzer, 1981). Other factors, such as the format and organization of the passage and the clarity of the sentences, have a greater effect on readability than do word length and sentence length (Huckin, 1983).

The problem with readability formulas is that they attempt to evaluate a portion of the communication—the words on paper—without considering the reader. A "difficult word" for a lawyer might not be difficult for a biologist, and vice versa. And someone who is interested in the subject being discussed will understand more than someone who isn't. Good writing has to be well thought out and carefully structured. The sentences must be clear, and the vocabulary must be appropriate for the readers. *Then* the writing will be readable.

A better way to assess the difficulty of the writing is to use the cloze procedure. The *cloze procedure* is an interactive measure of readability; that is, it examines people as they read the text. To perform the cloze procedure, randomly choose three passages from the text. In each passage, delete every fifth word, and then ask the reader to fill in the blanks. A person who scores 57 percent correct can understand 90 percent of the text. See Bormuth (1966) for more information on the cloze procedure.

Structuring Effective Sentences

Good technical communication is characterized by clear, correct, and graceful sentences that convey information without calling attention to themselves. This section consists of seven guidelines for structuring effective sentences, as shown in Figure 15.2.

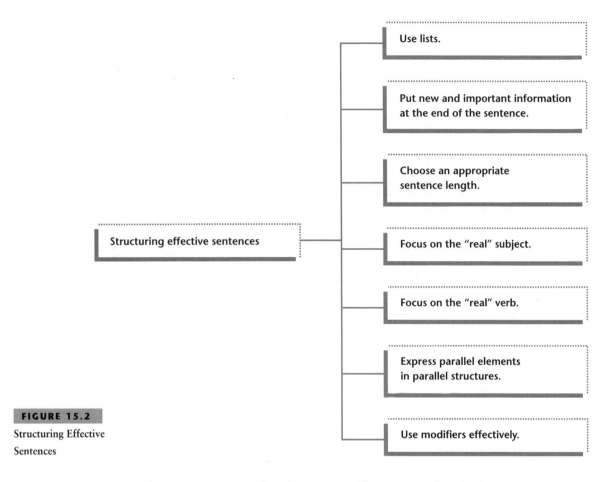

FIGURE 15.2

Structuring Effective
Sentences

[Figure boxes, top to bottom, branching from "Structuring effective sentences":]

Use lists.

Put new and important information at the end of the sentence.

Choose an appropriate sentence length.

Focus on the "real" subject.

Focus on the "real" verb.

Express parallel elements in parallel structures.

Use modifiers effectively.

Use Lists

list

Chapter 14 contained a discussion of lists as a technique for organizing pararaph-length passages of text. This section discusses lists as a technique for communicating shorter passages.

Many sentences in technical communication are long and complicated:

> We recommend that more work on heat-exchanger performance be done with a larger variety of different fuels at the same temperature, with similar fuels at different temperatures, and with special fuels such as diesel fuel and shale-oil-derived fuels.

Here readers cannot concentrate fully on the information because they are trying to remember all the *with* phrases following "done." If they could "see" how many phrases they had to remember, their job would be easier.

Revised as a list, the sentence is easier to follow:

> We recommend that more work on heat-exchanger performance be done:
> - with a larger variety of different fuels at the same temperature
> - with similar fuels at different temperatures
> - with special fuels such as diesel fuels and shale-oil-derived fuels

In this version, the placement of the words on the page reinforces the meaning. The bullets direct the reader's eyes to the three items in a series. And the fact that each item begins at the same left margin helps, too.

If you don't have enough space to list the items vertically, or if you are not permitted to do so, number the items within the sentence:

> We recommend that more work on heat-exchanger performance be done (1) with a larger variety of different fuels at the same temperature, (2) with similar fuels at different temperatures, and (3) with special fuels such as diesel fuels and shale-oil-derived fuels.

As you create lists, follow these six guidelines:

- *Indent the items in the list.* The amount of indentation depends on the length of the items. Single words or short phrases might be indented so that the list appears centered; longer items might be indented only two or three characters.

- *Set off each listed item with a number, a letter, or a symbol (usually a bullet).*

  – Use numbered lists to suggest sequence (as in the numbered steps in a set of instructions) or priority (the first item is the most important). Sometimes writers number a list to emphasize the total number of items (as in the "Seven Warning Signals of Cancer" from the American Cancer Society). For sublists, use lowercase letters:

    1. Item
        a. subitem
        b. subitem

    2. Item
        a. subitem
        b. subitem

  – Use bullets when you do not wish to suggest sequence or priority. Make sure you don't use numbers for lists of people; everyone except number 1 gets offended. For sublists, use hyphens.

    - Item
      – subitem
      – subitem

  – Use an open (unshaded) box (□) for checklists. See Chapter 12 for more information on checklists.

- *Break up long lists.* Because most people can remember only 5–9 items, break up lists that contain 10 or more items. Figure 15.3 shows how to break up a long list.

Original List	Revised List
Tool Kit:	Tool Kit:
• hand saw	• Saws
• coping saw	– hand saw
• hacksaw	– coping saw
• compass saw	– hacksaw
• adjustable wrench	– compass saw
• box wrench	
• Stillson wrench	• Wrenches
• socket wrench	– adjustable wrench
• open-end wrench	– box wrench
• Allen wrench	– Stillson wrench
	– socket wrench
	– open-end wrench
	– Allen wrench

FIGURE 15.3

Breaking Up a Long List

Carliner (1987) recommends breaking up a bulleted list of more than 5 items and a numbered list of more than 10. (The numbers give readers extra cues to help them remember the information.)

- *Present the items in a parallel structure.* Parallelism is discussed later in this chapter. Figure 15.4 shows how to revise an unparallel list.

  In Figure 15.4, the unparallel list is sloppy, a mixture of noun phrases (items 1, 3, 4, and 5), a verb phrase (item 2), and a participial phrase preceded by a dependent clause (item 6). The revision uses parallel verb phrases and deemphasizes the dependent clause in item 6 by placing it after the verb phrase.

- *Structure and punctuate the lead-in correctly.* Although standards vary from organization to organization, the most common is to use a grammatically complete lead-in followed by a colon, as shown in the following examples:

Nonparallel	Parallel
Here is the sequence we plan to follow:	Here is the sequence we plan to follow:
1. construction of the preliminary proposal	1. write the preliminary proposal
2. do library research	2. do library research
3. interview with the Bemco vice president	3. interview the Bemco vice president
4. first draft	4. write the first draft
5. revision of the first draft	5. revise the first draft
6. after we get your approval, typing of the final draft	6. type the final draft, after we receive your approval

FIGURE 15.4

Parallelism in Lists

Following are the three main assets:

The three main assets are as follows:

The three main assets are the following:

If you cannot use a grammatically complete lead-in, use a colon, a dash, or no punctuation at all:

The committee found that the employee
- did not cause the accident
- acted properly immediately after the accident
- reported the accident properly

- *Punctuate the list correctly.* Rules for punctuating lists vary. Like most stylistic issues in technical communication, there is no accepted standard, so you should find out whether your organization has a preference. If it does not, you can generally punctuate lists as follows:

– If the items are sentence fragments, use a lowercase letter at the start and do not use a period or a comma at the end.

The new facility will offer three advantages:
- lower leasing costs
- easier commuting distance
- a greater pool of potential workers

The last item in a list of fragments is generally *not* followed by a period. The white space to the right of the last listed item, as well as the white space that separates the list from the next line, clearly indicates the end of the list. Some writers, however, prefer to add a period after the last item.

– If the items are complete sentences, use an uppercase letter at the start and a period at the end.

The new facility will offer three advantages:
- The leasing costs will be lower.
- The commuting distance for most of the employees will be shorter.
- The pool of potential workers will be larger.

– If the items are fragments followed by complete sentences, begin the fragment with an uppercase letter and end with a period. Then begin the complete sentences with uppercase letters and end them with periods.

The new facility will offer three advantages:
- Lower leasing costs. The lease will cost $1,800 per month; currently we pay $2,300.
- Easier commuting distance. According to a recent questionnaire, our workers now spend an average of 18 minutes traveling to work. At the new location, the average would drop to 14 minutes.

- A greater pool of potential workers. In the last decade, the population has begun to shift westward, to the area of the new facility. We would be able to increase our potential work force, especially in the semiskilled and managerial categories.

– If the list consists of two different kinds of items—some fragments, some fragments followed by complete sentences—punctuate all the items the same way: with uppercase letters and periods.

The new facility will offer three advantages:
- Lower leasing costs.
- Easier commuting distance. According to a recent questionnaire, our workers now spend an average of 18 minutes traveling to work. At the new location, the average would drop to 14 minutes.
- A greater pool of potential workers. In the last decade, the population has begun to shift westward, to the area of the new facility. We would be able to increase our potential work force, especially in the semiskilled and managerial categories.

Note that in the design of lists, the second and subsequent lines in an item, called *turnovers*, are indented. Turnovers are aligned under the first letter of the first line. In other words, the bullet or the number at the start of the list extends out to the left of the text. This form, called *hanging indentation*, highlights the bullet or the number and thereby helps the reader see the organization of the passage.

Put New and Important Information at the End of the Sentence

new

Chapter 14 discussed the given/new concept: we learn by adding new information to the information we already know. This principle applies not only to large elements in a text but also to individual sentences. In general, your sentences will be easier to understand and more emphatic if you put the new information at the end of the sentence.

For instance, your company is experiencing labor problems, and you want to describe the possible results. Structure the sentence like this:

Because of the labor problems, we anticipate a three-week delay.

In this case, the *three-week delay* is the new information.

If, however, your readers know that there will be a three-week delay but don't know the reason for it, structure the sentence this way:

We anticipate the three-week delay in the production schedule because of labor problems.

Here, *labor problems* is the new information. Notice how the articles *the* and *a* signal the old and new information. In the second sentence, the definite article *the* is used with given information; in the first sentence, the indefinite article *a* is used with new information.

Try not to end the sentence with qualifying information that blunts the impact of the new information.

WEAK: The joint could fail under special circumstances.

IMPROVED: Under special circumstances, the joint could fail.

Also, put new or difficult terms at the end of the sentence.

WEAK: You use a wired glove to point to objects.

IMPROVED: To point to objects, you use a wired glove.

Choose an Appropriate Sentence Length

len

Sometimes sentence length affects the quality of your writing. In revising a draft, you might want to compute the average sentence length for a representative passage of writing. (Many software programs do this for you.)

No firm guidelines for appropriate sentence length exist. In general, however, an average length of 15–20 words is effective for most technical communication. A succession of 10-word sentences would be choppy; a series of 35-word sentences would probably be too demanding. And any succession of sentences of approximately the same length can distract the reader.

Avoid Overly Long Sentences

How long is too long? There is no simple answer, because the ease of reading a sentence depends on its length, vocabulary, and structure; the reader's knowledge of the topic; and the purpose of the communication.

Yet often your draft will include sentences such as the following:

The construction of the new facility is scheduled to begin in March, but it might be delayed by one or even two months by winter weather conditions, which can make it impossible or nearly impossible to begin excavating the foundation.

This 40-word sentence can be tiring to read. To make it more readable, divide it into two sentences:

The construction of the new facility is scheduled to begin in March. However, construction might be delayed until April or even May by winter weather conditions, which can make it impossible or nearly impossible to begin excavating the foundation.

Sometimes an overly long sentence can be fixed by creating a list:

WEAK: To connect the CD player to the amplifier, first be sure that the power is off on both units, then insert the plugs firmly into the jacks (the red plug into the right-channel jack and the black plug into the left-channel jack), making sure that you leave a little slack in the connecting cord to prevent shock or vibration.

IMPROVED: To connect the CD player to the amplifier, follow these steps:

1. Be sure that the power is off on both units.
2. Insert the plugs firmly into the jacks. The red plug goes into the right-channel jack, and the black plug into the left-channel jack.

Make sure that you leave a little slack in the connecting cord to prevent shock or vibration.

As this revision suggests, sometimes the best way to communicate technical information is to switch to a more visually oriented format by using lists or graphics.

Avoid Overly Short Sentences

Just as sentences can be too long, they can also be too short, as in the following example:

> The fan does not oscillate. It is stationary. The blade is made of plastic. This is done to increase safety. Safety is especially important because this design does not include a guard around the blade. A person could be seriously injured by putting his or her hand into the turning blade.

The problem is not that the word count of these sentences is too low, but rather that the sentences are choppy and contain too little information. The best way to revise a passage such as this is to combine sentences, as in the following example:

> The fan is stationary, not oscillating, with a plastic blade. Because the fan does not have a blade guard, the plastic blade is intended to prevent injury if a person accidentally touches it when it is moving.

One symptom of excessively short sentences is that they needlessly repeat key terms. Consider combining sentences, as in the following examples:

SLUGGISH: Computronics, a medium-sized consulting firm, consists of many diverse groups. Each group handles and develops its own contracts.

BETTER: Computronics, a medium-sized consulting firm, consists of many diverse groups, each of which handles and develops its own contracts.

SLUGGISH: I have experience working with various microprocessor-based systems. Some of these systems include the Z80, 6800 RCA 1802, and the AIM 6502.

BETTER: I have experience working with various microprocessor-based systems, including the Z80, 6800 RCA 1802, and the AIM 6502.

Focus on the "Real" Subject
rs

The conceptual or "real" subject of the sentence should also be the grammatical subject, and it should be prominent. Don't bury the real subject in a prepositional phrase following a useless or phantom grammatical subject. In the following examples, notice how the limp subjects disguise the real subjects. (The grammatical subjects are italicized.)

WEAK: The *use* of this method would eliminate the problem of motor damage.

STRONG: This *method* would eliminate the problem of motor damage.

WEAK: The *presence* of a six-membered lactone ring was detected.

STRONG: A six-membered lactone *ring* was detected.

Another way to make the subject of the sentence prominent is to reduce the number of grammatical *expletives: it is, there is,* and *there are.* In most cases, these constructions just waste space.

WEAK:	There is no alternative for us except to withdraw the product.
STRONG:	We have no alternative except to withdraw the product.

WEAK:	It is hoped that testing the evaluation copies of the software will help us make this decision.
STRONG:	I hope that testing the evaluation copies of the software will help us make this decision.

Often, as in this second example, the expletive *it is* is used along with the passive voice. See the discussion of the passive voice later in this chapter.

Expletives are not errors. Rather, they are conversational expressions that can sometimes help the reader by emphasizing the information that follows them.

WITH THE EXPLETIVE:	It is hard to say whether the recession will last more than a few months.
WITHOUT THE EXPLETIVE:	Whether the recession will last more than a few months is hard to say.

The version without the expletive is a little more difficult to understand because the reader has to read and remember a long subject—"Whether the recession will last more than a few months"—before getting to the verb—"is." However, the sentence could also be rewritten in other ways to make it easy to understand and to eliminate the expletive.

I don't know whether the recession will last for more than a few months.

Nobody really knows whether the recession will last more than a few months.

Use the search function of your word-processing program to find most weak subjects: usually they're right before the word *of.* Expletives are also easy to find.

Focus on the "Real" Verb

rv

A "real" verb, like a "real" subject, should stand out in every sentence. Few stylistic problems weaken a sentence more than nominalizing verbs. Writers nominalize verbs by changing them into nouns and then adding another verb, usually a weaker one, to clarify the meaning. *To install* becomes *to effect an installation; to analyze* becomes *to conduct an analysis.* Notice how nominalizing the verbs makes the following sentences both awkward and unnecessarily long. (The nominalized verbs are italicized.)

WEAK:	Each *preparation* of the solution is done twice.
STRONG:	Each solution is prepared twice.

WEAK:	An *investigation* of all possible alternatives was undertaken.
STRONG:	All possible alternatives were investigated.

WEAK:	*Consideration* should be given to an *acquisition* of the properties.
STRONG:	We should consider acquiring the properties.

Like expletives, nominalizations are not errors. Many common nouns are nominalizations; in addition, writers can often use nominalizations effectively to summarize an idea from a previous sentence. The nominalizations in the following example are italicized.

> The new *legislation* could delay our *entry* into the HDTV *market*. This *delay* could cost us millions.

Some software programs search for the most common nominalizations. With any word-processing program, you can identify most of the nominalizations if you search for character strings such as *tion, ment, sis, ence, ing,* and *ance.* If you search for the word *of,* you will also find many nominalizations.

Express Parallel Elements in Parallel Structures //

A sentence is parallel if its coordinate elements are expressed in the same grammatical form: that is, its clauses are either passive or active, its verbs are either infinitives or participles, and so on. Creating and sustaining a recognizable pattern makes the sentence easier to follow.

Notice how faulty parallelism weakens the following sentences.

NONPARALLEL:	Our present system *is costing* us profits and *reduces* our productivity. (unparallel verbs)
PARALLEL:	Our present system is *costing* us profits and *reducing* our productivity.

NONPARALLEL:	The dignitaries *watched* the launch, and the crew *was applauded.* (unparallel voice [see the discussion of voice later in this chapter])
PARALLEL:	The dignitaries *watched* the launch and *applauded* the crew.

NONPARALLEL:	*The typist should follow* the printed directions; *do not change* the originator's work. (unparallel mood: the first clause is subjunctive; the second, imperative)
PARALLEL:	*The typist should* follow the printed directions and *not change* the originator's work.

When creating parallel constructions, make sure that parallel items in a series do not overlap, thus changing or confusing the meaning of the sentence:

CONFUSING:	The speakers will include partners of law firms, businesspeople, and civic leaders.

CLEAR: The speakers will include businesspeople, civic leaders, and partners of law firms.

The problem with the original sentence is that "partners" appears to apply to "businesspeople" and "civic leaders." The revision solves the problem by rearranging the items so that "partners" cannot apply to the other two groups in the series.

CONFUSING: We need to buy more lumber, hardware, tools, and hire the subcontractors.

CLEAR: We need to buy more lumber, hardware, and tools, and we need to hire the subcontractors.

The problem with the confusing sentence is that the writer has two different ideas joined inappropriately. The first idea is that we need to buy three things: lumber, hardware, and tools. The second idea is that we need to hire the subcontractor. In the clear revision, each of these two ideas has its own subject and verb.

Use Modifiers Effectively   Technical communication is full of *modifiers*—words, phrases, and clauses that describe other elements in the sentence. To make your meaning clear, you must communicate to your readers whether a modifier provides necessary information about the word or phrase it refers to (its *referent*) or simply provides additional information. Further, you must make sure that the referent itself is always clearly identified.

Distinguish between Restrictive and Nonrestrictive Modifiers

mod

A *restrictive modifier,* as the term implies, restricts the meaning of its referent: that is, it provides information necessary to identify the referent. In the following examples, the restrictive modifiers are italicized:

The aircraft *used in the exhibitions* are slightly modified.

Please disregard the notice *you just received from us.*

In most cases, the restrictive modifier doesn't require a relative pronoun, such as *that* or *which*. If you choose to use a pronoun, however, use *that*:

The aircraft *that* are used in the exhibits are slightly modified.

(If the pronoun refers to a person or persons, use *who*.) Notice that restrictive modifiers are not set off by commas.

A *nonrestrictive modifier* does not restrict the meaning of its referent: the information it provides is not necessary to identify the referent. In the following examples, the nonrestrictive modifiers are italicized:

The Hubble telescope, *intended to answer fundamental questions about the origin of the universe,* was repaired in 1994.

When you arrive, go to the Registration Area, *which is located on the second floor.*

Like the restrictive modifier, the nonrestrictive modifier usually does not require a relative pronoun. If you use one, however, choose *which* (*who* or *whom* when referring to a person). Note that nonrestrictive modifiers are separated from the rest of the sentence by commas.

Avoid Misplaced Modifiers

mm

The placement of the modifier often determines the meaning of the sentence. Notice, for instance, how the placement of *only* changes the meaning in the following sentences.

Only Turner received a cost-of-living increase last year.
  (Meaning: Nobody else received one.)

Turner received *only* a cost-of-living increase last year.
  (Meaning: He didn't receive a merit increase.)

Turner received a cost-of-living increase *only* last year.
  (Meaning: He received a cost-of-living increase as recently as last year.)

Turner received a cost-of-living increase last year *only.*
  (Meaning: He received a cost-of-living increase in no other year.)

Misplaced modifiers—those that appear to modify the wrong referent—are common in technical communication. In general, the solution is to place the modifier as close as possible to its intended referent. Frequently, the misplaced modifier is a phrase or a clause:

MISPLACED: The subject of the meeting is the future of geothermal energy *in the downtown Webster Hotel.*

CORRECT: The subject of the meeting *in the downtown Webster Hotel* is the future of geothermal energy.

MISPLACED: *Jumping around nervously in their cages,* the researchers speculated on the health of the mice.

CORRECT: The researchers speculated on the health of the mice *jumping around nervously in their cages.*

A *squinting modifier* is a special kind of misplaced modifier. A squinting modifier is placed ambiguously between two potential referents, so that the reader cannot tell which one is being modified:

UNCLEAR: We decided *immediately* to purchase the new system.
            (Did we decide immediately, or did we decide to make the purchase immediately?)

CLEAR: We *immediately* decided to purchase the new system.

CLEAR: We decided to purchase the new system *immediately.*

> UNCLEAR: The people who worked on the Eagle assembly line *reluctantly* picked up their last paychecks.
> (Did they work reluctantly, or did they pick up their last checks reluctantly?)
>
> CLEAR: The people who worked *reluctantly* on the Eagle assembly line picked up their last paychecks.
>
> CLEAR: The people who worked on the Eagle assembly line picked up their last paychecks *reluctantly.*

A subtle form of misplaced modification often occurs with the *correlative constructions,* such as *either . . . or, neither . . . nor,* and *not only . . . but also:*

> NONPARALLEL: The new refrigerant not only decreases energy costs but also spoilage losses.
>
> PARALLEL: The new refrigerant decreases not only energy costs but also spoilage losses.

In this example, "decreases" applies to both "energy costs" and "spoilage losses." Therefore, the first half of the correlative construction should follow "decreases." Note that if the sentence contains two different verbs, the first half of the correlative construction should precede the verb:

> The new refrigerant not only decreases energy costs but also reduces spoilage losses.

Avoid Dangling Modifiers

dm

A *dangling modifier* has no referent in the sentence:

> DANGLING: Trying to solve the problem, the instructions seemed unclear.

In this sentence, the writer has not identified the person doing the trying. To correct the problem, rewrite the sentence to put the clarifying information either *within* the modifier or *next to* the modifier:

> CORRECT: *As I was trying to solve the problem,* the instructions seemed unclear.
>
> CORRECT: Trying to solve the problem, *I thought* the instructions seemed unclear.

Sometimes you can correct a dangling modifier by switching from the *indicative mood* (a statement of fact) to the *imperative mood* (a request or command):

> DANGLING: To initiate the procedure, the BEGIN button should be pushed.
>
> CORRECT: To initiate the procedure, push the BEGIN button.

In the imperative, the referent—in this case, *you*—is understood.

Choosing an Appropriate Level of Formality

lev

Although no standard definition of levels of formality exists, most experts agree that there are three levels:

INFORMAL: The Acorn 560 is a real screamer. With 100 meg of pure computing power, it slashes through even the thickest spreadsheets before you can say 2 + 2 = 4.

MODERATELY FORMAL: With its 100 MHz microprocessor, the Acorn 560 can handle even the most complicated spreadsheet problems quickly.

HIGHLY FORMAL: With a 100 MHz microprocessor, the Acorn 560 is a high-speed personal computer designed for computation-intensive applications such as large spreadsheets.

In general, technical communication is written in either a moderately formal or a formal style.

As you revise your writing to achieve an appropriate tone, think about your audience, subject, and purpose:

- *Audience.* You would use a more formal style in writing to a group of retired executives than to a group of college students; you would use a more formal style in writing to the company vice president than to your subordinates.

- *Subject.* You would use a more formal style in writing about a serious subject—safety regulations, important projects—than about preparations for the office Christmas party.

- *Purpose.* You would use a more formal style in a report to shareholders than in a company newsletter. Instructions tend to be relatively informal, often using the second person and the imperative mood. As discussed in Chapter 18, sometimes instructions are quite informal; they use contractions, the second person, and (occasionally) humor.

In general, it is better to err on the side of more formality rather than less. Avoid using an informal style in any writing that you do at the office, for two reasons:

- *Informal writing tends to be imprecise.* In the example about the computer, what exactly is a *screamer,* and what does it mean to *slash through* a problem?

- *Informal writing can be embarrassing.* If your boss unexpectedly sees an e-mail you wrote to a colleague, you might wish that it didn't begin, "How ya doin', loser?"

Choosing the Right Words and Phrases

Effective technical communication consists of the right words and phrases in the right places. The following section includes the three guidelines shown in Figure 15.5.

Be Clear and Specific

Follow these seven guidelines to make your writing clear and specific:

- Use active and passive voice appropriately.
- Be specific.
- Avoid unnecessary jargon.
- Use positive constructions.
- Avoid long noun strings.
- Avoid clichés.
- Avoid euphemisms.

Use Active and Passive Voice Appropriately

voice

In the active voice, the subject of the sentence performs the action expressed by the verb. In the passive voice, the subject receives the action. (In the following examples, the subjects are italicized.)

ACTIVE:	*Brushaw* drove the launch vehicle.
PASSIVE:	The launch *vehicle* was driven by Brushaw.
ACTIVE:	Many *physicists* support the big-bang theory.
PASSIVE:	The big-bang *theory* is supported by many physicists.

In most cases, the active voice works better than the passive voice. A sentence in the active voice more clearly emphasizes the agent. In addition, the active-voice sentence is shorter, because it does not require a form of the *to be* verb and the past participle, as the passive-voice sentence does. In the active ver-

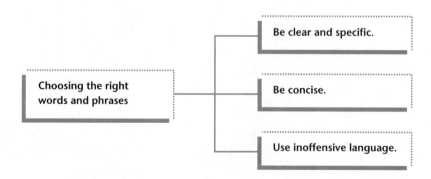

FIGURE 15.5

Choosing the Right
Words and Phrases

sion of the second example, for instance, the verb is "support," rather than "is supported," and "by" is unnecessary.

The passive voice, however, is generally more appropriate in four cases:

- The agent is clear from the context.

  Students are required to take both writing courses.

  The context makes it clear that the college requires that students take both writing courses.

- The agent is unknown.

  The comet was first referred to in an ancient Egyptian text.

  We don't know *who* referred to the comet.

- The agent is less important than the action.

  The documents were hand-delivered this morning.

  It doesn't matter *who* the messenger was.

- A reference to the agent is embarrassing, dangerous, or in some other way inappropriate.

  Incorrect data were recorded for the flow rate.

  It might be inappropriate to say *who* recorded the incorrect data; it might be tactful to avoid pointing a finger. However, it is unethical to use the passive voice to avoid responsibility for an action. In addition, the passive voice can help you maintain the focus of your paragraph.

  LANs have three major advantages. First, they are inexpensive to run. Second, they can be expanded easily. . . .

Many people who otherwise take little interest in grammar have strong feelings about voice. Some believe that the active voice is inappropriate because it emphasizes the person who does the work rather than the work itself, robbing the writing of objectivity. In many cases, this idea is valid. Why write "I analyzed the sample for traces of iodine"? If there is no ambiguity about who did the analysis, or if it is not necessary to identify who did the analysis, a focus on the action being performed is appropriate. But be conservative; if in doubt, use the active voice.

Supporters of the active voice argue that the passive voice creates a double ambiguity. When you write "The sample was analyzed for traces of iodine," your reader is not quite sure who did the analysis (you or someone else) or when it was done (as part of the project being described or sometime previously). Even though a passive-voice sentence can contain all the information found in its active-voice counterpart, often the writer omits the actor.

The best approach to the active-passive problem is to recognize how the two voices differ and use them appropriately. In the following examples, the writer mixes active and passive voice for no good reason.

AWKWARD: He lifted the cage door, and a white mouse was seen.

BETTER: He lifted the cage door and saw a white mouse.

AWKWARD: The new catalyst produced good-quality foam, and a flatter mold was caused by the new chute-opening size.

BETTER: The new catalyst produced good-quality foam, and the new chute-opening size resulted in a flatter mold.

A number of style programs can help you find the passive voice in your writing. Beware, however, of the advice some style programs offer about the passive voice: that it is undesirable, almost an error. This advice is wrongheaded. Use passive voice when it works better than the active voice; don't be bullied into changing to the active voice.

Any word-processing program allows you to search for *is, are, was,* and *were,* the forms of the verb *to be* that are most commonly used in passive-voice expressions. In addition, searching for *ed* will isolate the past participles, which also appear in most passive-voice expressions.

Be Specific

spec

Being specific involves using precise words, providing adequate detail, and avoiding ambiguity.

- *Use precise words.* A Ford Taurus is an automobile, but it is also a vehicle, a machine, and a thing. In describing the Ford Taurus, *automobile* is better than *vehicle,* because the less specific word also refers to pickup trucks, trains, hot-air balloons, and other means of transport. As words become more abstract—from *machine* to *thing,* for instance—the chances for misunderstanding increase.

- *Provide adequate detail.* Readers probably know less than you do about your subject. What might be perfectly clear to you might be too vague for them.

  VAGUE: An engine on the plane experienced some difficulties. *(Which engine? What plane? What kinds of difficulties?)*

  CLEAR: The left engine on the Jetson 411 unaccountably lost power during flight.

- *Avoid ambiguity.* Don't let readers wonder which of two meanings you are trying to convey.

  AMBIGUOUS: After stirring by hand for 10 seconds, add three drops of the iodine mixture to the solution. *(Stir the iodine mixture or the solution?)*

CLEAR: Stir the iodine mixture by hand for 10 seconds. Then add three drops to the solution.

CLEAR: Stir the solution by hand for 10 seconds. Then add three drops of the iodine mixture.

What should you do if you don't have the specific data? You have two options: to approximate—and clearly tell readers you are doing so—or to explain why the specific data are unavailable and indicate when they will become available.

VAGUE: The leakage in the fuel system is much greater than we had anticipated.

CLEAR: The leakage in the fuel system is much greater than we had anticipated; we estimate it to be at least 5 gallons per minute, not 2.

Several style programs isolate common vague terms and suggest more precise alternatives.

Avoid Unnecessary Jargon

jarg

Jargon is shoptalk. To a banker, *CD* means certificate of deposit; to an audiophile, it means compact disc. To the general reader, *ATM* means automated teller machine; to an electrical engineer, it means asynchronous transfer mode. Although jargon is often held up to ridicule, it is a useful and natural kind of communication in its proper sphere. Two software designers would find it hard to talk to each other about their craft if they couldn't use terms such as *SCSI port* and *WYSIWYG*.

However, using unnecessary jargon is inappropriate in four ways.

- *It can be imprecise.* The best current example is the degree to which electronics terminology has crept into everyday English. In offices, we ask employees to provide *feedback* after considering a proposal. What are we asking for: a facial expression, some body language, a phone call, or a written evaluation?

- *It can be confusing.* If we ask a computer novice to boot the system, he or she might have no idea what we're talking about.

- *It is often seen as condescending.* Many people react as if the writer is showing off—displaying a level of expertise that excludes most readers. While the readers are concentrating on how much they dislike the writer, they are missing the message.

- *It is often intimidating.* People feel somehow inadequate or stupid because they do not know what the writer is talking about. Obviously, this feeling hurts communication.

If you are addressing a technically knowledgeable audience, feel free to use appropriate jargon. (Sometimes, however, a technical term can have very different meanings, even when used by technical professionals with similar backgrounds.) If you are addressing an audience that includes managers or the general public, avoid jargon. If your document has separate sections for different audiences—as in the case of a technical report with an executive summary—use jargon accordingly. A glossary (list of definitions) is a useful addition if you suspect that managers will read the technical sections. See Chapter 11 for a discussion of these elements.

Use Positive Constructions

pos

The term *positive construction* has nothing to do with cheerfulness or an optimistic outlook on life. Rather, it indicates that the writer is describing what something is instead of what it is not. In the sentence "I was sad to see this project completed," "sad" is a positive construction. The negative construction would be "not happy."

Here are a few other examples of positive and negative constructions:

Positive Construction	*Negative Construction*
most	not all
few	not many
on time	not late, not delayed
positive	not negative
inefficient	not efficient
reject	cannot accept
impossible	not possible

Why should you try to use positive rather than negative constructions? Because readers understand positive constructions more quickly and more easily. And when a writer uses several negative constructions in the same sentence, readers must work much harder to untangle the meaning. Consider the following examples:

DIFFICULT: Because the team did not have sufficient time to complete the project, it was not unsurprising that it was unable to prepare a satisfactory report.

SIMPLER: Because the team had too little time to complete the project, it produced an unsatisfactory report.

DIFFICULT: Without an adequate population of krill, the entire food chain of the Antarctic region would be unable to sustain itself.

SIMPLER: If the krill population were too low, the entire food chain of the Antarctic region would be destroyed.

Avoid Long Noun Strings

str

A *noun string* is a phrase consisting of a series of nouns (or nouns and adjectives and adverbs), all of which modify the last noun. For example, in the phrase *parking-garage regulations,* the first two words modify *regulations.* Noun strings save time, and if your readers understand them, they are fine. A *passive-restraint system* is easier to write than *a restraint system that is passive,* and it won't confuse readers.

Hyphens help clarify some noun strings by linking two or more words that go together. For example, in the phrase *self-locking washer,* the hyphen links *self* and *locking;* together they modify *washer.* In other words, it is not a self washer, or a locking washer, but a self-locking washer. For more information on hyphens, see Appendix A.

However, noun strings are sometimes so long or so complex that hyphens won't ensure clarity. Consider untangling the phrases and restoring prepositions, as in the following examples:

UNCLEAR: preregistration procedures instruction sheet update

CLEAR: an update of the instruction sheet for preregistration procedures

UNCLEAR: operator-initiated default-prevention technique

CLEAR: a technique for preventing defaults that are initiated by the operator

An additional danger is that sometimes noun strings are merely pompous. If you are writing about a smoke detector, there is no reason to call it *a smoke-detection device.*

Avoid Clichés

cl

Good writing is original. Rather than use a cliché, just say what you want to say in plain English. Instead of "It's a whole new ball game," write "The situation has changed completely." Don't write "I am sure the new manager can cut the mustard"; write "I am sure the new manager can do his job effectively." Newer clichés have joined these hallowed examples: *movers and shakers, couch potato, fast lane,* and *paradigm shift.* But the advice is the same: if you're used to hearing and reading a phrase, avoid it.

Sometimes, writers further embarrass themselves by getting their clichés wrong: expressions become so timeworn that users forget what the words mean. The phrase "I could care less" often is used when the writer means just the opposite. The best solution to this problem is, of course, not to use clichés.

Following are a cliché-filled sentence and a translation into plain English:

TRITE: Afraid that we were *between a rock and a hard place,* we decided to *throw caution to the winds* with a *grandstand play* that would *catch our* competition *with its pants down.*

PLAIN: Afraid that we were in a dilemma, we decided on a risky, aggressive move that would surprise our competition.

Avoid Euphemisms

eu

A euphemism is a polite way to say something that makes people uncomfortable—the more uncomfortable the subject, the more euphemisms. Dozens of euphemisms deal with drinking, bathrooms, sex, and death. David Lord (as quoted in Fuchsberg, 1990) lists 48 euphemisms for firing someone, including the following:

personnel surplus reduction	dehiring
work force imbalance correction	decruiting
degrowing	redundancy elimination
indefinite idling	career-change-opportunity creation

It's fun to think of euphemisms like these, but don't write them. It's an ethical issue. People suffer when they get fired. Don't use language to cloud reality.

Be Concise   Technical communication should be concise. Follow these six guidelines:

- Avoid obvious statements.
- Avoid meaningless modifiers.
- Avoid unnecessary prepositional phrases.
- Avoid wordy phrases.
- Avoid redundant expressions.
- Avoid pompous words.

Avoid Obvious Statements

ob

Writing can become sluggish if it overexplains. The italicized words in the example below are sluggish:

SLUGGISH: The market for *the sale of* flash memory chips is dominated by *two chip manufacturers:* Intel and Advanced Micro Systems. These two *chip manufacturers* are responsible for 76 percent of the $1.3 billion market *in flash memory chips* last year.

IMPROVED: The market for flash memory chips is dominated by Intel and Advanced Micro Systems, two companies that claimed 76 percent of the $1.3 billion industry last year.

You don't need to state that the chips were made by chip manufacturers, nor do you need to explain that the $1.3 billion market is in flash memory chips. If it were in something else, the writer wouldn't be mentioning it in this context.

Avoid Meaningless Modifiers

w

Sometimes we include meaningless modifiers, such as the following:

basically	kind of
certain	sort of
essentially	various

BLOATED: *I think that, basically,* the board felt *sort of* betrayed, *in a sense,* by the *kind of* behavior the president displayed.

BETTER: The board felt betrayed by the president's behavior.

Of course, modifiers are not necessarily meaningless. For instance, it is appropriate to use *I think* or *it seems to me* to suggest an awareness that not everyone shares your view.

BLUNT: Next year we will face unprecedented challenges to our dominance of the market.

LESS BLUNT: In my view, next year we will face unprecedented challenges to our dominance of the market.

Of course, a sentence that is blunt to one reader is self-confident to another. You have to think about your audience's preferences.

Avoid Unnecessary Prepositional Phrases

prep

A prepositional phrase consists of a preposition followed by a noun or a noun equivalent. Here are some examples of prepositional phrases:

in the summary

on the engine

under the heading

Unnecessary prepositional phrases, often used along with abstract nouns and nominalizations, make writing long and boring. In the following examples, the prepositions are italicized:

LONG: The increase *in* the number *of* students enrolled *in* the materials engineering program *at* Lehigh University is a testament *to* the regard *in* which that program is held *by* the university's new students.

SHORTER: The growth of Lehigh University's materials engineering program suggests that the university's new students think it is a good program.

Avoid Wordy Phrases

w

Wordy phrases make technical communication long and boring. Sometimes people write phrases such as *on a weekly basis* rather than writing *weekly*. The long phrase rolls off the tongue easily and appears to carry the weight of scientific truth. But the humble *weekly* says the same thing, more concisely.

Figure 15.6 lists commonly used wordy phrases and their concise equivalents. Notice that almost all the wordy phrases contain prepositions, while many others contain nominalizations.

Following are a wordy sentence and a concise translation:

WORDY: I am of the opinion that, in regard to profit achievement, the statistics pertaining to this month will appear to indicate an upward tendency.

CONCISE: I think this month's statistics will show an increase in profits.

Avoid Redundant Expressions

redun

Avoid redundant expressions, such as *collaborate together, end result, any and all, each and every, still remain, completely eliminate,* and *very unique.* Be content to say something once.

REDUNDANT: We initially began our investigative analysis with a sample that was spherical in shape and heavy in weight.

BETTER: We began our analysis with a heavy, spherical sample.

FIGURE 15.6

Wordy Phrases and Their
Concise Equivalents

Wordy phrase	Concise phrase	Wordy phrase	Concise phrase
a majority of	most	in view of the fact that	because
a number of	some, many	it is often the case that	often
at an early date	soon	it is our opinion that	we think that
at the conclusion of	after, following	it is our recommendation that	we recommend that
at the present time	now	it is our understanding that	we understand that
at this point in time	now	make reference to	refer to
based on the fact that	because	of the opinion that	think that
despite the fact that	although	on a daily basis	daily
due to the fact that	because	on the grounds that	because
during the course of	during	prior to	before
during the time that	during, while	relative to	regarding, about
have the capability to	can	so as to	to
in connection with	about, concerning	subsequent to	after
in order to	to	take into consideration	consider
in regard to	regarding, about	until such time as	until
in the event that	if		

Avoid Pompous Words

pomp

Writers sometimes try to impress their readers by using pompous words, such as *initiate* for *begin*, *perform* for *do*, and *prioritize* for *rank*. When asked why they use big words where small ones will do, writers say they want to make sure their readers know they have a strong vocabulary, that they are well educated. In technical communication, plain talk is best. If you know what you're talking about, be direct and simple. Even if you're not so sure of what you're talking about, say it plainly; big words won't fool anyone for more than a few seconds.

Following are a few pompous sentences translated into plain English:

POMPOUS: The purchase of a database program will enhance our record-maintenance capabilities.

PLAIN: Buying a database program will help us maintain our records.

POMPOUS: It is the belief of the Accounting Department that the predicament was precipitated by a computational inaccuracy.

PLAIN: The Accounting Department thinks a math error caused the problem.

Figure 15.7 lists commonly used fancy words and their plain equivalents.

Several style programs isolate pompous words and expressions. Of course, you can use any word-processing program to search for terms you tend to misuse.

In the long run, your readers will be more impressed with clarity and accuracy. Don't waste your time thinking up fancy words.

FIGURE 15.7

Fancy Words and Their Plain Word Equivalents

Fancy word	Plain word	Fancy word	Plain word
advise	tell	impact (verb)	affect
ascertain	learn, find out	initiate	begin
attempt (verb)	try	manifest (verb)	show
commence	start, begin	parameters	variables, conditions
demonstrate	show	perform	do
employ (verb)	use	prioritize	rank
endeavor (verb)	try	procure	get, buy
eventuate (verb)	happen	quantify	measure
evidence (verb)	show	terminate	end, stop
finalize	end, settle, agree, finish	utilize	use
furnish	provide, give		

Use Inoffensive Language

Writing inoffensively is not merely a matter of politeness. It is a matter of perception. If language not only reflects people's attitudes but also helps form them, writing inoffensively helps to break down stereotypes, so that we see people as individuals.

Use Nonsexist Language

sexist

Sexist language favors one sex at the expense of the other. Although sexist language can shortchange men—as in some writing about female-dominated professions such as nursing—in most cases it victimizes women. Common examples include nouns such as *workman* and *chairman*. When writers use male pronouns to represent both males and females, as in the sentence "Each worker is responsible for his work area," they are also using sexist language. Some writers sidestep the problem of sexist language by simply claiming their innocence: "The use of the pronoun *he* does not in any way suggest a male bias." Many readers find this argument unpersuasive or even insincere.

In most organizations, people try to avoid sexist language. Although eliminating all gender bias from writing is not easy, many organizations have formulated guidelines in an attempt to reduce sexist language.

Six major techniques are used to eliminate sexist writing:

- *Replacing the male-gender words with non-gender-specific words. Chairman,* for instance, is replaced by *chairperson* or *chair. Firemen* are *firefighters, policemen* are *police officers.*
- *Switching to a different form of the verb.*

  SEXIST: The operator must pass a rigorous series of tests before *he is promoted* to supervisor.

  NONSEXIST: The operator must pass a rigorous series of tests before *being promoted* to supervisor.

- *Switching to the plural.*

  SEXIST: The operator must pass a rigorous series of tests before *he* is promoted to supervisor.

  NONSEXIST: Operators must pass a rigorous series of tests before *they* are promoted to supervisor.

In some organizations, you will see plural pronouns with singular nouns, particularly in informal documents such as memos:

If *an employee* wishes to apply for tuition reimbursement, *they* should consult Section 14.5 of the *Employee Manual.*

This construction is still considered a grammar error by careful writers and readers.

- *Switching to* he or she, he/she, s/he, *or* his or her. Sometimes, switching to the plural can make the sentence unclear:

  UNCLEAR: Operators are responsible for their operating manuals.
  (Does each operator have one operating manual or more than one?)

  CLEAR: Each operator is responsible for *his or her* operating manual.

  He or she, his or her, and the related constructions are awkward, especially if overused, but at least they are clear and inoffensive.
- *Addressing the reader directly.* If you use *you* and *your* or the understood *you,* you have avoided the problem.
- *Alternating* he *and* she. Joseph Williams (1994) and many other language authorities recommend alternating *he* and *she.*

If you use a word processor, search for *he, man,* and *men,* the words and parts of words most commonly associated with sexist writing. Some style programs search out the most common sexist terms and suggest nonsexist alternatives. But use your common sense. I have read the following sentence: "Every employee is responsible for the cost of his or her gynecological examination."

For a full discussion of nonsexist writing, see *The Handbook of Nonsexist Writing* (Miller & Swift, 1980).

Use Inoffensive Language When Referring to People with Disabilities

dis

According to the National Organization on Disability, one in six Americans—more than 40 million—has a physical, sensory, emotional, or mental impairment that interferes with daily life (Tyler, 1990).

In writing about people with disabilities, use the "people first" approach. This approach calls for writing about the person as someone with a disability, not as someone defined by that disability. The disability is a condition the person has, not what the person is.

Figure 15.8 on page 410, based on Tyler (1990, p. 65), lists some of the basic tenets of the people-first approach.

Understanding Simplified English for Nonnative Speakers

Because English is the language of over half of the world's scientific and technical communication, millions of nonnative speakers of English read technical communication written in English (Peterson, 1990).

Efforts to create a simplified language for these readers began as early as 1932, when Charles K. Ogden created an 850-word Basic English dictionary, complete with a set of simple grammar rules. The language used the active voice only and contained 18 "operators": verbs such as *put* and *get.*

- Refer to the person first, the disability second. Write "people with mental retardation," not "the mentally retarded."

- Don't confuse *handicap* with *disability*. *Disability* refers to the impairment or condition; *handicap* refers to the interaction between the person and his or her environment. A person can have a disability without being handicapped.

- Don't refer to victimization. Write "a person with AIDS," not "an AIDS victim" or "an AIDS sufferer."

- Don't refer to a person as "wheelchair bound" or "confined to a wheelchair." People who use wheelchairs to get around are not confined.

- Don't refer to people with disabilities as abnormal. They are atypical, not abnormal.

FIGURE 15.8

Principles of the
"People-First" Approach

Since that time, other companies and associations have created their own versions of *Simplified English,* as this general category of languages is known. All Simplified English languages share four characteristics:

- Each consists of a basic vocabulary of about 1,000 words.

- Each is based on the premise that a word has only one meaning. There are no synonyms.

- Each consists of a set of basic grammar rules.

- Each is geared to a specific discipline. For example, "AECMA Simplified English" is intended for aerospace workers.

Here is a sample of text translated into Simplified English.

BEFORE TRANSLATION INTO SIMPLIFIED ENGLISH:
Before filling the gas tank, it is necessary to turn off the propane line to the refrigerator. Failure to do so significantly increases the risk of explosion.

AFTER TRANSLATION INTO SIMPLIFIED ENGLISH:
Before you pump gasoline into the gas tank, turn off the propane line to the refrigerator. If you do not turn off the propane tank, it could explode.

For background information on Simplified English, see Peterson (1990). For information on a project to create a software program that assists in translating standard English into Simplified English, see Thomas, Jaffe, Kincaid, and Stees (1992).

Understanding Structured Writing

Structured writing is a general term referring to different strategies of imposing a visible structure on writing. For example, in one structure, used by many companies that write proposals, graphics are displayed on one page and the accompanying text is displayed on the facing page. (See Chapter 21.)

Like Simplified English, structured writing is intended to place restrictions on the use of the language in order to simplify and standardize it. Structured writing is not meant exclusively for nonnative speakers of English, although to the extent that it simplifies the language it assists them as well as native speakers.

A leading example of structured writing is Information Mapping®, a technique developed by Robert Horn in the mid-1960s. Figure 15.9, from an article by Horn (1985), shows the same information displayed first in traditional paragraphs and then using his technique.

TRANSFERS

There are more and more requests for transfers as the company expands and the key work force adopts a more flexible lifestyle. The company supervisor is a key person in facilitating such transfers and in determining whether they would be in the best interest of the company and the employees. This memorandum covers company policy that has been in effect for the past year and continues to be our policy. It outlines each supervisor's responsibilities when an employee under his or her supervision requests a transfer.

First, it is in the company's interest to retain employees who are performing satisfactorily; therefore, we will try to help employees to move to an area or job that they find more desirable. This is what you should do. Provide an employee who comes to talk about a transfer or to request one with Form 742, Application for Transfer, and tell him or her to fill it out as soon as possible.

If the employee is applying for a new job and not just a new location and if there are any parts of the new job that you as the supervisor consider may disqualify the employee, then you should discuss with the employee those areas immediately. Remember, it is the company policy that if an employee wishes to be transferred, the company will make every effort to find a job acceptable to the employee. So you should not discourage any request for transfer, even if it would disturb the completion of projects or goals in your department.

At the bottom of the form, you should fill out the supervisor's comment. Be brief and to the point. When you have finished that, you should make a photocopy of the employee's latest Performance Evaluation and attach it to the form.

If the employee's current performance rating is unsatisfactory, then your signature and your immediate supervisor's signature on Part C of the form are required. If the current performance rating is outstanding, then attach a copy of any letters of commendation. If the current performance rating is satisfactory, you do not have to attach anything.

Send a copy of the blue copy of the form to the company Placement Bureau and a pink copy to your Department File. The yellow copy should be given to the employee.

FIGURE 15.9

Information Mapping®

(Figure 15.9 continued)

How to Handle Transfer Requests

Introduction This procedure outlines each supervisor's responsibilities when an employee under your supervision requests a transfer.

Procedure table

STEP	ACTION
1	When employees request transfers, provide them with form 742, Application for Transfer.
2	Discuss with the employee any areas in the new job that you consider may disqualify the employee. Remember, it is our policy that if an employee wishes to be transferred, the company will make every effort to find a job acceptable to the employee. So you may in *no* way discourage a request for transfer.
3	At the bottom of the form, fill out the supervisor's comments.
4	Attach a copy of the latest Performance Evaluation.
5	<table><tr><td>IF the current performance rating is . . .</td><td>THEN</td></tr><tr><td>Unsatisfactory</td><td>Your signature and your immediate supervisor's signature on Part C of the form are required.</td></tr><tr><td>Outstanding</td><td>Attach a copy of the letter of recommendation.</td></tr></table>
6	<table><tr><td>Send this copy of the form</td><td>TO</td></tr><tr><td>Blue Copy</td><td>Placement Bureau</td></tr><tr><td>Pink Copy</td><td>Departmental File</td></tr><tr><td>Yellow Copy</td><td>Employee</td></tr></table>

WRITER'S CHECKLIST

Revision Software

1. Have you used a spell checker but also proofread your document for spelling errors?

2. If you have used a style program or a thesaurus, have you thought carefully about the advice offered by the software, and have you also revised the document without the software?

Lists

3. Is each list of the appropriate kind: bulleted, numbered, or checklist?

4. Does each list contain an appropriate number of items?

5. Are all the items in each list grammatically parallel?

6. Is the lead-in to each list structured and punctuated properly?

7. Are the items in each list punctuated properly?

Sentences

8. Are the sentences structured so that the new or important information comes near the end?

9. Are the sentences the appropriate length: neither short and choppy nor long and difficult to understand?

10. Does each sentence focus on the "real" subject?

11. Have you reduced the number of expletives used as sentence openers?

12. Does each sentence focus on the "real" verb, without including unnecessary nominalizations?

13. Have you eliminated parallelism problems from your sentences?

14. Have you used restrictive and nonrestrictive modifiers appropriately?

15. Have you eliminated misplaced modifiers, squinting modifiers, and dangling modifiers?

Level of Formality

16. Have you chosen an appropriate level of formality for your audience, subject, and purpose?

Words and Phrases

17. Have you used active and passive voice appropriately?

18. Have you used precise words?

19. Have you provided adequate detail?

20. Have you avoided ambiguity?

21. Have you avoided unnecessary jargon?

22. Have you used positive rather than negative constructions?

23. Have you avoided long noun strings?

24. Have you avoided clichés?

25. Have you avoided stating the obvious?

26. Have you avoided meaningless modifiers?

27. Have you avoided unnecessary prepositions?

28. Have you used the most concise phrases?

29. Have you avoided redundancy?

30. Have you avoided pompous words?

31. Have you avoided euphemisms?

32. Have you used nonsexist language?

33. Have you used people-first vocabulary in referring to people with disabilities?

EXERCISES

1. The information contained in the following sentences could be conveyed better in a list. Rewrite each sentence in the form of a list.

    a. The causes of burnout can be studied from three perspectives: physiological—the roles of sleep, diet, and physical fatigue; psychological—the roles of guilt, fear, jealousy, and frustration; environmental—the role of the physical surroundings at home and at work.

    b. There are many problems with the on-line registration system currently used at Dickerson. First, lists of closed sections cannot be updated as often as necessary. Second, students who want to register in a closed section must be assigned to a special terminal. Third, the computer staff is not trained to handle the student problems. Fourth, the Computer Center's own terminals cannot be used on the system; therefore, the university has to rent 15 extra terminals to handle registration.

2. In the following exercises, determine whether the new or important information is emphasized appropriately. If it is not, revise the sentence.

    a. Asynchronous transmission mode is likely to become the dominant communication mode over the next few years.

b. Multimedia interfaces are most successful when they are designed by engineers, computer scientists, psychologists, and applications experts, all working together.

c. The program must be carefully designed if it is to succeed.

3. The following sentences might be too long for some readers. Break each sentence into two or more sentences.

a. If we get the contract, we must be ready by June 1 with the necessary personnel and equipment to get the job done, so with this in mind a staff meeting, which all group managers are expected to attend, is scheduled for February 12.

b. Once we get the results of the stress tests on the 125-Z fiberglass mix, we will have a better idea of where we stand in terms of our time constraints, because if the mix isn't suitable we will really have to hurry to find and test a replacement by the Phase 1 deadline.

c. Although we had a frank discussion with Backer's legal staff, we were unable to get them to discuss specifics on what they would be looking for in an out-of-court settlement, but they gave us a strong impression that they would rather settle out of court.

4. The following examples contain choppy, abrupt sentences. Combine sentences to create a smoother prose style.

a. I need a figure on the surrender value of a policy. The policy number is A4399827. Can you get me this figure by tomorrow?

b. The program obviously contains an error. We didn't get the results we anticipated. Please ask Paul Davis to go through the program.

c. The supervisor is responsible for processing the outgoing mail. He is also responsible for maintaining and operating the equipment.

5. In the following sentences, the real subjects are buried in prepositional phrases or obscured by expletives. Revise the sentences so that the real subjects appear prominently.

a. There has been a decrease in the number of students enrolled in our training sessions.

b. It is on the basis of recent research that I recommend the new CAD system.

c. The use of in-store demonstrations has resulted in a dramatic increase in business.

6. In the following sentences, unnecessary nominalization has obscured the real verb. Revise the sentences to focus on the real verb.

a. Pollution constitutes a threat to the Wilson Wildlife Preserve.

b. Evaluation of the gumming tendency of the four tire types will be accomplished by comparing the amount of rubber that can be scraped from the tires.

c. Reduction of the size of the tear-gas generator has already been completed.

7. The following examples contain faulty parallelism. Revise the sentences to eliminate the errors.

a. The next two sections of the manual discuss how to analyze the data, the conclusions that can be drawn from your analysis, and how to decide what further steps are needed before establishing a journal list.

b. With our new product line, you would not only expand your tax practice, but your other accounting areas as well.

c. Sections 1 and 2 will introduce the entire system, while Sections 3 and 4 describe the automatic application and step-by-step instructions.

8. The following sentences contain punctuation or pronoun errors related to the use of modifiers. Revise the sentences to eliminate the errors.

a. The Greeting-Record Button records the greeting which is stored on a microchip inside the machine.

b. This problem that has been traced to manufacturing delays, has resulted in our losing four major contracts.

c. Please get in touch with Tom Harvey who is updating the instructions.

9. The following sentences contain misplaced modifiers. Revise the sentences to eliminate the errors.

a. Over the past three years it has been estimated that an average of eight hours per week are spent on this problem.

b. Information provided by this program is displayed at the close of the business day on the information board.

c. The computer provides a printout for the management team that shows the likely effects of the action.

10. The following sentences contain dangling modifiers. Revise the sentences to eliminate the errors.
    a. By following these instructions, your computer should provide good service for many years.
    b. To examine the chemical homogeneity of the plaque sample, one plaque was cut into nine sections.
    c. The boats in production could be modified in time for the February debut by choosing this method.

11. The following sentences are informal. Revise them to make them moderately formal.
    a. The learning modules were put together by a couple of professors in the department.
    b. The biggest problem faced by multimedia designers is that users freak if they don't see a button—or, heaven forbid, if they have to make up their own buttons!
    c. If the University of Arizona can't figure out where to dump its low-level radioactive waste, Uncle Sam could pull the plug on millions of dollars of research grants.

12. In the following sentences, the passive voice is used inappropriately. Rewrite the sentences to remove the inappropriate usages.
    a. Most of the information you need will be gathered as you document the history of the journals.
    b. When choosing multiple programs to record, be sure that the proper tape speed has been chosen.
    c. During this time I also cowrote a manual on the Roadway Management System. Frequent trips were also made to the field.
    d. Mistakes were made.
    e. Come to the reception desk when you arrive. A packet with your name on it can be picked up there.

13. The following sentences are vague. Revise them by replacing the vague elements with specific information. Make up any reasonable details.
    a. The results won't be available for a while.
    b. The fire in the lab caused extensive damage.
    c. A soil analysis of the land beneath the new stadium revealed an interesting fact.

14. The following sentences addressed to general readers contain unnecessary jargon. Revise the sentences to remove the jargon.

    a. Please submit your research assignment in hard-copy mode.
    b. The perpetrator was apprehended and placed under arrest directly adjacent to the scene of the incident.
    c. The new computer lab supports both platforms.

15. In the following sentences, convert the negative constructions to positive constructions.
    a. Williams was accused by management of making predictions that were not accurate.
    b. We must make sure that all our representatives do not act unprofessionally to potential clients.
    c. The shipment will not be delayed if Quality Control does not disapprove any of the latest revisions.

16. The following sentences contain long noun strings that the general reader might find awkard or difficult to understand. Rewrite the sentences to eliminate the long noun strings.
    a. The corporate-relations committee meeting has been scheduled for next Thursday.
    b. The research team discovered a glycerin-initiated, alkylene-oxide-based, long-chain polyether.
    c. We are considering purchasing a digital-imaging capable, diffusion-pump equipped, tungsten-gun SEM.

17. The following sentences contain clichés. Revise the sentences to eliminate the clichés.
    a. We hope the new program will positively impact all our branches.
    b. If we are to survive this difficult period, we are going to have to keep our ears to the ground and our noses to the grindstone.
    c. DataRight will be especially useful for those personnel tasked with maintaining the new system.

18. The following sentences contain euphemisms. Revise the sentences to eliminate the euphemisms.
    a. Downsizing will enable our division to achieve a more favorable cash-flow profile.
    b. Of course, accident statistics can be expected to show a moderate increase in response to a streamlining of the training schedule.
    c. Should the aircraft experience a water landing, your seat cushion will function as a personal flotation device.

19. The following sentences are verbose because they state the obvious. Revise the sentences to eliminate the obvious material.
    a. To register to take a course offered by the university, you must first determine whether the university will be offering that course that semester.
    b. The starting date of the project had to be postponed for a certain period of time due to a delay in the necessary authorization from the Project Oversight Committee.
    c. After you have installed DataQuick, please spend a few minutes responding to the questions about the process, then take the card to a post office and mail it to us.

20. The following sentences contain meaningless modifiers. Revise the sentences to remove them.
    a. It would seem to me that the indications are that the project has been essentially unsuccessful.
    b. For all intents and purposes, our company's long-term success depends to a certain degree on various factors that are in general difficult to foresee.
    c. This aspect of the presentation was quite well received overall, despite the fact that the meter readings were rather small.

21. The following sentences contain unnecessary prepositional phrases. Revise the sentences to eliminate them.
    a. Another advantage of the approach used by the Alpha team is that interfaces of different kinds can be combined.
    b. The complexity of the module will hamper the ability of the operator in the diagnosis of problems in equipment configuration.
    c. The purpose of this test of your aptitudes is to help you with the question of the decision of which major to enroll in.

22. The following sentences contain wordy phrases. Revise the sentences to make them more concise.
    a. The instruction manual for the new copier is lacking in clarity and completeness.
    b. The software packages enable the user to create graphic displays with a minimum of effort.
    c. We remain in communication with our sales staff on a daily basis.

23. The following sentences contain redundant expressions. Revise the sentences to remove the redundancies.
    a. In grateful appreciation of your patronage, we are pleased to offer you this free gift as a token gesture of our gratitude.
    b. An anticipated major breakthrough in storage technology will allow us to proceed ahead in the continuing evolution of our products.
    c. During the course of the next two hours, you will see a demonstration of our new speech-recognition system that will be introduced for the first time in November.

24. The following sentences contain pompous words. Revise the sentences to eliminate the pomposity.
    a. This state-of-the-art soda-dispensing module is to be utilized by the personnel associated with the Marketing Department.
    b. It is indeed a not unsupportable inference that we have been unsuccessful in our attempt to forward the proposal to the proper agency in advance of the mandated date by which such proposals must be in receipt.
    c. Deposit your newspapers and other debris in the trash receptacles located on the station platform.

25. The following sentences contain sexist language. Revise the sentences to eliminate the sexism.
    a. Each doctor is asked to make sure he follows the standard procedure for handling Medicare forms.
    b. Policemen are required to live in the city in which they work.
    c. Professor Harry Larson and Ms. Anita Sebastian—two of the university's distinguished professors—have been elected to the editorial board of *Modern Chemistry.*

26. The following sentences contain language that is offensive to some people with disabilities. Revise the sentences to eliminate the offensive language.
    a. This year, the number of female lung-cancer victims is expected to rise because of increased smoking.
    b. Mentally retarded people are finding greater opportunities in the service sector of the economy.
    c. This bus is specially equipped to accommodate the wheelchair-bound.

REFERENCES

Battison, R., & Goswami, D. (1981). Clear writing today. *Journal of Business Communication, 18*(4), 5–16.

Bormuth, J. R. (1966). Readability: A new approach. *Reading Research Quarterly, 2,* 79–132.

Carliner, S. (1987). Lists: The ultimate organizer for engineering writing. *IEEE Transactions on Professional Communication, 30*(4), 218–221.

Fuchsberg, G. (1990, December 7). Well, at least "terminated with extreme prejudice" wasn't cited. *Wall Street Journal,* p. B1.

Grammatik Windows version 2.0 user's guide. (1991). San Francisco: Reference Software International.

Horn, R. E. (1985). Results with structured writing using the Information Mapping® writing service standards. In T. M. Duffy and R. Waller (Eds.), *Designing Usable Texts.* Orlando, FL: Academic Press.

Huckin, T. N. (1983). A cognitive approach to readability. In P. V. Anderson, R. J. Brockmann, & C. Miller (Eds.), *New essays in technical and scientific communication: Research, theory, practice.* Farmingdale, NY: Baywood.

Miller, C., & Swift, K. (1980). *The handbook of nonsexist writing.* New York: Lippincott & Crowell.

Peterson, D. A. T. (1990). Developing a Simplified English vocabulary. *Technical Communication, 37*(2), 130–133.

Selzer, J. (1981). Readability is a four-letter word. *Journal of Business Communication, 18*(4), 23–34.

Thomas, M., Jaffe, G., Kincaid, J. P., & Stees, Y. (1992). Learning to use Simplified English: A preliminary study. *Technical Communication, 39*(1), 69–73.

Tyler, L. (1990). Communicating about people with disabilities: Does the language we use make a difference? *Bulletin of the Association of Business Communication, 53*(3), 65–67.

Williams, J. (1994). *Style: Ten lessons in clarity & grace* (4th ed.). New York: HarperCollins.

Electronic Mail, Memos, and Letters

This chapter discusses three short forms of communication used often in the working world: electronic mail (e-mail), memos, and letters.

Chapter 5 contained a brief discussion of the Internet as a tool for gathering information. That chapter pointed out that e-mail is the function performed most often on the Internet. E-mail is also an informal medium of communication between persons who are on a local area network (LAN). The discussion in this chapter will cover the basics of netiquette—e-mail etiquette—whether you are sending messages to another floor or across the country.

This chapter also discusses traditional memos: brief in-house reports. The discussion explains basic concepts of structuring memos and provides guidelines for writing common kinds of memos.

The final section of the chapter covers letters, the most traditional and formal of the three formats. Although letters are sometimes sent from one person to another within the same organization, letters are used most often when people in different organizations communicate. In addition to showing the basic formats, the chapter discusses tone and strategy in letter writing as well as common kinds of letters.

Electronic Mail

E-mail is the communication of brief messages on a computer network. The Congressional Office of Technology Assessment has estimated that by the year 2000, some 140 billion pieces of mail—two-thirds of the nation's total—will be sent by e-mail, not by the U.S. Postal Service.

E-mail offers four chief advantages over the interoffice delivery of paper documents or the U.S. Postal Service.

- *E-mail is fast.* On a LAN, delivery usually takes less than a second.
- *E-mail is cheap.* Once the network is in place, it doesn't cost anything to send a message to one person or a thousand. You don't pay per copy, as you do with photocopies, because you're not using any paper.
- *E-mail is easy to use.* Once you learn how to use your particular e-mail system, it is easy to send mail either to one person or to a group. You create an address book with the names and addresses of the people or groups you write to, and assign each one a nickname. Then, if you want to send the message to, say, a group of coworkers scattered across the country, you just type in the group's nickname, such as "AdCom," and the message automatically goes to each person in the group. The message sits in the mailbox until the recipient turns on the computer and reads it. Responding to an e-mail message or forwarding it to a third party—or many third parties—is also simple.
- *E-mail is digital.* E-mail can be read and erased or printed out, but it can also be stored like any other electronic file. Therefore, it can be used in other documents. For this reason, e-mail is a convenient way for people in different places to work collaboratively.

Because of these advantages, e-mail is quickly replacing many interoffice memos as well as some phone calls. Keep in mind, however, that there will always be a place for hard-copy interoffice mail, for two reasons:

- In some environments, people do not have access to a LAN or do not know how to use one.
- For legal reasons, people sometimes need original hard copies, on letter-head stationery or memo forms, complete with signatures.

Guidelines for Writing E-Mail

As you write e-mail, follow two guidelines:

- *Use appropriate formality.* In some organizations, managers expect e-mail messages to be as formal as paper-based documents; in other organizations, managers expect them or even want them to be quite informal. Learn the expectations in your organization. If the e-mail messages you read sound just like memos or letters, you know that the company considers e-mail as a practical way to communicate relatively formal messages. However, if you see writers using emoticons (also called *smiley faces*, of which :-) is the most popular) at the end of comments that are meant to be taken ironically or as a joke, or using abbreviations (such as *BTW* for *by the way*), you know that the organization sees e-mail not only as a way to communicate but also as a way to foster group cohesiveness. Before you send e-mail messages, read those written by other people.

But don't confuse informality with inarticulateness. Just because e-mail is an informal medium doesn't mean that unclear writing is acceptable. It isn't. Don't send your message until it is clear. Keep revising it until it says what you want it to, just as you would with any other kind of communication.

- *Realize that e-mail, like print, is permanent.* Many people do not realize that most networks are archived. That is, all the activity on the network is backed up on some kind of tape or disk system. Therefore, the message that you send to your colleague is probably stored somewhere, even if the recipient has deleted it. For this reason, do not write in an e-mail message anything that you would not write in a letter or memo. A simple test: would you be unhappy if your message appeared in tomorrow's newspaper? Companies are springing up that know how to search companies' computer archives to retrieve "erased" e-mail messages and other kinds of computer files.

  E-mail is being treated just like other forms of written communication in court cases. And the court has ruled that the organization that established the e-mail network may look at all e-mail traffic on that network without violating the privacy rights of employees using it (Crowe, 1994, p. 31).

Netiquette   To use e-mail effectively, you should know the basics of *netiquette*—etiquette on a network. Follow these six guidelines:

- *Take some care with your writing.* E-mail is informal, but messages shouldn't be sloppy. Although text-editing functions on most e-mail systems are much more limited than on a word processor, don't embarrass yourself by sending messages that you haven't at least proofread.

- *Don't flame.* Scorching a reader with scathing criticism, usually in response to something the person said in a previous message, is known as *flaming*. Flaming is rude. It is also embarrassing to the flamer who later realizes it is appropriate to apologize. When you are really angry and want to give your reader a piece of your mind, keep your hands away from the keyboard.

- *Use the subject line.* Readers like to be able to decide whether they want to read the message. The subject line that appears on their list of messages helps them decide.

- *Make your message easy on the eyes.* Use uppercase and lowercase letters as you do in other forms of correspondence, and skip lines between paragraphs. Don't use italics or underlining; they will appear as bizarre characters. Instead, use uppercase letters for emphasis. Keep the line length to under 65 characters so that your lines are not broken up if your reader has a smaller screen. For important messages, many people compose on their word processor, which makes it easy to set a short line length and to revise, then upload the message and send it.

- *Don't repost a message without the writer's permission.* If you receive a message from someone, don't post it to another person without the permission of the writer. Such a posting might be illegal—the courts haven't decided yet—but it is certainly unethical.

- *Don't send a message unless you have something to say.* Resist the temptation to write a message that says, in effect, that you agree with another message. If you can add something new, fine, but don't send a message just to be part of the conversation.

Figure 16.1 shows an e-mail that violates some of these guidelines. The writer is a technical professional working for a microchip manufacturer. Figure 16.2 is a revised version of this e-mail message.

The writer does not make clear in the first paragraph what his purpose is.

The writer uses an accusatory tone throughout the e-mail. Obviously, he is angry.

The writer has not proofread the e-mail.

The writer slips into the passive voice (see Chapter 15) when giving instructions for his readers to follow.

The writer does not clearly tell his readers who is to receive this information.

FIGURE 16.1

E-Mail Message That Violates Guidelines

Lately, we have been missing laser repair files for our 4meg wafers. After brief investigation, I have found the main reason for the missing data.

Occasionally, some of you have wrongly probed the wafers under the correlate step and the data is then copied into the Nonprod step using the QTR program. This is really stupid. When date is copied this way the repair data is not copied, it remains under the correlate step.

To avoid this problem, first probe the wafers the right way. If a wafer must be probed under a different step, the wafer in the CHANGE file must be renamed to the *.* format.

Editing the wafer data file should be used only as a last resort. If this becomes a common problem, we could have more problems with invalid data that there are now.

Supers and leads: please pass this information along to those who need to know.

Lately, we have been missing laser repair files for our 4meg wafers. In this e-mail I want to briefly describe the problem and recommend a method for solving it.

Here is what I think is happening. Some of the wafers have been probed under the correlate step; this method copies the data into the Nonprod step, and leaves the repair data uncopied. It remains under the correlate step.

To prevent this problem, please use the probing method outlined in Spec 344-012. If a wafer must be probed using a different method, rename the wafer in the CHANGE file to the *.* format. Edit the wafer data file only as a last resort.

I'm sending along copies of Spec 344-012. Would you please pass this e-mail and the spec to all of your operators.

Thanks. Please get in touch with me if you have any questions.

The first paragraph provides a clear statement of the writer's purpose.

The writer has removed the errors throughout the message.

The writer's diagnosis of the problem is stated more diplomatically.

The correct method is stated more clearly.

The writer promises to provide copies of the relevant specification.

The writer concludes politely.

FIGURE 16.2

E-Mail That Adheres to Guidelines

Memos

Although e-mail is replacing hard copy as the preferred medium for in-house communication, the memorandum remains an important form of technical communication. Each day, the average employee may receive a half-dozen memos and send out another half-dozen. Whereas most memos convey routine news addressed to several readers, some memos are used for brief technical reports.

This section concentrates not on the "FYI" memo—the simple communication addressed "For Your Information"—but on the more substantive brief technical reports written in memo form, such as directives, responses to inquiries, trip reports, and field reports.

Like all technical communication, memos should be ethical, clear, accurate, comprehensive, concise, accessible, professional, and correct. But they need not be ceremonious.

The printed forms on which memos are usually written include a subject heading—a place for the writer to define the subject. If you use this space efficiently, you can define both your subject and your purpose and thus begin to communicate immediately.

In the body of the memo, headings help your readers understand your message. For example, the simple heading *Results* enables them to decide whether or not to read the paragraph. (See Chapters 13 and 14 for more detailed discussions of headings.)

Writing a memo is essentially like writing any other form of technical communication. First you have to understand your audience and purpose. Then you gather your information, create some sort of outline, write a draft, and revise it. Making the memo look like a memo—adding the structural features that your readers will expect—is relatively simple. Your software has templates, or you can build the structure into your outline or shape the draft at some later stage.

Understanding the Elements of the Memo

The memo is made up of five components, as shown in Figure 16.3.

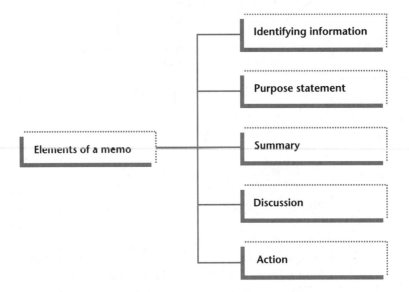

FIGURE 16.3

Elements of a Memo

Identifying Information

The top of a memo should identify the writing situation as efficiently as possible. Name yourself, the audience, and the subject, ideally with some indication of the purpose of the memo. Most writers put their initials or signature next to their typed names or at the end of the memo; the initials or signature shows that the writer has reviewed the memo and is accepting responsibility for sending it. Replace the inside address—the mailing address of the reader—with a department name or an office number, generally listed after the person's name. Sometimes you give no "address" at all.

Almost all memos have five elements at the top:

- the logo or a brief letterhead of the organization
- the "to" line
- the "from" line

- the "subject" line
- the "date" line

Some organizations have a "copies" or "c" (copy) heading as well. "Memo," "Memorandum," or "Interoffice" might be printed on the forms.

Organizations sometimes specify how to fill out the headings. Some prefer full names of the writer and reader; others want only the first initials and the last names. Some prefer job titles; some do not. If your organization does not object, include your job title and your reader's. The memo will then be informative for someone referring to it after you or your reader have moved on to a new position as well as for readers elsewhere in the organization who might not know either of you. List the names of persons receiving photocopies of the memo, either alphabetically or in descending order of organizational rank. In listing the date, write out the month (March 4, 19XX or 4 March 19XX). Do not use the all-numeral format (3/4/XX); people from other countries use different notations and therefore could be confused.

The subject heading—the title of the memo—deserves special mention. Don't be *too* concise. Avoid naming only the subject, such as *Tower Load Test;* rather, specify what aspect of the test you wish to address. For instance, *Tower Load Test Results* or *Results of Tower Load Test* would be much more informative than *Tower Load Test,* which does not tell the reader whether the memo is about the date, the location, the methods, the results, or any number of other factors related to the test. In short, each subject line should be informative, accurate, and unique. (See Chapter 5 on purpose and Chapters 13 and 14 on titles.)

Figure 16.4 on page 426 shows several common styles of filling in the identifying information of a memo.

Type the second and all subsequent pages of memos on plain paper rather than on letterhead. Include the following information in the upper left-hand corner of each page:

- the name of the recipient
- the date of the memo
- the page number

You might even define the communication as a memo and repeat the primary names. A typical second page of such a memo begins like this:

Memo to: J. Alders   April 6, 19XX
from: R. Rossini   Page 2

Purpose Statement

Memos are reproduced very freely. The majority of those received might be only marginally relevant to the reader. After starting to read their incoming memos,

AMRO   MEMO

To: B. Pabst
From: J. Alonso
Subject: MIXER RECOMMENDATION FOR PHILLIPS
Date: 11 June 19XX

INTEROFFICE

To: C. Cleveland   c: B. Aaron
From: H. Rainbow   K. Lau
Subject: Shipment Date of Blueprints   J. Manuputra
          to Collier   W. Williams
Date: 2 October 19XX

NORTHERN PETROLEUM COMPANY   INTERNAL CORRESPONDENCE

Date: January 3, 19XX
To: William Weeks, Director of Operations
From: Helen Cho, Chemical Engineering Dept.
Subject: Trip Report—Conference on Improved Procedures
          for Chemical Analysis Laboratory

FIGURE 16.4

Identifying Information
in a Memo

many readers ask, "Why is the writer telling me this?" The first sentence of the body—the purpose statement—should answer that question. Following are a few examples of purpose statements:

I want to tell you about a problem we're having with the pressure on the main pump, because I think you might be able to help us.

The purpose of this memo is to request authorization to travel to the Brownsville plant Monday to meet with the other quality inspectors.

This memo presents the results of the internal audit of the Phoenix branch, an audit that you authorized March 13, 19XX.

I want to congratulate you on the quarterly record of your division.

Notice that the best purpose statements are concise and direct. Make sure your purpose statement has a verb that clearly communicates what you want the memo to accomplish, such as *to request, to explain,* or *to authorize.* (See Chapter 5 for a more detailed discussion of purpose.) Some students of logical argument object to a direct statement of purpose—especially when the writer is asking for something, as in the example about requesting authorization to travel to the Brownsville plant. Rather than beginning with a direct statement of purpose, an argument for such a request would open with the reasons that the trip is necessary, the trip's potential benefits, and so on. Then the writer would conclude the memo with the actual request: "For these reasons, I am requesting authorization to. . . ." Although some readers would rather have the reasons presented first, far more would prefer to know immediately why you have written. There are two basic problems with using the standard argument structure:

- Some readers will toss the memo aside if you seem to be rambling on about the Brownsville plant without getting to the point.
- Some readers will suspect that you are trying to manipulate them into doing something they don't want to do.

The purpose statement sacrifices subtlety for directness.

Summary

Along with the purpose statement, the summary forms the core of the memo. It has three main goals:

- to help all readers follow the subsequent discussion
- to enable executive readers to skip the rest of the memo if they so desire
- to remind readers of the main points

Following are some examples of summaries:

The conference was of great value. The lectures on new coolants suggested techniques that might be useful in our Omega line, and I met three potential customers who have since written inquiry letters.

The analysis shows that lateral stress was the cause of the failure. We are now trying to determine why the beam did not sustain a lateral stress weaker than that it was rated for.

In March, we completed Phase I (Design) on schedule. At this point, we anticipate no delays that will jeopardize our projected completion date.

The summary should reflect the length and complexity of the memo. It might range in length from one simple sentence to a long and technical paragraph. If possible, the summary should reflect the sequence of the information in the body of the memo.

Discussion

The discussion contains the basic arguments of the memo. Generally, the discussion begins with a background paragraph. Even if you think the reader will be familiar with the background, include a brief statement, just to be safe. Also, the background will be valuable to a reader who refers to the memo later.

Each background discussion is, of course, unique; however, some basic guidelines are useful. If the memo defines a problem—for example, a flaw detected in a product line—the background might discuss how the problem was discovered or present the basic facts about the product line: what the product is, how long it has been produced, and in what quantities. If the memo reports the results of a field trip, the background might discuss why the trip was undertaken, what its objectives were, who participated, and so on.

Following is a background paragraph from a memo requesting authorization to have a piece of equipment retooled:

Background

The stamping machine, a Curtiss Model 6143, is used in the sheet-metal shop. We bought it in 1986 and it is in excellent condition. However, since we switched the size of the tin rolls last year, the stamping machine no longer performs to specifications.

After the background comes the detailed discussion. Here you give your readers a clear and complete idea of what you have to say. You might divide the detailed discussion into the subsections of a more formal report: materials, equipment, methods, results, conclusions, and recommendations. Or you might give it headings that pertain specifically to the subject you are discussing. You might include small tables or figures but should attach larger ones as appendices (see Chapter 12).

The discussion section of the memo can be developed according to any of the basic patterns for structuring arguments, including chronological, spatial, more important to less important, and cause and effect. These patterns are discussed in detail in Chapter 8.

Figure 16.5 is the detailed-discussion section from a memo written by a salesman working for the XYZ Company, which makes electronic typewriters. The XYZ salesman met an IBM salesman by chance one day, and they talked about the XYZ typewriters. The XYZ salesman is writing to his supervisor, telling her what he learned from the IBM salesman and also what XYZ's research and development (R&D) department told him in response to the comments of the IBM salesman.

Action

Many memos present information that will eventually be used in major projects or policies. These memos will become part of the files on these projects or poli-

> ### DISCUSSION
>
> In this section I relate the salesman's comments to me and the response from our R&D department.
>
> **Salesman's Comments:**
> In our conversation, he talked about the strengths of our machines and then mentioned two problems: excessive ribbon consumption and excessive training time.
>
> In general, he had high praise for the XYZ machines. In particular, he liked the idea of the rotary and linear stepping motor. Also, he liked having all the options within the confines of the typewriter. He said that although he knows we have some reliability problems, the machines worked well while he was training on them.
>
> The <u>major problem</u> with the XYZ machines, he said, is excessive ribbon consumption. According to his customers who have XYZ machines, the $5 cartridge lasts only about two weeks. This adds up to about $120 a year, about a third the cost of our basic Model A machine.
>
> The <u>minor problem</u> with the machines, he said, is that most customers are used to the IBM system. Since our language is very different, customers are spending more time learning our system than they had anticipated. He didn't offer any specifics on training-time differences.
>
> **R&D's Response:**
> I relayed these comments to Steve Brown in R&D. Here is what he told me.
> Ribbon Consumption: A recent 20 percent price reduction in the 4.2" cartridge should help. In addition, in a few days our 4.9" correctable cartridge—with a 40 percent increase in character output—should be ready for shipment. R&D is fully aware of the ribbon-consumption problem and will work on further improvements, such as thinner ribbon, increased diameter, and new cartridges.
> Training Time: New software is being developed that should reduce the training time.
>
> If I can answer any questions about the IBM salesman's comments, please call me at x1234.

The argument is presented chronologically: the writer describes first his discussion with the IBM salesman and then the response from R&D. Within each of these two subsections, the basic pattern is more important to less important: the ribbon-consumption problem is more serious than the training-time problem, so ribbon consumption is discussed first.

In this section, the writer describes how he conveyed the problem to his technical colleague. The colleague responded that the company is working on solutions to the problem.

FIGURE 16.5

Detailed Discussion of a Memo

cies. Some reports, however, require follow-up action more immediately, by either the writer or the readers. For example, a memo addressing a group of supervisors might discuss a problem and end by stating what the writer is going to do about it. A supervisor might use the action component of a memo to delegate tasks for other employees. In writing the action component of a memo, be sure to define clearly *who* is to do *what* and *when*. Following are two examples of action components:

Action:
I would appreciate it if you would work on the following tasks and have your results ready for the meeting on Monday, June 9.

1. Henderson to recalculate the flow rate.
2. Smith to set up meeting with the regional EPA representative for sometime during the week of February 13.
3. Falvey to ask Armitra in Houston for his advice.

Action:
To follow up these leads, I will do the following this week:

1. Send the promotional package to the three companies.
2. Ask Customer Relations to work up a sample design to show the three companies.
3. Request interviews with the appropriate personnel at the three companies.

In the first example, notice that although the writer is the supervisor of his readers, he uses a polite tone in this introductory sentence.

Understanding Common Types of Memos

Each memo is written by a specific writer to a specific audience for a specific purpose. Every memo is unique. In one, for instance, you might try to persuade your readers that although your idea for improving the public image of your company will cost money in the short run, the campaign is necessary and will more than pay for itself in the long run. There is no magic formula for writing this kind of memo; you have to figure out the best way to convince your readers that your idea makes sense.

Even though no two memos are the same, there are four broad categories of memos according to the functions they fulfill, as shown in Figure 16.6.

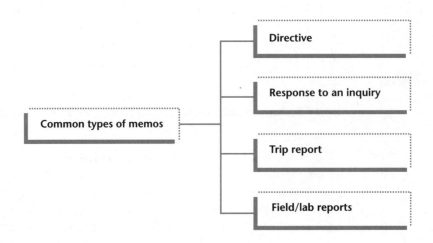

FIGURE 16.6

Common Types of
Memos

Notice as you read about each type of memo how the purpose-summary-discussion-action strategy is tailored to the occasion. Pay particular attention to the headings, lists, and indentation used to highlight structure.

The Directive

In a *directive memo,* you state a policy or procedure you want your readers to follow. If possible, explain the reason for the directive; otherwise, it might seem like an irrational order rather than a thoughtful request. For short memos, the explanation should precede the directive, to prevent the appearance of bluntness. For longer memos, the actual directive might precede the detailed explanation, to ensure that readers will not overlook the directive. Of course, the body of the memo should begin with a polite explanatory note, such as the following:

> The purpose of this memo is to establish a uniform policy for dealing with customers who fall more than 60 days behind in their accounts. The policy is defined below under the heading "Policy Statement." Following the statement is our rationale.

Figure 16.7 shows an example of a directive. Notice that the directive is stated as a request, not an order. A polite tone works best.

Note that in spite of its brevity and simplicity, this example follows, without headings, the purpose-summary-discussion-action structure. The subject line identifies the purpose, the first paragraph is a combination of summary and discussion (extensive discussion is hardly necessary in this situation), and the second paragraph dictates the specific action to be taken.

FIGURE 16.7

Directive

Quimby   Interoffice

Date: March 19, 19XX
To: All supervisors and sales personnel
From: D. Bartown, Engineering
Subject: Avoiding customer exposure to sensitive information
                    outside Conference Room B

It has come to our attention that customers meeting in Conference Room B have been allowed to use the secretary's phone directly outside the room. This practice presents a problem: the proposals that the secretary is working on are in full view of the customers. Proprietary information such as pricing can be jeopardized unintentionally.

In the future, would you please escort any customers or non-Quimby personnel needing to use a phone to the one outside the Estimating Department? Your cooperation in this matter will be greatly appreciated.

The Response to an Inquiry

Often a colleague might ask you to provide information that cannot be communicated on the phone because of its complexity or importance. When you respond to such an inquiry, the purpose-summary-discussion-action strategy is particularly useful. The purpose of the memo is simple: to provide the reader with the information requested. The summary states the major points of the subsequent discussion and calls the reader's attention to any parts of it that might be especially important. The action section (if it is necessary) defines any relevant steps that you or some other personnel are taking or will take. Figure 16.8 shows an example of a response to an inquiry.

NATIONAL INSURANCE COMPANY

TO: J. M. Sosry, Vice President
FROM: G. Lancasey, Accounting
SUBJECT: National's Compliance with the Federal Pay Standards
DATE: February 2, 19XX

Purpose: This memo responds to your request for an assessment of our compliance with the Federal Non-Inflationary Pay and Price Behavior Standards.

Summary: 1. We are in compliance except for a few minor violations.
2. Legal Affairs feels we are exercising "good faith," a measure of compliance with the Standards.
3. Data Processing is currently studying the costs and benefits of implementing data processing of the computations.

The following discussion elaborates on these three points.

Discussion: Here is my assessment of our compliance with the new federal standards.
1. We are in compliance with the Standards except for the following details related to fringe benefits.
a. The fringe benefits of terminated individuals have not yet been eliminated from our calculations. The salaries have been eliminated.
b. The fringe benefits associated with promotional increases have not yet been eliminated from our calculations.

Notice how the numbered items in the summary section correspond to the numbered items in the discussion section. This parallelism helps readers find quickly the discussion they want.

FIGURE 16.8

Response to an Inquiry

(Figure 16.8 continued)

   c. The fringe benefits of employees paid $14,800 or less have not yet been eliminated from our calculations.
2. I met with Joe Brighton of Legal Affairs last Thursday to discuss the question of compliance. Joe is aware of our minor violations. His research, including several calls to Washington, suggests that the Standards define "good faith" efforts to comply for various-size corporations, and that we are well within these guidelines.
3. I talked with Ted Goldstein of Data Processing last Friday. They have been studying the costs/benefits of implementing data processing for the calculations. Their results won't be complete until next week, but Ted predicts that it will take up to two months to write the program internally. He is talking to representatives of computer service companies this week, but he doubts if they can provide an economical solution.

As things stand now—doing the calculations manually—we will need three months to catch up, and even then we will always be about two weeks behind.

Action: I have asked Ted Goldstein to send you the results of the cost/ benefits study when they are in. I have also asked Joe Brighton to keep you informed of any new developments with the Standards.

If I can be of any further assistance, please let me know.

The Trip Report

A *trip report* is a record of a business trip written after the employee returns to the office. Most often, a trip report takes the form of a memo. The key to writing a good trip report is to remember that your reader is less interested in an hour-by-hour narrative of what happened than in a carefully structured discussion of what was important. If, for instance, you attended a professional conference, don't list all the presentations—simply attach the agenda or program if you think your reader will be interested. Communicate the important information you learned, or describe the important questions that didn't get answered. If you traveled to meet a client (or a potential client), focus on what your reader is interested in: how to follow up on the trip and maintain a good business relationship with the client.

In most cases, the standard purpose-summary-discussion-action structure is appropriate for this type of memo. Briefly mention the purpose of the trip—even if your reader might already know its purpose. By doing this, you will provide a

The writer and reader appear to be relatively equal in rank: the informal tone of the "Recommendation" section suggest that they have worked together before. Despite this familiarity, however, the memo is clearly organized to make it easy to read and refer to later, or to pass on to another employee who might follow up on it.

Dynacol Corporation

INTEROFFICE COMMUNICATION

To:	G. Granby, R&D
From:	P. Rabin, Technical Services
Subject:	Trip Report—Computer Dynamics, Inc.
Date:	September 20, 19XX

Purpose:

This memo presents my impressions of the Computer Dynamics technical seminar of September 18. The purpose of the seminar was to introduce their new PQ-500 line of computers.

Summary:

In general, I was not impressed with the new line. The only hardware that might be of interest to us is their graphics terminal, which I'd like to talk to you about.

Discussion:

Computer Dynamics offers several models in its 500 series, ranging in price from $11,000 to $45,000. The top model has a 64 M memory. Although it's very fast at matrix operations, this feature would be of little value to us. The other models offer nothing new.

I was disturbed by some of the answers offered by the Computer Dynamics representatives, which everyone agreed included misinformation.

The most interesting item was the graphics terminal. It is user-friendly. Integrating their terminal with our system could cost $4,000 and some 4–5 person-months. But I think that we want to go in the direction of graphics terminals, and this one looks very good.

Recommendation:

I'd like to talk to you, when you get a chance, about our plans for the addition of graphics terminals. I think we should have McKinley and Rossiter take a look at what's available. Give me a call (x3442) and we'll talk.

FIGURE 16.9

Trip Report

complete record for future reference. In the action section, list either the pertinent actions you have taken since the trip or what you recommend that your reader do. Figure 16.9 provides an example of a typical trip report.

Field and Lab Reports

Many organizations use memos to report on inspection and maintenance procedures. These memos, known as *field* or *lab reports,* include the same information that high-school lab reports do—the problem, methods, results, and conclusions—but they deemphasize the methods and can include a recommendations section.

A typical field or lab report, therefore, has the following structure:

1. purpose of the memo
2. summary
3. problem leading to the decision to perform the procedure
4. methods
5. results
6. conclusions
7. recommendations

A single word—"visual" —constitutes the discussion of the inspection procedure in the purpose section. Nothing else needs to be said.

FIGURE 16.10

Field Report

> ### *Lobate Construction*
> MEMO
>
> To: C. Amalli
> From: W. Kabor
> Subject: Inspection of Chemopump after Run #9
> Date: 6 January 19XX
> c: A. Beren
>   S. Dworkin
>   N. Mancini
>
> **Purpose:**
>   This memo presents the findings of my visual inspection of the Chemopump after it was run for 30 days on Kentucky #10 coal. The memo also requests authorization to carry out follow-up procedures.
>
> **Problem:**
>   The inspection was designed to determine if the new Chemopump is compatible with Kentucky #10, our lowest-grade coal. In preparation for the 30-day test run, the following three modifications were made:

(Figure 16.10 continued)

1. New front bearing housing buffer plates of tungsten carbide were installed.
2. The pump casting volute liner was coated with tungsten carbide.
3. New bearings were installed.

Summary:

A number of small problems with the pump were observed, but nothing serious and nothing surprising. Normal break-in accounts for the wear. The pump accepted the Kentucky #10 well.

Findings:

The following problems were observed:

1. The outer lip of the front-end bell was chipped along two thirds of its circumference.
2. Opposite the pump discharge, the volute liner received a slight wear groove along one-third of its circumference.
3. The impeller was not free-rotating.
4. The holes in the front-end bell were filled with insulating mud.

The following components showed no wear:

1. The 5½" impeller.
2. The suction neck liner.
3. The discharge neck liner.

Conclusions:

The problems can be attributed to normal break-in for a new Chemopump. The Kentucky #10 coal does not appear to have caused any extraordinary problems. The new Chemopump seems to be operating well.

Recommendations:

I would like authorization to modify the pump as follows:

1. Replace the front-end bell with a tungsten carbide-coated front-end bell.
2. Replace the bearings on the impeller.
3. Install insulation plugs in the holes in the front-end bell.

I recommend that the pump be reinspected after another 30-day run on Kentucky #10.

If you have any questions, please call me at x241.

In the findings section, the writer lists the bad news—the problems—before the good news. This is a logical order, because the bad news means more to the readers.

Sometimes several sections are combined. Purpose and problem often are discussed together, as are results and conclusions.

The lab report shown in Figure 16.10 illustrates some of the possible variations on this standard report structure.

Letters

The letter is the basic means of communication between two organizations: close to 100 million business letters are written each workday. Letters remain a basic link, because they provide documentary records. Often, phone conversations and transactions are immediately written up as letters, to become a part of the files of both organizations. The fax machine is used routinely to speed up the delivery of the letter.

Even as a new employee, you can expect to write letters regularly. And as you advance to positions of greater responsibility, you will write even more letters, for you will be representing your organization more often. Writing a letter is much like writing any other technical document. First you have to analyze your audience and determine your purpose. Then you have to gather your information, create an outline, write a draft, and revise it.

Projecting the "You Attitude"

Like any other type of technical communication, the letter should be ethical, clear, concise, comprehensive, accessible, correct, professional, and accurate. It must convey information in a logical order. It should not contain small talk; the first paragraph should get directly to the point without wasting words. And to enable the reader to locate information quickly and easily, a topic sentence should appear at the start of each paragraph in the body. Often, letters use headings and indentation just as reports do. In fact, some writers use the term *letter report* to describe a technical letter of more than, say, two or three pages. In substance, it is a report; in form, it is a letter, containing all of a letter's traditional elements.

Moreover, because it is a communication from one person to another, a letter must also convey a courteous, positive tone. The key is the "you attitude," which means looking at the situation from your reader's point of view and adjusting the content, structure, and tone to meet the person's needs. The you attitude is largely common sense. If, for example, you are writing to a supplier who has failed to deliver some merchandise on the agreed-upon date, the you attitude dictates that you not discuss problems you are having with other suppliers—those problems don't concern your reader. Rather, you should concentrate on explaining clearly and politely that your reader has violated your agreement and that not having the merchandise is costing you money. Then you should propose ways to expedite the shipment.

ENGINEER JOHN MOLLER ON WRITING BRIEF DOCUMENTS

on the kinds of short documents he writes routinely

My responsibility is business development, so I routinely write letters to prospective clients outlining my company's capabilities, proposals, meeting minutes that record what the people assigned to a project are supposed to do (and what they're *not* supposed to do), and brochures.

on planning the kind of information that goes into a document

There are certain items that go into each kind of document. For instance, minutes include information about where and when the meeting took place, who attended, what was discussed, and what was resolved. But the most important thing I think about when I sit down to write any kind of document is the needs of the people who will read it.

on using existing documents as a template

I start with an existing example of the document I'm going to write, then make the necessary changes. All our documents are in electronic form, so this is the easiest way. In addition, using an existing document as a template ensures that I don't leave out any necessary kinds of information.

on the characteristics he tries to achieve in his short documents

The first sentence or paragraph is usually a summary. The key is to get the summary into one or two sentences. Also, I try to keep everything to one page; it increases the probability that someone will read it. Big contracts and proposals, of course, have to be longer, but for routine documents I strive for clarity and conciseness.

on his education

I never took a technical writing course. Except for a few lab courses, writing wasn't really built into the engineering curriculum. At one company I worked for, I think the owner got fed up with the quality of the documents we produced, and he hired a writing consultant; the entire company took his course, discussing writing and doing a lot of exercises. And it worked. We got a lot better.

on the amount of writing he does

I write almost 100 percent of the time.

on the importance of writing to his career

It's absolutely critical to my career and to the success of this company. If we can't communicate clearly with our customers, we're in big trouble.

John Moller is vice president of Laramore, Douglass and Popham, a consulting engineering firm in New York City.

Looking at things from the other person's point of view would be simple if both parties always saw things the same way. They don't, of course. Sometimes the context of the letter is a dispute. Nevertheless, good letter writers always maintain a polite tone. Civilized behavior is good business, as well as a good way to live.

Following are examples of thoughtless sentences, each followed by an improved version that exhibits the you attitude:

EGOTISTICAL: Only our award-winning research and development department could have devised this revolutionary new sump pump.

BETTER: Our new sump pump features significant innovations that you may appreciate.

BLUNT: You wrote to the wrong department. We don't handle complaints.

BETTER:	Your letter has been forwarded to the Customer Service Division.
ACCUSING:	You must have dropped the engine. The housing is badly cracked.
BETTER:	The badly cracked housing suggests that your engine must have fallen onto a hard surface from some height.
SARCASTIC:	You'll need two months to deliver these parts? Who do you think you are, the Post Office?
BETTER:	Surely you would find a two-month delay for the delivery of parts unacceptable in your business. That's how I feel too.
BELLIGERENT:	I'm sure you have a boss, and I doubt if he'd like to hear about how you've mishandled our account.
BETTER:	I'm sure you would prefer to settle the account between ourselves rather than having it brought to your supervisor's attention.
CONDESCENDING:	Haven't you ever dealt with a major corporation before? A 60-day payment period happens to be standard.
BETTER:	Perhaps you were not aware of the standard 60-day payment period.
OVERSTATED:	Your air-filter bags are awful. They're all torn. We want our money back.
BETTER:	You will doubtless be surprised to learn that 19 of the 100 air-filter bags we purchased are torn. We hope you agree that refunding the purchase price of the 19 bags—$190.00—is the fair thing to do.

After you have drafted a letter, look back through it. Put yourself in your reader's place. How would you react if you received it? A calm, respectful, polite tone always makes the best impression and therefore increases your chances of achieving your goal.

Avoiding Letter Clichés

Related to the you attitude is the issue of letter clichés. Over the decades, a set of words and phrases has come to be associated with letters; one common example is *as per your request*. For some reason, many people think that these phrases are required. They're not. They make the letter sound stilted and insincere. If you would feel awkward or uncomfortable saying these clichés to a friend, avoid them in your letters. Figure 16.11 on page 440 is a list of some of the common clichés and their more natural equivalents.

Figures 16.12 and 16.13 on pages 440–441 contain two versions of the same letter: one written in letter clichés, the other in plain language. The letter in Figure 16.13 not only avoids the clichés but also shows a much better under-

Letter Clichés	Natural Equivalents
attached please find	attached is
cognizant of	aware that
enclosed please find	enclosed is
endeavor (verb)	try
herewith ("We herewith submit . . .)	(None. "Herewith" doesn't say anything. Skip it.)
hereinabove	previously, already
in receipt of ("We are in receipt of . . .")	"We have received . . ."
permit me to say	(None. Permission granted. Just say it.)
pursuant to our agreement	as we agreed
referring to your ("Referring to your letter of March 19, the shipment of pianos . . .")	"As you wrote in your letter of March 19, the . . ." or subordinate the reference at the end of your sentence.
same (as a pronoun: "Payment for same is requested . . .")	(Use the noun instead: "Payment for the merchandise is requested . . .")
wish to advise ("We wish to advise that . . .")	(The phrase doesn't say anything. Just say what you want to say.)
the writer ("The writer believes that . . .")	"I believe . . ."

FIGURE 16.11

Letter Clichés

Dear Mr. Kim:

Referring to your letter regarding the problem encountered with your new Eskimo Snowmobile. Our Customer Service Department has just tendered its report.

It is their conclusion that the malfunction is caused by water being present in the fuel line. It is our unalterable conclusion that you must have purchased some bad gasoline. We trust you are cognizant of the fact that while we guarantee our snowmobiles for a period of not less than one year against defects in workmanship and materials, responsibility cannot be assumed for inadequate care. We wish to advise, for the reason mentioned hereinabove, that we cannot grant your request to repair the snowmobile free of charge.

FIGURE 16.12

Letter Containing Letter Clichés

(Figure 16.12 continued)

Permit me to say, however, that the writer would be pleased to see that the fuel line is flushed at cost, $30. Your Eskimo would then give you many years of trouble-free service.

Enclosed please find an authorization card. Should we receive it, we shall endeavor to perform the above-mentioned repair and deliver your snowmobile forthwith.

Sincerely yours,

Dear Mr. Kim:

Thank you for writing to us about the problem with your new Eskimo Snowmobile.

Our Customer Service Department has found water in the fuel line. Apparently some of the gasoline was bad. While we guarantee our snowmobiles for one year against defects in workmanship and materials, we cannot assume responsibility for problems caused by bad gasoline. We cannot, therefore, grant your request to repair the snowmobile free of charge.

However, no serious harm was done to the snowmobile. We would be happy to flush the fuel line at cost, $30. Your Eskimo would then give you many years of trouble-free service.

If you will authorize us to do this work, we will have your snowmobile back to you within four working days. Just fill out the enclosed authorization card and drop it in the mail.

Sincerely yours,

FIGURE 16.13

Letter Written in Natural Language

standing of the you attitude. Rather than building the letter around the violation of the warranty, as the first writer does, the second writer presents the message as good news: the snowmobile is not ruined, and it can be returned in less than a week for a low cost.

Understanding the Elements of the Letter

Almost every letter has a heading, inside address, salutation, body, complimentary close, signature, and reference initials. In addition, some letters contain one or more of the following notations: attention, subject, enclosure, and copy.

For short, simple letters, you can compose the elements in the sequence in which they will appear. For more complex letters, however, use the strategy discussed with all the other technical-communication applications: start with the body and continue to the end. Then go back and add the preliminary elements.

In the following paragraphs, the elements of the letter are discussed in the order they would usually appear. Six common types of letters will be discussed in detail later in this chapter.

Heading

The typical organization has its own stationery, with its name, address, phone number, and perhaps a logo—the letterhead—printed at the top. The letterhead and the date the letter will be sent (printed two lines below the letterhead) make up the heading. When printing on blank paper, use your address (without your name) and date as the heading. Print only the first page of any letter on letterhead stationery. Print the second and all subsequent pages on blank paper, with the name of the recipient, the page number, and the date in the upper left-hand corner. For example:

Mr. John Cummings
Page 2
July 3, 19XX

Do not number the first page of any letter.

Inside Address

The inside address is your reader's name, position, organization, and business address. If your reader has a courtesy title, such as *Dr., Professor,* or—for public officials—*Honorable,* use it. If not, use *Mr.* or *Ms.* (unless you know the reader prefers *Mrs.* or *Miss*). If your reader's position can fit conveniently on the same line as his or her name, add it after a comma; otherwise, place it on the line below. Spell the name of the organization the way the organization itself does: for example, International Business Machines calls itself IBM. Include the complete mailing address: the street, city, state, and zip code.

Attention Line

Sometimes you will be unable to address the letter to a particular person. If you don't know (and cannot easily find out) the person's first name or don't know the person's name at all, use the attention line:

Attention: Technical Director

Use the attention line if you want to make sure that the organization you are writing to responds even if the person you write to is unavailable. In this case, put the name of the organization or of one of its divisions on the first line of the inside address:

Operations Department
Haverford Electronics
117 County Line Road
Haverford, MA 01765

Attention: Charles Fulbright, Director

Subject Line

On the subject line, put either a project number (for example, "Subject: Project 31402") or a brief phrase defining the subject of the letter (for example, "Subject: Price Quotation for the R13 Submersible Pump").

Operations Department
Haverford Electronics
117 County Line Road
Haverford, MA 01765

Attention: Charles Fulbright, Director
Subject: Purchase Order #41763

Salutation

If you have no attention line or subject line, put the salutation two lines below the inside address. The traditional salutation is *Dear* followed by the reader's courtesy title and last name. Use a colon after the name, not a comma. If you are fairly well acquainted with your reader, use *Dear* followed by the first name. When you do not know the reader's name, use a general salutation:

Dear Technical Director:
Dear Sir or Madam:

When you are addressing a group of people, use one of the following salutations:

Ladies and Gentlemen:
Gentlemen: (if all the readers are male)
Ladies: (if all the readers are female)

Or you can tailor the salutation to your readers:

Dear Members of the Restoration Committee:
Dear Members of Theta Chi Fraternity:

This same strategy is useful for sales letters without individual inside addresses:

Dear Home Owner:
Dear Customer:

Body

The body is the substance of the letter. In most cases, you will have at least three paragraphs: an introductory paragraph, a concluding paragraph, and one or more body paragraphs. For information on how to develop the argument of the body, see Chapter 8.

Complimentary Close

After the body of the letter, include one of the traditional closing expressions: *Sincerely, Sincerely yours, Yours sincerely, Yours very truly, Very truly yours.* Capitalize only the first word in the complimentary close, and follow all such expressions by a comma. Today, all these phrases have lost whatever particular meanings and connotations they once possessed. They are interchangeable.

Signature

Type your full name on the fourth line below the complimentary close. Sign the letter, in ink, above the typewritten name. Most organizations prefer that you add your position, beneath your typed name. For example:

Very truly yours,

José Santos

José Santos
Personnel Manager

Reference Line

If someone else types your letters, the reference line identifies, usually by initials, both you and the typist. It appears a few spaces below the signature line, along the left margin. Generally, the writer's initials—which always come first—are capitalized, and the typist's initials are lowercase. For example, if Marjorie Chu wrote a letter that Alice Wagner typed, the standard reference notation would be MC/aw.

Enclosure Line

If the envelope contains any documents other than the letter itself, identify the number of enclosures:

FOR ONE ENCLOSURE:	Enclosure
	or
	Enclosure (1)
FOR MORE THAN ONE ENCLOSURE:	Enclosures (2)
	Enclosures (3)

In determining the number of enclosures, count only the separate items, not the number of pages. A 3-page memo and a 10-page report constitute only two enclosures. Some writers like to identify the enclosures by name:

Enclosure: 1995 Placement Bulletin

Enclosures (2): "This Year at Ammex"
              1995 Annual Report

Copy Line

If you want the reader to know that other people are receiving a copy, use the symbol *c* (for "copy") or *pc* (for "photocopy") followed by the names of the other recipients (listed either alphabetically or according to organizational rank). If you do not want your reader to know about other copies, type *bc* ("blind copy") on the copies only—not on the original.

Learning the Format of the Letter

Three popular formats are used for letters: modified block, modified block with paragraph indentations, and full block. Figures 16.14, 16.15, and 16.16 (pages 446–448) show diagrams of letters written in these three formats.

Understanding Common Types of Letters

Dozens of kinds of letters exist for specific occasions. This section focuses on the six types written most frequently in the technical world, as shown in Figure 16.17 on page 449. Two other types of letters are discussed in this book: the transmittal letter in Chapter 11 and the job-application letter in Chapter 17. For a more detailed discussion of business letters, consult one of the several full-length books on the subject.

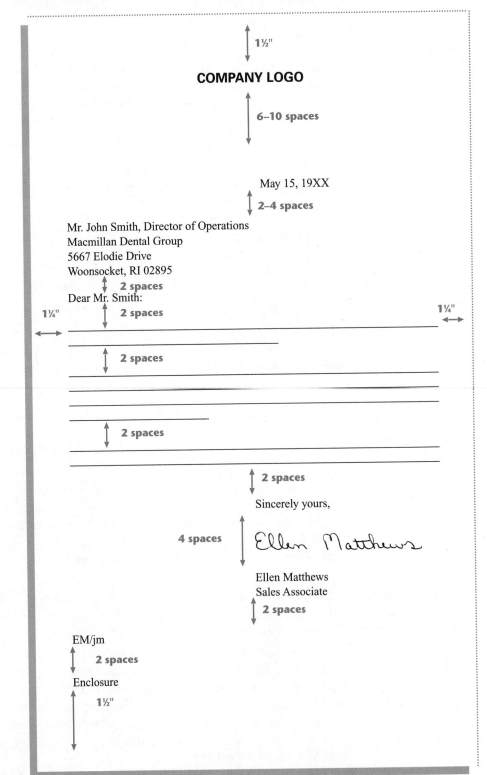

1½"

COMPANY LOGO

6–10 spaces

May 15, 19XX

2–4 spaces

Mr. John Smith, Director of Operations
Macmillan Dental Group
5667 Elodie Drive
Woonsocket, RI 02895

2 spaces

Dear Mr. Smith:

2 spaces

1¼"   **1¼"**

2 spaces

2 spaces

2 spaces

Sincerely yours,

4 spaces   *Ellen Matthews*

Ellen Matthews
Sales Associate

2 spaces

EM/jm

2 spaces

Enclosure

1½"

FIGURE 16.14

Modified Block Format

446

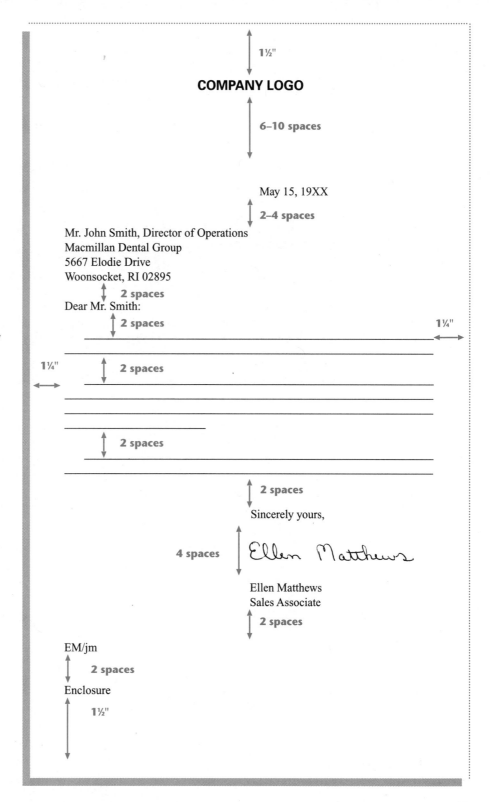

COMPANY LOGO

1½"

6–10 spaces

May 15, 19XX

2–4 spaces

Mr. John Smith, Director of Operations
Macmillan Dental Group
5667 Elodie Drive
Woonsocket, RI 02895

2 spaces

Dear Mr. Smith:

2 spaces

Paragraphs are generally indented five characters.

1¼"

1¼"

2 spaces

2 spaces

2 spaces

Sincerely yours,

4 spaces

Ellen Matthews

Ellen Matthews
Sales Associate

2 spaces

EM/jm

2 spaces

Enclosure

1½"

FIGURE 16.15

Modified Block Format
with Paragraph
Indentation

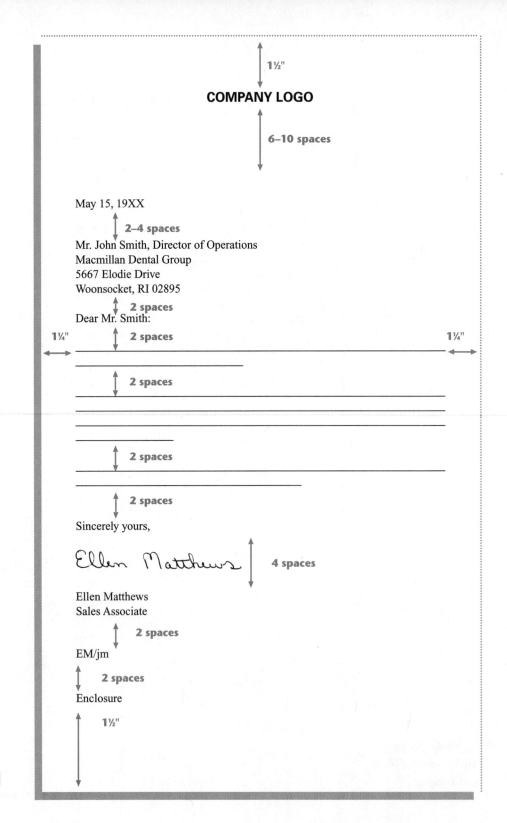

COMPANY LOGO

1½"

6–10 spaces

May 15, 19XX

2–4 spaces

Mr. John Smith, Director of Operations
Macmillan Dental Group
5667 Elodie Drive
Woonsocket, RI 02895

2 spaces

Dear Mr. Smith:

2 spaces

1¼"   1¼"

2 spaces

2 spaces

2 spaces

Sincerely yours,

4 spaces

Ellen Matthews
Sales Associate

2 spaces

EM/jm

2 spaces

Enclosure

1½"

FIGURE 16.16

Full-Block Format

448

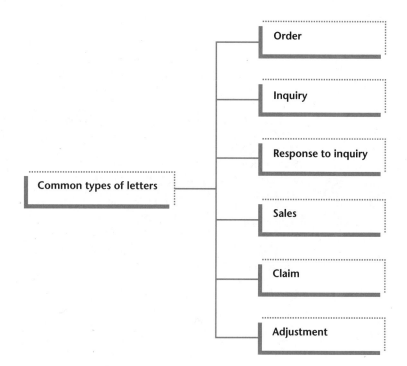

FIGURE 16.17

Common Types of
Letters

The Order Letter

Perhaps the most basic form of business correspondence is the order letter, written to a manufacturer, wholesaler, or retailer. When writing an order letter, include all the information your reader will need to identify the merchandise: the quantity, model number, dimensions, capacity, material, price, and any other pertinent details. Also specify the method of delivery. A typical order letter is shown in Figure 16.18 on page 450. (Notice that the writer uses an informal table to describe the parts she orders.)

Many organizations have preprinted forms, called purchase orders, for ordering products or services. A purchase order calls for the same information that appears in an order letter.

The Inquiry Letter

Your purpose in writing an inquiry letter is to obtain information from your reader. Writing an inquiry letter is more or less difficult, depending on whether the reader is expecting your letter.

- If the reader is expecting the letter, your task is easy. For example, if a company that makes institutional furniture has advertised that it will send its 48-page, full-color brochure to prospective clients, you need write merely a one-sentence letter: "Would you please send me the

WAGNER AIRCRAFT
116 North Miller Road
Akron, OH 44313

September 4, 19XX

Mr. Frank DiFazio
Franklin Aerospace Parts
623 Manufacturer's Blvd.
Bethpage, NY 11741

Dear Mr. DiFazio:

Would you please send us the following parts by parcel post? All page numbers refer to your 19XX catalog.

Quantity	Model No.	Page	Catalog Description	Price
2	36113-NP	42	Seal fins	$ 34.95
1	03112-Bx	12	Turbine bearing support	19.75
5	90135-QN	102	Turbine disc	47.50
1	63152-Bx	75	Turbine bearing housing	16.15
			Total Price:	$118.35

Yours very truly,

Christine O'Hanlon
Christine O'Hanlon
Purchasing Agent

FIGURE 16.18

Order Letter

brochure advertised in *Higher Education Today,* May 13, 19XX?" The manufacturer knows why you're writing and, naturally, wants to receive letters such as yours, so no explanation is necessary. You can also ask a technical question, or set of questions, about any product or service for sale. Your inquiry letter might begin, "We are considering purchasing your new X-15 work stations for an office staff of 35 and would like some further information. Would you please answer the following questions?" The detail about the size of the potential order is not necessary, but it does make the inquiry seem serious and the potential sale substantial. An inquiry letter of this kind will inspire a prompt and gracious reply.

- If the reader is not expecting your letter, your task is more difficult. If the reader will not directly benefit from supplying information, you must ask a favor. Only careful, persuasive writing will make your reader *want* to respond despite the absence of the profit motive.

Follow these four guidelines in writing an inquiry letter:

- *In the first paragraph of this kind of letter, state why you are writing to this person or organization.* You might use subtle flattery at this point— for example, "I was hoping that, as the leader in solid-state electronics, your company might be able to furnish some information about. . . ." Then you might explain why you want the information. Obviously, a company will not furnish information to a competitor. You have to show that your interests are not commercial—for instance, "I will be using this information in a senior project in agronomy at Illinois State University. I am trying to devise a. . . ." If you need the information by a certain date, this might be a good place to mention it: "The project is to be completed by April 15, 19XX."

- *List your questions.* Readers are understandably annoyed by thoughtless requests to send "everything you have" on a topic. They much prefer a set of carefully thought out technical questions showing that the writer has already done substantial research. "Is your Model 311 compatible with Norwood's Model B12?" is obviously much easier to respond to than "Would you please tell me about your Model 311?" If your questions can be answered in a small space, simply leave room for your reader's reply after each question or in the margin.

- *Offer something in return, because you are asking someone to do something for you.* In many cases, all you can offer are the results of your research. If possible, state that you would be happy to send a copy of the report you are working on. And finally, express your appreciation. Don't say "Thank you for sending me this information." Such a statement is presumptuous, because it assumes that the reader is both willing and able to meet your request. A statement such as the following is more effective: "I would greatly appreciate any help you could give me in answering these questions." Finally, if the answers will be brief, enclose a stamped self-addressed envelope for your reader's reply.

- *Write a brief thank-you note to someone who has responded to your inquiry letter.*

Figure 16.19 on page 452 shows an example of a letter of inquiry.

14 Hawthorne Ave.
Belleview, TX 75234

November 2, 19XX

Dr. Andrew Shakir
Director of Technical Services
Orion Corporation
721 West Douglas Avenue
Maryville, TN 31409

Dear Dr. Shakir:

I am writing to you because of Orion's reputation as a leader in the manufacture of adjustable X-ray tables. I am a graduate student in biomedical engineering at the University of Texas, and I am working on an analysis of diagnostic equipment for a seminar paper. Would you be able to answer a few questions about your Microspot 311?

1. Can the Microspot 311 be used with lead oxide cassettes, or does it accept only lead-free cassettes?
2. Are standard generators compatible with the Microspot 311?
3. What would you say is the greatest advantage, for the operator, in using the Microspot 311? For the patient?

My project is due on January 15. I would greatly appreciate your assistance in answering these questions. Of course, I would be happy to send you a copy of my report when it is completed.

Yours very truly,

Albert K. Stern

Albert K. Stern

FIGURE 16.19

Inquiry Letter

The Response to an Inquiry

If you ever receive an inquiry letter, keep the following suggestions in mind. If you wish to provide the information the writer asks for, do so graciously. If the questions are numbered, number your responses correspondingly. If you cannot answer the questions, either because you don't know the answers or because you cannot divulge proprietary information, explain the reasons and offer to assist with other requests. Figure 16.20 shows a response to the inquiry letter in Figure 16.19.

ORION

721 WEST DOUGLAS AVE.   (615) 619-8132
MARYVILLE, TN 31409   TECHNICAL SERVICES

November 7, 19XX

Mr. Albert K. Stern
14 Hawthorne Ave.
Belleview, TX 75234

Dear Mr. Stern:

I would be pleased to answer your questions about the Microspot 311. We think it is the best unit of its type on the market today.

1. The 311 can handle lead oxide or lead-free cassettes.
2. At the moment, the 311 is fully compatible only with our Duramatic generator. However, special wiring kits are available to make the 311 compatible with our earlier generator models—the Olympus and the Saturn. We are currently working on other wiring kits.
3. For the operator, the 311 increases the effectiveness of the radiological procedure while at the same time cutting down the amount of film used. For the patient, it cuts down the number of repeat exposures and therefore reduces the total dose.

I am enclosing a copy of our brochure on the Microspot 311. If you would like additional copies, please let me know. I would be happy to receive a copy of your analysis when it is complete. Good luck!

Sincerely yours,

Andrew Shakir M.D.

Andrew Shakir, M.D.
Director of Technical Services

AS/le

Enclosure
c: Robert Anderson, Executive Vice President

FIGURE 16.20

Response to an Inquiry

The Sales Letter

A large, sophisticated sales campaign costs millions of dollars—for marketing surveys and consulting fees, printing, postage, and promotions. This kind of campaign is beyond the scope of this book. However, you may well have to draft a sales letter for a product or service.

The you attitude is crucial in a sales letter. Your readers don't care why you want to sell your product or service. They want to know why they should buy it. Because you are asking them to spend valuable time studying the letter, you must provide clear, specific information to help them understand what you are selling and how it will help them. Be upbeat and positive, but never forget that readers want facts.

Most writers of sales letters use a four-part strategy:

1. *Gain the reader's attention.* Unless the opening sentence seems either interesting or important, the reader will put the letter aside. To attract the reader, use interesting facts, quotations, or questions. In particular, try to identify a problem that will interest your reader. A few examples of effective openings follow:

    How much have construction costs risen since your plant was built? Do you know how much it would cost to rebuild at today's prices?

    The Datafix copier is better than the Xerox—and it costs less, too. We'll repeat: it's better and it costs less!

    If you're like most training directors, we bet you've seen your share of empty promises. We've heard all the stories, too. And that's why we think you'll be interested in what *Fortune* said about us last month.

2. *Describe the product or service you are trying to sell.* What does it do? How does it work? What problems does it solve?

    The Datafix copier features automatic loading, so your people don't waste time watching the copies come out. Datafix copies from a two-sided original—automatically! And Datafix can turn out 90 copies a minute—which is 25 percent faster than our fastest competitor. . . .

3. *Convince your reader that your claims are accurate.* Refer to users' experience, testimonials, or evaluations performed by reputable experts or testing laboratories.

    In a recent evaluation conducted by *Office Management Today,* more than 85 percent of our customers said they would buy another Datafix. The next best competitor: 71 percent. And Datafix earned a "Highly Reliable" rating, the highest recommendation in the reliability category. All in all, Datafix scored *higher* than any other copier in the desktop class. . . .

4. *Tell your reader how to find out more about your product or service.* If possible, provide a postcard that the reader can use to request more informa-

tion or arrange for a visit from one of your sales representatives. Make it easy to proceed to the next step in the sales process.

Figure 16.21 shows an example of a sales letter.

DAVIS TREE CARE
1300 Lancaster Avenue
Berwyn, PA 19092

May 13, 19XX

Dear Home Owner:

Do you know how much your trees are worth? That's right—your trees. As a recent purchaser of a home, you know how much of an investment your house is. Your property is a big part of your total investment.

Most people don't know that even the heartiest trees need periodic care. Like shrubs, trees should be fertilized and pruned. And they should be protected against the many kinds of diseases and pests that are common in this area.

At Davis Tree Care, we have the skills and experience to keep your trees healthy and beautiful. Our diagnostic staff is made up of graduates of major agricultural and forestry universities, and all of our crews attend special workshops to keep current with the latest information on tree maintenance. Add to this our proven record of 43 years of continuous service in the Berwyn area, and you have a company you can trust.

May we stop by to give you an analysis of your trees—absolutely without cost or obligation? A few minutes with one of our diagnosticians could prove to be one of the wisest moves you've ever made. Just give us a call at 555-9187 and we'll be happy to arrange an appointment at your convenience.

Sincerely yours,

Jasmine Brown
President

FIGURE 16.21
Sales Letter

The Claim Letter

A claim letter is a polite and reasonable complaint. If as a private individual or a representative of an organization you purchase a defective or falsely advertised product or receive inadequate service, your first recourse is a claim letter.

The purpose of the claim letter is to convince your reader that you are a fair and honest customer who is justifiably dissatisfied. If it does, your chances of receiving an equitable settlement are good. Most organizations today pay attention to reasonable claims, because they realize that unhappy customers are bad business. In addition, claim letters indicate the weak points in their product or service.

Writing a claim letter calls for a four-part strategy:

1. *Identify the product or service.* List the model numbers, serial numbers, sizes, and any other pertinent data.

2. *Explain the problem.* State the symptoms clearly and specifically. What function does not work? What exactly is wrong with the service?

3. *Propose an adjustment.* Define what you want the reader to do: for example, refund the purchase price, replace or repair the item, improve the service.

4. *Conclude courteously.* Say that you trust the reader, in the interest of fairness, to abide by your proposed adjustment.

The you attitude is just as important as content in a claim letter. You must project a calm and rational tone. A complaint such as "I'm sick and tired of being ripped off by companies like yours" will hurt your chances of an easy settlement. If, however, you write "I am very disappointed in the performance of my new Eversharp Electric Razor," you sound like a responsible adult. There is no reason to show anger in a claim letter, even if the other party has made an unsatisfactory response to an earlier one. Calmly explain what you plan to do, and why. Your reader will then more likely see the situation from your perspective. Figure 16.22 shows an example of a claim letter that the writer faxed to the reader.

The Adjustment Letter

In an adjustment letter, you respond to a claim letter and tell the customer how you plan to handle the situation. Whether you are granting the customer everything proposed in the claim letter, part of it, or none of it, your purpose remains the same: to show that your organization is fair and reasonable, and that you value the customer's business.

If you can grant the request, the letter will be simple to write. Express your regret about the situation, state the adjustment you are going to make, and end

ROBBINS CONSTRUCTION, INC.
255 Robbins Place   Centerville, MO 65101 (417) 555-1850

August 19, 19XX

Mr. David Larsen
Larsen Supply Company
311 Elmerine Avenue
Anderson, MO 63501

Dear Mr. Larsen:

As steady customers of yours for over 15 years, we came to you first when we needed a quiet pile driver for a job near a residential area. On your recommendation, we bought your Vista 500 Quiet Driver, at $14,900. We have since found, much to our embarrassment, that it is not substantially quieter than a regular pile driver.

We received the contract to do the bridge repair here in Centerville after promising to keep the noise to under 90 db during the day. The Vista 500 (see enclosed copy of bill of sale for particulars) is rated at 85 db, maximum. We began our work and, although one of our workers said the driver didn't seem sufficiently quiet to him, assured the people living near the job site that we were well within the agreed sound limit. One of them, an acoustical engineer, marched out the next day and demonstrated that we were putting out 104 db. Obviously, something is wrong with the pile driver.

I think you will agree that we have a problem. We were able to secure other equipment, at considerable inconvenience, to finish the job on schedule. When I telephoned your company that humiliating day, however, a Mr. Meredith informed me that I should have done an acoustical reading on the driver before I accepted delivery.

I would like you to send out a technician—as soon as possible—either to repair the driver so that it performs according to specifications or to take it back for a full refund.

Yours truly,

Jack Robbins

Jack Robbins, President

JR/lr

Enclosure

FIGURE 16.22

Claim Letter

on a positive note by encouraging the customer to continue doing business with you.

If you cannot grant the request, try to salvage as much goodwill as you can. Obviously, your reader will not be happy. If your letter is carefully written, however, it can show that you have acted reasonably. In denying a request, you attempt to explain your side of the matter, thus educating your reader about how the problem occurred and how to prevent it in the future.

This more difficult kind of adjustment letter generally has a four-part structure:

1. *Attempt to meet the customer on some neutral ground.* Consider an expression of regret but not an apology. You might even thank the customer for bringing the matter to the attention of the company. If you admit that the customer is right in this kind of adjustment letter—such as by writing "We are sorry that the engine you purchased from us is defective"—the customer would have a good case against you if the dispute ended up in court.

2. *Explain why your company is not at fault.* Most often, you explain to the customer the steps that led to the failure of the product or service. Do not say "You caused this." Instead, use the less blunt passive voice: "The air pressure apparently was not monitored."

3. *Clearly state that your company is denying the request, for the reasons you have noted in the letter.* This statement must come late in the letter. If you begin with it, most readers will not finish the letter, and therefore you will not be able to achieve your twin goals of education and goodwill.

4. *Try to create goodwill.* You might, for instance, offer a special discount on another, similar product. A company's profit margin on any one item is almost always large enough to permit attractive discounts as an inducement to continue doing business.

Figures 16.23 and 16.24 on pages 459 and 460 show examples of "good news" and "bad news" adjustment letters. The first letter is a reply to the claim letter shown in Figure 16.22.

Larsen Supply Company
311 Elmerine Avenue
Anderson, MO 63501

August 21, 19XX

Mr. Jack Robbins, President
Robbins Construction, Inc.
255 Robbins Place
Centerville, MO 65101

Dear Mr. Robbins:

I was very unhappy to read your letter of August 19 telling me about the failure of the Vista 500. I regretted most the treatment you received from one of my employees when you called us.

Harry Rivers, our best technician, has already been in touch with you to arrange a convenient time to come out to Centerville to talk with you about the driver. We will of course repair it, replace it, or refund the price. Just let us know your wish.

I realize that I cannot undo the damage that was done on the day that a piece of our equipment failed. To make up for some of the extra trouble and expense you incurred, let me offer you a 10 percent discount on your next purchase or service order with us, up to $1,000 total discount.

You have indeed been a good customer for many years, and I would hate to have this unfortunate incident spoil that relationship. Won't you give us another chance? Just bring in this letter when you visit us next, and we'll give you that 10 percent discount.

Sincerely,

Dave Larsen

Dave Larsen, President

FIGURE 16.23
"Good News"
Adjustment Letter

Quality Video Products

February 3, 19XX

Ms. Maya Raphael
1903 Highland Avenue
Glenn Mills, NE 69032

Dear Ms. Raphael:

Thank you for writing us about the videotape you purchased on January 11, 19XX.

You used the videotape to record your daughter's wedding. While you were playing it back last week, the tape jammed and broke as you were trying to remove it from your VCR. You are asking us to reimburse you $500 because of the sentimental value of that recording.

As you know, our videotapes carry a lifetime guarantee covering parts and workmanship. We will gladly replace the broken videotape. However, the guarantee states that the manufacturer and the retailer will not assume any incidental liability. Thus we are responsible only for the retail value of the blank tape.

However, your wedding tape can probably be fixed. A reputable dealer can splice tape so skillfully that you will hardly notice the break. It's a good idea to make backup copies of your valuable tapes.

Attached to this letter is a list of our authorized dealers in your area, any of which would be glad to do the repairs for you. We have already sent out your new videotape. It should arrive within the next two days.

Please contact us if we can be of any further assistance.

Sincerely yours,

Paul Blackwood

Paul R. Blackwood, Manager
Customer Relations

FIGURE 16.24
"Bad News" Adjustment Letter

E-mail

1. Is the tone appropriate?

2. Have you written the message carefully and revised it?

3. Have you avoided flaming?

4. Have you used the subject line?

5. Have you used uppercase and lowercase letters?

6. Have you skipped lines between paragraphs?

7. Have you kept the line length to under 65 characters?

8. Have you checked with the writer before reposting his or her message?

Memos

1. Does the identifying information include
   ❏ the names and (if appropriate) the job positions of both you and your readers?
   ❏ a sufficiently informative subject heading?
   ❏ the date?

2. Does the purpose statement clearly tell the readers why you are asking them to read the memo?

3. Does the summary
   ❏ briefly state the major points developed in the body of the memo?
   ❏ reflect the structure of the memo?

4. Does the discussion section include
   ❏ a background paragraph?
   ❏ headings to clarify the structure and content?

5. Does the action section clearly and politely identify tasks that you or your readers will carry out?

Letter Format

1. Is the first page typed on letterhead stationery?

2. Is the date included?

3. Is the inside address complete and correct? Is the appropriate courtesy title used?

4. If appropriate, is an attention line included?

5. If appropriate, is a subject line included?

6. Is the salutation appropriate?

7. Is the complimentary close typed with only the first word capitalized? Is the complimentary close followed by a comma?

8. Is the signature legible and the writer's name typed beneath the signature?

9. If appropriate, are the reference initials included?

10. If appropriate, is an enclosure line included?

11. If appropriate, is a copy line included?

12. Is the letter typed in one of the standard formats?

Types of Letters

1. Does the order letter
   ❏ include the necessary identifying information, such as quantities and model numbers?
   ❏ specify, if appropriate, the terms of payment?
   ❏ specify the method of delivery?

2. Does the inquiry letter
   ❏ explain why you chose the reader to receive the inquiry?
   ❏ explain why you are requesting the information and to what use you will put it?
   ❏ specify by what date you need the information?
   ❏ list the questions clearly and, if appropriate, provide room for the reader's response?
   ❏ offer, if appropriate, the product of your research?

3. Does the response to an inquiry letter answer the reader's questions or explain why they cannot be answered?

4. Does the sales letter
   ❏ gain the reader's attention?
   ❏ describe the product or service?
   ❏ convince the reader that the claims are accurate?
   ❏ encourage the reader to find out more about the product or service?

5. Does the claim letter
   ❏ identify specifically the unsatisfactory product or service?
   ❏ explain the problem(s) clearly?
   ❏ propose an adjustment?
   ❏ conclude courteously?

6. Does the "good-news" adjustment letter
   ❏ express your regret?
   ❏ explain the adjustment you will make?
   ❏ conclude on a positive note?

7. Does the "bad-news" adjustment letter
   ❏ meet the reader on neutral ground, expressing regret but not apologizing?
   ❏ explain why the company is not at fault?

❑ clearly deny the reader's request?
❑ attempt to create goodwill?

EXERCISES

1. Revise the following e-mail message to make it more effective and more professional. The writer is a technician responding to a request from a colleague for information about machines called coat tracks.

COAT TRACKS ARE THE MACHINES USED IN THE FIRST STEP OF THE PHOTO PROCESS. THE WAFERS COEM TO COAT TO HAVE A LAYER OF A PHOTO-SENSITIVE RESIST APPLYED. THIS REQUIRES SEVERAL OPERATIONS TO BE DONE BY THE SAME MACHINE. FIRST THE WAFER IS COATED WITH A LAYER OF PRIMER OR HMDS. THIS ENSURES THAT THE RESIST WILL ADHERE TO THE WAGER. THE WAFER IS THEN CAOTED WITH RESIST. THERE ARE 5 DIFFERENT TYPES OF RESIST IN USE, EACH HAS ITS WON CHARACTERISTICS, AND ALL ARE USED ON DIFFERENT LEVELS AND PART TYPES. I DONT REALLY HAVE TIME TO GO INTO ALL THE DETAILS NOW. THE PHOTO RESIST MUST BE APPLIED INA UNIFORM LAYER AS THE UNFIOROMITY OF THE RESIST CAN EFFECT SEVERAL OTHER STEPS TO INCLUDE EXSPOSURE ON THE STEPPER TO THE ETCH RATE ON A LAM. TO INSURE THE PROPER UNFIROMITY AND RESIST VOLUMES THE TRACKS ARE INSPECTED AT 24 HOUR INTERVALS, ALL THE FUNCTIONS ARE CHECKED AND PARTICAL MONITORS ARE RAN TO ENSURE PROPER OPERATION AND CLEANLYNESS. AFTER THE WAFER IS COATED IT IS SOFT BAKED, THIS RIDS THE WAFER OF SOLVENTS IN THE PRIMER AND RESIST AND ALSO HARDENS THE RESIST. THE WAFERS ARE THEN READY TO GO TO THE NEXT STEP AT THE P&E OR THE STEPPER.

2. As the manager of Lewis Auto Parts Store, you have noticed that some of your salespeople smoke in the showroom. You have received several complaints from customers. Write a memo defining a new policy: salespeople may smoke in the employees' lounge but not in the showroom.

3. There are 20 secretaries in the six departments at your office. Although they are free to take their lunch hours whenever they wish, sometimes several departments have no secretarial coverage between 1:00 and 1:30 P.M. Write a memo to the secretaries, explaining why this lack of coverage is undesirable and asking for their cooperation in staggering their lunch hours.

4. You are a senior with an important position in a school organization, such as a technical society or the campus newspaper. The faculty adviser to the organization has asked you to explain, for your successors, how to carry out the responsibilities of the position. Write a memo in response to the request.

5. The boss at the company where you last worked has phoned you, asking for your opinions on how to improve the working conditions and productivity there. Using your own experiences, write a memo responding to the boss's inquiry.

6. If you have attended a lecture or presentation in your area of academic concentration, write a trip report memo to an appropriate instructor assessing its quality.

7. Write up a recent lab or field project in the form of a memo to the instructor of the course.

8. The memos on pages 463-465 could be improved in tone, substance, and structure. Revise them to increase their effectiveness, adding any reasonable details.

a.

KLINE MEDICAL PRODUCTS

Date: 1 September 19XX
To: Mike Framson
From: Fran Sturdiven
Subject: Device Master Records

The safety and efficiency of a medical device depends on the adequacy of its design and the entire manufacturing process. To ensure that safety and effectiveness are manufactured into a device, all design and manufacturing requirements must be properly defined and documented. This documentation package is called by the FDA a "Device Master Record."

The FDA's specific definition of a "Device Master Record" has already been distributed.

Paragraph 3.2 of the definition requires that a company define the "compilation of records" that makes up a "Device Master Record." But we have no such index or reference for our records.

Paragraph 6.4 says that any changes in the DMR must be authorized in writing by the signature of a designated individual. We have no such procedure.

These problems are to be solved by 15 September 19XX.

b.

Diversified Chemicals, Inc.
Memo

Date: August 27, 19XX
To: R. Martins
From: J. Speletz
Subject: Charles Research Conference on Corrosion

The subject of the conference was high-temperature dry corrosion. Some of the topics discussed were

1–thin film formation and growth on metal surfaces. The lectures focused on the study of oxidation and corrosion by spectroscopy.

(continued)

2–the use of microscopy to study the microstructure of thick film formation on metals and alloys. The speakers were from the University of Colorado and MIT.

3–one of the most interesting topics was hot corrosion and erosion. The speakers were from Penn State and Westinghouse.

4–future research directions for high-temperature dry corrosion were discussed from five viewpoints.

1–university research
2–government research
3–industry research
4–European industry research
5–European government research

5–corrosion of ceramics, especially the oxidation of Si_3N_4. One paper dealt with the formation of Si ALON, which could be an inexpensive substitute for Si_3N_4. This topic should be pursued.

c.

FREEMAN, INC.   INTEROFFICE

To:	C. F. Ortiz
From:	R. C. Nedden
Subject:	Testing of Continuous Solder Strip Alternative for Large-Scale Integrated Terminals
c:	J. A. Jones
	M. H. Miller

We ordered samples of continuous solder strips in three thicknesses for our testing: 1.5 mils, 2.5 mils, and 4.0 mils. Then we manufactured each thickness into terminals to test for pull strength.

The 1.5 mil material had an average pull strength of 1.62 pounds, which is above our goal of 1.5 pounds. But 30 percent of these terminals did not meet the goal. The 2.5 mil material had an average pull strength of 2.4 pounds, with a minimum force of 1.65 pounds. The 4.0 mil material had an average pull strength of 2.6 pounds per terminal, with a minimum of 1.9. Even though there was 60 percent more solder available than with the 2.5 mils material, the average pull strength increased by only 8 percent.

We concluded from this that the limit to the pull strength of the terminal is dependent on the geometry of the terminal, not on the amount of solder.

(continued)

For this reason, we believe that the 2.5 mils material would be the most cost-effective solution to the problem of inadequate pull strength in our LSI terminals.

Please let me know if you have any questions.

9. Write an order letter to John Saville, general manager of White's Electrical Supply House (13 Avondale Circle, Los Angeles, CA 90014). These are the items you want: one SB11 40-ampere battery backup kit, at $73.50; twelve SW402 red wire kits, at $2.50 each; ten SW400 white wire kits, at $2.00 each; and one SB201 mounting hardware kit, at $7.85. Invent any reasonable details about methods of payment and delivery.

10. Secure the graduate catalog of a university offering a graduate program in your field. Write an inquiry letter to the appropriate representative, asking at least three questions about the program the university offers.

11. You are the marketing director of the company that publishes this book. Draft a sales letter that might be sent to teachers of the course you are presently taking.

12. You are the marketing director of the company that makes your bicycle (or calculator, stereo set, running shoes, etc.). Write a sales letter that might be sent to retailers to encourage them to sell the product.

13. You are the recruiting officer for your college. Write a letter that might be sent to juniors in local high schools to encourage them to apply to the college when they are seniors.

14. You purchased four "D"-size batteries for your cassette player, and they didn't work. The package they came in says that the manufacturer will refund the purchase price if you return the defective items. Inventing any reasonable details, write a claim letter asking for not only the purchase price but other related expenses.

15. A thermos you just purchased for $8.95 has a serious leak. The grape drink you put in it ruined a $15.00 white tablecloth. Inventing any reasonable

details, write a claim letter to the manufacturer of the thermos.

16. The gasoline you purchased from New Jersey Petroleum contained water and particulate matter; after using it, you had to have your automobile tank flushed at a cost of $50. You have a letter signed by the mechanic explaining what he found in the bottom of your tank. As a credit-card customer of New Jersey Petroleum, write a claim letter. Invent any reasonable details.

17. As the recipient of one of the claim letters described in exercises 14–16, write an adjustment granting the customer's request.

18. You are the manager of a private swim club. A member has written to say that she lost a contact lens (value $55) in your pool. She wants you to pay for a replacement. The contract that all members sign explicitly states that the management is not responsible for loss of personal possessions. Write an adjustment letter denying the request. Invent any reasonable details.

19. As manager of a stereo equipment retail store, you guarantee that you will not be undersold. If a customer who buys something from you can prove within one month that another retailer sells the same equipment for less money, you will refund the difference in price. A customer has written to you, enclosing an ad from another store showing that it is selling the equipment he purchased for $26.50 less than he paid at your store. The advertised price at the other store was a one-week sale that began five weeks after the date of his purchase from you. He wants his $26.50 back. Inventing any reasonable details, write an adjustment letter denying his request. You are willing, however, to offer him a blank cassette tape worth $4.95 for his equipment if he would like to come pick it up.

20. The letters below and on pages 467-469 could be improved in both tone and substance. Revise them to increase their effectiveness, adding any reasonable details.

REFERENCE

Crowe, E. P. (1994). *The electronic traveller: Exploring alternative online systems.* New York: Windcrest/McGraw-Hill.

a.

GUARDSMAN PROTECTIVE EQUIPMENT, INC.
3751 PORTER STREET
NEWARK, DE 19304

April 11, 19XX

Dear Smith Family:

A rose is a rose is a rose, the poet said. But not all home protection alarms are alike. In a time when burglaries are skyrocketing, can you afford the second-best alarm system?

Your home is your most valuable possession. It is worth far more than your car. And if you haven't checked your house insurance policy recently, you'll probably be shocked to see how inadequate your coverage really is.

The best kind of insurance you can buy is the Watchdog Alarm System. What makes the Watchdog unique is that it can detect intruders before they enter your home and scare them away. Scaring them away while they're still outside is certainly better than scaring them once they're inside, where your loved ones are.

At less than two hundred dollars, you can purchase real peace of mind. Isn't your family's safety worth that much?

If you answered yes to that question, just mail in the enclosed postage-paid card, and we'll send you a 12-page, fact-filled brochure that tells you why the Watchdog is the best on the market.

Very truly yours,

b.

19 Lowry's Lane
Morgan, TN 30610

April 13, 19XX

Sea-Tasty Tuna
Route 113
Lynchburg, TN 30563

Gentlemen:

I've been buying your tuna fish for years, and up to now it's been OK.

But this morning I opened a can to make myself a sandwich. What do you think was staring me in the face? A fly. That's right, a housefly. That's him you see taped to the bottom of this letter.

What are you going to do about this?

                                              Yours very truly,

c.

Handyman Hardware, Inc.   **Millersville, AL 61304**

December 4, 19XX

Hefty Industries, Inc.
19 Central Avenue
Dover, TX 76104

Gentlemen:

I have a problem I'd like to discuss with you. I've been carrying your line of hand tools for many years.

Your 9" pipe wrench has always been a big seller. But there seems to be something wrong with its design. I have had three complaints in the last few months about the handle snapping off when pressure is exerted on it. One user cut his hand seriously enough to require 19 stitches.

Frankly, I'm hesitant to sell any more of the 9" pipe wrenches, but I still have over two dozen in inventory.

Have you had any other complaints about this product?

Sincerely yours,

d.

Sea-Tasty Tuna
Route 113
Lynchburg, TN 30563

April 21, 19XX

Mr. Seth Reeves
19 Lowry's Lane
Morgan, TN 30610

Dear Mr. Reeves:

We were very sorry to learn that you found a fly in your tuna fish.

Here at Sea-Tasty we are very careful about the hygiene of our plant. The tuna are scrubbed thoroughly as soon as we receive them. After they are processed, they are inspected visually at three different points. Before we pack them, we rinse and sterilize the cans to ensure that no foreign material is sealed in them.

Because of these stringent controls, we really don't see how you could have found a fly in the can. Nevertheless, we are enclosing coupons good for two cans of Sea-Tasty tuna.

We hope this letter restores your confidence in us.

Truly yours,

e.

Hefty Industries, Inc.
19 Central Avenue
Dover, TX 76104

December 11, 19XX

Mr. Peter Arlen, Manager
Handyman Hardware, Inc.
Millersville, AL 61304

Dear Mr. Arlen:

Thank you for bringing this matter to our attention.

In answer to your question—yes, we have had a few complaints about the handle snapping on our 9" pipe wrench.

Our engineers brought the wrench back to the lab and discovered a design flaw that accounts for the problem. We have redesigned the wrench and have not had any complaints since.

We are not selling the old model anymore because of the risk. Therefore we have no use for your two dozen. However, since you have been a good customer, we would be willing to exchange the old ones for the new design.

We trust this will be a satisfactory solution.

Sincerely,

| CHAPTER 17 | Job-Application Materials |

For most of you, the first nonacademic test of your technical writing skills will come when you create your job-application materials. These materials will inform employers about your academic and employment experience, personal characteristics, and reasons for applying. But they will provide much more information, too. Employers have learned that one of the most important skills an employee can bring to a job is the ability to communicate effectively. Therefore, potential employers look carefully for evidence of writing skills. Job-application materials pose a double hurdle: the tasks of showing employers both what you can do and how well you can communicate.

Some students think that once they get a satisfactory job, they will never again have to worry about résumés and application letters. Statistics suggest otherwise. The typical professional changes jobs more than five times. Although this chapter pays special attention to the student's first job hunt, the skills and materials discussed here also apply to established professionals who wish to change jobs.

Planning for the Job Search

Planning for the job search is a lot of work, stretching over weeks and months, not days. Three main tasks are involved in planning the search, as shown in Figure 17.1.

- *Doing a self-inventory.* Before you can start thinking of where you want to work, you need to answer some questions about yourself:
  - *What are your strengths and weaknesses?* Are your skills primarily technical? Do you work well with people? Do you work best with supervision or without?
  - *What are your preferences?* Think about what you have or have not liked about your previous positions and your college courses. In addition to a salary, what do you want to get out of a job?
  - *What kind of organization would you like to work for?* Profit or nonprofit? Government or private industry? Small or large?
  - *What are your geographical preferences?* If you are free to move, where would you like to live? How do you feel about commuting?

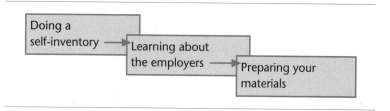

FIGURE 17.1

Planning for a Job Search

- *Learning about the employers.* Don't base your job search exclusively on the information provided in a job advertisement in the paper. Learn as much as you can about the organizations through other means as well:

  - *Attend job fairs.* Your college and perhaps your community hold job fairs, where different employers provide information about their organizations.
  - *Find out about trends in your field.* Read the *Occupational Handbook*, published by the U.S. Department of Labor, for information about your field and related fields. Talk with professors in your field and with the people at your job-placement office.
  - *Research the companies you are interested in.* Write to them for information. Scan the index of the *Wall Street Journal* for articles. Study their annual reports, many of which are collected in your college library.

- *Preparing your materials.* You know you will need to write application letters and résumés and that you will go on interviews. Start planning early by obtaining materials from the career-placement office. Talk with friends who have gone through the process successfully; study their application materials. Read some of the books on different aspects of job searching.

     One more very important part of preparing your materials: make up a *portfolio,* a collection of your best work. You'll want to be able to give a prospective employer a copy of the portfolio as an example of the kind of work you can do. For a technical communicator, the portfolio will include a variety of documents you made in courses and in previous positions. For technical professionals, the portfolio will include documents such as project reports, as well as other materials, including computer simulations or output. A portfolio is generally presented in a loose-leaf notebook, with each item preceded by a descriptive and evaluative statement. Often, a portfolio contains a table of contents and an introductory statement.

Understanding the Seven Ways to Look for a Position

Once you have done your planning, you can start to look for a position. The seven major ways to find a job are shown in Figure 17.2.

- *Through a college or university placement office.* Almost all colleges and universities have placement offices, which bring companies and students together. Generally, students submit a résumé—a brief list of credentials—to the placement office. The résumés are then made available to representatives of business, government, and industry, who use the place-

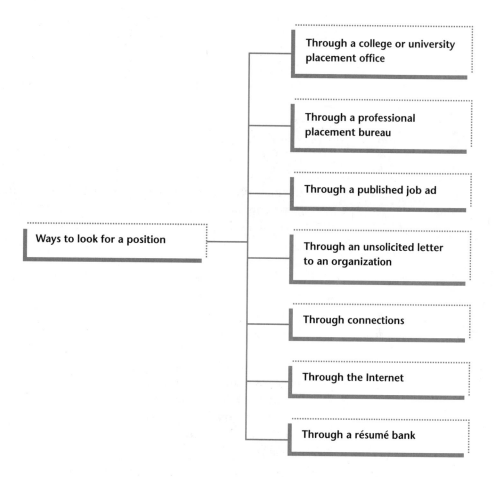

Ways to look for a position

- Through a college or university placement office
- Through a professional placement bureau
- Through a published job ad
- Through an unsolicited letter to an organization
- Through connections
- Through the Internet
- Through a résumé bank

FIGURE 17.2

Seven Ways to Look for a Position

ment office to arrange on-campus interviews with selected students. Those who do best in the campus interviews are then invited by the representatives to visit the organization for a tour and another interview. Sometimes a third interview is scheduled; sometimes an offer is made immediately or shortly thereafter. The advantage of this system is twofold: it is free and it is easy. You merely deliver a résumé to the placement office and wait to be contacted.

- *Through a professional placement bureau.* A professional placement bureau offers essentially the same service as a college placement office, but it charges either the employer (or, rarely, the new employee) a fee—often 10 percent of the first year's salary—when the client accepts a position. Placement bureaus cater primarily to more advanced professionals who are looking to change jobs.

- *Through a published job ad.* Published job ads generally offer the best opportunity for both students and professionals. Organizations advertise in three basic kinds of publications: public-relations catalogs (such as *College Placement Annual*), technical journals, and newspapers. Check the major technical journals in your field and the large metropolitan newspapers, especially the Sunday editions. In responding to an ad, you must include with the résumé a job-application letter that highlights the crucial information on the résumé.

- *Through an unsolicited letter to an organization.* Instead of waiting for a published ad, consider writing unsolicited letters of application to organizations you would like to work for. The disadvantage of this technique is obvious: there might not be an opening. Yet many professionals favor this technique, because fewer people are competing for those jobs that do exist, and sometimes organizations do not advertise all available positions. Occasionally, an impressive unsolicited application prompts an organization to create a position.

   Before you write an unsolicited application, make sure you learn as much as you can about the organization (see Chapters 6 and 7): current and anticipated major projects, hiring plans, and so on. You should be knowledgeable about any organization you are applying to, of course, but when you are submitting an unsolicited application you have no other source of information on which to base your résumé and letter. The business librarian at your college or university will be able to point out sources of information, such as the Dun and Bradstreet guides, the *F&S Index of Corporations,* and the indexed newspapers such as the *New York Times,* the *Washington Post,* the *Los Angeles Times,* and the *Wall Street Journal.*

- *Through connections.* A relative or an acquaintance who can exert influence or at least point out a new position can help you get a job. Other good contacts include employers from your past summer jobs and faculty members in your field. Become active in the student chapter of your field's professional organization, through which you can meet professionals in your area.

- *Through the Internet.* The Internet and several major commercial access services have special news services and bulletin boards specifically devoted to job searching. While the postings on these services sometimes duplicate newspaper ads, you would not have an opportunity to check the ads published in thousands of local newspapers otherwise. And many of the postings are unique. Contact your access service for information on how to find the employment services. If as a student you have Internet access, your college placement office should be able to help you.

   In addition, some people find work by posting their résumés to Usenet newsgroups and bulletin boards. See Chapter 6 for more information about the Internet.

- *Through a résumé bank.* A résumé bank is a company that collects and stores résumés in a database, then sells them to organizations seeking to hire people. Some résumé banks specialize in certain kinds of positions, such as managers or clerical workers; other résumé banks specialize in workers in a certain price salary range, such as $20,000–$50,000.

Writing the Résumé

Many students ask whether they should write the résumé themselves or use a résumé-preparation agency. I think it's best to make your own résumé, for three reasons:

- *You know yourself better than anyone else does.* No matter how thorough and professional the work of a résumé-preparation agency, you can do a better job communicating the important information about yourself.
- *Employment officers can recognize the style of the local agencies.* Readers who realize that you did not write your own résumé might wonder what kinds of deficiencies you are trying to hide.
- *If you write your own résumé, you will more likely vary it to meet the needs of different situations.* You are much less likely to go back to a résumé-preparation agency and pay them an additional fee to make a minor revision.

Whether you write the résumé by yourself or get help, it communicates in two ways: by its appearance and by its content.

Appearance of the Résumé

Because potential employers normally see your résumé before they see you, it has to make a good first impression. Employers believe—often correctly—that the appearance of the résumé reflects the writer's professionalism. A sloppy résumé implies that you would do sloppy work. A neat résumé implies that you would do professional work. When employers look at a résumé, they see the documents they will be reading if they hire you.

Some colleges and universities advise students to have their résumé professionally printed. A printed résumé is attractive, and that's good—provided, of course, that the information on it is consistent with its professional appearance. Most employers agree, however, that a neatly word-processed résumé, printed on a laser printer and photocopied on good-quality paper, is just as effective.

People who photocopy a word-processed résumé are more likely to tailor different versions to the needs of the organizations to which they apply—a good strategy. People who go to the trouble and expense of a professional printing job are far less likely to make up different résumés. The résumé looks so good that they don't want to tinker with it. This strategy is dangerous, for it encourages the writer to underestimate the importance of tailoring the résumé to a specific audience.

However they are reproduced, résumés should appear neat and professional. They should have

- *Generous margins.* Leave a 1-inch margin on all four sides.
- *Clear type.* Use a good-quality laser printer.
- *Balance.* Arrange the information so that the page has a balanced appearance.
- *Clear organization.* Use adequate white space. Make sure the line spacing *between* items is greater than the line spacing *within* an item. That is, the line spacing between your education section and your employment section should be greater than that within one of those sections. You should be able to see the different sections clearly if you stand while the résumé is on the floor in front of your feet.

Use indentation clearly. When you arrange items in a vertical list, indent turnovers a few spaces. (*Turnovers* are the second and subsequent lines of any item.) Notice, for example, that the following list from the computer-skills section of a résumé can be confusing:

Computer Experience
Systems: IBM, Macintosh, UNIX, Andover AC-256, Prime 360
Software: Lotus 1-2-3, DBase V, PlanPerfect, Micrografx Designer, Aldus Pagemaker, Microsoft Word
Languages: Fortran, Pascal, C

With the second line of an entry indented, the arrangement is much easier to understand:

Computer Experience
Systems: IBM, Macintosh, UNIX, Andover AC-256, Prime 360
Software: Lotus 1-2-3, DBase V, PlanPerfect, Micrografx Designer, Aldus Pagemaker,
    Microsoft Word
Languages: Fortran, Pascal, C

If you are submitting your résumé to a résumé bank, follow these suggestions on formatting (based on Wright, 1994):

- Use a single-column design. Scanners have trouble with multicolumn designs.
- Use type sizes between 10 point and 14 point (see Chapter 13 for a discussion of type sizes).
- Do not italicize or underline words.
- Use standard typefaces such as Helvetica, Futura, Optima, Times Roman, New Century Schoolbook, Courier, Univers, and Bookman.
- Use a laser printer.

See Chapter 13 for more information on page design.

Content of the Résumé

Although different experts advocate different approaches to résumé writing, everyone agrees on three things:

- *The résumé must be honest.* No accurate statistics exist on how many résumés contain lies, but the figure is probably considerable. One source ("Looking," 1994, p.10) estimates the figure at one third. You have probably read about employees who have been caught including not only exaggerations but outright lies on their résumés. Naturally, they are fired immediately. Many employers today routinely check candidates' credentials. So, for practical as well as ethical reasons, tell the truth.

- *The résumé must be completely free of errors.* Grammar, punctuation, usage, and spelling errors cast doubt on the accuracy of the information in the résumé. Ask for assistance after you have written the draft, and proofread the finished product at least twice.

- *The résumé must provide clear, specific information, without generalizations or self-congratulation.* Your résumé is a sales document, but you are both the salesperson and the product. You cannot say "I am a terrific job candidate" as if the product were a toaster or an automobile. Instead, you have to provide the specific details that will lead the reader to the conclusion that you are a terrific job candidate. You must *show* the reader. Telling the reader is graceless and, worse, unconvincing.

A résumé should be long enough to include all the pertinent information but not so long that it bores or irritates the reader. Generally, keep it to one page. If, however, you have special accomplishments—such as journal articles, patents, or substantial service in student government—a two-page résumé is appropriate. If your information takes up just over one page, either eliminate or condense some material to make it fit onto one page, or modify the layout so that it fills a substantial part of the second page. According to a recent study (Harcourt, Krizan, & Merrier, 1991), summarized in Table 17.1, most hiring officials prefer that a new college graduate's résumé be brief.

The two major styles of résumés are *chronological* and *analytical*. In the chronological résumé, you present your information chronologically using time as the organizing factor for each section of the résumé, including education and

TABLE 17.1	*Preferred Length for a Résumé*
Preferred Length	**Hiring Officials (Percent)**
No longer than one page	23.6
No longer than two pages	41.8
Depends on applicant's information	32.7
Other	01.8

experience. In the analytical résumé, you present a separate section called *skills,* which is organized according to skills, not time.

Recent graduates commonly use the chronological résumé because in most cases they haven't had an opportunity to build the record of skills and accomplishments needed to fill out an analytical résumé. However, if you have had professional work experience, you might consider the analytical style.

Elements of the Chronological Résumé

Almost every chronological résumé has the six basic elements shown in Figure 17.3. However, your résumé should reflect one particular person: you. Many people have some special skills or background that could be conveyed in additional sections.

Identifying Information

Place your full name, address, and phone number at the top of the page. Generally, you should present your name in boldface letters, centered at the top. Use your complete address, with the zip code. Use the two-letter state abbreviations used by the Postal Service. Also give your complete phone number, with the area code.

If your address during the academic year differs from your home address, list both and identify them clearly. An employer might call during an academic holiday to arrange an interview.

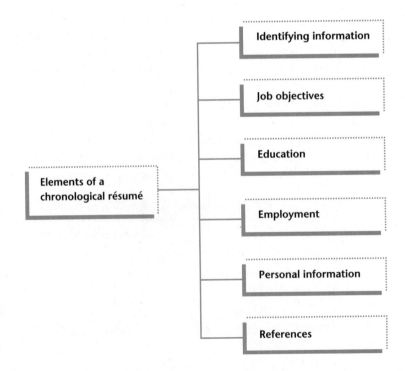

FIGURE 17.3

Elements of a
Chronological Résumé

Job Objectives

After the identifying information, add a statement of objectives, in the form of a brief phrase or sentence—for example, "Objective: Entry-level position as a dietitian in a hospital." According to a recent study, 88.5 percent of managers making personnel decisions consider a statement of objectives important, because it gives the appearance that the writer has clear sense of direction and goals (Harcourt, Krizan, & Merrier, 1991). In drafting the statement, follow two guidelines:

- *State only goals or duties explicitly mentioned in the job advertisement, or at least clearly implied.* If you unintentionally suggest that your goals are substantially different from the job responsibilities, the organization might infer that you would not be happy working there and not consider you further.
- *Avoid meaningless generalities.* You accomplish nothing by writing, "Position offering opportunities for advancement in the field of health science, where I can use my communication and analytical skills."

Education

If you are a student or a recent graduate, add the education section after the identifying information. If you have substantial professional experience, place the employment experience section before the education section.

Include at least the following information in the education section:

- *The degree.* After the degree abbreviation (such as B.S., B.A., A.A., or M.S.), list your academic major (and, if you have one, your minor)—for example, "B.S. in Materials Engineering, minor in General Business."
- *The institution.* Identify the institution by its full name: "The Pennsylvania State University," not "Penn State."
- *The location of the institution.* Include the city and state.
- *The date of graduation.* If your degree has not yet been granted, write "Anticipated date of graduation" or "Degree expected in" before the month and year.

Include one additional item:

- *Information about other schools you attended.* List any other institutions you attended beyond the high-school level—even those at which you did not earn a degree. Employers are generally impressed to learn that a student began at a community college or junior college and was able to transfer to a four-year college or university. The listing for other institutions attended should include the same information as the main listing. Arrange the entries in reverse chronological order: that is, list first the school you attended most recently.

The basic information about the educational experience is necessary, but it is not enough. It implies that you merely endured an institution and received a degree. The following four strategies can be useful in elaborating on your educational background:

- *List your grade-point average.* If your grade-point average is significantly above the median for the graduating class, list it. Or list your average in the courses in your major, if that is more impressive.

- *Create a list of courses.* Choose courses that will be of particular interest to your reader. Advanced courses in an area of your major concentration might be appropriate, especially if the potential employer has mentioned that area in the job ad. Also useful would be a list of communications courses—technical communication, public speaking, organizational communications, and the like. A listing of business courses on an engineer's résumé—or the reverse—also shows special knowledge and skills. The only kind of course listing that is not particularly helpful is one that merely names the traditional required courses for your major. When you list courses, include the substantive title; employers won't know what "Chemistry 450" is. Call it "Chemistry 450. Organic Chemistry."

- *Describe a special accomplishment.* If you did a special senior design or research project, for example, a two- or three-line description would be informative. Include in this description the title and objective of the project, any special or advanced techniques or equipment you used, and—if you know them—the major results. Such a description might be phrased as follows: "A Study of Composite Substitutes for Steel—a senior design project intended to formulate a composite material that can be used to replace steel in car axles." Even a traditional academic course in which you conducted a sophisticated project and wrote a sustained report can be described profitably. A project discussion makes you look less like a student—someone who takes courses—and more like a professional—someone who designs and carries out projects.

- *List any honors and awards you received.* Scholarships, fellowships, and academic awards offer evidence of exceptional ability. If you have received a number of such honors, or some that were not exclusively academic, you might list them separately (in a section called *Honors* or *Awards*) rather than in a subsection of the education section. Often, some information could logically be placed in two or even three different locations, and you must decide where it will make the best impression.

The education section is the easiest part of the résumé to adapt for different positions. For example, a student majoring in electrical engineering who is applying for a position that requires strong communications skills can list communications courses in one version of the résumé. The same student can list advanced electrical engineering courses in a résumé directed to another potential employer. As you compose the education section of your résumé, carefully emphasize those aspects of your background that meet the requirements for the particular job.

Employment

The employment section, like the education section, conveys at least the basic information about each job you've held:

- dates of employment
- the organization's name and location
- your position or title

However, such a skeletal list would not be very informative or impressive. As in the education section, provide carefully selected details.

What readers want to know, after they have learned where and when you were employed, is what you actually did. Therefore, provide at least a two- to three-line description for each position. For particularly important or relevant jobs, give a longer description. Focus the description on one or more of the following factors:

- *Documents*. What kinds of documents did you write or assist in writing? List, especially, various governmental forms and any long reports, manuals, or proposals.

- *Clients*. What kinds of, and how many, clients did you transact business with as a representative of your organization?

- *Skills*. What kinds of technical skills did you learn or practice in doing the job?

- *Equipment*. What kinds of technical equipment did you operate or supervise? Mention, in particular, any computer skills you demonstrated, for they can be useful in almost every kind of position.

- *Money*. How much money were you responsible for? Even if you considered your bookkeeping position fairly simple, the fact that the organization grossed, say, $2 million a year shows that the position involved real responsibility.

- *Personnel*. How many personnel did you supervise? Students sometimes supervise small groups of other students or technicians. Naturally, supervision shows maturity and responsibility.

Whenever possible, emphasize *results*. If you reorganized the shifts of the weekend employees you supervised, cite the results:

Reorganized the weekend shift, resulting in a cost savings of over $3,000 per year.

Wrote and produced (with desktop publishing) a parts catalog that is still used by the company and that increased our phone inquiries by more than 25 percent.

When you describe your positions, use the active voice—"supervised three workers"—rather than the passive voice—"three workers were supervised by me." The active voice emphasizes the verb, which describes the action. In thinking about your functions and responsibilities in your various positions, keep in mind the strong action verbs that clearly communicate your activities. (See Chapter 15 for more information on strong verbs.) Also note that résumés often omit the *I* at the start of sentences. Rather than write "I prepared bids," many people simply write "Prepared bids." Whichever style you use, be consistent. Figure 17.4 lists some strong verbs.

Here is a sample listing:

June–September 19XX: Millersville General Hospital, Millersville, TX. Student Dietitian. Gathered dietary histories and assisted in preparing menus for a 300-bed hospital. Received "excellent" on all items in evaluation by head dietitian.

In just a few lines, you can show that you sought and accepted responsibility and that you acted professionally. Do not write "I accepted responsibility"; rather, present facts that lead the reader to that impression.

Naturally, not all jobs entail such professional skills and responsibilities. Many students find summer work as laborers, salesclerks, short-order cooks, and so on. If you have not held a professional position, list the jobs you have held, even if they were completely unrelated to your career plans. If the job title is self-explanatory—such as waitress or service-station attendant—don't elaborate. Every job is valuable. You learn that you are expected to be someplace at a specific time, wear appropriate clothes, and perform some specific duties. Also, every job helps pay college expenses. If you can say that you earned, say, 50 percent of your annual expenses through a job, employers will be impressed by your self-reliance. Most of them probably started out with nonprofessional positions. And any job you have held can yield a valuable reference.

FIGURE 17.4

Strong Action Verbs for Use in Résumés

administered	coordinated	evaluated	maintained	provided
advised	corresponded	examined	managed	purchased
analyzed	created	expanded	monitored	recorded
assembled	delivered	hired	obtained	reported
built	developed	identified	operated	researched
collected	devised	implemented	organized	solved
completed	directed	improved	performed	supervised
conducted	discovered	increased	prepared	trained
constructed	edited	instituted	produced	wrote

One further word: if you have held a number of nonprofessional positions as well as several professional positions, you can group the nonprofessional ones together in one listing:

Other Employment: cashier (summer, 1989), salesperson (part-time, 1989), clerk (summer, 1988).

This technique prevents the nonprofessional positions from drawing the reader's attention away from the more important positions.

List your jobs in reverse chronological order on the résumé, to highlight those positions you have held most recently.

Personal Information

Most résumés today no longer include such information as the writer's height, weight, date of birth, and marital status. One explanation for this change is that federal legislation prohibits organizations from requiring this information. Perhaps more important, most employers now feel that such personal information is irrelevant to a person's ability.

The personal information section of the résumé *is* the appropriate place for a few items about your outside interests:

- participation in community-service organizations—such as Big Brothers—or volunteer work in a hospital
- hobbies related to your career interests—for example, amateur electronics for an engineer
- sports, especially those that might be socially useful in your professional career, such as tennis, racquetball, and golf
- university-sanctioned activities, such as membership on a team, participation on the college newspaper, or election to a responsible position in an academic organization or a residence hall

Do not include activities that might create a negative impression: hunting, gambling, performing in a rock band. And always omit such activities as meeting people and reading—everybody meets people and reads.

References

Potential employers will want to learn more about you from your professors and your previous employers. In providing these references, follow three steps:

- *Decide how you want to present the references.* You may list the names of three or four referees—people who have written letters of recommendation or who have agreed to speak on your behalf—on your résumé. Or you may simply say that you will furnish the names of the referees on request. The length of your résumé sometimes dictates which approach to use. If the résumé is already long, the abbreviated form might be preferable. If the résumé does not fill out the page, the longer form might be the one to use. However, each style has advantages and disadvantages that you should consider carefully.

Furnishing the referees' names appears open and forthright. It shows that you have already secured your referees and have nothing to hide. If one or several of the referees are prominent people in their fields, the reader is likely to be impressed. And, perhaps most important, the reader can easily phone the referees or write them a letter. Listing the referees makes it easy for the prospective employer to proceed with the hiring process. The only disadvantage is that it takes up space on the résumé that might be needed for other information.

Writing "References will be furnished on request" takes up only one line, however, and it leaves you in a flexible position. You can still secure referees after you have sent out the résumé. Also, you can send selected letters of reference to prospective employers according to your analysis of what they want. Using different references for different positions is sometimes just as valuable as sending different résumés. However, some readers will interpret the lack of names and addresses as evasive or secretive or assume that you have not yet asked prospective referees. A greater disadvantage is that if the readers are impressed by the résumé and want to learn more about you, they cannot do so quickly and directly.

What do personnel officers prefer regarding references? According to the study referred to earlier (Harcourt, Krizan, & Merrier, 1991), 58.4 percent of the hiring officials surveyed want to see the full references, including the name, title, organization, mailing address, and phone number for each referee. For example:

Dr. Robert Ariel
Assistant Professor of Biology
Central University
1910 Westerly Parkway
Portland, OR 97202
(503) 555-5746

- *Choose your referees carefully.* Solicit references only from persons who know your work best and for whom you have done the best work—for instance, a previous employer whom you worked with closely, or a professor from whom you have received A's. It is unwise to ask prominent professors who do not know your work well, for the advantage of having a famous name on the résumé will be offset by the referee's brief and uninformative letter. Often, a less well known professor can write the most informative letter or provide the best recommendation.

- *Give the potential referee an opportunity to decline gracefully.* Sometimes the person has not been as impressed with your work as you think. Also, if you simply ask "Would you please write a reference letter for me?" the potential referee might accept and then write a lukewarm letter. It is better to follow up the first question with "Would you be able to write an enthusiastic letter for me?" or "Do you feel you know me well enough to write a strong recommendation?" If the potential referee shows any signs of hesitation or reluctance, you can withdraw the invitation at that point. The scene is a little embarrassing, but it is better than receiving a half-hearted recommendation.

Other Elements

So far, the discussion has concentrated on the sections that appear on virtually everyone's résumé. Other sections either are discretionary or are appropriate only for certain writers.

- *Military experience.* If you are a veteran of the armed forces, include a military service section on the résumé. Define your military service as if it were any other job, citing the dates, locations, positions, ranks, and tasks. Often a serviceperson receives regular evaluations from a superior; these evaluations can work in your favor.

- *Language ability.* If you have a working knowledge of a foreign language, your résumé should include a *Language Skills* section. Language skills are particularly relevant if the potential employer has international interests and you could be useful in translation or foreign service.

- *Willingness to relocate.* If you are willing to relocate, state that fact explicitly. Many organizations today will find you a more attractive candidate if they know you are willing to move around as you learn the business.

Elements of the Analytical Résumé

The analytical résumé differs from the chronological in that it conveys job skills and experience not in the experience section but in a separate section, usually called *Skills* or *Skills and abilities*. Figure 17.5 on page 486 shows the seven basic elements of an analytical résumé. The employment section of an analytical résumé becomes a briefer list of identifying information about the candidate's employment history: name of the company, dates of employment, and name of the position. The skills section is usually placed prominently near the top of the résumé.

Following is an example of a skills section.

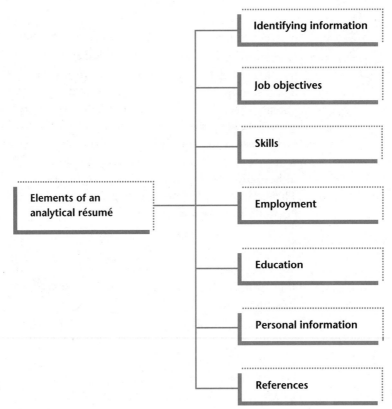

FIGURE 17.5

Elements of an Analytical
Résumé

Skills and Abilities
Management
  Served as weekend manager of six employees in the retail clothing business. Also
  trained three summer interns at a health-maintenance organization.
Writing and Editing
  Wrote status reports, edited performance appraisals, participated in assembling and
  producing an environmental impact statement using desktop-publishing.
Teaching and Tutoring
  Tutored in the University Writing Center. Taught a two-week course in electronics for
  teenagers. Coach youth basketball.

In a skills section, you choose the headings, the arrangement, and the level of detail. Your goal, of course, is to highlight those skills that the employer is seeking.

Figures 17.6, 17.7, and 17.8 on pages 487–489 show effective résumés. Figure 17.6 is a chronological résumé for Carlos Rodriguez, a student seeking a full-time job. He is what is known as a traditional student: he entered college right after high school and has proceeded on schedule. He has no dependents and is willing to relocate, so he makes that point clearly. Because he has some

CARLOS RODRIGUEZ

3109 Vista Street   Philadelphia, PA 19136   (215) 555-3880

Objective   Entry-level position in signal processing

Education   Bachelor of Science in Electrical Engineering
Drexel University, Philadelphia, PA
Anticipated June 1996

Grade-Point Average: 3.67 (of 4.0)

Senior Design Project: "Enhanced Path-Planning Software for Robotics"

Advanced Engineering Courses

Digital Signal Processing	Computer Hardware
Introduction to Operating	Systems Design
Systems I, II	Computer Logic Circuits I, II
Digital Filters	

Employment

6/93–1/94   RCA Advanced Technology Laboratory, Moorestown, NJ
Developed and tested new LISP software to implement the automated production of VLSI standard cell family databooks. Assisted senior engineer in designing and testing two Gate Array (2700 and 5000 gates) VLSI integrated chips.

6/91–1/92   RCA Advanced Technology Laboratory, Moorestown, NJ
Verified and documented several integrated circuit designs. Used CAD software and hardware to simulate, check, and evaluate these designs. Gained experience on the VAX and Applicon.

Honors, Awards, and Organizations

Eta Kappa Nu (Electrical Engineering Honor Society)
Tau Beta Pi (General Engineering Honor Society)
Institute of Electrical and Electronics Engineers, Inc.

Willing to relocate.

References

Ms. Anita Feller	Mr. Fred Borelli	Mr. Sam Shamir
Engineering Consultant	Unit Manager	Comptroller
700 Church Road	RCA Corporation	RCA Corporation
Cherry Hill, NJ 08002	Route 38	Route 38
(609) 555-7836	Moorestown, NJ 08057	Moorestown, NJ 08057
	(609) 555-2435	(609) 555-7849

FIGURE 17.6

Chronological Résumé of
a Traditional Student

Alice P. Linder   1781 Weber Road
Warminster, PA 18974
(215) 555-3999

Objective: A position in molecular research that uses my computer skills

Education: Harmon College, West Yardley, PA
Major: Bioscience and Biotechnology
Expected Graduation Date: June 1997

Related Course Work
General Chemistry I, II, III   Biology I, II, III
Organic Chemistry I, II   Statistical Methods for Research
Physics I, II   Technical Communication
Calculus I, II

Employment Experience:
June 1994–present   Smith Kline & French Laboratory, Upper Merion, PA
(20 hours per week)   Analyze molecular data on E & S PS300, Macintosh, and
                       IBM PCs. Write programs in C, and wrote a user's guide
                       for an instructional computing package. Train and con-
                       sult with scientists and deliver in-house briefings.

June 1980–Jan. 1984   Anchor Products, Inc., Ambler, PA
                      Managed 12-person office in a $1.2 million company.
                      Also performed general bookkeeping and payroll.

Oct. 1991–present   Children's Hospital of Philadelphia, Philadelphia, PA
                      Volunteer in the physical therapy unit. Assist therapists
                      and guide patients with their therapy. Use play therapy to
                      enhance strengthening progress.

Honors: Awarded three $5,000 tuition scholarships (1993–1996) from the
Gould Foundation.

Additional Information:
Member, Harmon Biology Club, Yearbook Staff
Raised three school-age children
Tuition 100% self-financed

References: Available on request.

FIGURE 17.7

Chronological Résumé of
a Nontraditional Student

Alice P. Linder	1781 Weber Road
	Warminster, PA 18974
	(215) 555-3999

Objective: A position in molecular research that uses my computer skills

Skills and Abilities: Laboratory Skills

Analyze molecular data on E & S PS300, Macintosh, and IBM PCs. Write programs in C.

Have taken 12 credits in biology and chemistry labs.

Communication Skills

Wrote a user's guide for an instructional computing package.

Train and consult with scientists and deliver in-house briefings.

Management Skills

Managed 12-person office in a $1.2 million company.

Education: Harmon College, West Yardley, PA

Major: Bioscience and Biotechnology

Expected Graduation Date: June 1997

Related Course Work

General Chemistry I, II, III	Biology I, II, III
Organic Chemistry I, II	Statistical Methods for
Physics I, II	Research
Calculus I, II	Technical Communication

Employment Experience:

June 1994–present Smith Kline & French Laboratory, Upper Merion, PA
(20 hours per week) Laboratory Assistant Grade 3

June 1980–Jan. 1984 Anchor Products, Inc., Ambler, PA
Office Manager

Oct. 1991–present Children's Hospital of Philadelphia, Philadelphia, PA
Volunteer in the physical therapy unit. Assist therapists and guide patients with their therapy. Use play therapy to enhance strengthening progress.

Honors: Awarded three $5,000 tuition scholarships (1993–1996) from the Gould Foundation.

Additional Information:

Member, Harmon Biology Club, Yearbook Staff
Raised three school-age children
Tuition 100% self-financed

References: Available on request.

FIGURE 17.8

Analytical Résumé of a
Nontraditional Student

high-powered referees from his former employers, he chooses to list their names at the end of the résumé.

Figure 17.7 is a chronological résumé written by Alice Linder, a nontraditional student: a single mother who has returned to school after having raised three children. She is applying for an internship position. Figure 17.8 is an analytical résumé by Alice Linder.

A note about Alice Linder's résumés: there are two different strategies for presenting the information that a candidate is a nontraditional student. Alice Linder feels that her nontraditional status is an asset: her maturity and experience will make her a more effective employee than the traditional student. For this reason, she mentions this information on her résumé. Other people would deemphasize this information because they feel it is irrelevant and draws attention away from important credentials. As a compromise between these two strategies, a student could mention the nontraditional status briefly in the job application letter but omit it from the résumé.

Writing the Job-Application Letter

The job-application letter is crucial because it is the first thing your reader sees. If the letter is ineffective, the reader probably will not bother to read the résumé.

If students had infinite time and patience, they would create a different résumé for each prospective employer, highlighting their appropriateness for that one job. But because they don't have unlimited time and patience, they usually make one or two versions of their résumés. As a result, the typical résumé makes a candidate look only relatively close to the ideal the employer has in mind. Thus the job-application letter must appeal directly and specifically to the needs and desires of the particular employer.

Appearance of the Job-Application Letter

Like the résumé it introduces, the letter must be error-free and professional looking. See Chapter 16 for a discussion of letter formats.

Content of the Job-Application Letter

Like the résumé, the job-application letter is a sales document. Its purpose is to convince the reader that you are an outstanding candidate who should be invited for an interview. Of course, you accomplish this purpose through hard evidence, not empty self-praise.

The job-application letter is *not* an expanded version of everything in the résumé. The key to a good application letter is selectivity. In the letter, choose from the résumé two or three points of greatest interest to the potential employer and develop them into paragraphs. Emphasize results, such as improved productivity or quality or decreased costs. If one of your previous part-time positions called for specific skills that the employer needs, make that position the subject of a substantial paragraph in the letter, even though the résumé devotes only a few lines to it. If you try to cover every point on your résumé, the letter will be fragmented. The reader then will have a hard time forming a clear impression of you, and the purpose of the letter will be thwarted.

In most cases, a job-application letter should fill up the better part of a page. Like all business correspondence, it should be single spaced, with double spaces between paragraphs. For more experienced candidates, the letter may be longer, but most students find that they can adequately describe their credentials in one page. If you write at length on a minor point, you end up being boring. Worse still, you appear to have poor judgment. Employers always seek candidates who can say much in a short space.

Elements of the Job-Application Letter

Among the mechanical elements of the job-application letter, the inside address—the name, title, organization, and address of the recipient—is most important. If you know the correct form of this information from an ad, there is no problem. However, if you are uncertain about any of the information—the recipient's name, for example, might have an unusual spelling—verify it by phoning the organization. Many people are sensitive about such matters, so do not risk beginning the letter with a misspelling or incorrect title.

When you do not know who should receive the letter, do not address it to a department of the company—unless the job ad specifically says to do so—because no one in that department might feel responsible for dealing with it. Instead, phone the company to find out who manages the department. If you are unsure of the appropriate department or division to write to, address the letter to a high-level executive, such as the president. The letter will be directed to the right person. Also, because the application includes both a letter and a résumé, type the enclosure notation (see Chapter 16) in the lower left corner of the letter.

The four-paragraph letter that will be discussed here is only a basic model. Because every substantial job-application letter has an introductory paragraph, two body paragraphs, and a concluding paragraph, the minimum number of paragraphs for the job-application letter has to be four. But there is no reason that an application letter cannot have five or six paragraphs. The four basic paragraphs are shown in Figure 17.9 on page 492.

Because this is such an important letter, plan it carefully. Choose information from your background that best responds to the needs of the potential employer. Draft the letter and then revise it. Let it sit for a while, and then revise it again. Spend as much time on it as you can. Make each paragraph a unified, functional part of the whole letter. Supply clear transitions from one paragraph to the next. (See Chapter 14 for more information on paragraphs.)

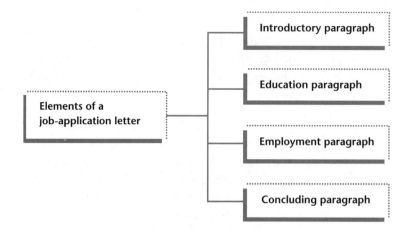

FIGURE 17.9

Elements of a Job-
Application Letter

The Introductory Paragraph

The introductory paragraph establishes the tone for the rest of the letter and captures the reader's attention. It also has four specific functions:

- *It identifies your source of information.* In an unsolicited application, all you can do is ask if a position is available. For most applications, however, your source of information is a published advertisement or an employee already working for the organization. If your source is an ad, identify specifically the publication and its date of issue. If an employee told you about the position, identify that person by name and title.

- *It identifies the position you are interested in.* Often, an organization has a number of positions advertised; without the title, your reader might not know which position you seek.

- *It states that you wish to be considered for the position.* Although the context makes your wish obvious, you should mention it because the letter would be awkward without it.

- *It forecasts the rest of the letter.* Choose a few phrases that forecast the body of the letter, so that the letter flows smoothly.

These four aspects of the introductory paragraph need not appear in any particular order, nor need each be covered in a single sentence. The following examples of introductory paragraphs demonstrate different ways to provide the necessary information.

RESPONSE TO
A JOB AD:

I am writing in response to your notice in the May 13 *New York Times*. I would like to be considered for the position in system programming. I hope you find that my studies at Eastern University in computer science, along with my programming experience at Airborne Instruments, would qualify me for the position.

UNSOLICITED: My academic training in hotel management and my experience with Sheraton International have given me a solid background in the hotel industry. Would you please consider me for any management trainee position that might be available?

UNSOLICITED PERSONAL CONTACT: Mr. Howard Alcott of your Research and Development Department has suggested that I write. He thinks that my organic chemistry degree and my practical experience with Brown Laboratories might be of value to you. Do you have an entry-level position in organic chemistry for which I might be considered?

As these sample paragraphs indicate, the important information can be conveyed in any number of ways. The difficult part of the introductory paragraph—and of the whole letter—is to achieve the proper tone: quiet self-confidence. Because the letter will be read by someone who is professionally superior to you, the tone must be modest, but it should not be self-effacing or negative. Never say, for example, "I do not have a very good background in computers but I'm willing to learn." The reader will take this kind of statement at face value and probably stop reading right there. Show pride in your education and experience, while at the same time suggesting by your tone that you have much to learn.

The Education Paragraph

Most students should place the education paragraph before the employment paragraph because the content of the former will be stronger. If, however, your work experience is more pertinent than your education, discuss your work first.

In devising your education paragraph, take your cue from the job ad (if you are responding to one). What aspect of your education most directly fits the job requirements? If the ad stresses versatility, you might structure your paragraph around the range and diversity of your courses. Also, you might discuss course work in a field related to your major, such as business or communication skills. Extracurricular activities are often valuable; if you were an officer in a student organization in your field, discuss the activities and programs that you coordinated. Perhaps the most popular strategy for developing the education paragraph is to discuss skills and knowledge gained from advanced course work in the major field.

Whatever information you provide, the key to the education paragraph is to develop one unified idea, rather than to toss a series of unrelated facts onto the page. Notice how each of the following education paragraphs develops a unified idea:

At Eastern University, I have taken a wide range of courses in the sciences, but my most advanced work has been in chemistry. In one laboratory course, I developed a new aseptic brewing technique that lowered the risk of infection by over 40 percent. This new technique was the subject of an article in the *Eastern Science Digest*. Representatives from three national breweries have visited our laboratory to discuss the technique with me.

To broaden my education at Southern, I took eight business courses in addition to my requirements for the Civil Engineering degree. Because your ad mentions that the position will require substantial client contact, I believe that my work in marketing, in particular, would be of special value. In an advanced marketing seminar, I used Pagemaker to produce a 20-page sales brochure describing Oppenheimer Properties' different kinds of building structures for sale to industrial customers in our section of the city. That brochure is now being used at Oppenheimer Properties, where I am serving as an intern.

The most rewarding part of my education at Western University took place outside the classroom. My entry in a fashion-design competition sponsored by the university won second place. More important, through the competition I met the chief psychologist at Eastern Regional Hospital, who invited me to design clothing for people with physical handicaps. I have since completed six different outfits. These designs are now being tested at the hospital. I hope to be able to pursue this interest once I start work.

Notice that each of these paragraphs begins with a topic sentence—a forecast of the rest of the paragraph—and uses considerable detail and elaboration to develop the main idea. Notice, too, a small point: if you haven't already specified your major and college or university in the introductory paragraph, be sure to do so in the education paragraph.

The Employment Paragraph

Like the education paragraph, the employment paragraph should begin with a topic sentence and then elaborate a single idea. That idea might be that you have had a broad background or that a single job has given you special skills that make you particularly suitable for the available job. The following examples show effective employment paragraphs.

For the past three summers and part-time during the academic year, I have worked for Redego, Inc., a firm that specializes in designing and planning industrial complexes. I began as an assistant in the drafting room. By the second summer, I was accompanying a civil engineer on field inspections. Most recently, I have used CAD to assist an engineer in designing and drafting the main structural supports for a 15-acre, $30 million chemical facility.

Although I have worked every summer since I was 15, my most recent position, as a technical editor, has been the most rewarding. I was chosen by Digital Systems, Inc., from among 30 candidates because of my dual background in computer science and writing. My job was to coordinate the editing of computer manuals. Our copy editors, who are generally not trained in computer science, need someone to help verify the technical accuracy of their revisions. When I was unable to answer their questions, I was responsible for interviewing our systems analysts to find out the correct answer and making sure the computer novice could follow it. This position gave me a good understanding of the process by which operating manuals are created.

I have worked in merchandising for three years as a part-time and summer salesperson in men's fashions and accessories. I have had experience running inventory control software and helped one company switch from a manual to an on-line system. Most recently, I assisted in clearing $200,000 in out-of-date men's fashions: I coordinated a campaign to sell half of the merchandise at cost and was able to convince the manufacturer's representative to accept the other half for full credit. For this project, I received a certificate of appreciation from the company president.

Notice how these writers carefully define their duties to give their readers a clear idea of the nature and extent of their responsibilities.

Although you will discuss your education and experience in separate paragraphs, try to link these two halves of your background. If an academic course led to an interest that you were able to pursue in a job, make that point clear in the transition from one paragraph to the other. Similarly, if a job experience helped shape your academic career, tell the reader about it.

The Concluding Paragraph

The concluding paragraph of the job-application letter, like that of any sales letter, is intended to stimulate action. In this case, you want the reader to invite you for an interview. In the preceding paragraphs of the letter, you provided the information that should have convinced your reader to give you another look. In the last paragraph, you want to make it easy for him or her to do so. The concluding paragraph contains three main elements:

- a reference to your résumé
- a request for an interview
- your phone number

If you have not yet referred to the enclosed résumé, do so at this point. Then, politely but confidently request an interview—making sure to use the phrase *at your convenience*. Don't make the request sound as if you're asking a personal favor. And be sure to include your phone number and the time of day you can be reached, even though the phone number is also on your résumé. Many employers will pick up a phone and call promising candidates personally, so make it as easy as possible for them to do so.

Following are examples of effective concluding paragraphs:

The enclosed résumé provides more information about my education and experience. Could we meet at your convenience to discuss further the skills and experience I could bring to Pentamax? You can leave a message for me anytime at (303) 555-5957.

More information about my education and experience is included on the enclosed résumé, but I would appreciate the opportunity to meet with you at your convenience to discuss my application. You can reach me after noon on Tuesdays and Thursdays at (212) 555-4527.

Figures 17.10 and 17.11 on pages 496–497 show examples of effective job-application letters corresponding to the résumés presented in Figure 17.6 and 17.7.

3109 Vista Street
Philadelphia, PA 19136

January 19, 19XX

Ms. Stephanie Spencer, Director of Personnel
Department 411
Boeing Naval Systems
103 Industrial Drive
Wilmington, DE 20093

Dear Ms. Spencer:

I am writing in response to your advertisement in the January 16 *Philadelphia Inquirer.* Would you please consider me for the position in Signal Processing? I believe that my academic training at Drexel University in Electrical Engineering, along with my experience with RCA Advanced Technology Laboratory, would qualify me for the position.

My education at Drexel University has given me a strong background in computer hardware and system design. I have concentrated on digital and computer applications, developing and designing computer and signal-processing hardware in two graduate-level engineering courses I was permitted to take. For my senior-design project, I am working with four other undergraduates in using OO programming techniques to enhance the path-planning software for an infrared night-vision robotics application.

While working at RCA I was able to apply my computer experience to the field of the VLSI design. In one project I used my background in LISP to develop and test new LISP software used in the automated production of VLSI standard cell family databooks. In another project, I used CAD software on a VAX to evaluate IC designs.

The enclosed résumé provides an overview of my education and experience. Could I meet with you at your convenience to discuss my qualifications for this position? Please write to me at the above address or leave a message anytime at (215) 555-3880.

Yours truly,

Carlos Rodriguez

Carlos Rodriguez

Enclosure (1)

FIGURE 17.10

Job-Application Letter

1781 Weber Road
Warminster, PA 18974

January 17, 19XX

Mr. Harry Gail
Fox Run Medical Center
399 N. Abbey Road
Warminster, PA 18974

Dear Mr. Gail:

Last April I contacted you regarding the possibility of an internship as a laboratory assistant at your center. Your assistant, Mary McGuire, told me then that you might consider such a position this year. With the experience I have gained since last year, I believe that I would be a valuable addition to your center in many ways.

At Harmon College, I have earned a 3.7 GPA in 36 credits in chemistry and biology; all but two of these courses have had laboratory components. One important skill stressed here is the ability to communicate effectively in writing and orally. Our science courses have extensive writing and speaking requirements; my portfolio includes seven research papers and lab reports of more than 20 pages each, and I have delivered four oral presentations, one of 45 minutes, to classes.

At Smith Kline & French, where I currently work part-time, I analyze molecular data on an E&S PS300, a Macintosh, and an IBM PC. I have tried to remain current with the latest advances; my manager at Smith Kline has allowed me to attend two different two-day in-house seminars on computerized data analysis using SAS.

Having been out of school for more than a decade, I am well aware of how much the technology has changed. However, as the manager of a 12-person office for four years, I believe that I have acquired skills that would benefit Fox Run. In addition, as a single mother of three I know something about time management.

More information about my education and experience is included on the enclosed résumé, but I would appreciate the opportunity to meet with you at your convenience to discuss my application. If you would like any additional information about me or Harmon's internship program, please write to me at the above address or call me at (215) 555-3999.

Very truly yours,

Alice P. Linder

Alice P. Linder

Enclosure

The letter in Figure 17.11 needs some comment. Earlier I discussed different strategies for discussing a student's non-traditional status. The writer of the letter in Figure 17.11 decided to discuss her nontraditional background—she is a single mother who has been out of school for more than 10 years—without overemphasizing it. Accordingly, the first two paragraphs of the body discuss her qualifications for the internship position. She wants to make sure that the reader evaluates her as he would any candidate. Then she adds a paragraph that attempts to explain how her additional experience—as an office manager and a single mother—has enabled her to develop skills that would be of value to anyone in any field. She exploits her situation gracefully, always appealing to the reader's needs, not to his sympathy.

Writing Follow-Up Letters

If you do get an interview, write two of the four types of follow-up letters: the first type plus one of the other three:

- *The letter of appreciation after an interview.* Your purpose in writing a follow-up letter is to thank the representative for taking the time to see you and to emphasize your particular qualifications. You can also take this opportunity to restate your interest in the position. The follow-up letter can do more good with less effort than any other step in the job application procedure, because so few candidates take the time to write it. Figure 17.12 is the body of a letter of appreciation.
- *The letter of acceptance of a job offer.* This one is easy: express appreciation, show enthusiasm, and repeat the major terms of your employment. Figure 17.13 is the body of a letter of acceptance of a job offer.

Thank you for taking the time yesterday to show me your facilities and to introduce me to your colleagues.

Your advances in piping design were particularly impressive. As a person with hands-on experience in piping design, I can appreciate the advantages your design will have.

The vitality of your projects and the good fellowship among your employees further confirm my initial belief that Cynergo would be a fine place to work. I would look forward to joining your staff.

FIGURE 17.12

Body of a Letter of Appreciation after an Interview

Thank you very much for the offer to join your staff. I accept.

I look forward to joining your design team on Monday, July 19. The salary, as you indicate in your letter, is $31,250.

As you have recommended, I will get in touch with Mr. Matthews in Personnel to begin the paperwork.

I appreciate the trust you have placed in me, and I assure you that I will do what I can to be a productive team member at Cynergo.

FIGURE 17.13

Body of a Letter of
Acceptance of a Job Offer

- *The letter of rejection in response to a job offer.* This one is also easy. If you decide not to accept a job offer, write a polite letter, expressing your appreciation and, if appropriate, explaining why you decline the offer. Remember, you might want to work for this company sometime. Figure 17.14 shows the body of this kind of letter.
- *The letter acknowledging a job rejection.* This one is hard to write. Why should you write back after you have been rejected for a job? To maintain good relations with the company. You just might get a phone call the next week, explaining that the person who accepted the job has had to change her plans and offering you the position. Figure 17.15 on page 500 shows the body of this kind of letter.

I appreciate very much the offer to join your staff.

Although I am certain that I would benefit greatly from working at Cynergo, I have decided to take a job with a firm in Pittsburgh, where I have been accepted at Carnegie-Mellon to pursue my master's degree at night.

Again, thank you for your generous offer.

FIGURE 17.14

Body of a Letter of
Rejection in Response to
a Job Offer

I was disappointed to learn that I will not have a chance to join your staff, for I feel that I could make a substantial contribution. However, I appreciate that job decisions are complex, involving many candidates and many factors.

Thank you very much for the courtesy you have shown me. I have long believed—and I still believe—that Cynergo is a first-class organization.

FIGURE 17.15

Body of a Letter of Acknowledgment of a Job Rejection

WRITER'S CHECKLIST

Résumé

1. Does the résumé respond to the needs of its readers?

2. Is the résumé honest?

3. Does the résumé have a professional appearance, with generous margins, a symmetrical layout, adequate white space, and effective indentation?

4. Is the résumé free of errors?

5. Does the identifying information section contain your name, address(es), and phone number(s)?

6. Have you included a clear statement of your job objectives?

7. Does the education section include your degree, your institution and its location, and your anticipated date of graduation, as well as any other information that will help your reader appreciate your qualifications?

8. Does the employment section include, for each job, the dates of employment, the organization's name and location, and (if you are writing a chronological résumé) your position or title, as well as a description of your duties and accomplishments?

9. Does the personal information section include relevant hobbies or activities, as well as extracurricular interests?

10. Does the references section include the names, job title, organization, mailing address, and phone number of three or four referees? If you are not listing this information, does the strength of the rest of the résumé offset the omission?

11. Does the résumé include any other appropriate sections, such as military service, language skills, or honors?

Job-Application Letter

1. Does the letter respond to the needs of its readers?

2. Is the letter honest?

3. Does the letter look professional?

4. Does the introductory paragraph identify your source of information and the position you are applying for, state that you wish to be considered, and forecast the rest of the letter?

5. Does the education paragraph respond to your reader's needs with a unified idea introduced by a topic sentence?

6. Does the employment paragraph respond to your reader's needs with a unified idea introduced by a topic sentence?

7. Does the concluding paragraph include a reference to your résumé, a request for an interview, and your phone number?

8. Does the letter include an enclosure notation?

Letter of Appreciation

1. Does the letter of appreciation for a job interview thank the interviewer and briefly restate your qualifications?

EXERCISES

Several of the following exercises call for you to write memos. See Chapter 16 for a discussion of memos.

1. In a newspaper or journal, find a job advertisement for a position in your field for which you might be qualified. Write a résumé and a job-application letter in response to the ad. Include the job ad or a photocopy of it. You will be evaluated not only on the content and appearance of the materials, but also on how well you have targeted the materials to the job ad.

2. You are a human-resources specialist for XYZ Corporation. Your primary responsibility is to screen incoming job applications, rejecting those of applicants who seem inappropriate for a particular position and forwarding those of applicants who seem qualified. Your supervisor is interested in knowing more about how you do the work you do; she has provided you with the résumé that appears below and asked you to comment on it. Write a 500-word memo to your supervisor evaluating the strengths and weaknesses.

The writer is responding to an ad describing the following duties: "Research and develop key technology and system concepts for spectrally efficient digital radio frequency data networks such as digital cellular mobile radio telephones, public safety trunked digital radio systems, and satellite communications."

3. You are a human-resources specialist for XYZ Corporation. Your primary job duty is to screen incoming job applications, rejecting those of applicants

Rajiv Siharath
2319 Fifth Avenue
Waverly, CT 01603
Phone: 555-3356

Personal Data: 22 years old
                   Height 5'11"
                   Weight 176 lbs.

Education: B.S. in Electrical Engineering
                   University of Connecticut,
                   June 19XX

Experience: 6/XX–9/XX Falcon Electronics
                   Examined panels for good wiring. Also, I revised several
                   schematics.
                   6/XX–9/XX MacDonalds Electrical Supply Co.
                   Worked parts counter.
                   6/XX–9/XX Happy Burger
                   Made hamburgers, fries, shakes, fish sandwiches, and fried
                   chicken.
                   6/XX–9/XX Town of Waverly
                   Outdoor maintenance. In charge of cleaning up McHenry
                   Park and Municipal Pool picnic grounds. Did repairs on some
                   electrical equipment.

Backround: Born and raised in Waverly.
                   Third baseman, Fisherman's Rest softball team.
                   Hobbies: jogging, salvaging and repairing appliances, reading
                   magazines, politics.

References: Will be furnished on request.

who seem inappropriate for a particular position and forwarding those of applicants who seem qualified. Your supervisor is interested in knowing more about how you do the work you do; she has provided you with the letter that appears below and asked you to comment on it. Write a 500-word memo to your supervisor evaluating the strengths and weaknesses.

The writer is responding to an ad describing the following duties: "Buyer for a high-fashion ladies' dress shop. Experience required."

April 13, 19XX

Marilyn Grissert
Best Department Store
113 Hawthorn
Atlanta, Georgia

Dear Ms. Grissert:

As I was reading the Sunday *Examiner,* I came upon your ad for a buyer. I have always been interested in learning about the South, so would you consider my application?

I will receive my degree in fashion design in one month. I have taken many courses in fashion design, so I feel I have a strong background in the field.

Also, I have had extensive experience in retail work. For two summers I sold women's accessories at a local clothing store. In addition, I was a temporary department head for two weeks.

I have enclosed a résumé and would like to interview you at your convenience. I hope to see you in the near future. My phone number is 555-6103.

Sincerely,

Brenda Sisneros

REFERENCES

Harcourt, J., Krizan, A. C., & Merrier, P. (1991). Teaching résumé content: Hiring officials' preferences versus college recruiters' preferences. *Business Education Forum, 45*(7), 13–17.

Looking at job applications? Remember—It's hirer beware. (1994, May/June). *Canadian Banker, 101*(3), 10.

Wright, J. W. (1994). *Résumés for people who hate to write résumés.* Livermore, CA: Shastar Press.

CHAPTER 18 Instructions and Manuals

The people who answer the customer-support phones at Dell Computers can tell you that no matter how hard their technical communicators try, their manuals will never please everyone. One customer called, asking how to install batteries in her new laptop. When told the instructions were on page one of the manual, she replied, "I just paid $2,000 for this [deleted] thing, and I'm not going to read a book" ("Befuddled," 1994, p. B1). Instructions and manuals have acquired a bad reputation over the years, and for that reason many people don't even bother trying to read them. Perhaps that is why many technical communicators use the phrase *RTFM*, which (roughly translated) stands for Read the Manual. But many people apparently don't. This situation is regrettable, because instructions and manuals are fundamentally important to carrying out procedures and using products safely and effectively.

Chapter 10 discussed process descriptions, which explain how a process occurs. The present chapter discusses instructions, which are process descriptions written to help the reader perform a specific task. For instance, a set of instructions explains and shows how to install a water heater in your home.

This chapter also discusses manuals, which are documents consisting primarily of instructions. Often manuals are printed and bound, like books. Manuals can be classified according to function. One common type is a user's manual. Like a set of instructions, its function is to instruct. For example, it explains how to use a software program, maintain inventory, or operate a piece of machinery. Other types include installation manuals, maintenance manuals, and repair manuals.

Effective instructions and manuals are challenging to write. You must make sure that your audience will want to read your document and be able to understand and follow it easily. You also must make sure that in performing the tasks they won't damage any equipment or, more important, injure themselves or other people.

Three other aspects of instructional writing are also discussed in this chapter: minimal manuals, writing instructions and manuals for nonnative speakers of English, and hypertext.

Understanding the Role of Instructions and Manuals

Instructions and manuals are central to technical communication. If you are a technical professional such as an engineer, you will probably be asked to write or contribute to instructions and manuals often in your career. If you are a technical communicator, you will write them more often than any other major kind of technical document.

Whereas just a decade or two ago little attention was paid to the quality of instructions and manuals, today the goal is to make systems safe and "user friendly."

- *Safety* is an important issue largely because of liability suits. As discussed in Chapter 2, a company can lose millions if a court finds that its instructions and manuals failed to explain how to use the product properly or to alert the user about dangers of using its product.
- *User-friendliness* is critical, because if the product is hard to use, it could fail in the marketplace. Major advertising campaigns for all kinds of products stress clear, simple, easy-to-use manuals.

Instructions and manuals are no longer an afterthought but rather an integral part of the planning process. At the most progressive companies, the people who create the documentation—the documents that come with the product—are part of the research and development team.

Analyzing Your Audience

When instructions and manuals are ineffective, chances are that the writer has inaccurately assessed the audience. Performing a function, such as assembling a backyard swing or maintaining a conveyor belt, is easy for the expert but not necessarily so easy for the person reading the documentation. A reader who doesn't know what a self-locking washer is or what the calibrations on a timing control mean might not be able to complete the process.

Before you start to write a set of instructions or a manual, think carefully about the background and skill level of your audience. If you are writing to people who are experienced in the field, use technical vocabulary and concepts. But if you are addressing general readers, define technical terms and give your directions in more detail. Don't be content to write "Make sure the tires are rotated properly." Instead, define proper rotation and describe how to achieve it.

The best way to make sure you have assessed your audience effectively is to find people whose backgrounds resemble those of your intended readers and then test the effectiveness of the documentation. This process, called usability testing, is discussed in detail in Chapter 19.

Planning for Safety

Your most important responsibility in writing documentation is to make sure you do everything you can to ensure the safety of your readers. Even though some kinds of tasks do not involve safety risks, many do. Planning for safety includes three steps:

- writing the safety information
- creating a design for the safety information
- placing the safety information in appropriate locations

Writing the Safety Information

Be clear and concise. Avoid complicated sentences.

COMPLICATED: It is required that safety glasses be worn when inside this laboratory.
SIMPLE: You must wear safety glasses in this laboratory.
SIMPLE: Wear safety glasses in this laboratory.

Sometimes a phrase works better than a sentence: "Safety Glasses Area."

Because a typical set of instructions or manual can contain dozens of comments—both safety and nonsafety comments—experts have created different terms to indicate the seriousness of the advice. Unfortunately, this terminology is not consistent.

For instance, the American National Standards Institute (ANSI) and the U.S. military's MILSPEC publish definitions that differ significantly, and many private companies have their own definitions that don't conform with either ANSI or MILSPEC. The following explanation of four commonly used terms points out the significant differences between ANSI and MILSPEC. The four terms are presented here from most serious to least.

- *Danger.* MILSPEC does not use this term, but for ANSI and many companies *danger* warns the reader of an immediate and serious hazard that will likely be fatal.

DANGER. EXTREMELY HIGH VOLTAGE. STAND BACK.

Often, safety warnings are written in all-uppercase letters.

- *Warning.* For MILSPEC, *warning* is the most serious level, indicating an action that could result in serious injury or death. For ANSI, it also suggests the potential for serious injury or death. Among different companies, *warning* ranges from serious injury or death to serious damage to equipment.

WARNING: TO PREVENT SERIOUS INJURY TO YOUR ARMS AND HANDS, MAKE SURE THE ARM RESTRAINTS ARE IN PLACE BEFORE YOU OPERATE THIS MACHINE.

- *Caution.* For MILSPEC, *caution* warns of the potential for both equipment damage and long-term health hazards. For ANSI, it indicates the potential for minor or moderate injury. Among companies, *caution* signals the potential for anything from moderate injury to serious equipment damage or destruction.

Caution: Do not use nonrechargeable batteries in this charging unit; they could damage the charging unit.

- *Note.* A note is a tip or suggestion to help the readers carry out the procedure successfully.

  Note: Two different kinds of washers are provided: regular and locking. Be sure to use the locking washers here.

If your organization does not have its own guidelines for safety labeling, I recommend the following definitions:

- *Danger.* Likelihood of serious injury, including death.
- *Warning.* Potential for minor, moderate, or serious injury.
- *Caution.* Potential for damage to equipment.
- *Note.* A suggestion on how to carry out a task.

Creating a Design for the Safety Information

Whether placed in a document or on machinery or equipment, safety information should be prominent and easy to read. Many organizations use visual symbols to accompany their different levels of safety comments, but these symbols are not yet standardized. Determine whether your organization already has a set of symbols that you can use in your document. If it doesn't, create different designs for each kind of comment. For instance, you could present warnings in 18-point type, boldfaced, within a box. Of course, the more critical the safety comment, the larger and more emphatic it should be. Safety information is often printed in color. Text is presented against a yellow, orange, or red background. Symbols are printed in color, too. Flames are presented in red, for instance. See the discussion of color in Chapter 12 for more information.

The clip art in Figure 18.1 on page 508, from *Arts and Letters* (1990), is typical of symbols used in safety information. The three symbols represent fire danger, electrical danger, and the need to wear safety glasses. Note that the triangle indicates that the danger is severe, whereas the image inside the triangle conveys the specific danger.

Placing the Safety Information in Appropriate Locations

What are appropriate locations? This question has no easy answer because you cannot control how your audience reads your document. Be conservative: put in safety information wherever you think the reader is likely to see it, and don't be afraid to repeat yourself. Naturally, you don't want to repeat the same piece of advice in front of each of 20 steps, because that will merely teach your readers to stop paying attention to you. But a reasonable amount of repetition—such as including the same safety comment at the top of each page—might be effective. If your company's procedure format calls for a safety section near the beginning of the document, put that information there *and* right before the appropriate step in the step-by-step section.

The Occupational Safety and Health Administration Guidelines (Chapter XVII, Sections 1910.145 and 1926.155) describe proper standards for placing

safety messages on products and manuals. These standards address the following questions:

- Is the message prominently displayed so that the user sees it?
- Is the message large enough and clearly legible, under operating conditions?
- Are the graphics and the words of the message clear and informative?

Figure 18.2, from an operator's manual for a John Deere lawnmower, shows one company's approach to placing safety warnings on machinery.

Drafting Effective Instructions

Instructions can be brief—a small sheet of paper—or extensive, up to 20 pages or more. Brief instructions might be produced by one or two people: a writer, or a writer and an artist. Sometimes a subject-matter expert—an expert in the technical subject being written about—is added to the team. For more extensive instructions, other people—marketing and legal personnel, for example—might be added. The team could consist of 10 or even 20 professionals working with a budget of many thousands of dollars.

As is the case with process descriptions, you will probably find it easiest to write instructions sequentially; that is, in the order in which they will appear. As they write instructions, many writers perform the task they are explaining, a process that enables them to detect errors and omissions.

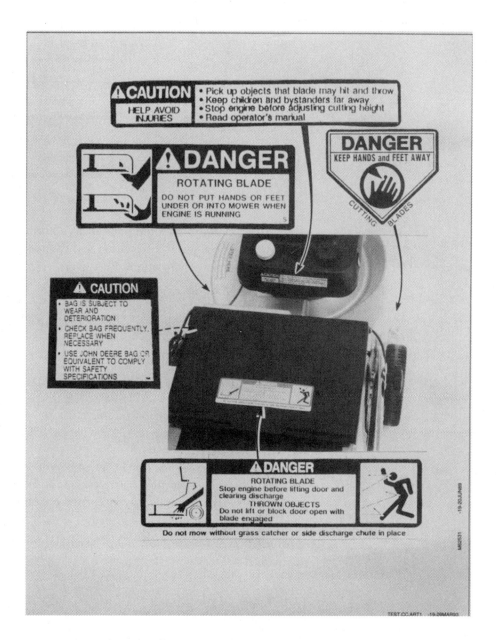

Safety Information on
Machinery

Regardless of the size of the project, most instructions are structured like process descriptions (see Chapter 10). The main difference is that the conclusion of a set of instructions is less a summary than an explanation of how to make sure the reader has followed the instructions correctly. Most sets of instructions contain the three components shown in Figure 18.3 on page 510.

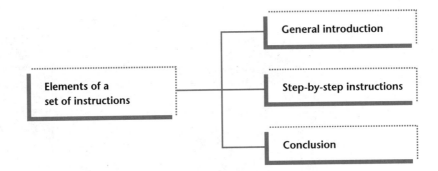

FIGURE 18.3

Elements of a Set of
Instructions

**The General
Introduction**

The general introduction provides the preliminary information that readers will need in order to follow the instructions safely and easily. In writing the introduction, answer these three questions about the process:

- *Why should the reader carry out this task?* Often, you do not need to answer this question because the reason is obvious. For example, you don't need to explain why a backyard barbecue grill should be assembled. Sometimes, however, you do need to explain, as in the case of many preventive-maintenance chores, such as changing radiator antifreeze every two years.

  If appropriate, answer two more questions:

  – *When should the reader carry out the task?* Some tasks, such as rotating tires or planting crops, need to be performed at particular times or at particular intervals.
  – *Who should carry out the task?* Sometimes you need to describe the person or persons who are to carry out a task. Some kinds of aircraft maintenance, for example, may be carried out only by persons certified to do that maintenance.

- *What safety measures or other concerns should the reader understand?* In addition to stating the safety measures that apply to the whole task, state any tips that will make your reader's job easier. For example:

  NOTE: For ease of assembly, leave all nuts loose. Give only 3 or 4 complete turns on bolt threads.

- *What tools and materials will the reader need?* Include the list of necessary tools and equipment in the introduction so that readers will not have to interrupt their work to hunt for something. If you think your readers might not be able to identify the tools or materials easily, include drawings next to the names (see Chapter 12).

  Following is a list of tools and materials from a set of instructions on replacing broken window glass:

You will need the following tools and materials:

Tools	Materials
glass cutter	putty
putty knife	glass of proper size
window scraper	paint
chisel	hand cleaner
electric soldering iron	work gloves
razor blade	linseed oil
pliers	glazier's points
paint brush	

The Step-by-Step Instructions

Follow these five guidelines when writing step-by-step instructions:

- *Number the instructions.* For long, complex instructions, use two-level numbering, such as a decimal system:

  1
      1.1
      1.2
  2
      2.1
      2.2
  etc.

- *Put the right amount of information in each step.* Each step should define a single task that the reader can carry out easily, without having to refer back to the instructions.

  TOO MUCH INFORMATION:
  1. Mix one part of the cement with one part water, using the trowel. When the mixture is a thick consistency without any lumps bigger than a marble, place a strip of the mixture about 1" high and 1" wide along the face of the brick.

  TOO LITTLE INFORMATION:
  1. Pick up the trowel.

  RIGHT AMOUNT OF INFORMATION:
  1. Mix one part of the cement with one part water, using the trowel, until the mixture is a thick consistency without any lumps bigger than a marble.
  2. Place a strip of the mixture about 1" high and 1" wide along the face of the brick.

- *Use the imperative mood.* For example, "Attach the red wire. . . ." The imperative is more direct and economical than the indicative mood ("You should attach the red wire . . ." or "The operator should attach the red wire . . ."). Make sure your sentences are grammatically parallel. Avoid the passive voice ("The red wire is attached . . ."), because it can be ambiguous: Is the red wire already attached?

- *Include graphics.* When appropriate, accompany each step with a photograph or drawing that shows the reader what to do. Some kinds of activities—such as adding two drops of a reagent to a mixture—do not need illustration, but they might be clarified by charts or tables.

    Figure 18.4 shows the extent to which a set of instructions can integrate words and graphics. This excerpt is from the operating-instructions booklet for a video-cassette recorder. See Chapter 12 for a discussion of graphics.

- *Do not omit the articles* (a, an, the) *to save space.* Omitting the articles makes the instructions hard to read and, sometimes, unclear. In the sentence "Locate midpoint and draw line," for example, the reader cannot tell if "draw line" is a noun ("the draw line") or a verb and its object ("draw the line").

The Conclusion   Instructions often conclude with *maintenance tips.* Another popular conclusion for a set of instructions is a *troubleshooter's guide,* usually in the form of a table, that identifies common problems and explains how to solve them.

Following is a portion of the troubleshooter's guide included in the operating instructions for a lawnmower.

Problem	Cause	Correction
THE MOWER DOES NOT START.	1. The mower is out of gas. 2. The gas is stale. 3. The spark plug wire is disconnected from the spark plug.	1. Fill the gas tank. 2. Drain the tank and refill it with fresh gas. 3. Connect the wire to the plug.
THE MOWER LOSES POWER.	1. The grass is too high. 2. The air cleaner is dirty. 3. There is a buildup of grass, leaves, or trash.	1. Set the mower to the "higher cut" position. 2. Replace the air cleaner. 3. Disconnect the spark plug wire, attach it to the retainer post, and clean the underside of the mower housing.

Some troubleshooter's checklists refer the reader to the page where the recommended action is described. For example, the Correction column in the lawn-mower troubleshooting checklist might say, "1. Fill the gas tank. See page 4."

Figure 18.5 on page 514 is the set of instructions for an anti-tip bracket that comes with a stove. Note the following six points about these instructions:

- The safety information includes a graphic that illustrates the hazard that the device is intended to prevent.
- The description of the parts includes drawings.

EASY OPERATION GUIDE

First   Refer to BASIC OPERATION.

To watch TV normally

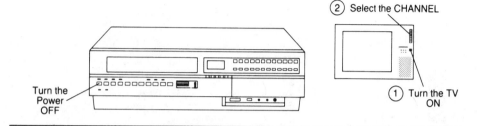

Recording

To watch and record the same TV program

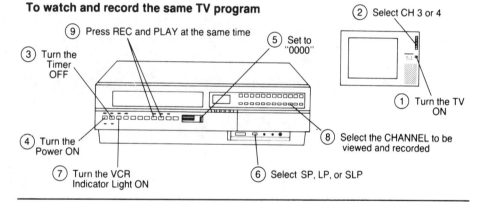

To watch one TV program while recording another

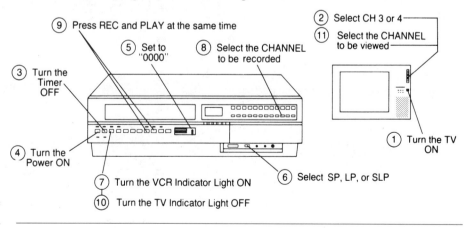

FIGURE 18.4

Instructions that Integrate
Words and Graphics

Anti-Tip Bracket

Read and follow Range Installation Instruction sheet first.

Complete this set of steps to secure the range to the floor before you move the range into final operating position.

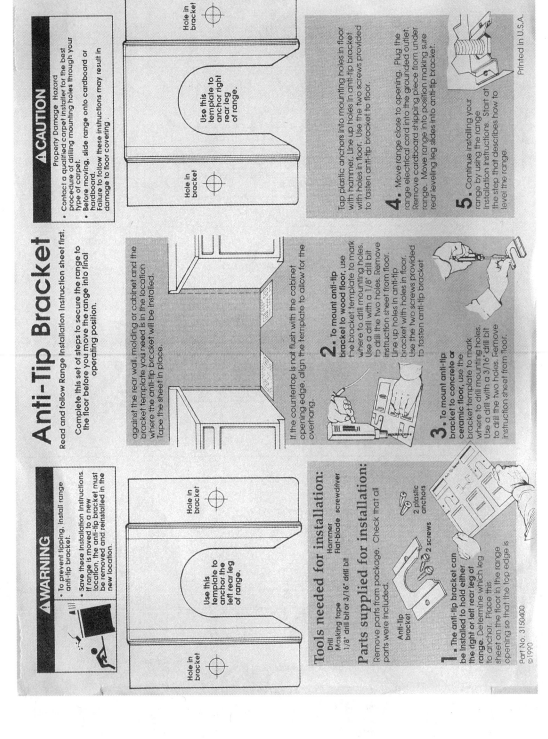

Hole in bracket

Use this template to anchor the left rear leg of range.

Hole in bracket

Tools needed for installation:
Drill Hammer
Masking tape Flat-blade screwdriver
1/8" drill bit or 3/16" drill bit

Parts supplied for installation:
Remove parts from package. Check that all parts were included.

Anti-Tip bracket

2 plastic anchors

2 screws

1. The anti-tip bracket can be installed to hold either the right or left rear leg of range. Determine which leg to anchor. Place this sheet on the floor in the range opening so that the top edge is

against the rear wall, molding or cabinet and the bracket template you need is in the location where the anti-tip bracket will be installed. Tape the sheet in place.

If the countertop is not flush with the cabinet opening edge, align the template to allow for the overhang.

2. To mount anti-tip bracket to wood floor, use the bracket template to mark where to drill mounting holes. Use a drill with a 1/8" drill bit to drill the two holes. Remove instruction sheet from floor. Line up holes in anti-tip bracket with holes in floor. Use the two screws provided to fasten anti-tip bracket

3. To mount anti-tip bracket to concrete or ceramic floor, use the bracket template to mark where to drill mounting holes. Use a drill with a 3/16" drill bit to drill the two holes. Remove instruction sheet from floor.

Part No. 3150400
©1990

Hole in bracket

Hole in bracket

Use this template to anchor right rear leg of range.

Tap plastic anchors into mounting holes in floor with hammer. Line up holes in anti-tip bracket with holes in floor. Use the two screws provided to fasten anti-tip bracket to floor.

4. Move range close to opening. Plug the range electrical cord into the grounded outlet. Remove cardboard shipping piece from under range. Move range into position making sure rear leveling leg slides into anti-tip bracket.

5. Continue installing your range by using the range installation instructions. Start at the step that describes how to level the range.

Printed in U.S.A.

FIGURE 18.5

Instructions

514

- The instructions include templates to help the reader carry out the procedure.

- The step-by-step instructions include drawings to help the reader understand the procedure.

- Although these instructions are generally well done, they would be easier to follow if steps 1 and 3 were not broken up and presented in two columns. Step 1, in particular, is difficult to follow because it starts at the bottom of the left column and continues near the top of the middle column.

- Some readers would find that each numbered step contains too much information; they would want to add more steps.

Drafting Effective Manuals

Most of what has been said in this chapter about instructions applies to manuals. For example, you have to analyze your audience, explain procedures clearly, and complement your descriptions with graphics.

However, because manuals are usually much more ambitious projects than instructions, they require more planning. A bigger investment is at stake for the organization, not only in the costs of writing and producing the manual but also in the potential effects of the published manual. In addition to reducing the possibility of injuries and liability risks, a good manual attracts customers and reduces costs, for the organization needs fewer customer-support people. A poor manual is expensive to produce because it must be revised more often and alienates customers.

Drafting a manual, like drafting a book, is so complex that the following discussion can provide only an introduction. If you are involved in creating a manual, consult the sources listed in the bibliography (Appendix D).

Writing a manual is almost always a collaborative project. A full-size manual is simply too big and requires too many skills for one person to write: technical skills in the subject area, writing skills, graphics skills, production skills, as well as the knowledge of contracts and law that is required to prevent lawsuits. In addition, obtaining other people's perspectives can prevent little problems from becoming big problems.

The following discussion comments on the three stages of writing a manual.

- *Prewriting.* The most important stage of manual writing is prewriting. First, you must analyze your audience and purpose.

   The documentation for a sophisticated procedure or system almost always accommodates a diverse audience. Often you will decide to produce a set of manuals—one for the user, one for the manager, one for the installer, one for the maintenance technicians, and so on. Or you might decide on a main manual and some other, related documents, such as brochures, flyers, and workbooks.

While your primary purpose varies with the kind of manual, you also have a secondary purpose: to motivate your readers. The mere fact that the manual might be 200 pages or even longer means that you have to persuade people to read it. In many cases, readers are uncomfortable with the new system. Your job is to make the task of learning it seem less overwhelming. A well-designed manual, with plenty of white space and graphics, will help considerably.

In planning a manual, follow the guidelines discussed in Chapters 4–7 of this text:

Chapter 4. Analyzing Your Audience

Chapter 5. Determining Your Purpose and Strategy

Chapter 6. Performing Secondary Research

Chapter 7. Performing Primary Research

- *Drafting.* Drafting a manual is much the same as drafting any other kind of technical document. Sometimes one person collects information from all the people on the documentation team and creates a draft from it. Sometimes different subject-matter experts write their own drafts, which a writer or a writing team then revises or rewrites. In drafting a manual, you also need to create and integrate graphics, and to design the whole manual and each page.

- *Revising.* Because manuals are bigger and more complex than most other kinds of technical documents, revision is a more complicated process. In fact, revision is a series of different kinds of checks: from checks of the technical information down to checks of structure, organization, emphasis, and style. See the discussion of usability testing in Chapter 19.

The Front Matter   The front matter, which helps readers understand the content of the manual and the best way to use it, consists of everything before the body. Most manuals have a cover or title page and a table of contents. In addition, most manuals contain a preface and a how-to-use-this-manual section. Sometimes, however, these last two items are simply combined in an introduction.

To decide whether to use a cover or just a title page, consider the manual's size and intended use. Manuals that will receive some wear and tear—such as maintenance manuals for oil-rigging equipment—need a hard cover, usually made of a water-resistant material. Manuals used around an office usually don't need hard covers, unless they are so big that they require extra strength. See Chapter 13 for more information on covers.

The title page, usually designed by a graphic artist, contains at least the title of the manual. An extensive table of contents is especially important in a manual because people refer to manuals for specific information; they don't read them straight through. The headings in an effective table of contents are clear and focus on the task the reader wants to accomplish. See Chapters 13 and 14 for more information on phrasing headings.

The other introductory information, which sometimes precedes and sometimes follows the table of contents, takes a number of forms. The first page of text in the manual, for instance, can be called "Introduction" or "Preface" or have no heading at all. Another popular strategy is to use several "about" phrases: "About Product X," "About the Product X Documentation Set," and so on.

Regardless of what you call this introductory information, it must answer five basic questions for your reader:

- Who should use this manual?
- What product, procedure, or system does the manual describe?
- What is the manual's purpose?
- What are the manual's major components?
- How should the manual be used?

These questions need not be answered in separate paragraphs or even in separate sentences.

Some manuals need to answer one other question: what does the typography signify? If you use type creatively in the manual, and you want your readers to understand your conventions, define it in the front matter.

Figure 18.6 on page 518 (Microsoft, 1993, p. xii) shows the conventions section from a software manual.

Figure 18.7 on pages 520–521 (*Using*, 1991, pp. 5–6) is an example of an effective introduction to a manual. The first page of this introduction provides an overview of the parts of the manual and directs the reader to the opening chapters. Then it describes the background information that readers will need in order to understand the manual. The second page of the introduction provides a selected bibliography and then explains the typographic conventions used in the manual.

For more information on front matter, see Chapter 11.

The Body   The body of a manual might look like the body of a traditional report, or it might look radically different. Its structure, style, and use of graphics will depend on its purpose and audience. For instance, the body of a manual might include summaries and diagnostic tests to help readers determine whether they have understood the discussion. A long manual might have more than one "body"; that is, each chapter might be a self-contained unit with its own introduction, body, and conclusion.

Follow these four guidelines when drafting the body of a manual:

Document Conventions

To help you locate and interpret information easily, this guide uses consistent visual cues and a standard key combination format. These conventions are explained as follows.

This	Represents
bold	Commands and the switches that follow them. You must type commands and their switches exactly as they appear.
italic	Placeholders that represent information you must provide. Italic type also signals a new term. An explanation precedes or follows the italicized term.
ALL CAPITALS	The names of computers, printers, directories, and files.

Microsoft documentation uses the term *MS-DOS* to refer to the MS-DOS and IBM® Personal Computer DOS operating systems.

Keyboard Conventions

Key combinations and key sequences appear in the following format:

Notation	Meaning
KEY1+KEY2	Hold down the first key while you press the second key. For example, "Press CTRL+C" means that you press CTRL and hold it down while you press C.
KEY1, KEY2	A comma (,) between key names means you must press the keys in sequence—for example, "Press ALT, F10" means that you press and release the ALT key, and then press and release the F10 key.

FIGURE 18.6

Conventions

- *Structure the body according to the way the reader will use it.* For example, if the reader is supposed to carry out a process, structure the body chronologically. If the reader is supposed to understand a concept, structure the body from more important elements to less important elements. The various argument patterns discussed in Chapter 8 are appropriate, but, as always, the writing situation might call for another pattern.
- *Write clearly.* Simple, short sentences work best. Use the imperative when giving instructions.
- *Be informal if appropriate.* For some kinds of manuals, especially those used by readers untrained in the subject, an informal style is appropriate. Contractions and everyday vocabulary are common. A caution about informality: safety warnings and information about serious subjects such as disease or war are usually presented in a formal style.
- *Use graphics.* They break up the text and improve comprehension.

Whenever you want your readers to perform an action with their hands, include a drawing or photograph showing the action being performed. Where appropriate, use tables and figures. Deere & Co., the farm-equipment manufacturer, uses the term *illustruction* to describe the writing style of its owner/operator manuals: each step is an illustration with a brief explanatory comment. See Chapter 12 for more information on graphics.

Figure 18.8 on pages 522–523 (*Series 935*, 1994) consists of two pages from the body of a user's manual for an industrial control product. The first page is intended for a new user, as the tab states. The second page is for an experienced user.

Figure 18.9 on page 524 (*LaserJet*, 1994, pp. 4–6) is a page from the user's manual for a computer printer.

The Back Matter   The audience and purpose of the manual will determine what makes up the back matter. However, two items are common: a glossary and an index.

A glossary is an alphabetized list of definitions of important terms used in the document. An index is common for most manuals of 20–30 pages or more. Glossaries and indexes are discussed in Chapter 11.

Appendices to manuals contain many different kinds of information. Procedures manuals often have flowcharts or other graphics that picture the processes described in the body of the manual. Sometimes, these graphics are produced so that they can be removed from the manual and taped to the office wall for frequent reference. User's guides often have diagnostic tests and reference materials—error messages and sample data lists for the computer system, troubleshooting guides, and the like. For a further discussion of appendices, see Chapter 11.

Drafting Revisions   In high-tech industries, a new generation of a product might come along as
of Manuals   often as every 18 months. As products evolve, technical communicators need to revise the manuals that go with them. When a new version of the product is released, you can take one of two approaches to revising.

- *Publishing a "new" manual.* Although a new manual will likely contain elements from the old manual, it makes no explicit reference to the old manual. Someone who has not used the previous version of the product or seen its manual would not be aware that they even existed.

  The advantage of a new manual is that it can reflect the look of the new product. It encourages people to take out their checkbooks. The disadvantage of a new manual is that people who are switching from the old product to the new will have to start all over again; they cannot distinguish the new information from the old.

- *Publishing a "revised" manual.* Revised manuals, like new ones, contain a lot of information from the old version, but the new information is flagged so that users can find it easily. The new information is usually marked with an icon in the margin or with a *change bar:* a vertical bar in the margin. Sometimes a section in the front matter describes the changes to this version of the product.

The advantage of a revised manual is that it is easier for previous users to work with. The disadvantage is that it can look like a patch job.

Some software companies make a point of reminding the reader to study the descriptions of the new material: the text that appears on the screen during installation contains a reference to the new material; if the

---

About this manual

This manual is organized in four parts:

- Part I, "Introduction to the ToolBook Development Environment," describes the ToolBook development environment and takes a look at the scripts in DayBook, the personal information manager provided with ToolBook.
- Part II, "Elements of OpenScript," covers the major elements of the Open-Script language. Each chapter provides examples that apply the concepts discussed to real-world programming problems.
- Part III, "Building ToolBook Applications," covers programming tasks associated with application development, such as initialization and data management.
- Part IV, "Reference," provides an alphabetic reference to OpenScript, including both general topics and specific terms, appendices on OpenScript keywords and ToolBook's Dialog utility, a glossary, and an index.

To get started quickly, read Chapter 1, "Overview" for important concepts, then read Chapter 2, "A Guided Tour of DayBook," for a detailed look at the Open-Script programming in a finished ToolBook application.

What you should already know

To get the most out of this manual, you need some experience with:

- A programming or macro language. An object-oriented programming language is helpful but not essential.
- The Windows user interface.
- ToolBook's user interface, including its menu commands, palettes, and script-building tools.
- OpenScript Help, an online reference to the OpenScript language.

If you encounter an unfamiliar term, check the glossary in the back of this manual for a definition. Important terms appear in bold type and are defined the first time they occur. For details about ToolBook's user interface and how to use OpenScript Help, see *Using ToolBook.* For details about the Windows user interface, see the *Windows User's Guide.*

FIGURE 18.7

Introduction to a Manual

(Figure 18.7 continued)

Other recommended reading

For advanced technical information or insights on application design, we recommend the following:

- *The ToolBook Companion* by Joe Pierce (Microsoft Press, 1990)
- *Windows Guide to Programming* and the *Windows Programmer's Reference* (Microsoft Windows Software Developer's Kit and Microsoft Press, 1990)
- *The Art of Human-Computer Interface Design* (Addison-Wesley, 1990)
- *The Design of Everyday Things* by Donald A. Norman (Doubleday, 1988)
- *The Visual Display of Quantitative Information* by Edward Tufte (Graphics Press, 1983)

Conventions

Document conventions

Type style	Meaning
italic text	OpenScript keywords within paragraphs. In step-by-step procedures, italic text is also used for literal characters you should type.
bold text	First mention of a term defined in the glossary.
sans serif text	OpenScript statements in syntax lines and code examples.

OpenScript syntax conventions

Symbol	Meaning
< >	Parameters standing for a literal value or expression
( )	Required parentheses
[ ]	Optional words and parameters
|	Any one of the separated items can be used
. . .	The preceding parameter can be repeated

Notice that this page contains a lot of white space and several simple drawings, and that it contains cross-references to more detailed sections of the manual.

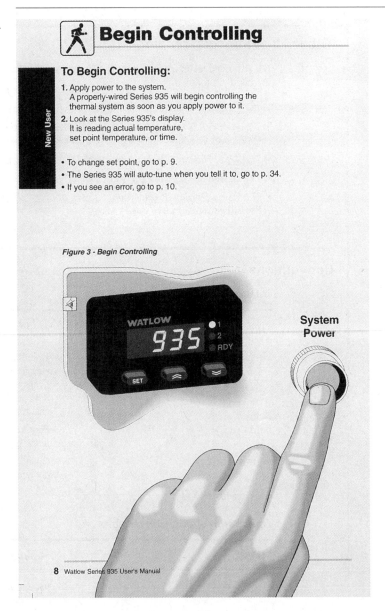

Begin Controlling

New User

To Begin Controlling:

1. Apply power to the system.
   A properly-wired Series 935 will begin controlling the thermal system as soon as you apply power to it.

2. Look at the Series 935's display.
   It is reading actual temperature, set point temperature, or time.

• To change set point, go to p. 9.
• The Series 935 will auto-tune when you tell it to, go to p. 34.
• If you see an error, go to p. 10.

Figure 3 - Begin Controlling

WATLOW

935

● 1
● 2
● RDY

SET

System Power

8   Watlow Series 935 User's Manual

FIGURE 18.8

Pages from the Body of a Manual

reader does not acknowledge the reference by hitting a particular key, the installation procedure stops.

The Minimal Manual

Researchers studying ways to improve the effectiveness of manuals have learned that in many cases the goals of the writers and the readers do not match. Writers

(Figure 18.8 continued)

This page begins with a brief introduction to the Configuration Menu and a step-by-step guide to using it. Notice that Table 6 shows the different images that will appear on the screen in the Configuration menu.

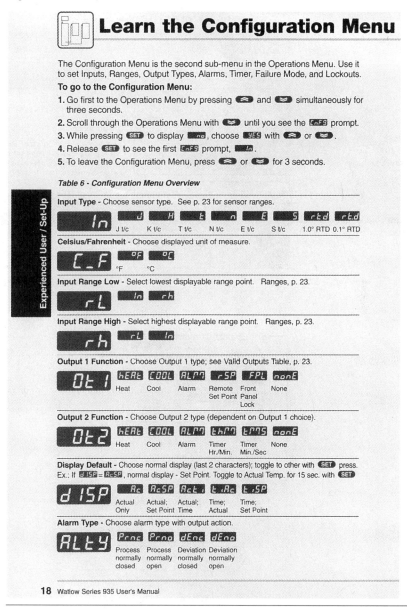

Learn the Configuration Menu

The Configuration Menu is the second sub-menu in the Operations Menu. Use it to set Inputs, Ranges, Output Types, Alarms, Timer, Failure Mode, and Lockouts.

To go to the Configuration Menu:

1. Go first to the Operations Menu by pressing ⬆ and ⬇ simultaneously for three seconds.
2. Scroll through the Operations Menu with ⬇ until you see the `CnFg` prompt.
3. While pressing `SET` to display `no`, choose `YES` with ⬆ or ⬇.
4. Release `SET` to see the first `CnFg` prompt, `In`.
5. To leave the Configuration Menu, press ⬆ or ⬇ for 3 seconds.

Table 6 - Configuration Menu Overview

Experienced User / Set-Up

Input Type - Choose sensor type. See p. 23 for sensor ranges.

`In`	`J`	`H`	`t`	`n`	`E`	`S`	`rtd`	`rt.d`
	J t/c	K t/c	T t/c	N t/c	E t/c	S t/c	1.0° RTD	0.1° RTD

Celsius/Fahrenheit - Choose displayed unit of measure.

`C_F`	`°F`	`°C`
	°F	°C

Input Range Low - Select lowest displayable range point.   Ranges, p. 23.

`rL`	`In`	`rh`

Input Range High - Select highest displayable range point.   Ranges, p. 23.

`rh`	`rL`	`In`

Output 1 Function - Choose Output 1 type; see Valid Outputs Table, p. 23.

`Ot1`	`hEAt`	`COOL`	`ALM`	`rSP`	`FPL`	`nonE`
	Heat	Cool	Alarm	Remote Set Point	Front Panel Lock	None

Output 2 Function - Choose Output 2 type (dependent on Output 1 choice).

`Ot2`	`hEAt`	`COOL`	`ALM`	`thrM`	`tMS`	`nonE`
	Heat	Cool	Alarm	Timer Hr./Min.	Timer Min./Sec	None

Display Default - Choose normal display (last 2 characters); toggle to other with `SET` press.
Ex.: If `dISP` = `Ac-SP`, normal display - Set Point. Toggle to Actual Temp. for 15 sec. with `SET`

`dISP`	`Ac`	`AcSP`	`Acti`	`tiAc`	`tiSP`
	Actual Only	Actual; Set Point	Actual; Time	Time; Actual	Time; Set Point

Alarm Type - Choose alarm type with output action.

`ALtY`	`Prnc`	`Prno`	`dEnc`	`dEno`
	Process normally closed	Process normally open	Deviation normally closed	Deviation normally open

18 Watlow Series 935 User's Manual

want to explain all the features of the product. Readers just want to learn how to perform a task. Researchers developed the concept of the minimal manual to address this conflict.

As articulated by John M. Carroll and his colleagues (1988), you should apply three principles when writing a minimal manual:

Printing a Self Test

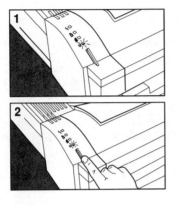

The page begins by posing a question: "Why print a self-test page?" If the reader doesn't learn the answer to that question, the rest of the information on that page will be wasted.

The three numbered steps in the explanation of how to print a self-test page correspond to the drawings.

Why Print a Self Test Page?

It is useful to print a self test page to troubleshoot printer problems or see what fonts have downloaded to the printer.

The data in the Printer Information section of the self test page resets every time the self test page prints, the printer's settings are changed, or when the printer is disconnected from its power source.

To Print a Self Test Page

1 Make sure the green Ready light is on, and all other lights are off. (If necessary, reset the printer.)

2 Briefly press the front panel button.

3 After about 6 seconds, the Data light comes on and the Ready light blinks. Then the self test page prints.

Usually, only one page prints. If you have downloaded fonts or macros, additional pages print as needed.

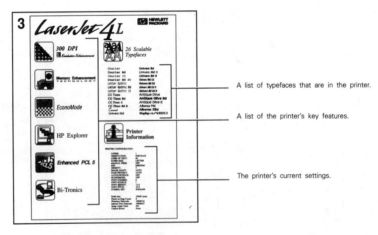

A list of typefaces that are in the printer.

A list of the printer's key features.

The printer's current settings.

FIGURE 18.9

Page from a User's Manual for a Printer

- *Focus on real tasks and activities.* Instead of telling readers how to use all the functions of the product, tell them how to do a real task. For example, in a manual for a word-processing program, don't explain the dozens of features; rather, have readers create a letter or memo. Introduce the major features, such as cutting and pasting, along the way.
- *Slash the verbiage.* The goal of a minimal manual is to reduce the length of a manual by as much as 75 percent. People don't want to read manu-

als; they want to accomplish a task. In describing a task, the minimal manual provides only basic guidelines, forcing the reader to experiment. For example, a manual might include a sentence such as the following:

You can delete what you have typed by using the **backspace** key or the **delete** key. Try these two keys.

This brevity is meant not only to save space but also to force the reader to experiment with the system rather than rely on the manual.

- *Support error recognition and error recovery.* People make mistakes when they experiment, so the minimal manual helps readers recognize and recover from them.

To indent an entire paragraph, use the **indent** key. If only the first line is indented, you have hit **tab** instead of **indent**. Use **backspace** to delete the **tab,** then try again.

Several experiments suggest that readers learn faster and make fewer errors when using minimal manuals than when using traditional manuals. However, some readers prefer more traditional manuals that provide greater structure and more detailed information.

The most basic idea behind minimalism—that people learn best when performing realistic tasks—is now reflected in most manuals. Commercial software, for instance, routinely comes with a tutorial that lets the user perform real tasks, such as setting up a spreadsheet, entering data, and printing it. And many software programs now come with "concise" manuals and an invitation to request (or buy) the full manual. Some concise manuals are truly minimalist; other, less helpful ones just leave out a lot of information to save printing and shipping costs.

Writing Instructions and Manuals for Nonnative Speakers of English

International trade is fundamental to the financial health of many American companies. When an American company wishes to provide a set of instructions or a manual to a non-English-speaking audience or to an audience whose first language is not English, it can either translate the document into the reader's native language or try to make the English easy to understand.

Translation is sometimes the best or the only alternative, but companies often use a simplified form of English, particularly when readers need to acquire a basic competence for long-term use of English. For instance, thousands of non-English-speaking aircraft technicians train for months in the United States to learn how to maintain U.S.-made civilian and military aircraft purchased by their countries. The manufacturers would find it prohibitively expensive to translate the many volumes of documentation into all the different languages used by their customers; therefore, they use a simple form of English.

The different forms of simple English include Simplified English, Controlled English, and Fundamental English. Some of these forms have been devised by scholars, others by companies such as Caterpillar Tractor and Kodak. They share three basic characteristics:

- *A limited vocabulary.* Some forms use vocabularies of as few as 500 words. Every word has only one meaning: *right* is the opposite of *left,* not of *wrong;* and *correct* is the opposite of *incorrect* (Sanderlin, 1988).

- *A limited grammar and sentence structure.* Sentences are short and simple. The imperative mood dominates. Following is an example of a sentence that has been revised from standard English into a simple English.

  ORIGINAL: After visually inspecting the gap to ensure the gap is no more than 0.025 inches, replace the housing.

  REVISED: Use the gauge to make sure the gap is no more than 0.025 inches. Then replace the housing.

- *A reliance on graphics.* As discussed in Chapter 12, graphics are used extensively in communicating with speakers whose first language is not English. If you can use a drawing, diagram, or photograph instead of words, do so. But be careful—graphics easily project a cultural bias that interferes with the communication or actually offends your readers.

The length of instruction in different forms of Simplified English ranges from 30 hours to 3 months.

For more information, see the following article:

Sanderlin, S. (1988). Preparing instruction manuals for non-English readers. *Technical Communication, 35,* 96–100.

Also see Chapter 15 for a discussion of Simplified English.

Hypertext and Hypermedia

The word *hypertext* was coined in 1965 by Ted Nelson, whose vision is to create a "docuverse," an electronically linked document universe in which every piece of information is linked to every other piece of information. Hypertext refers to a *multilinear* technique for presenting information.

A book is linear not only in the obvious way—the pages are numbered—but also in the most essential way: the author has laid out a clear path for the reader to follow. Even though you can skip around in the book by reading chapters out of order (as you may be doing with this textbook in your class), the idea behind paper-based documents is that the author has created a linear structure for presenting the information and the reader is intended to work within that structure.

In contrast, hypertext is multilinear. Readers are encouraged to jump from one piece of information to another, according to their interests and needs. Readers create their own paths through the information. Two concepts—nodes and links—are central to understanding hypertext:

- A *node* is a unit of information. It can be a sentence, a paragraph, a longer passage, or a whole document; it can be a photograph, a diagram, or a some other sort of graphic. In a *hypermedia* application, it can also be a video or film clip, animation, music, or spoken words. (Because hypertext and hypermedia differ only in the range of media used in the nodes, the word *hypertext* is used throughout this discussion.)

   Nodes are created by the author, but in some hypertext systems the readers can also make their own nodes by adding information.

- A *link* is a software connection between two nodes. For instance, a hypertext training manual for electronics technicians might include a section on how to solder. Readers who don't know what soldering is might use the link to jump to another node that defines soldering. But readers who already know what soldering is will not use that link. These more knowledgeable readers might jump to a video clip that shows a person demonstrating how to solder correctly. Readers activate these links by putting the cursor on a *button* and clicking the mouse, hitting a special key on the keyboard, or, in some cases, touching the screen.

   Like nodes, links are created by the author, but some hypertext applications let the reader create additional links. Some hypertext systems use one-directional links; that is, you can only go forward. Others use two-directional links; you can go forward or retrace your steps. Still others use both one- and two-directional links.

If you have used the help function on a software program, you have used a hypertext document. As Figure 18.10 on page 528 illustrates, the reader of hypertext participates in creating the flow of information much more actively than if he or she were reading a paper document. Notice in the figure the mixture of one-directional and two-directional links. Notice, too, that a link can be anchored to an entire node or to only a portion of a node.

Figure 18.11 on page 529 (Hall, 1994, p. 65) shows a screen from a hypertext application.

Hypertext Applications   Although hypertext is still a relatively new technology, we are beginning to understand its potential and its limitations. Obviously, it cannot be used without a computer. And it should not be used for brief documents that a person is intended to read from cover to cover. It is most effective when the writer needs to present a large, complex body of information to numerous readers, each of whom has different needs and goals.

For this reason, hypertext is used most commonly as a means of presenting three basic kinds of instructional material:

- *Instructional material for computer users.* The most obvious role for hypertext is in helping people use computers. Software tutorial manuals appear in it, as do on-line help functions, error messages, and reference manuals.

- *Instructional material for technicians.* Hypertext is used commonly in training manuals, particularly for repair and maintenance personnel. Flight mechanics, auto mechanics, and other technicians use hypertext manuals to progress from the novice to the expert level.

- *Advertising and consumer information.* Hypertext is popular in the following applications:

  – point-of-sale demonstrations (the computerized presentations you see set up in department stores) for consumer and industrial products
  – trade show presentations such as those intended for automobile dealers introducing the new year's product line
  – electronic kiosks used in museums and shopping malls to help users find their way, do educational learning exercises, look up product information, and order goods in self-service stores

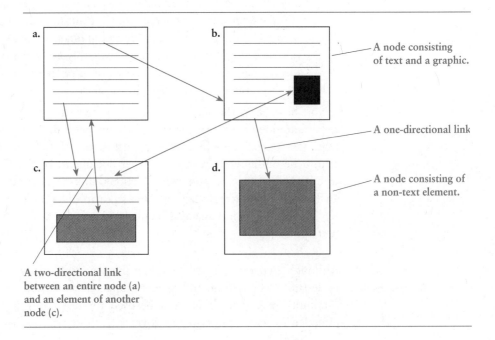

a.

b.
— A node consisting of text and a graphic.

— A one-directional link

c.

d.
— A node consisting of a non-text element.

A two-directional link between an entire node (a) and an element of another node (c).

FIGURE 18.10

The Basic Concept of Nodes and Links

The text on the right appears on the screen because the user has chosen the link "1. Introduction and Schema."

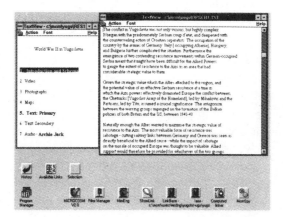

a

The user has read the text, then pulled down the Action menu and selected "Follow Link." In other words, the user is following the path suggested by the designers of the system. The system recommends that the user select "Mihailovic."

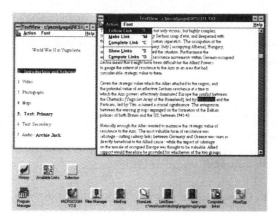

b

The Available Links box has opened, and the user has selected a portrait of Mihailovic.

FIGURE 18.11

Hypertext Screens

c

Advantages of Hypertext and Hypermedia

Although we don't fully understand the promise and limitations of hypertext, scholars are fairly confident that it offers the following seven advantages:

- *It accommodates different kinds of readers.* The chief advantage of hypertext is that readers create their own documents to match their own aptitudes, interests, and needs. Information can be layered, so that readers who need only a superficial understanding of the topic can skim along the top layer of nodes, whereas those who need more detailed information can delve into deeper layers.

- *It enables readers to customize the information.* Some hypertext systems keep track of the path the reader has chosen, enabling the reader to reproduce that path on a later occasion. And because some hypertext systems allow readers to create their own nodes and links, they can write notes for their own future reference. Authoring software—software that lets you create hypertext documents—is widely available now; much of it is suitable even for nonspecialists.

- *It saves space.* According to Nielsen (1990, pp. 13–14), the documentation for an F-18 fighter is 300,000 pages long and consumes 68 cubic feet of space. In hypertext on a CD-ROM, it requires 0.04 cubic feet.

- *It is relatively inexpensive to produce.* The cost of producing a CD-ROM, the most common delivery medium for hypertext, is less than a dollar, using a machine that costs about $6,000. If your business involves printing and sending large quantities of information to many people, hypertext documents can be much cheaper to produce than paper documents. The savings increase if the documents have to be updated frequently: you just produce and send out new disks.

- *It can be used like other digital information.* For example, documents can be put on networks so that they can be used by many people simultaneously. And they can be revised to make new documents.

- *It can reduce training costs.* Hypertext manuals can reduce the need for some kinds of training that require human instructors, because readers study the hypertext document at their own pace.

- *It can reduce training hazards.* Some kinds of training—involving electricity, explosives, and some chemicals, for instance—are inherently dangerous. The more knowledgeable the readers become through hypertext training, the less danger they face.

Disadvantages of Hypertext and Hypermedia

Hypertext, however, has the following six disadvantages:

- *Readers can become "lost in hyperspace."* Some readers find themselves unsure of where they are or have been, how to find the information they want, or how to get out of the system. Hypertext authors are trying to solve this problem by providing different kinds of maps and hierarchical diagrams of the links and nodes.

- *Readers can be swamped by information.* Using hypertext media, readers learn by association, jumping from one idea to another according to their needs and desires. The writer therefore does not have the opportunity to create a frame of reference; the reader is in control. Consequently, readers can be overwhelmed by new information and not learn as effectively as they would if the writer exerted more control over the pace and sequence.

- *Readers can skip important information.* Because readers are on their own, writers cannot ensure that they read important information.

- *Hypertext is expensive to write.* Writing hypertext documents is quite different from writing traditional documents. Few people are practiced at it. For this reason, good hypertext documents can be very expensive to create.

- *Hypertext is a new medium.* In most cases, companies subcontract the creation of hypertext documents. But there is no assurance that the subcontractor you hire today will exist tomorrow. Investments can be lost.

- *Hypertext software and hardware is not yet standardized.* One piece of hypertext software might not be compatible with another. This can create logistical complications when you try to use a hypertext document on a system other than the one on which it was created.

Software for creating hypertext documents is becoming easier to use and less expensive every year. And colleges and universities are offering courses about how to create hypertext. Departments of technical communication, instructional technology, media studies, communication, and art are most likely to be involved in educating students about hypertext.

For more information on hypertext, consult the following books:

Horton, W. (1994). *Designing and writing online documentation: Hypermedia for self-supporting products* (2nd ed.). New York: Wiley.

Nielsen, J. (1990). *Hypertext and hypermedia.* Boston: Academic Press.

Instructions

1. Does the introduction to the set of instructions
   - ❑ state the purpose of the task?
   - ❑ describe safety measures or other concerns that the readers should understand?
   - ❑ list necessary tools and materials?

2. Are the step-by-step instructions
   - ❑ numbered?
   - ❑ expressed in the imperative mood?
   - ❑ simple and direct?

3. Are appropriate graphics included?

4. Does the conclusion
   - ❑ include any necessary follow-up advice?
   - ❑ include, if appropriate, a troubleshooter's guide?

Manuals

5. Does the manual include, if appropriate, a cover?

6. Does the title page provide all the necessary information to help readers understand whether they are reading the appropriate manual?

7. Is the table of contents clear and explicit? Are the items phrased to indicate clearly the task that readers are to carry out?

8. Does the other front matter clearly indicate
   - ❑ the product, procedure, or system the manual describes?
   - ❑ the purpose of the manual?
   - ❑ the major components of the manual?
   - ❑ the best way to use the manual?

9. Is the body of the manual organized clearly?

10. Are appropriate graphics included?

11. Is a glossary included, if appropriate?

12. Is an index included, if appropriate?

13. Are all other appropriate appendix items included?

14. Is the writing style clear and simple throughout the manual?

EXERCISES

Several of the following exercises call for you to write a memo. See Chapter 16 for a discussion of memos.

1. Write a set of instructions for one of the following activities or for a process used in your field. Include appropriate graphics. In a brief note preceding the instructions, indicate your audience and purpose.
   a. how to change a bicycle tire
   b. how to parallel-park a car
   c. how to study a chapter in a text
   d. how to light a fire in a fireplace
   e. how to make a cassette-tape copy of a compact disc
   f. how to tune up a car

2. You work in the customer-relations department of a company that makes plumbing supplies. The products-development head has just handed you the draft of a set of installation instructions for a sliding tub door that you see on page 533. He has asked you to comment on their effectiveness. Write a 500-word memo to him, evaluating the instructions and suggesting improvements.

3. Write a brief manual for some process you are familiar with. Consider writing a procedures manual for a school activity or part-time job, such as your work as the business manager of the school newspaper or as a tutor in the Writing Center.

4. Write a 500-word memo to your instructor evaluating the front matter of a manual. Does it explain the purpose of the manual and how to use it? Is it welcoming and professional in tone? How might it be improved? Attach to your memo copies of the pages you are evaluating.

REFERENCES

Anti-tip bracket installation instructions. Whirlpool Corporation, 1990.

Arts and Letters Clip Art Handbook. 15th ed. 1990. Dallas: Computer Support Corporation.

Befuddled PC users flood help lines, and no question seems to be too basic. (1994, March 1). *Wall Street Journal,* sec. B, p. 1.

Carroll, J. M., Smith-Kerker, P. L., Ford, J. R., & Mazur-Rimetz, S. A. (1988). The minimal manual. In S. Doheny-Farina (Ed.), *Effective documentation: What we have learned from research,* (pp. 73–102). Cambridge, MA: MIT Press.

Hall, W. (1994). Ending the tyranny of the button. *IEEE MultiMedia, 1*(1), 60–68.

LaserJet 4L printer user's manual. (1994). Boise, ID: Hewlett-Packard Company.

McKnight, C., Dillon, A., & Richardson, J. (1991).

INSTALLATION INSTRUCTIONS

CAUTION: SEE BOX NO. 1 BEFORE CUTTING ALUMINUM HEADER OR SILL

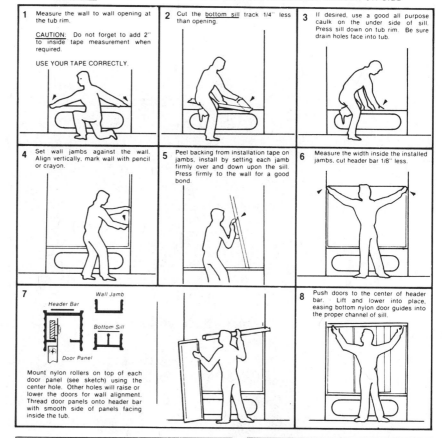

1 Measure the wall to wall opening at the tub rim.

CAUTION: Do not forget to add 2" to inside tape measurement when required.

USE YOUR TAPE CORRECTLY.

2 Cut the bottom sill track 1/4" less than opening.

3 If desired, use a good all purpose caulk on the under side of sill. Press sill down on tub rim. Be sure drain holes face into tub.

4 Set wall jambs against the wall. Align vertically, mark wall with pencil or crayon.

5 Peel backing from installation tape on jambs, install by setting each jamb firmly over and down upon the sill. Press firmly to the wall for a good bond.

6 Measure the width inside the installed jambs, cut header bar 1/8" less.

7 *Wall Jamb* *Header Bar* *Bottom Sill* *Door Panel*

Mount nylon rollers on top of each door panel (see sketch) using the center hole. Other holes will raise or lower the doors for wall alignment. Thread door panels onto header bar with smooth side of panels facing inside the tub.

8 Push doors to the center of header bar. Lift and lower into place, easing bottom nylon door guides into the proper channel of sill.

TRIDOR MODEL ONLY:

To reverse direction of panels, raise panels out of bottom track and slide catches past each other thereby reversing direction so that shower head does not throw water between the panels.

HARDWARE KIT CONTENTS
TUDOR MODEL
4 nylon bearings
4 ball bearing screws # 8-32 × 3/8"
TRIDOR MODEL
6 nylon bearings
6 ball bearing screws # 8-32 × 3/8"

Hypertext in context. Cambridge, UK: Cambridge University Press.

Microsoft Corporation. (1993). *Concise user's guide: Microsoft® MS-DOS® 6 for the MS-DOS operating system.* Redmond, WA: Microsoft Corporation.

Nielsen, J. (1990). *Hypertext and hypermedia.* Boston: Academic Press.

Sanderlin, S. (1988). Preparing instruction manuals for non-English readers. *Technical Communication, 35,* 96–100.

Series 935 user's manual. (1994). Winona, MN: Watlow Controls.

Using OpenScript. (1991). Bellevue, WA: Asymetrix Corporation.

CHAPTER 19

Usability Testing of Instructions and Manuals

Writers of instructions and manuals have always tried to write clearly. They have used checklists, style programs, and spell checkers. And they have had the documents reviewed by subject-matter experts and by other kinds of information specialists, such as legal counsel and marketing specialists. These reviews are helpful in pointing out potential problems.

In recent years, however, many organizations have also begun to do usability testing on their instructions and manuals. *Usability testing* is the process of performing experiments with people who represent real users, to see how easily they can use a product or carry out a process. Usability testing has been performed for decades on all kinds of machinery and consumer products. However, it has become popular as a way to test documents only in the last decade. Usability testing is different from other forms of review in that it allows you to examine the document from the perspective of the reader, the person who will be using the document.

Throughout this chapter, examples will be drawn from tests of common computer software and documentation, because technical communicators work on these products more than on any others. And the word *product* will refer to the items being tested, whether they are hardware or software, accompanying documents, or some combination.

This chapter covers five main topics:

- the goals of usability testing
- the basic concepts of usability testing
- preparing for a usability test
- conducting a usability test
- interpreting and reporting the data from a usability test

The information in this section is based on two excellent recent books:

Rubin, J. (1994). *Handbook of usability testing: How to plan, design, and conduct effective tests.* New York: Wiley.

Dumas, J. S., & Redish, J. C. (1993). *A practical guide to usability testing.* Norwood, NJ: Ablex.

If your organization is interested in a program of usability testing, I strongly recommend these two books; they will help you get the best results from your tests and save you a lot of time and frustration.

USABILITY TESTER STEPHANIE ROSENBAUM ON USABILITY TESTING

on the kinds of companies that do usability testing

All kinds do it. But the companies that do it most are the ones that realize that their customers are having problems—that's why you see so much usability testing in the computer industries—and those companies with mission-critical products and systems—for example, the banking and insurance industry. When the systems in place must be used correctly, and if they are so complicated that people make a lot of mistakes, then it's really important to do usability testing.

on the methodology of usability testing

There are quite a few accepted methods, but many occasions arise for creating new methods. Among experienced usability engineers, there is quite a bit of thinking and research, but when you are just starting, you begin with accepted methods.

on whether to leave it to professionals

It's comparable to technical communication: an untrained person could buy a few books about writing a manual, and if you studied them and took it seriously you could write a better manual than if you hadn't read the books, but you wouldn't be as skilled as if you had gone through a good university program or a good apprenticeship program. When you're just getting started, it's best to work with experienced usability professionals. You don't need to make every mistake that has been made for the last 15 years.

on teaching companies to create their own usability programs

Maybe a third to a half of our projects are skill-transfer projects: the purpose is to teach the client's own staff how to do that kind of project, so that over the course of several projects and over time they will build their own skills. It's the same process we do internally: the people we hire often require training. They read the research literature, go to seminars, observe projects in progress, and assist experienced practitioners.

on the biggest challenge of usability programs

Many people do not realize that study design and administration must be rigorous if you want useful results. It's awfully easy to lead the witness to the answer you are hoping to get rather than an objective answer—all without realizing it. Also, when planning a study if you don't recognize, for example, that people who have used Product X will perform differently on Product Y from people who haven't used Product X, you're going to get biased results. Designing studies that collect valid and reliable information about relevant issues is the biggest challenge.

on doing quick-and-dirty tests

People always talk about doing a quick-and-dirty test. It doesn't have to be quick and dirty; it can be quick and clean. You just have to narrow the scope of the study. It doesn't do you any good to get information that you cannot trust.

return on investment

If you establish a usability testing program early in a product-development project, it will reduce your development and support costs. But no single test is going to do everything; you need an ongoing program. One test can identify obvious usability problems, but there's always going to be another product or another version. A lot of what you learn in usability testing can't be applied to the current product but can be applied only to future products. You're never done with usability testing, just as you're never done with development or support.

on the future of usability testing

There's going to be a lot more usability work in the future. As more of this work moves earlier in the development cycle, as it should, there might be less "testing" but more usability projects. Testing will become more a validation of the user-centered design you did early on. All usability work pays for itself many times over. What we have to do is collect enough data and communicate the benefits to management. But it would be hard for anyone who has observed user studies—and seen the difference they make in the success of the product or system—not to be a believer.

Stephanie Rosenbaum is president of Tec Ed, a company specializing in usability testing and other technical-communication services.

The Goals of Usability Testing

The military tests a gas mask to make sure it functions correctly and that soldiers know how to use it effectively so they won't die in combat. In most cases, however, the goal of usability testing is not a matter of life and death—it is a matter of money. The ultimate goal of most usability testing is to increase the chances that the product will sell.

Usability testing has six main goals:

- *To help a company design safe products.* Usability testing helps a company see whether users can carry out tasks safely, effectively, and quickly.
- *To help a company market its products.* Usability testing helps the company understand and exploit the product's competitive advantages.
- *To help reduce the expense of service calls and customer-support phone lines.* Sophisticated products are often sold with maintenance and technical support included for a certain period, such as 90 days. The less often a technician is sent out to the customer, or the less often the customer calls for help, the lower the costs.
- *To help reduce the number of updates to the product.* Computer and software companies do usability testing to kill the bugs before the product is released, so that it doesn't have to be updated often.
- *To help a company learn how to improve other products.* The insights gained from usability testing improve not only the product being tested but also the process of making all similar products.
- *To help increase customer satisfaction with the product.* Unhappy customers share two characteristics: they try very hard not to buy another product from the same company, and they tell everyone they can think of about why they are unhappy.

The Basic Concepts of Usability Testing

Why is it so common for consumers to bring home a product and be unable to figure out how to use it, or even how to put it together? Sometimes it seems as if the people who designed and built the product weren't even thinking that someone would actually try to use it. Modern high-tech products can be baffling to the average person for two main reasons:

- *High-tech products today are complicated.* For $45 you can buy an internal fax modem. For $1,000 you can buy a computer more complex and powerful than the ones used by NASA to send people to the moon.
- *Many manufacturers don't think much about usability.* When personal computers were introduced about 15 years ago, users had to search to

find the on/off switch. It tended to be near the back of the box, because that was the most convenient place for the engineers to put it: that's where the electrical power system was. No one asked the user whether that was the most convenient place.

What does the complexity of high-tech products have to do with technical communication? Most obviously, technical communicators play the major role in writing the documentation that explains these products: the user's guides, reference manuals, installation guides, maintenance manuals, and tutorials. In the most successful companies, however, technical communicators also act as the customer's advocate. They work alongside the technical professionals to help the company make a product that meets the customer's needs.

The process of testing a product for usability is really only the visible part of a much larger concept of usability: the idea that the organization's mission is to offer products that real people working in real-life environments will find easy to use. There are five basic principles of usability testing:

- *Usability testing is an idea that permeates product development, not a single task that must be accomplished.* Usability testing involves a systematic program of testing the product rigorously and often to make sure it is easy to use. The organization devotes substantial resources to training and equipping the testers and to enabling the designers to use the information generated by the tests.

- *Usability testing involves teamwork by different kinds of specialists.* Most high-technology products are so complicated that no one person could understand every aspect of how they work. A usability test involves the collaborative efforts of a number of people with different areas of expertise, including product experts, usability experts, technical communicators, and video-camera operators.

- *Usability testing involves studying real users as they use the product.* The people who designed the product are not the best people to test it because they know it too well. Usability testing requires real users, or people who are as close to real users as can be found. A company will always learn important information from real users that it would not have learned from people within the organization.

- *Usability testing involves setting measurable goals and determining whether the product meets those goals.* Usability testing involves determining in advance what the user is supposed to be able to do. For instance, in testing a user's guide for a word-processing program, the testers might decide that the user should be able to find the section on saving a file and carry out that task successfully in less than one minute. With a clear, measurable goal, the testers can study a number of participants and record meaningful data.

- *Usability testing involves repeatedly revising the product.* If the test shows that a large percentage of participants are having trouble with a task, testers examine all the information—including questionnaires the participants filled out, oral comments they made while performing the task, and videotapes of the test—to determine what went wrong. Then the testers try to fix the problem; test more participants, and fix the problem; and so on. They keep at it, until they run out of time or money.

Preparing for a Usability Test

Usability testing requires careful planning over a period of weeks or months. According to Kantner (1994), planning accounts for one-half to three-quarters of the total time devoted to testing. Eight main tasks must be accomplished in planning a usability test, as shown in Figure 19.1.

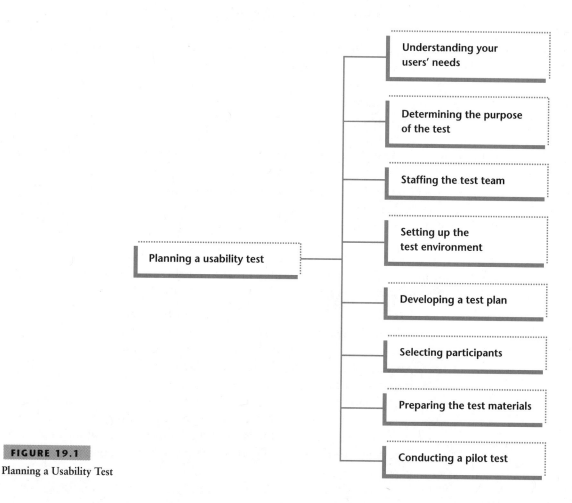

FIGURE 19.1

Planning a Usability Test

Understanding Your Users' Needs

Because the purpose of usability testing is to shift the focus from making the product to using it, the most critical step in preparing for a test is to understand your users' needs. Companies conducting usability tests frequently use the following four techniques:

- *Conducting focus groups.* A *focus group* is a group of people brought together for a few hours to talk about a product or an issue. In politics, a candidate convenes a focus group to learn about the voters' concerns, attitudes, and behaviors, in order to create or refine campaign strategy. In the corporate world, a skilled interviewer representing the company convenes a group of about a dozen people who represent potential users of the company's product, such as nurses who would operate a piece of medical equipment.

- *Testing existing products.* One way to prepare to test a new product is to test a similar existing product—either the company's own or a competitor's. Testers might find, for example, that their competitors all have quick-start guides for their software, whereas they do not.

- *Having experts review the product.* Subject-matter experts, such as software developers, can evaluate the usability of the product by looking at it and thinking about it, but currently there is no clear consensus that this technique is effective. Usability experts appear to uncover more problems, and more important problems, than software experts and other subject-matter experts (Dumas & Redish, 1993, p. 67). Not surprisingly, the best results are obtained by people with expertise in both engineering and usability, especially when they work in groups of four or five.

- *Conducting on-site interviews and observations.* It makes sense to visit real users in the workplace, to observe them doing their work and to interview them.

Determining the Purpose of the Test

Because *usability* is such a broad term, it is not sufficient to say that you want to test the usability of an item. Rubin (1994) describes four kinds of tests, each of which has a different purpose:

- *Exploratory test.* When a company creates a new product or completely redesigns an existing one, it starts with an *exploratory test,* intended to determine how well the participants understand the broad outlines of the product. For instance, in a word-processing program with pull-down menus, do the participants understand the basic terminology and know how to carry out such basic tasks as opening a new file or retrieving an existing one? Because an exploratory test occurs early in the development cycle of a product, often the product does not even exist at the time of testing. For example, pull-down menus might be represented on paper.

- *Assessment test.* Often called an *information-gathering test,* an *assessment test* occurs later in the development cycle than does the exploratory test. The product already exists, even if only as a rough prototype, and participants actually carry out tasks rather than discuss how they would carry them out. In an assessment test, testers collect data on performance.

- *Validation test.* Shortly before the item is introduced, a *validation test,* sometimes called a *verification test,* is carried out. At this point, the product exists, and the usability standards it is intended to meet are known. For instance, the company might have decided that since its main competitor's version of the spell checker checks a 10,000-word passage in 50 seconds with no more than four correctly spelled words flagged, its own product must at least match that benchmark.

- *Comparison test.* At any stage in the development of the product, testers might want to conduct a *comparison test,* in which the participant is presented with two versions of the same product. The goal is to determine whether one version works better than the other. For instance, the company might want to know whether users of a word-processing program prefer to have the spell checker work automatically as they draft, or at the user's command. Rubin (1994) points out that comparison tests often reveal that users would prefer a third alternative, or some combination of the two you have presented.

Staffing the Test Team Extensive programs in usability testing involve a number of specialists, each of whom does one, and only one, job. In many companies, however, only two or three people do all the tasks involved in the tests. The following description covers six functions that must be performed during a usability test:

- *Test administrator.* This person is in charge of the whole test: ensuring that all the personnel are present on time and prepared, the equipment is set up, the test materials are ready, and the test participant shows up. During the test, the administrator is responsible for dealing with any problems that arise. After the test, the administrator oversees the process of debriefing the participant, analyzing the data, and writing the report.

- *Briefer.* The briefer interacts with the participant. The briefer greets the participant, provides the necessary orientation or training, makes sure the participant fills out any forms and questionnaires, and, after the test, pays the participant and escorts him or her from the building.

- *Camera operator.* The camera operator sets up the equipment, runs it during the test, and makes sure all the tapes are clearly and accurately labeled. Often the camera operator is responsible for creating a *highlights tape,* which shows the most important events from the tests of several participants.

- *Data recorder.* The data recorder is responsible for logging the data—either on paper or on software—for each test. After the test, the data recorder compiles the data and performs the statistical analyses.
- *Help-desk operator.* The help-desk operator plays the role of the technical adviser who is available to help users once the product is on the market.
- *Product expert.* Because the product being tested is often a prototype, it can and does fail. The product expert is there to fix it or get a replacement.

Setting Up the Test Environment

Many people immediately think of the word *laboratory* when they hear the phrase *usability testing.* But usability tests can be conducted without a lab, or with only a modest one. Rubin recommends beginning with a minimal setup, then adding equipment and devoting space to usability as needed (1994, p. 49). The following discussion outlines some of the basic approaches to creating a test environment.

The equipment used in testing is fairly basic:

- VCRs to record the test
- video monitors (if the test observers are in another room)
- an intercom system with microphones (if the participant is to sit alone in the test room)
- videotape editing machines to create a single tape from several source tapes

This equipment can be supplemented with sophisticated devices that show two images on the video screen at the same time and that add text to videotapes.

Usability testers use four basic environments. Here they are discussed from least expensive to most expensive:

- *Portable lab.* The equipment for a portable lab can be moved from one location to another within the company, or to a remote location such as a meeting room at a local hotel or a customer's site. Figure 19.2 shows the basic floor plan for setting up the equipment in a room. The obvious advantages of a portable lab are that it is inexpensive and flexible. The disadvantage is that the test participant can be intimidated or made self-conscious by the presence and actions of the testers.
- *One-room lab.* A one-room lab is set up like a portable lab, but because the lab is in a permanent location, the camera can be mounted on the wall rather than on a tripod, and a second camera can be added to record the test participant's face and hands. The advantage of the one-room lab is that you can make the participant more comfortable, but the disadvantage is the same as with a portable lab: the testers are present in the room.

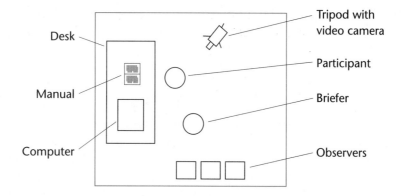

FIGURE 19.2

Floor Plan of a Portable
Lab

- *Two-room lab.* In a two-room lab, people can observe the test from an adjoining room by watching through a one-way mirror or by watching video monitors. The briefer can sit in the room with the participant or in the adjoining room. If the briefer is not with the participant, communication takes place through an intercom. Figure 19.3 shows the floor plan for a two-room lab. The advantages of the two-room lab are that fewer people are in the test room to distract the participant, and that the observers can talk among themselves or come and go as they wish, without interfering with the test.
- *Three-room lab.* The extra room in the three-room lab is for people observing the test but not participating in it directly. The three-room lab gives you the most comfortable, versatile setup, but obviously it is the most expensive.

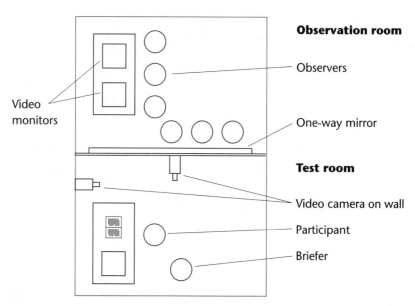

FIGURE 19.3

Floor Plan of a Two-
Room Lab

Developing a Test Plan   Rubin (1994) recommends that testers write a test plan, a document that describes and justifies what they plan to do. The test plan is a kind of proposal, a statement requesting approval and resources (see Chapter 21 on proposals). The test plan is also the document that serves as the link between the developers and the testers and as the basis for the script for the test itself. A basic test plan consists of the following seven sections:

- *Purpose.* Why does the company need to carry out the test? Perhaps the company had an unexpectedly large number of problems with the previous model, or a new competitor with a different technology is gaining market share.

- *Test objectives.* Usually, test objectives are stated as questions the company wants to answer, such as "Will users be able to determine the appropriate manual to consult for installing the program?" or "Will users prefer on-line help to printed help?"

- *User profile.* The user profile characterizes the users of the product. Table 19.1 shows a user profile that might be appropriate for a test of a user's guide for a new word-processing program.

- *Experimental design.* The heart of a test plan is the experimental-design section, which describes what the testers want to do and how to do it. This section describes how many participants are needed and what they will do. The general principle is to have as many participants as possible, to increase the chances of obtaining valid results from the tests. Experts recommend testing at least 4 or 5 people, a number that will uncover 80 percent of the problems (Rubin, 1994, p. 93). For even better results, testers use 10 or 12 participants.

TABLE 19.1	*User Profile*
Criterion	**Profile**
Word-processing experience	None: 30% 1–2 years: 30% 2+ years: 40%
Education	High school: 50% College: 40% Graduate school: 10%
Age	Adult: 100%
Sex	Female: 60% Male: 40%
Platform	Mac: 30% DOS: 70%

Often the user profile will suggest that testing should involve, for example, experienced and novice users and managers and technical people. In such cases, testers have to work out the different permutations and hire as many participants as possible. For instance, testers might assemble a group consisting of

– novice managers
– experienced managers
– novice technical people
– experienced technical people

- *Task description.* This section lists each task the participant will perform and defines criteria for successful completion of the task. For instance, if the company is testing whether the user understands a description of how to turn spreadsheet data into a bar graph, the task description might look like Table 19.2. In most cases, the task description includes numerous tasks. For instance, in testing the manual for the spreadsheet, the company will probably want to see how successfully the user customizes the bar chart by modifying bar width and adding data, labels, a legend, and a title. The test plan should present each of these tasks separately so that meaningful data can be generated.

  Harris (1991) recommends starting with relatively simple tasks to put the participant at ease.

- *Equipment.* This section specifies what equipment will be needed, including the products to be tested, such as a personal computer and a set of manuals.

- *Evaluation data.* The evaluation data are the data to be collected for each participant for each task in each test. In general, evaluation data fall into two categories: performance data and attitudinal data. Table 19.3 on page 546 lists typical evaluation data. Performance data are quantitative; that is, they are measured in numbers. Some attitudinal data are quantitative; others are qualitative. Attitudinal data are derived from questionnaires and interviews (see Chapter 7).

TABLE 19.2	*Excerpt from a Task Description*
Task	**Description**
Turn spreadsheet data into a bar graph.	The participant will be presented with a simple spreadsheet and told that the manual describes how to perform the task.
	Successful completion: creating a bar graph within five minutes.

TABLE 19.3	*Typical Evaluation Data*
Performance Data	**Attitudinal Data**
• Time to complete the task • Number and percentage of tasks completed successfully and unsuccessfully • Number of correct and incorrect choices (such as incorrect menu selections) • Number of calls to the help desk • Number of repetitions of a particular choice	• Value of the product overall • Value of individual elements of the product, such as the index of a manual • Ease of setup or installation • Ease of use of the product and its individual elements • Helpfulness of help desk personnel

Selecting Participants

Selecting the right participants for a usability test is essential. This section discusses five factors in choosing participants.

- *Defining participants' ideal characteristics.* As mentioned earlier, the test plan includes a section profiling the users. Each participant's technical background and amount of experience with the product being tested should match that of a typical user. In addition, a sufficient number of participants should be recruited to obtain valid statistics in the tests.

- *Finding the right participants.* For reasons of convenience, the test administrator is often tempted to use company personnel as test participants. This tactic is risky: they might know too much about the product to be representative of the real user.

  When testers go outside their own organization to recruit, they think first of the obvious places. If the product will be used by corporate trainers, testers contact corporate trainers. If the product will be used by accountants, they contact accountants. If the users are not specific to one industry, the job is somewhat easier, for participants can be recruited from temporary agencies, professional organizations, universities, and computer user's groups. Testers also post notices on bulletin boards—physical ones and electronic ones—and place ads in the classified section of the newspaper.

- *Writing an introductory letter to potential participants.* To maintain a consistent approach in their recruiting, testers draft a letter to send to all potential participants. The letter accomplishes the following five tasks.

  – It briefly describes the product:

    We are testing the user's manual of a database program that has an improved graphical user interface.

– It describes the nature of their participation:

> You will be asked to come to our testing center at a mutually convenient time in early July and perform various tasks with this new program. While you do this, we will videotape you and may ask you why you made certain choices in performing the tasks. After the session, we will ask you more questions and request that you fill out a questionnaire.

– It indicates how long the session will last:

> The session will last no more than three hours.

– It describes any incentive the participant will be receiving:

> In appreciation for your taking the time to help us improve our new product, we will give you a check for $75 at the end of the session.

– It tells the potential participant to get in touch with the testers to ask questions:

> Please call me at 555-3088 if you have any questions.

- *Having the potential participant respond to a questionnaire.* To make sure the potential participants match the user profile, testers have them fill out a questionnaire, either in writing or over the phone.

  The questionnaire, which can be enclosed with the introductory letter or sent later, usually contains 5-10 questions about the potential participant's knowledge of the kind of product to be tested. Following are a few sample questions from a questionnaire used in conjunction with a test for a new word-processing program:

How long have you used a Windows-based word-processing program?
_____ less than six months
_____ between six months and a year
_____ more than a year

Is your computer linked to a network?
_____ yes
_____ no

If the questionnaire is administered over the telephone, it should be carefully scripted to ensure consistency. For example, when a particular answer calls for the questioner to skip down to a particular follow-up question or to terminate the interview, this information should be indicated clearly, as in the following question:

How long have you used a Windows-based word-processing program?
_____ less than six months **REJECT**
_____ between six months and a year **RECRUIT AS NOVICE**
_____ more than a year **RECRUIT AS EXPERIENCED**

The script tells the questioner how to respond to the answers given.

After receiving the questionnaire results, testers determine which participants to recruit. From this point on, notifying the participants, scheduling their tests, and ensuring that they arrive on time are largely clerical tasks.

- *Writing a confirmation letter to the participants.* Kantner (1994) recommends sending each participant a confirmation letter, timed to arrive three days before the test. The letter should thank them for their willingness to assist you, indicate the time and place of the test, enclose a map, and reaffirm the compensation they will receive.

Preparing the Test Materials

The most time-consuming task in preparing for a usability test is to create the test materials. Most tests use six basic kinds of test materials:

- *Legal forms.* To protect the rights of the participant and the organization, usability experts recommend creating appropriate legal forms and having them approved by the organization's legal counsel. Dumas and Redish (1993, p. 205) explain that the concept underlying usability testing is called *minimal risk.* According to the Federal Register, minimal risk means that "the probability and magnitude of harm or discomfort anticipated in the test are not greater, in and of themselves, than those ordinarily encountered in daily life or during the performance of routine physical or psychological examinations or tests." If testers can't abide by this principle, they shouldn't do the test.

- *Orientation script.* Before the test begins, the briefer reads an orientation script to the participant. The script ensures that each participant is treated exactly the same way. It introduces the members of the test team and repeats the purpose of the test and the participant's role, emphasizing that the participant is not being tested but that the product is.

    The briefer indicates that the participant may ask questions at any time and offers to answer any at this time. Finally, before the test begins, the briefer asks the participant to fill out the background questionnaire.

- *Background questionnaire.* A background questionnaire is a more detailed version of the questionnaire that was administered by letter or on the phone. The purpose is to make sure the testers have all the information they need to understand the participant's behavior during the test. Following are some examples of questions that might be included in a background questionnaire.

    At work, is a printer connected directly to your computer?
    _____ Yes
    _____ No

    If yes, what is the brand and model of the printer?
    _____

Did you connect the computer to the printer, or did someone else do it?

_____ I did it.

_____ Someone else did it.

Testers often want to determine the participant's attitudes toward the product being tested. They could, for example, ask the participant to respond to a Likert-scale statement that reads "It would be very useful if the word-processing software enabled me to print the address from a letter onto an envelope, without having to write a macro." (See Chapter 7 for a discussion of kinds of questions used in questionnaires.)

- *Task scenarios*. A task scenario is a set of instructions given to the participant for each task to be performed. Task scenarios are usually given piecemeal; that is, the briefer gives the first scenario, and the participant does the first task. Then the briefer gives the second scenario, and so on. Task scenarios can be written and given to the participant, or read to the participant. Either way, the briefer should invite the participant to ask questions when something in the scenario is unclear.

    The task scenario should be sufficiently challenging to test whether the participant understands the concept behind the task, as in the following excerpt:

    Your supervisor has read your proposal to expand your department's office space; your proposal is on your screen now. She would prefer to have the Recommendations section, which is now on page 14, at the end of the executive summary. You may use the user's guide or not, as you please. Tell me when you have finished the task. Do you have any questions?

    Harris (1991) suggests avoiding vocabulary that might give the participant a clue about how to perform the task. The scenario quoted here, for example, does not say to "move" the Recommendations section, because *move* might tell the participant what action to take or how to find the appropriate discussion in the index.

- *Data logs*. A data log can be electronic or manual. For each task, a data log includes the task scenario and space to record the appropriate performance and attitudinal data. For the task scenario involving moving text within a document, for example, the data log might include the following information:

Performed task correctly? y/n _____
Total time to complete: _____
Incorrect commands executed: _____
                              _____
                              _____
Referred to index?   y/n _____
Referred to contents?   y/n _____
Called help desk?   y/n _____

Participant's comments

Test team's comments

- *Post-task and post-test questionnaires.* To learn the participant's opinions and attitudes about the product, testers use questionnaires administered after each task and questionnaires administered after the whole test is complete. Rubin (1994, p. 200) recommends using post-task questionnaires to break up tests of more than two hours; otherwise, the participant might forget useful information.

Conducting a Pilot Test A pilot test is, in effect, a usability test for the usability test. A pilot test can uncover four basic categories of problems:

- *Problems with the equipment.* The system can fail for some unknown reason or make one of the tasks impossible to complete. Equipment problems must be solved before testers bring in a real test participant and waste time and money—and use up a qualified participant.
- *Problems with the instructions or manual.* Information can be missing or out of sequence, or passages can be unclear or contain company-specific jargon. These problems could frustrate the participant and prevent testers from discovering the information the company needs.
- *Problems with the test materials.* Questionnaires can be unclear or too ambiguous to yield useful information. All kinds of other problems can afflict the other test documents: missing materials, missing passages, contradictions, or unnecessary redundancy from one document to another.
- *Problems with the test design.* The test can be too long, so the participant does not finish, or the test can be too short because a lengthy task was omitted by mistake. Maybe the tasks are not presented in a logical order, or the participant performs a task in a way the testers did not anticipate, causing the system to fail or avoiding the procedure they wanted to test.

Conducting a Usability Test

Although every tester makes mistakes every time, the goal is to minimize the number and severity of the mistakes so that as much as possible can be learned from each test. The three aspects of conducting the test are shown in Figure 19.4.

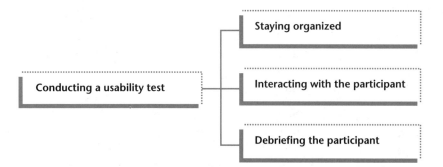

FIGURE 19.4
Conducting a Usability
Test

Staying Organized   Because a usability test involves dozens of details, and because the testers will be nervous, they need to do everything they can to stay organized. Experienced testers recommend making up a checklist and a time schedule for the test day. The schedule should cover everything from before the first participant arrives until after the last one is escorted out—including packing the equipment away in preparation for the next day. In addition, experts recommend keeping each participant's materials in a separate folder.

Interacting with the Participant   Interacting with the participant is perhaps the most challenging part of usability testing. Testers can never predict what will happen; there are no routine usability tests. Following are two suggestions about interacting with the participant:

- *Consider using a think-aloud protocol.* Most of the information derived from a test will come from what the testers see: the videotapes, the data logs, and impressions of the participant's body language. But many testers use an additional technique: the *think-aloud protocol*. In this technique, the participant speaks what he or she is thinking—for instance, "I guess I'm supposed to press ENTER here, but I'm not sure because the manual didn't say to do it." If the participant feels comfortable thinking aloud, the information can prove invaluable, for the participant is, in effect, writing the data log. However, many people become very self-conscious when thinking aloud. In addition, some testers mistrust the technique because they feel it makes the participants slow down and act more carefully than they normally would.

- *Don't bias the participant.* When testers ask a participant a question, they try not to reveal the answer they want. They do not say, "Well, that part of the test was pretty simple, wasn't it?" Regardless of whether the participant thought it was simple or difficult, his or her impulse will be to answer yes. Dumas and Redish (1993, p. 298) recommend using neutral phrasing, such as "How was it performing that procedure?" or "Did you find that procedure easy or difficult?" In responding to questions, testers are indirect. If the participant asks "Should I press ENTER now?" they respond, "Do you think you should?" or "I'd like to see you decide."

A related problem occurs when the participant has trouble performing a task and becomes frustrated. Harris (1991) recommends waiting five minutes before intervening. However, Rubin (1994, pp. 222–223) suggests that you intervene immediately in any of the following three situations:

– The participant is hopelessly confused or so frustrated that he or she is about to quit and go home.
– The participant feels uncomfortable performing a task because it seems too risky or likely to jeopardize the test.
– A malfunction occurs that is unrelated to the participant's action.

When testers do intervene, they should not blame the participant. Even a person who did something that caused the system to fail has done you a favor by alerting you to a potentially catastrophic problem.

Debriefing the Participant

After the test is completed, testers usually have a number of questions about the participant's actions. For this reason, they debrief—that is, interview—the participant. The debriefing is critically important, for once the participant walks out the door, it is difficult and expensive to ask any questions, and the participant likely will have forgotten the details. Consequently, the debriefing can take as long as the test itself did.

While the participant is filling out the post-test questionnaire, the test team quickly looks through the data log and notes the most important areas to investigate. Their purpose in debriefing is to obtain as much information as possible about what occurred during the test; their purpose is not to think of ways of redesigning the product to prevent future problems. Testers want to concentrate on analyzing what happened; they don't want to use up their time with the participant discussing an issue about which he or she probably cannot offer much useful information. In addition, they want to hold off on thinking of ways to fix the problem, so that they can let the information from the test sink in. That way, they increase the chances that the redesign ideas will be successful.

Rubin (1994, p. 246) suggests beginning the debriefing with a neutral question such as "So, what did you think?" This kind of question encourages the participant to start off with an important suggestion or impression. During the debriefing session, testers probe high-level concerns before getting to the smaller details. They try not to get sidetracked by a minor problem.

Interpreting and Reporting the Data from a Usability Test

After completing a usability test, testers have a great deal of data, including notes, questionnaires, and videotapes. The three steps in turning that data into useful information are shown in Figure 19.5.

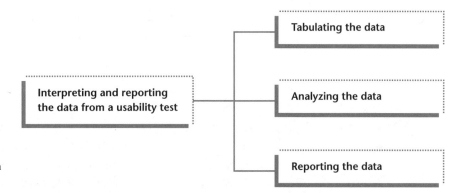

FIGURE 19.5

Interpreting and
Reporting the Data from
a Usability Test

**Tabulating the
Data**

Tabulating the data involves gathering all the information, both quantitative and qualitative, and, when appropriate, performing statistical procedures on it. The quantitative data include performance measures, such as how long it took a participant to complete a task, and some attitude measures, such as data on how easy the participant found the task to perform. The qualitative data include non-numerical information, such as notes taken by the test team and quotations from participants during or after the test.

Simple data can be tabulated with pen and paper. More complicated quantitative data can be tabulated using a spreadsheet, which will automatically calculate basic descriptive statistics, such as totals, means, and standard deviations. Although these descriptive statistics are the kind used most often in usability testing, inferential statistics—designed to establish causal relationships among different factors in the test—are sometimes used. For more information about inferential statistics, see Spyridakis (1992).

After the data are tabulated, they are transcribed and summarized. Handwritten notes and audiotapes are entered on a word processor. Once all this information is electronic, the testers can summarize it and extract quotations to support their generalizations. Then they compile the data according to logical categories.

Dumas and Redish (1993, p. 313) emphasize that it is important to pay attention to aberrant data, scores that are significantly different from the mean. If the mean time to complete a task was about 5 minutes but one participant took 32 minutes, there could be several explanations. Perhaps this participant was having a bad day; if this is the case, most or all of his or her data should be aberrant. Perhaps the participant was not sufficiently experienced to handle the test (and should have been eliminated during the screening process).

But if nothing was "wrong" with the participant, the testers go back to the data to try to figure out why this participant took so long to complete the task. If the aberration can be explained by determining that the data logger forgot to indicate a system failure—fine. But testers need to think carefully about whether something about the product itself caused the participant to work so slowly; one of only four participants might represent 25 percent of the real users.

Analyzing the Data   Once testers have compiled all the data, they analyze it to understand and fix the usability problems. Naturally, they want to concentrate on the most important ones. Two guidelines apply here:

- *Determine the importance of each usability problem.* Importance can be measured by two factors: the *severity* and the *frequency* of the problem. Table 19.6, based on Rubin's classification scheme (1994, p. 277), shows the severity measure. Table 19.7 shows the frequency measure.

  How important is a usability problem? To answer this question, assign each problem a severity ranking and a frequency ranking, and then total the two rankings. For instance, if you rank a problem 4 in severity and 4 in frequency, that's 8, as important as it gets. Obviously, a total of 2 is minor. But notice that an irritant that occurs all the time yields a 5, just as does a catastrophic problem that occurs only rarely.

  Adding two single-digit numbers is easy enough; the difficult part is to figure out how to rank the problem in the first place. Who decides whether a problem is catastrophic or merely severe? Test participants' responses to post-test questionnaires and comments during debriefing are the best source for this information. Testers have to resist the temptation to overrule the participants.

TABLE 19.6   *Severity Measure*

Ranking	Severity Measure	Example
4	**Catastrophic.** The user is so frustrated that he or she refuses to use the document.	The manual has no index or table of contents.
3	**Severe.** The user becomes extremely frustrated because a function is too difficult to perform.	The instructions omit vital steps in the procedure.
2	**Moderate.** The user is annoyed by the inconvenience of performing a function.	The instructions do not list the tools the reader will need to perform the procedure.
1	**Irritant.** The problem is only a minor annoyance.	The manual has no headers, footers, tabs, or other navigational tools to indicate where the reader is.

TABLE 19.7	*Frequency Measure*
Ranking	**Frequency**
4	Occurs 91–100 percent of the time
3	Occurs 51–89 percent of the time
2	Occurs 11–50 percent of the time
1	Occurs 0–10 percent of the time

- *Redesign and retest.* On a good day, testers figure out right away what caused the problems, and they and the product specialists fix them easily. Often, however, they have to think long and hard to determine the causes. Then begins the process of redesigning the product. If the product is a manual, the redesign can be as simple as adding running heads or improving the table of contents, or as complex as rewriting it entirely to make it task-oriented rather than function-oriented.

  No matter how difficult the redesign process, testers can never be sure they have solved the problem until they retest the product. In many cases, they don't have the time and money to retest; but ideally, usability testing is a series of tests intended to make the product progressively easier to use.

Reporting the Data Time pressures are often so great that the testers can't retest. Frequently, testers can't even write the report before they have to present their findings to management. Testers sometimes have to give an impromptu oral report about the test before they have even had a chance to tabulate the data, or they have to provide a one- or two-page report the next day.

At some point, however, testers need to submit a written report. Even though writing it can require weeks and add 20 to 25 percent to the cost of the test (Kantner, 1994), the rigor required in compiling a report often leads the testers to insights they might not have achieved otherwise.

A traditional written report is usually required. Because people with different backgrounds, interests, and needs will read it, the report should be presented in a modular form, with a special section for the executives, a detailed table of contents, a clear introduction, and a detailed set of findings: the results, conclusions, and recommendations. Participants' quotations are an important component of usability reports, for managerial readers often find them more meaningful than tables of data. Transcripts and other data are either attached or made available to readers who want to read them. See Chapter 23 for a detailed discussion of reports. In addition to writing a report, many usability teams produce a highlights tape to circulate with the report.

Usability testing might seem like an extremely expensive and difficult undertaking, and certainly it is frustrating. But testers who are methodical, open-minded, and curious about how people use the product find that usability testing is the least expensive and most effective way to improve its quality.

WRITER'S CHECKLIST

1. In preparing for a usability test, have you
   - ❑ made sure you understand your users' needs?
   - ❑ decided on the purpose of the test?
   - ❑ assembled a test team?
   - ❑ set up the test environment?
   - ❑ developed a test plan?
   - ❑ selected participants who are similar to your real users?
   - ❑ prepared the test materials?
   - ❑ conducted a pilot test?

2. In conducting a usability test, have you
   - ❑ established procedures for staying organized?
   - ❑ avoided biasing the participant?
   - ❑ debriefed the participant?

3. In interpreting and reporting the data, have you
   - ❑ tabulated the data using appropriate methods?
   - ❑ analyzed the data to learn what you can about the product?
   - ❑ reported the data in a formal report and, if appropriate, a highlights tape?

EXERCISES

Exercises 1 and 2 call for you to write memos. See Chapter 16 for a discussion of memos.

1. Contact local manufacturing companies and computer hardware and software companies to see whether any of them perform usability testing. Interview a person who performs usability testing at one of these organizations. Then write a 1,000-word memo to your instructor describing how the process of conducting usability testing at this organization differs from that described in this chapter.

2. If a local company conducts usability testing, see whether you can sign up to be a test participant.

After the test, write a 1,000-word memo to your instructor describing the experience, focusing on what you learned about usability testing.

3. In your technical communication class, form a group of four or five students and conduct a usability test for the assembly or use of one of the following products:
   a. a piece of computer hardware such as a printer
   b. a piece of software (or a portion of a piece of software)
   c. a document that accompanies a piece of software (or a portion of one)
   d. a piece of equipment used in your major field of study
   e. a product such as a coffee grinder, a calculator, a voice-mail system, or an external modem

   Submit a usability report to your instructor.

REFERENCES

Dumas, J. S., & Redish, J. C. (1993). *A practical guide to usability testing.* Norwood, NJ: Ablex.

Harris, R. A. (1991). A do-it-yourself usability kit. *Journal of Technical Writing and Communication, 21,* 351–368.

Kantner, L. (1994). The art of managing usability tests. *IEEE Transactions on Professional Communication, 37,* 143–148.

Rubin, J. (1994). *Handbook of usability testing: How to plan, design, and conduct effective tests.* New York: Wiley.

Spyridakis, J. (1992). Conducting research in technical communication: The application of true experimental designs. *Technical Communication, 39,* 607–624.

Oral Presentations

So far, this book has focused on writing. In this chapter, the focus shifts to speaking. Even though documents and presentations are different, the techniques for preparing them are quite similar; you analyze your audience, gather information, organize it, and create graphics. The big difference between writing and speaking is, of course, the form of delivery.

There are four basic types of presentations:

- *Impromptu presentations.* You deliver an impromptu presentation without any advance notice. For instance, at a meeting, your supervisor calls on you to speak for a few minutes about a project you are working on. You did not know you were going to be asked; you did not prepare any materials or rehearse.

- *Extemporaneous presentations.* In an extemporaneous presentation, you might refer to notes or an outline, but you actually make up the sentences as you go along. Regardless of how much you have planned and rehearsed the presentation, you create it as you speak.

  The extemporaneous presentation is preferable for all but the most formal occasions. At its best, it is clear and seemingly spontaneous. If you have planned and rehearsed your presentation sufficiently, the information will be accurate, complete, organized, and easy to follow. And if you can think well on your feet, the presentation will have a naturalness that will help your audience concentrate on what you are saying.

- *Scripted presentations.* In a scripted presentation, you read a text that you (or someone else) wrote out completely in advance. You sacrifice naturalness for increased clarity and precision.

- *Memorized presentations.* In a memorized presentation, you speak without notes or a script. Memorized presentations are not appropriate for most technical subjects because of the difficulty involved in trying to remember technical data. In addition, few people other than trained actors can memorize presentations of more than a few minutes.

This chapter discusses extemporaneous and scripted presentations. It explains how to prepare an outline or speaking script, how to prepare graphics, and how to rehearse. Finally, the chapter addresses how to give the presentation and respond to questions afterward.

Understanding the Role of Oral Presentations

In certain respects, an oral presentation is inefficient. For the speaker, preparing and rehearsing it generally take more time than writing the same information in a document would take. For the audience, physical conditions during the presentation—such as noises, poor lighting or acoustics, or an uncomfortable temperature—can make concentration difficult.

Yet an oral presentation has one big advantage over a written one: it permits a dialogue between the speaker and the audience. Listeners can offer alternative explanations and viewpoints, or simply ask questions that help the speaker clarify the information. And the speaker and listeners can converse before and after the presentation.

Oral presentations are therefore a popular means of technical communication. You can expect to give oral presentations to four different types of audiences:

- *Clients and customers.* An oral presentation can be a valuable sales technique. Whether you are trying to interest clients in a silicon chip or a bulldozer, you will present its features and its advantages over the competition. Then, after the sale, you will likely provide detailed oral operating instructions and maintenance procedures to the users.

- *Colleagues in your organization.* If you are the resident expert on a mechanism, procedure, or technical subject, you will instruct your fellow workers, both technical and nontechnical. After you return from an important conference or an out-of-town project, your supervisors will want a briefing—an oral report. If you have an idea for improving operations at your organization, you probably will write a brief informal proposal and then present the idea orally to a small group of managers. Your presentation will help them determine whether it is prudent to devote resources to studying the idea.

- *Fellow professionals at technical conferences.* You might speak on your own research project or on a team project carried out at your organization. Or you may be invited to speak to professionals in other fields. If you are an economist, for example, you might be invited to speak to real-estate agents about interest rates.

- *The general public.* As you assume greater prominence in your field, you will receive more invitations to speak to civic organizations and governmental bodies. Your organization will encourage you to give these presentations, for they reflect positively on the organization.

You might not have had much experience in public speaking, and perhaps your few attempts have been difficult. Natural speakers who can talk off the cuff are rare; for most of us, an oral presentation requires deliberate and careful preparation.

HOSPITAL ADMINISTRATOR VINCENT KURAITIS ON ORAL COMMUNICATION

on the importance of oral communications skills to his job

They're really important. I have to build consensus, I need to persuade people, I need to get my ideas across, and much of that is done orally.

on where he learned to give presentations

It's an acquired skill over time. I debated in college and high school, so I had some very specific skill building from back then. And I took a speech course in college. In my first job there were ongoing training sessions, and public speaking was part of that, and then I guess on-the-job training over time is another way you build your skills.

on the need to practice giving presentations

Practice is extremely important. Through practice almost everybody can become more proficient at public speaking. I think Toastmasters is one of the most economical, practical ways to practice, and I've been involved in that and recommended it to other people. You get together and practice a wide variety of different types of speeches with other people who will give you some candid feedback.

on how to prepare to give an important presentation

It's really important to reflect on who the audience is and to tailor each presentation with the audience in mind. I address a wide diversity of audiences. On the one hand, I make presentations to hospital employees, and some of them look at their work as a job, not a career. They may have very little knowledge of what's going on in health care or in the industry, and I've got to be able to speak to that level. At the other extreme, I make presentations to our hospital board of directors and the people in our corporate office, and they are very knowledgeable about business and social issues. I think it's important to plan the presentation

according to what your audience knows, to the way they like information presented; some people like numbers, some people like pictures, some people like words.

on the need to consider your purpose in planning a presentation

You also need to think about how you can most effectively achieve your goal. Sometimes the goal is to inform, sometimes it's to persuade, sometimes it's a combination. You have to be conscious of what you're trying to achieve when you're talking to them.

on planning resources for making graphics

We use transparencies more often than slides. It varies, depending on the audience. If I'm going to make something for a presentation and use it once, I'm probably not going to bother to go out and spend $50 per slide for a presentation, but there are some presentations that we deliver to different groups of our employees over a period of time. So it makes a little more sense to invest some time and energy so that the presentations look a little more professional.

on dealing with nervousness

I look at nervousness as an opportunity to try and channel that adrenaline, to use it as something that will help me do my best rather than detract from my ability to be persuasive. So I can get a little bit more alert, a little bit more in touch with my audience, and I look at the adrenaline rush that comes with speaking in front of an audience as something to look forward to because it makes me do my best.

Vincent Kuraitis, J.D./M.B.A., is vice president of corporate development and specialty services at Saint Alphonsus Regional Medical Center in Boise, Idaho.

Preparing an Oral Presentation

Professional speakers, like professionals in every other field, make presentations look easy. But when you see an excellent 20-minute presentation, you are seeing only the last 20 minutes of a process that took many hours. How much time should you devote to preparing an oral presentation? Experts recommend devoting 20–60 minutes for each minute of the presentation (Smith, 1991, p. 6). At an average of 40 minutes, you would need over 13 hours to prepare a 20-minute presentation. Obviously, there are many variables, including your knowledge of the subject, your experience in creating presentation graphics, and your experience in giving previous presentations on the same subject. But the point is clear: good presentations don't just happen.

Preparing an oral presentation requires the five steps shown in Figure 20.1.

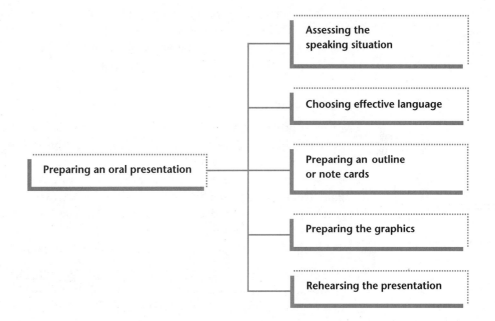

FIGURE 20.1

Preparing an Oral
Presentation

Assessing the Speaking Situation

The first step in assessing the speaking situation is to analyze your audience and purpose. Then you determine how much information you can deliver in the allotted time.

Analyzing Your Audience and Purpose

Consider audience and purpose, just as you would in writing a document.

- *Audience.* Who are the people in the audience? How much do they know about your subject? You must answer these questions to determine the level of technical vocabulary and concepts appropriate to the audience. Speaking over an audience's head puzzles them; oversimplifying makes you appear condescending and insulting.

  Ask yourself the same kinds of questions you would ask about readers: Why is the audience there? What do they want to accomplish in listening to your presentation? Are they likely to be hostile, enthusiastic, or neutral? A presentation on the virtues of free trade, for instance, will be received one way by conservative economists and another way by American steelworkers.

- *Purpose.* Are you attempting to inform your audience, or to both inform and persuade them? If you are explaining how windmills can be used to generate power, you have one type of argument. If you are explaining why *your* windmills are an economical means of generating power, you have another type.

Your analysis of your audience and purpose will affect the strategy—the content and the form—of your presentation. You might have to emphasize some aspects of your subject or ignore some altogether. You might have to arrange topics to accommodate the audience's needs.

Budgeting Your Time

At most professional conferences, the organizers clearly state a maximum time, such as 20 minutes, for each speaker. If the question-and-answer period is part of your allotted time, plan accordingly. Even at an informal presentation, you probably will have to work within an unstated time limit that you must determine from the speaking situation. Claiming more than your share of an audience's time is rude and egotistical, and eventually your listeners will start to resent you or simply stop paying attention.

For a 20-minute presentation, the time allotment in Table 20.1 is typical.

TABLE 20.1	*Time Allotment for a 20-Minute Presentation*
Task	**Time (minutes)**
• Introduction	2
• Body	
– First Major Point	4
– Second Major Point	4
– Third Major Point	4
• Conclusion	2
• Questions	4

In delivering a scripted presentation, most speakers need more than a minute to deliver a double-spaced page of text effectively.

Choosing Effective Language

Delivering an oral presentation is more challenging than writing a document, for two reasons:

- Listeners can't flip backward in the presentation to listen again to something they didn't understand.

- Because you are speaking live, you must maintain your listeners' attention, even if they are hungry or tired or the room is too hot. (Readers can be distracted, too, but they have more freedom to deal with the distractions.)

Using language effectively helps you meet these two challenges.

Using Language to Signal Advance Organizers, Summaries, and Transitions

Even if you use graphics effectively, listeners cannot "see" the organization of a presentation as well as readers can. For this reason, use language to alert your listeners to advance organizers, summaries, and transitions.

- *Advance organizers.* An *advance organizer* is a statement that tells the listener what you are about to say. You will use an advance organizer in the introduction, of course, when you tell your audience your purpose, scope, main points, and organization. In addition, you will use advance organizers when you introduce main ideas in the body of the presentation.

  Advance organizers have to be explicit:

  In the next 20 minutes, I'd like to discuss the implications of the new RCRA regulations on the long-range waste-management strategy for Radnor Township. I want to make three major points. First, that. . . . Second, that. . . . And third, that. . . . After the presentation, I'll be happy to field your questions.

  Notice that the speaker numbers his points. He can use this numbering system throughout the presentation to help his listeners follow him.

- *Summaries.* The major summary is in the conclusion, but you might also summarize at strategic points in the body of the presentation. For instance, after a three- to four-minute discussion of a major point, you might summarize it in one sentence before going on to the next major point. Here is a sample summary from a conclusion:

  Let me conclude by summarizing my three main points about the implications of the new RCRA regulations on the long-range waste-management strategy for Radnor Township. The first point: . . . . The second point: . . . . The third point: . . . . I hope this presentation will give you some ideas as you think about the RCRA. If you have any questions, I'd be happy to try to answer them at this time.

- *Transitions.* As you move from one point to the next, signal the transition clearly. Summarize the old point, and then announce that you are moving to the next point.

  It is clear, then, that the federal government has issued regulations without indicating how it wants county governments to comply with them. I'd like to turn now to my second main point: . . .

Using Memorable Language

If people doze off while reading a document you have written, you probably won't know it, at least until they complain to you later or you realize that the document failed to accomplish your purpose. But if they doze off while you are speaking to them, you'll know it right away. Effective presentations require memorable language.

Three techniques can help make your language more memorable.

- *Involve the audience.* People are more interested in their own concerns than in yours. Talk to the audience about their problems and their solutions. In the introduction, establish a link between your topic and the audience's interests. For instance, in the presentation to the Radnor Township Council about waste management, you might begin like this:

  Picture yourself on the Radnor Township Council two years from now. After exhaustive hearings, proposals, and feasibility studies, you still don't have a waste-management plan that meets federal regulations. What you do have is a mounting debt: the township is being assessed $1,000 per day until you implement an acceptable plan.

- *Refer to people, not to abstractions.* People remember specifics; they forget abstractions. When you want to make a point memorable, describe it in human terms.

  What could you do with that $365,000 every year? You could buy almost 200 personal computers; that's a computer for almost every classroom in every elementary school in Radnor Township. You could expand your school-lunch program to feed every needy child in the township. You could extend your after-school programs to cover an additional 3,000 students.

- *Use interesting facts, figures, and quotations.* Do your research and find interesting information about your subject. For instance, you might find a brief quotation from an authoritative figure in the field or a famous person not generally associated with the field (for example, Abraham Lincoln on waste management).

A note about humor: There are about 500 people in the United States who make a good living being funny. In comparison, there are millions of engineers, accountants, scientists, and technical communicators. Don't plan to tell a joke. If something happens in the context of the presentation that provides an opening for a witty remark, and you are good at it, fine. But don't *prepare* to be funny.

Preparing an Outline or Note Cards

After assessing your audience, purpose, and strategy, prepare an outline or a set of note cards. Keep in mind that an oral presentation should, in general, be simpler than a written version of the same material. For example, keep statistics and equations to a minimum.

Structure the presentation logically. The argument patterns presented in Chapter 8 are as useful for an oral presentation as for a document.

Write an outline or note cards just as you would for a document. Your own command of the facts—and your ability to remember them while you are under stress—will determine the degree of specificity necessary. Many people prefer a sentence outline (see Chapter 8) because of its specificity; others feel more comfortable with a topic outline, especially when they are using note cards. Figure 20.2 shows a combined sentence/topic outline, and Figure 20.3 on page 566 shows a topic outline. (Of course, the outline could be written on note cards.) The speaker is the specialist in waste-treatment facilities. The audience is a group of civil engineers interested in gaining a general understanding of new

---

Purpose: to describe, to a group of civil engineers, a new method of treatment and disposal of industrial waste.

1. Introduction
    1.1 The recent Resource Conservation Recovery Act places stringent restrictions on plant engineers.
    1.2 With neutralization, precipitation, and filtration no longer available, plant engineers will have to turn to more sophisticated treatment and disposal techniques.

2. The Principle Behind the New Techniques
    2.1 Waste has to be converted into a cementitious load-supporting material with a low permeability coefficient.
    2.2 Conversion Dynamics, Inc., has devised a new technique to accomplish this.
    2.3 The technique is to combine pozzolan stabilization technology with the traditional treatment and disposal techniques.

3. The Applications of the New Technique
    3.1 For new low-volume generators, there are two options.
        3.1.1 Discussion of the San Diego plant.
        3.2.2 Discussion of the Boston plant.
    3.2 For existing low-volume generators, Conversion Dynamics offers a range of portable disposal facilities.
        3.2.1 Discussion of the Montreal plant.
        3.2.2 Discussion of the Albany plant.
    3.3 For new high-volume generators, Conversion Dynamics designs, constructs, and operates complete waste-disposal management facilities.
        3.3.1 The Chicago plant now processes up to 1.5 million tons per year.
        3.3.2 The Atlanta plant now processes up to 1.75 million tons per year.
    3.4 For existing high-volume generators, Conversion Dynamics offers add-on facilities.
        3.4.1 The Roanoke plant already complies with the new RCRA requirements.
        3.4.2 The Houston plant will be in compliance within six months.

4. Conclusion
    The Resource Conservation Recovery Act will necessitate substantial capital expenditures over the next decade.

FIGURE 20.2

Sentence/Topic Outline for an Oral Presentation

Purpose: to describe, to a group of civil engineers, a new method of treatment and disposal of industrial waste.

Introduction
– Implications of the RCRA

Principle Behind New Technique
– reduce permeability of waste
– use pozzolan stabilization technology

Applications of New Technique
– for new low-volume generators (San Diego, Boston)
– for existing low-volume generators (Montreal, Albany)
– for new high-volume generators (Chicago, Atlanta)
– for existing high-volume generators (Roanoke, Houston)

Conclusion

FIGURE 20.3

Topic Outline for an Oral
Presentation

developments in industrial-waste disposal. The speaker's purpose is to provide this information and to suggest that his company is a leader in the field.

Notice the differences in content and form between these two versions of the same outline. Whereas some speakers prefer to have full sentences before them, others find that full sentences interfere with spontaneity. (For presentations read from manuscript, of course, the outline is only preliminary to writing the text.)

Preparing Presentation Graphics

Graphics fulfill the same purpose in an oral presentation that they do in a written one: they clarify or highlight important ideas or facts. Statistical data, in particular, lend themselves to graphical presentation, as do descriptions of equipment or processes. Research reported by Smith (1991) indicates that presentations that include transparencies are judged more professional, persuasive, and credible than those without them, and that audiences remember information better if it is accompanied by graphics. Smith (1991, p. 58) offers these figures:

	Retention after	
	3 hours	3 days
Without graphics	70%	10%
With graphics	85%	65%

If you have not had a lot of experience creating graphics for presentations, review Chapter 12; everything said there about graphics applies to graphics for presentations.

Characteristics of an Effective Graphic

Effective graphics have five characteristics:

- *Visibility.* The most common problem with presentation graphics is that they are too small. Many speakers mistakenly try to transfer information from an 8.5 × 11 inch page to a slide or transparency. As a general rule, text has to be presented in 24-point type or larger to be visible on the screen.

  To save space, reduce traditional sentences to brief phrases.

  TEXT IN A DOCUMENT:
  The current system has three problems:
  - It is expensive to maintain.
  - It requires nonstandard components.
  - It is not compliant with the new MILSPEC.

  SAME TEXT ON A SCREEN:
  Three problems:
  - expensive to maintain
  - nonstandard components
  - noncompliant with new MILSPEC

- *Legibility.* Use clear, legible lines for drawings and diagrams: black on white works best for transparencies. Use legible typefaces for text; a boldfaced sans serif typeface such as Helvetica is effective because it reproduces clearly on a screen. Avoid shadowed and outlined letters. See Chapter 13 for more information on type. And see Chapter 12 for information on color in presentation graphics and for examples of a black-and-white transparency and a color slide.

- *Simplicity.* Both text and drawings must be simple. Each graphic should present only a single idea. Remember that your listeners have not seen the graphic before and will not be able to linger over it.

- *Clarity.* Of course, the information has to make sense to your audience. In cutting verbiage and simplifying concepts and visual representations, make sure the point of the graphic remains clear.

- *Correctness.* Rare is the presentation that does not contain at least one graphic with a typo or some other error. Everyone makes mistakes, but mistakes are particularly embarrassing when they are 10 inches tall on a screen.

See Chapter 12 on graphics and Chapter 13 on design. These two chapters provide a detailed look at the principles of using visual elements to enhance communication. One point made in Chapter 12 is worth repeating: be careful when you use graphics templates in your software. Some of the templates violate basic principles of design. And be careful when you use clip art. If it helps you communicate, go ahead. But don't use clip art just to fill in a blank space on a transparency or slide.

Graphics and the Speaking Situation

To plan your graphics, analyze four aspects of the speaking situation:

- *Length of the presentation.* How many graphics should you have? A rule of thumb is to have a different graphic for every 30 seconds of the presentation. It is far better to have a series of simple graphics than to have one complicated one that stays on the screen for 10 minutes.

- *Audience aptitude and experience.* What kinds of graphics can your audience understand easily? You don't want to present scatter graphs if your listeners have had no experience interpreting them.

- *Size and layout of the room.* The graphics for a presentation in a small meeting room differ from those for a presentation in a 500-seat auditorium. Think first about the size of the images, then about the layout of the room. For instance, will a window create glare that you will have to consider as you plan the type or placement of the graphics?

- *Equipment.* Find out ahead of time what kind of equipment is available in the presentation room. Inquire about backups in case of equipment failure. If possible, bring your own equipment with you. That way, you know it works and you know how to use it. Some speakers bring graphics in two media just in case; that is, they have slides but they also have transparencies of the same graphics.

When experienced speakers make presentations away from the office, they often bring a set of supplies with them just in case. The following list, based on Smith (1991, pp. 148–149), specifies some of these items:

electrical plug adapter
extension cord
spare bulbs for overhead projector and slide projector
chalk and eraser
transparency pens
blank transparency sheets
transparent tape
scissors

Using Graphics to Signal the Organization of the Presentation

Used effectively, graphics can help you communicate the organization of the presentation. You can use the transition from one graphic to the next to indicate the transition from one point to the next.

As you create your graphics, try to reflect the basic organization of the presentation. Start with a title graphic that lists the title of the presentation, as well as your name, affiliation, and job title. Sometimes speakers also indicate the occasion for the presentation and the date. Figure 20.4 is a typical title graphic.

Next, display a graphic showing the first-level headings from your outline. For the presentation outlined in Figure 20.2, the first-level headings would appear as in Figure 20.5.

To move from one topic to the next, display a graphic for the new topic, complete with its subheadings. For instance, Figure 20.6 on page 570 shows the graphic that introduces the topic "Principles Behind the New Technique."

RCRA: Implications for the Future

Charles E. Harim

Department of Civil Engineering
The Johns Hopkins University

Radnor Township Council
Resource Committee

October 15, 1995

FIGURE 20.4
Title Graphic

Overview

1. **Introduction: Implications of the RCRA**

2. **Principles Behind the New Technique**

3. **Applications of the New Technique**

FIGURE 20.5
Graphic Showing First-Level Headings

4. **Conclusion**

2. Principles Behind the New Technique

2.1 Reduce permeability of waste

2.2 Use pozzolan stabilization technology

FIGURE 20.6

Graphic Showing
Subheadings

Notice that the outline numbering system helps listeners understand the organization of the talk. This system of layering the graphics can be extended down to a third level if needed.

For your last graphic, consider a summary of your main points or a brief set of questions. The questions give you an occasion to restate your main points as you prompt the audience to synthesize the information from the presentation.

Different Media Used for Graphics

Table 20.2 describes the basic media for graphics.

TABLE 20.2 *Basic Media for Oral Presentations*		
Medium	**Advantages**	**Disadvantages**
COMPUTER PRESENTATIONS: IMAGES APPEAR ON A COMPUTER SCREEN OR ON A PROJECTOR ATTACHED TO THE COMPUTER, OR CAN BE TRANSFERRED TO SLIDES, VIDEOTAPE, OR FILM.	• Very professional appearance. • You can produce any combination of static or dynamic images, from simple graphs to sophisticated, three-dimensional images. • You can control the rate at which the images change.	• The equipment is not available everywhere. • Preparing the graphics is very time consuming. • Presentations prepared on one system might not run on an incompatible system.
SLIDE PROJECTOR: PROJECTS PREVIOUSLY PREPARED SLIDES ONTO A SCREEN.	• Very professional appearance. • Versatile—can handle photographs or artwork, color or black-and-white. • With a second projector, you can eliminate the pause between slides. • During the presentation, you can easily advance and reverse the slides. • Graphics software lets you create small paper copies of your slides to distribute to the audience after the presentation.	• Slides can be expensive to produce. • Room has to be kept relatively dark during the slide presentation.

TABLE 20.2   *(continued)*

Medium	Advantages	Disadvantages
OVERHEAD PROJECTOR: PROJECTS TRANSPARENCIES ONTO A SCREEN.	• Transparencies are inexpensive and easy to create. • You can draw transparencies "live." • You can create overlays by placing one transparency over another. • Lights can remain on during the presentation. • You can face the audience. • Graphics software lets you create small paper copies of your transparencies to distribute to the audience after the presentation.	• Not as professional looking as slides. • Each transparency must be loaded separately by hand.
OPAQUE PROJECTOR: PROJECTS IMAGES ON PAPER ONTO A SCREEN.	• You can project single sheets or pages in a bound volume. • Requires no expense or advance preparation.	• Room must be kept dark during the presentation. • Cannot magnify sufficiently for a large auditorium. • Each page must be loaded separately by hand. • The projector is noisy.
POSTER: A GRAPHIC DRAWN ON OAK TAG OR OTHER PAPER PRODUCT.	• Inexpensive. • Posters can be drawn or modified "live."	• Too small for large rooms.
FLIP CHART: A SERIES OF POSTERS, BOUND TOGETHER AT THE TOP LIKE A LOOSE-LEAF BINDER; GENERALLY PLACED ON AN EASEL.	• Relatively inexpensive. • You can easily flip back or forward. • Posters can be drawn or modified "live."	• Too small for large rooms.
FELT BOARD: A HARD, FLAT SURFACE COVERED WITH FELT, ONTO WHICH PAPER CAN BE ATTACHED USING DOUBLED-OVER ADHESIVE TAPE.	• Relatively inexpensive. • Effective if you wish to rearrange the items on the board during the presentation. • Versatile—can handle paper, photographs, cutouts.	• Informal appearance.
CHALKBOARD OR OTHER HARD WRITING SURFACE.	• Almost universally available. • You have complete control—can add, delete, or modify the graphic easily.	• Complicated or extensive graphics are difficult to create. • Ineffective in large rooms. • Very informal appearance.
OBJECTS: MODELS OR SAMPLES OF MATERIAL THAT CAN BE HELD UP OR PASSED AROUND THROUGH THE AUDIENCE.	• Interesting for the audience. • They provide a good look at the object.	• Audience members might not listen while they are looking at the object. • It can take a long while to pass an object around a large room. • The object might not survive intact.
HANDOUTS: PHOTOCOPIES OF WRITTEN MATERIAL GIVEN TO EACH AUDIENCE MEMBER.	• Much material can be fit on the page. • Audience members can write on their copies and keep them.	• Audience members might read the handout rather than listen to the speaker.

Rehearsing the Presentation   Even the most gifted speakers need to rehearse. It is a good idea to set aside enough time to rehearse your speech thoroughly.

Rehearsing an Extemporaneous Presentation

Rehearse your extemporaneous presentation at least three times.

- *First rehearsal.* Don't worry about posture or voice projection. Just try to compose your presentation out loud with your outline before you. Your goal is to see if the speech makes sense—if you can explain all the points you have listed and can forge effective transitions from point to point. If you have any trouble, stop and try to figure out the problem. If you need more information, get it. If you need a better transition, create one.

  You might have to revise your outline or notes. This is very common and no cause for alarm. Pick up where you left off and continue through the presentation, stopping again where necessary to revise the outline. When you have finished your first rehearsal, put the outline away and do something else.

- *Second rehearsal.* Once you are rested, try the presentation again. This time, it should flow more easily. Make any necessary revisions to the outline or notes. When you have complete control over the organization and flow, check to see if you are within the time limits.

- *Third rehearsal.* After a satisfactory second rehearsal, try the presentation again, under more realistic circumstances—if possible, in front of people. The listeners might offer constructive advice about parts they don't understand or about your speaking style. If people aren't available, tape-record the presentation and then evaluate your own delivery. If you can visit the site of the presentation to get the feel of the room and rehearse there, you will find giving the actual speech a little easier.

Rehearse again until you are satisfied with your presentation. Then stop. Don't attempt to memorize it; if you do, you will surely panic the first time you forget a phrase. During the presentation, you must be thinking of your subject, not about the words you used during the rehearsals.

Rehearsing a Scripted Presentation

Rehearsing a scripted presentation is a combination of revising the text and rehearsing it. As you revise, read the script out loud. You want to hear how it sounds. Once you think the presentation says what you want to say, try reading it aloud into a tape recorder. Revise it until you are satisfied, and then rehearse in front of real people. Again, do not memorize. There is no need to: you will have your script there on the podium.

Giving the Oral Presentation

After all the preparation, the time to give the presentation finally arrives. In giving the presentation, you will concentrate on what you have to say. In addition, however, you will have three concerns, as shown in Figure 20.7.

Dealing with Nerves

Most professional actors freely admit to being nervous before a performance, so it is no wonder that most technical speakers are nervous. You might well fear that you will forget everything or that no one will be able to hear you. These fears are common. But keep in mind three facts about nervousness:

- *You are much more aware of your nervousness than the audience is.* They are farther away from your trembling hands.
- *Nervousness gives you energy and enthusiasm.* Without energy and enthusiasm, your presentation will almost surely fail. If you *seem* bored and listless, your audience will *be* bored and listless.
- *After a few minutes, your nervousness will pass.* You will be able to relax and concentrate on the subject.

This advice, however, is unlikely to make you feel much better if you are distracted by nerves as you wait to give your presentation. Experienced speakers suggest different points to keep in mind when you find yourself getting nervous before a presentation:

- *Realize that you are prepared.* If you have done your homework and rehearsed the presentation, you'll be fine. You are in control of the presentation.
- *Realize that the audience is there to hear you, not to judge you.* Your listeners want to hear what you have to say. They are much less interested in your nervousness than you are.

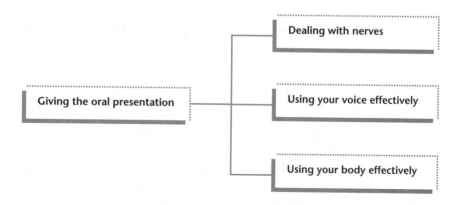

FIGURE 20.7
Giving the Oral
Presentation

- *Realize that your audience is made up of individual people who happen to be sitting in the same room.* If you tell yourself that the audience members are individuals like yourself who also get nervous before making presentations, you'll feel better.

And experienced speakers suggest several actions you can take to lessen the nerves:

- *Walk around.* A brisk walk of a minute or two can calm you down by dissipating some of your nervous energy.
- *Go off by yourself for a few minutes.* Some people find that getting away for a moment helps them compose their thoughts and realize that they can handle the nervousness.
- *Talk with someone for a few minutes.* For some speakers, distraction works best. Talk with someone who has come to the presentation a few minutes early.
- *Take several deep breaths, exhaling slowly.* Doing so will help you control your nerves.

When it is time to begin, don't jump up to the lectern and start speaking quickly. Walk up slowly and arrange your text, outline, or note cards before you. If water is available, take a sip. Look out at the audience for a few seconds before you begin. It is polite to begin formal presentations with "Good morning" (or "Good afternoon," or "Good evening") and to refer to the officers and dignitaries present. If you have not been introduced by the introductory speaker, identify yourself. In less formal contexts, just begin your presentation.

So that the audience will listen to you and have confidence in what you say, try to project the same attitude that you would in a job interview: restrained self-confidence. Show interest in your topic and knowledge about your subject. You can convey this sense of control through your voice and your body.

Using Your Voice Effectively

Inexperienced speakers encounter problems with five aspects of vocalizing.

- *Volume.* Acoustics vary greatly from room to room, so you won't know how well your voice will carry in a room until you have heard someone speaking there. In some well-constructed auditoriums, speakers can use a conversational volume. Other rooms require greater voice projection. Some rooms have an annoying echo.

  These special circumstances aside, more people speak too softly than too loudly. After your first few sentences, ask if the people in the back of the room can hear you. When people speak into microphones, they tend to bend down toward the microphone and end up speaking too loudly. Glance at your audience to see if you are having volume problems. The body language of the audience members will be clear.

- *Speed.* Nervousness makes people speak more quickly. Even if you think you are speaking at the right rate, you might be going a little too fast for some of your audience. Remember, you know what you will say; your listeners, however, are trying to understand new information. For particularly difficult points, slow down for emphasis. After finishing a major point, pause before beginning the next point.
- *Pitch.* In an effort to control their voices, many speakers end up flattening their pitch. The resulting monotone is boring—and, for some listeners, actually distracting. Try to let the pitch of your voice go up or down as it would in a normal conversation. In fact, experienced speakers often exaggerate pitch variations slightly.
- *Articulation.* The nervousness that goes along with an oral presentation tends to accentuate sloppy pronunciation. If you want to say *environment,* don't say *envirament.* Don't drop final gs. Say *trying,* not *tryin'.* A related pronunciation problem concerns technical words and phrases, especially the important ones. When a speaker uses a phrase over and over, it tends to get clipped and becomes difficult to understand. Unless you articulate carefully, *Scanlon Plan* will end up as *Scanluhplah.*
- *Nonfluencies.* Avoid such meaningless fillers as *you know, like, okay, right, uh,* and *um.* These phrases do not disguise the fact that you aren't saying anything; they call attention to it. A thoughtful pause is better than an annoying verbal tic.

Using Your Body Effectively

Besides listening to what you say, the audience will be looking at you. Effective speakers know how to use their bodies to help the listeners follow the presentation. Keep in mind four guidelines about physical movement:

- *Maintain eye contact.* It is only polite to look at your audience. This is called *eye contact.* For small groups, look at each listener randomly; for larger groups, be sure to look at each segment of the audience frequently during your speech. Do not stare at your notes, at the floor, at the ceiling, or out the window. Eye contact is more than politeness; it is a way for you to see how the audience is receiving the presentation. You will see, for instance, if listeners in the back are having trouble hearing you.
- *Use natural gestures.* When people talk, they tend to gesture with their hands. Most of the time, these gestures make the presentation look natural and improve listeners' comprehension. You can supplement your natural gestures by using your arms and hands to signal pauses and to emphasize important points. When referring to graphics, walk to the screen and point to direct the audience's attention.

  Avoid mannerisms—those physical gestures that serve no useful purpose. Don't play with your jewelry or the coins in your pocket. Don't tug at your beard or fix your hair. These nervous gestures can quickly distract an audience from what you are saying. Don't pace back and forth. Like verbal mannerisms, physical mannerisms are often unconscious. Constructive criticism from friends can help you pinpoint them.

- *Don't block the audience's view of the screen.* Don't stand at the overhead projector if doing so blocks some people's view of the screen. After you put on a transparency, step back so that you are standing to the side of the screen. Use a pointer to indicate key words or images on the screen.

- *Control the audience's attention.* People will listen to and look at anything that is interesting. Don't lose the audience's attention. If you hand out photocopies at the start of the presentation, some people will start to read them and stop listening to you. If you leave an image on the screen after you are done talking about it, some people will keep looking at it instead of listening to you. When you want the audience to look at you and listen to you, remove the graphics or turn off the projector.

Answering Questions after the Presentation

In some presentations, particularly informal ones, audience members ask questions throughout the presentation. This method of questioning helps you clarify points as you go along, but it can make it difficult for you to stay on track.

More often, an oral presentation is followed by a question-and-answer period. When you invite questions, don't abruptly say, "Any questions?" This phrasing suggests that you don't really want any questions. Instead, say something like this: "If you have any questions, I'll be happy to try to answer them now." If invited politely, people will be much more likely to ask; in that way, you will more likely communicate your information effectively.

When you respond to questions, five situations occur frequently:

- *You're not sure everyone heard the question.* If there is no moderator to ask if the question was audible, ask if people have heard the question. If they haven't, repeat or paraphrase it yourself, perhaps as an introduction to your response: "Your question about the efficiency of these three techniques. . . ." Some speakers always repeat the question; that way, they are sure that everyone hears it, and they get an extra moment to think about their answer.

- *You don't understand the question.* Ask for a clarification. After responding, ask if you have answered the question adequately.

- *You don't know the answer to the question.* Tell the truth. Only novices believe that they ought to know all the answers. If you have some ideas about how to find out the answer—by checking a certain reference source, for example—share them. If the question is obviously important to the person who asked it, you might offer to meet with him or her after the question-and-answer period to discuss ways for you to give a more complete response, such as through the mail.

- *You get a question that you have already answered in the presentation.* Restate the answer politely. Begin your answer with a phrase such as the following: "I'm sorry I didn't make that point clear in my talk. I wanted to explain how. . . ." Never insult the person by pointing out that you already covered that: "I already answered that question in my talk, but let me repeat it for you. . . ."

- *A belligerent member of the audience rejects your response and insists on restating his or her original point.* Offer politely to discuss the matter further after the session. This strategy will prevent the person from boring or annoying the rest of the audience.

If it is appropriate to stay after the session to talk individually with members of the audience, offer to do so. Remember to thank them for their courtesy in listening to you.

Sample Evaluation Form

FIGURE 20.8

Sample Evaluation Form

Figure 20.8 is a list of questions that can help you focus your thoughts as you watch and listen to a presentation.

Evaluation Form

To the left of each of the following statements, write a number from 1 to 5, with 5 signifying strong agreement and 1 signifying strong disagreement.

( ) 1. In the introduction, the speaker made an attempt to relate the topic to my concerns.

( ) 2. In the introduction, the speaker explained the main points he or she wanted to make in the presentation.

( ) 3. In the introduction, the speaker explained the organization of the presentation.

( ) 4. The speaker used interesting, clear language to get the points across.

( ) 5. The speaker made the transitions from one point to the next clearly.

( ) 6. The speaker used clear and distinct enunciation.

( ) 7. The speaker exhibited no distracting vocal mannerisms.

( ) 8. The speaker exhibited no distracting physical mannerisms.

( ) 9. The speaker used graphics effectively to reinforce and explain the main points.

( ) 10. The speaker summarized the main points effectively in the conclusion.

( ) 11. The graphics helped me understand the organization of the presentation.

( ) 12. Throughout the presentation, the speaker paid attention to the audience.

( ) 13. The speaker seemed to be enthusiastic throughout the presentation.

( ) 14. The speaker used the allotted time effectively.

( ) 15. The speaker invited questions politely.

( ) 16. The speaker answered questions effectively.

Answer the following two questions on the other side of this sheet.

17. What did you particularly like about this presentation?

18. What would you have done differently if you had been the speaker?

This checklist covers the steps involved in preparing to give an oral presentation.

1. Have you assessed the speaking situation—the audience and purpose of the presentation?

2. Have you determined how much information you can communicate in your allotted time?

3. Have you chosen language to help you signal advance organizers, summaries, and transitions?

4. Have you chosen language that is vivid and memorable?

5. Have you outlined your information on note cards?

6. Have you prepared graphics that are
   ☐ visible?
   ☐ legible?
   ☐ simple?
   ☐ clear?
   ☐ correct?

7. In planning your graphics, have you considered your audience's aptitude and experience, the size and layout of the room, and the equipment?

8. Have you planned your graphics so that they help the audience understand the organization of your presentation?

9. Have you chosen appropriate media for your graphics?

10. Have you made sure that the presentation room will have the necessary equipment for the graphics?

11. Do you have appropriate supplies with you in case you need them?

12. Have you rehearsed your presentation several times, with a tape recorder or a live audience?

EXERCISES

Exercise 4 calls for you to write a memo. See Chapter 16 for a discussion of memos.

1. Prepare a five-minute presentation, including graphics, on one of the topics listed here. For each presentation, your audience consists of the other students in your class, and your purpose is to introduce them to an aspect of your academic field.
   a. Define a key term or concept in your field.
   b. Describe how a particular piece of equipment is used in your field.
   c. Describe how to carry out a procedure common in your field.

   The instructor and the other students will evaluate the presentation by filling out the form in Figure 20.8. If your instructor wishes, this assignment can be done collaboratively.

2. Prepare a five-minute presentation based on your proposal for a completion-report topic. Your audience consists of the other students in your class, and your purpose is to introduce them to your topic. The instructor and the other students will evaluate the presentation by filling out the form in Figure 20.8. If your instructor wishes, this assignment can be done collaboratively.

3. After the report is written, prepare a five-minute presentation based on your findings and conclusions. Your audience consists of the other students in your class, and your purpose is to introduce them to your topic. The instructor and the other students will evaluate the presentation by filling out the form in Figure 20.8. If your instructor wishes, this assignment can be done collaboratively.

4. Write a 1,000-word memo to your instructor in which you describe and evaluate a recent oral presentation of a guest speaker at your college or a politician on television.

REFERENCE

Smith, T. C. (1991). *Making successful presentations: A self-teaching guide.* New York: Wiley.

Proposals

> The Format of the Proposal
> Sample Internal Proposal
> Writer's Checklist
> Exercises

Most projects undertaken by organizations, as well as most major changes made within organizations, begin with a proposal. A proposal is an offer to carry out research or to provide a product or service.

This chapter begins by describing the logistics of proposals, concentrating on external and internal proposals, as well as solicited and unsolicited proposals. Then it explains the "deliverables" of proposals: what you deliver at the end of the project. The chapter then discusses the need for persuasion in proposal writing. Finally, it discusses the components of a proposal and presents an example of an internal proposal.

The Logistics of Proposals

Proposals are classified as either external or internal and as solicited or unsolicited. Figure 21.1 shows the relationship among these four terms.

External and Internal Proposals   Proposals are either *external* (if submitted to another organization) or *internal* (if submitted to the writer's own organization).

External Proposals

An *external proposal* is a proposal submitted to an outside organization. No organization produces all the products or provides all the services it needs.

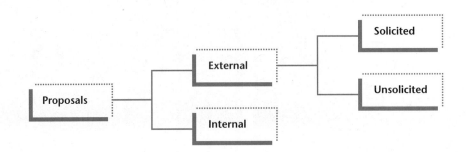

FIGURE 21.1

The Logistics of Proposals

Paper clips and company cars have to be purchased. Offices must be cleaned and maintained. Sometimes projects that require unusual expertise, such as sophisticated market analyses, have to be carried out. Any number of companies would love to provide the paper clips or the cars, and a few dozen consulting organizations would happily conduct the studies. For this reason, it is almost always a buyer's market. To get the best deal, most organizations require that their potential suppliers compete for the business by submitting a proposal, a document in which the supplier attempts to make the case that it deserves the contract.

Internal Proposals

One day, while you're working on a project in the laboratory, you realize that if you had a new centrifuge you could do your job better and faster. The increased productivity would save your company the cost of the equipment in a few months. You call your supervisor, who tells you to send a memo describing what you want, why you want it, what you're going to do with it, and what it costs; if your request seems reasonable, you'll likely get the money.

Your memo is an *internal proposal*—a persuasive argument, submitted within an organization, for carrying out an activity that will benefit the organization, generally by increasing productivity or quality or by reducing costs. An internal proposal can recommend that the organization conduct research, purchase a product, or change some aspect of its policies or procedures.

The scope of the proposal determines its format. A simple request might be conveyed orally, either in person or on the phone. A more ambitious request might require a brief memo. The most ambitious requests are generally conveyed in formal proposals. Organizations often use dollar figures to determine the format of the proposal. For instance, employees use a brief form to communicate proposals that would cost less than $1,000, whereas they use a report to communicate proposals that would cost more than $1,000.

Solicited and Unsolicited Proposals External proposals are either solicited or unsolicited. A *solicited proposal* originates with a request from a potential customer. An *unsolicited proposal* originates with the potential supplier.

Solicited Proposals

When an organization wants to purchase a product or service, it publishes one of two basic kinds of statements:

- An IFB—*information for bid*—is used for standard products. When an agency of the federal government needs office equipment, for instance, it lets suppliers know that it wants to purchase, say, 100 secretary's chairs of a particular type. The supplier that offers the lowest bid wins the contract.

- An RFP—*request for proposal*—is issued for customized products or services. Police cars are likely to differ from the standard consumer model: they might have different engines, cooling systems, suspensions, and upholstery. The police department's RFP might be a long and detailed set of technical specifications. The supplier that can provide the automobile most closely resembling the specifications—at a reasonable price—will probably win the contract. Sometimes the RFP is a more general statement of goals. The customer is in effect asking the suppliers to create their own designs or describe how they will achieve the specified goals. The supplier that offers the most persuasive proposal will probably win the contract.

Most organizations issue RFPs and IFBs in newspapers or send them in the mail to past suppliers. Government RFPs and IFBs are published in the journal *Commerce Business Daily*. Figure 21.2 shows a sample entry from that journal.

Unsolicited Proposals

An unsolicited proposal looks essentially like a solicited proposal, except, of course, that it does not refer to an RFP. Even though the potential customer never formally requested the proposal, in almost all cases the supplier was invited to submit the proposal after people from the two organizations met and discussed the project informally. Because proposals are expensive to write, suppliers are reluctant to submit them without assurances that the potential customer will study them carefully. Thus, the word *unsolicited* is only partially accurate.

External proposals—both solicited and unsolicited—can culminate in contracts of several different types: a flat fee for a product or a one-time service; a

---

Dept of Veteran Affairs, 800 Zorn Avenue, Mail Code (90C), Louisville, KY 40206 C-A/E SERVICES FOR DESIGN OF EXPAND AMBULATORY CARE SOL 603-33-95 DUE 100694 POC Contact Point, Marsha Caudill (502) 894-6113, Contracting Officer, Marsha Caudill (502) 894-6113. Services of an Architectural/Engineer (A/E) to provide fully developed preliminary design with the option to award final working design and construction period services for Project 603-075 "Expand Ambulatory Care at the VA Medical Center, Louisville, Kentucky." Project involves construction of a three story addition of approximately 18,000 GSF which will provide additional space for Ambulatory Care Service, and for relocation of the Mental Health Clinic, Prosthetics Service, and the Audiology and Speech Pathology Service. Drawings shall be accomplished on "CAD" and be compatible with VA Autocad system format software, Auto Cad Release 12. The A/E will be required to complete preliminary design within 60 days of contract award. If the option for final working design is exercised, completion will be required within an additional 120 days. Estimated construction cost is between $2,000,000 and $5,000,000. Area of consideration is restricted to businesses within 50 mile radius of VAMC, Louisville, KY. This solicitation is unrestricted in accordance with the Small Business Competitiveness Demo. Program. SIC 8712 applies to this solicitation. Interested firms must submit appropriate data described in Numbered Note 24 no later than 4:30 p.m. local time on 9/28/94. Subject to availability of funds. (0244)

FIGURE 21.2

Extract from Commerce Business Daily

leasing agreement; or a "cost-plus" contract, under which the supplier is reimbursed for the actual cost plus a profit set at a fixed percentage of the costs.

The "Deliverables" of Proposals

When people talk about *deliverables,* they are referring to what the supplier will deliver at the end of the project. Deliverables can be classified into two major categories, as shown in Figure 21.3.

Research Proposals   In submitting a research proposal, you are promising to provide a research report of some sort. Here are a few examples:

- A biologist working for a state bureau of land management writes a proposal to the National Science Foundation asking for resources to build a window-lined tunnel in the forest to study tree and plant roots and the growth of fungi. The biologist also wishes to understand better the relationship between plant growth and the activity of insects and worms. The deliverable will be a report submitted to the National Science Foundation and, perhaps, an article published in a professional journal.

- A manager of the technical-publications department at a manufacturing company writes a proposal to her supervisor asking for resources to study whether the company should convert its internal documents from paper to on-line. The deliverable will be a completion report (see Chapter 23) that contains her recommendations.

- A university sociologist writes to his state board of education proposing to study a nearby migrant community to determine the extent to which this community understands and uses the municipal services available to it. The deliverable will be a report, submitted to the state board of education and the municipal government, that presents his findings. If the findings are of interest to people beyond his immediate geographical area, he will also write a journal article.

A research proposal often leads to two other kinds of documents: progress reports and completion reports.

After the proposal has been approved and the researchers have begun work, they often submit one or more progress reports. A *progress report* tells the spon-

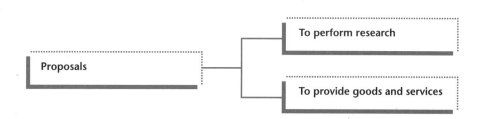

FIGURE 21.3

"Deliverables" of a
Proposal

sor of the project how the work is proceeding: Is it following the plan of work outlined in the proposal? Is it going according to schedule? Is it staying within budget? Progress reports are the subject of Chapter 22.

At the end of the project, researchers prepare a *completion report,* often called a *final report,* a *project report,* or simply a *report.* A completion report tells the readers the whole story of the research project, beginning with the problem or opportunity that motivated it, the methods that were used in carrying it out, and the findings: the results, conclusions, and recommendations. Completion reports are the subject of Chapter 23.

People perform research projects for three basic reasons:

- *Curiosity.* Researchers want to understand as much as they can about their field.
- *Professional advancement.* Many organizations require that their professional employees carry out research and publish in appropriate reports, journals, or books. Government researchers and university professors, for instance, are expected to remain active. Writing proposals is one way to get the resources—time as well as money for travel, equipment, and assistants—to carry out the research.
- *Personal satisfaction.* Performing research, writing, and publishing are fun.

Goods-and-Services Proposals   Whereas a research proposal leads to a report of some kind, a goods-and-services proposal leads to a tangible product (a fleet of automobiles), a service (building maintenance), or some combination of the two (the building of a house).

A vast network of goods-and-services contracts spans the working world. The U.S. government, the world's biggest customer, spent about $300 billion in 1993 on work performed by organizations that submitted proposals. The defense and aerospace industries, for example, are almost totally dependent on government contracts. But goods-and-services contracts are by no means limited to government contractors. One auto manufacturer buys engines from another, and a company that makes spark plugs buys its steel from another company. In fact, most products and services are purchased by contract. The world of work depends on goods-and-services proposals.

Persuasion and Proposals

Regardless of whether the supplier is a professor applying for a research grant, an employee proposing a project at the office, or an electronics company seeking a government contract to build a sophisticated radar device for a new jet aircraft, their proposals will be analyzed carefully and skeptically.

The agency reviewing the professor's proposal will want to see that the professor fully understands the important research questions pertaining to the subject, has a feasible plan for carrying out the project, and has a good track record in such projects. Many other professors will be competing for the same grant, and the agency wants to make sure it is spending its money wisely.

The supervisor of the employee writing the internal proposal will want to be satisfied that the employee has isolated a real problem and devised a feasible strategy for solving it, and that the employee has a good record of carrying through on similar projects.

The government officials reviewing the radar-device proposals will want to be satisfied that the supplier will live up to its promise: to build, on schedule, the best radar device at the best price. With perhaps a dozen suppliers competing for the contract, the government officials will know only that a number of companies want the work; they can never be sure—not even after the contract has been awarded—that they have made the best choice.

A proposal, then, is an argument. To be successful, it must be persuasive. The writers must convince the readers that the *future benefits* will outweigh the *immediate and projected costs*. Basically, proposal writers must clearly demonstrate that they

- understand the readers' needs
- are able to fulfill their own promises
- are committed to fulfilling their own promises

Understanding the Readers' Needs   The most crucial element of the proposal is the definition of the problem or opportunity to which the project is intended to respond. This would seem to be mere common sense: how can you expect to write a successful proposal if you don't show that you understand the readers' needs? Yet the people who evaluate proposals—whether government readers, private foundation officials, or managers in small corporations—agree that an inadequate or inaccurate understanding of the problem or opportunity is the most common weakness of the proposals they see.

Responding to Readers' Needs in an External Proposal

Sometimes the RFP fails to convey the problem or opportunity. More often, however, the writers of the proposal are at fault. The suppliers might not read the RFP carefully and simply assume that they understand the client's needs. Or perhaps the suppliers, in response to a request, know they cannot satisfy a client's needs but nonetheless prepare proposals detailing a project they can complete, hoping either that the readers won't notice or that no other supplier will come closer to responding to the real problem. Suppliers can easily concentrate on the details of what *they* want to do rather than on what the customers need.

But most readers will toss the proposal aside as soon as they realize that it does not respond to their needs. When you respond to an RFP, study it thoroughly. If you don't understand something in it, contact the organization that issued it; they will be happy to clarify it, for a bad proposal wastes everyone's time. Your first job as a proposal writer is to demonstrate your grasp of the problem.

When you write an unsolicited proposal, analyze your audience carefully. How can you define the problem or opportunity so that your readers will understand it? Keep in mind the readers' needs (even if the readers are oblivious to them) and, if possible, the readers' backgrounds. Concentrate on how the problem has decreased productivity or quality or on how your ideas would create new opportunities. When you submit an unsolicited proposal, your task in many cases is to convince your readers that a need exists. Even when you have reached an understanding with some of your customer's representatives, your proposal will still have to persuade other officials.

Responding to Readers' Needs in an Internal Proposal

An internal proposal also must respond to readers' needs. If you propose that your organization hire a new person, you have to make the case that the person is needed and would save or bring in more money than he or she costs. In addition, you have to make sure there is a place for the new person in your current facilities. Writing an internal proposal is both more simple and more complicated than writing an external proposal. It is simpler because you have more access to your readers than you would to external readers. And you can get more information, more easily.

However, you might find it more difficult to get a true sense of the situation in your own organization. Some coworkers might not be willing to tell you directly if your proposal is a long shot. Another danger is that when you identify the problem you want to solve, you are often criticizing—directly or indirectly—the person in your organization who instituted the system that needs to be revised or who failed to take necessary actions earlier. Therefore, before you write an internal proposal, discuss your ideas thoroughly with as many potential readers as you can. In this way you will more likely find out what the organization really thinks of your idea before you commit it to paper.

Describing What You Plan to Do Once you have shown that you understand why something needs to be done, describe what you plan to do. Convince your readers that you can respond to the situation you have just described. Discuss your approach to the subject: indicate the procedures and equipment you would use. If appropriate, justify your choices. After you say that you want to do ultrasonic testing on a structure, explain why, unless the reason is obvious.

Create a complete picture of how you would progress from the first day of the project to the last. Many inexperienced writers believe that they need only convince the reader of their enthusiasm and good faith. Unfortunately, few readers are satisfied with assurances, no matter how well intentioned. Most look for a detailed plan that shows that the writer has actually started to do the work.

No proposal can anticipate all of your readers' questions about what you plan to do, of course. But the more planning you have done before you submit the proposal, the greater the chances you will be able to do the work successfully if you get the go-ahead. A full discussion of your plan suggests to your readers that you are interested in the project itself, not just in winning the contract or in receiving authorization.

Demonstrating Your Professionalism

After showing that you understand the readers' needs and have a well-conceived plan, demonstrate that you are the kind of person—or yours is the kind of organization—that is committed to delivering what is promised. Many other people or organizations could probably carry out the project. You want to convince your readers that you have the pride, ingenuity, and perseverance to solve the problems that inevitably occur in any big undertaking. In short, you want to show that you are a professional.

Demonstrating your professionalism usually involves providing four kinds of information:

- *Credentials and work history.* Who are the people in your organization with the qualifications and experience to carry out the project? What equipment and facilities do you have that will enable you to do the work? What management structure will you use to maintain coordination and keep all the different activities running smoothly? What similar projects have you completed successfully? In short, make the case that you know how to make this project work because you have made similar projects work.

- *Work schedule.* Sometimes called a *task schedule,* this schedule—which usually takes the form of a graph or chart— shows when the various phases of the project will be carried out. In one sense, the work schedule is a straightforward piece of information that enables your readers to see how you would apportion your time. But in another sense, it reveals more about your attitudes toward your work than about what you will actually be doing on any given day. Events rarely proceed according to plan: some tasks take more time than anticipated, some take less. A careful and detailed work schedule is therefore really another way of showing that you have done your homework, that you have attempted to foresee any problems that might jeopardize the success of the project.

- *Quality control.* Your readers will want to see what procedures you have established to evaluate the effectiveness and efficiency of your work on the project. Sometimes quality-control procedures consist of technical evaluations carried out periodically by the project staff. Sometimes the writer will build into the proposal provisions for on-site evaluation by recognized authorities in the field or by representatives of the potential client. Quality control is also measured by progress reports (see Chapter 22).
- *Budget.* Most proposals conclude with a budget, a formal statement of how much the project will cost.

Writing a Proposal

In writing a proposal, you use the same basic techniques of prewriting, drafting, and revising that you use in any other kind of writing. However, a proposal can be such a big project that two aspects of the writing process—resource planning and collaboration—assume greater importance than they do in smaller documents.

The Role of Resource Planning

As discussed in Chapter 5, planning a project requires a lot of work. You need to see whether your organization can devote resources to two different activities:

- writing the proposal
- carrying out the project if the proposal is successful

Sometimes an organization writes a proposal, wins the contract, and then loses money because it doesn't have the resources to do the project and in effect must subcontract major portions of it.

The resources you need fall into three basic categories:

- *Personnel.* Will the necessary technical personnel, managers, and support people be available?
- *Facilities.* Will facilities be available to carry out the research and production, or can they be leased? Can you profitably subcontract portions of the job to companies that have the appropriate facilities?
- *Equipment.* Do you have the right equipment available? If not, can you buy it or lease it or subcontract the work? Some contracts provide for the purchase of equipment, but others don't.

Don't write the proposal unless you are confident that you can carry out the project if the proposal is successful.

The Role of Collaboration

Collaboration is critical in most proposals of more than a few pages because no one person has the time and expertise to do all the work. Writing major proposals calls for the expertise of technical personnel, writers, editors, graphic artists,

PROPOSAL WRITER JIM HILL ON PROPOSALS

on the importance of proposals

They are absolutely essential if you're going to do any government contract work or indeed any kind of contract work. You certainly hope your win ratio will be good; companies that don't have a good win ratio just go under.

on the team members who write proposals

The critical ones are the people who know the technology. Then you usually have a few people on a professional proposal team who are writers or editors or project managers who bring people together and outline the proposal. They assign it to the technical specialists to write their parts. And there are marketing people and legal staff involved in looking at a proposal before it goes out.

on the importance of planning

Preproposal work is all-important. Your marketing people must always be working with potential customers to find out their plans and explain your company's capabilities. No RFP [request for proposal] should be a surprise; if it is, you've already lost.

on developing a win strategy

At the very start you have to identify what the customer wants, and then figure out what it is that you have that will give you a competitive advantage over other companies bidding on the project.

on learning how to write winning proposals

You start with a good course in how to control stress. Second, you have to know your business. You can't lag in any aspect of your business. And you have to be able to work with people; you have to be able to keep things on track. A good proposal writer has to maintain the lines of communication with management and try to anticipate problems before putting all those resources into a proposal that's not going to fly.

on analyzing your audiences

You're writing to different audiences, and this is one of the things your marketing people have to find out. They try to find out how the proposal is going to be graded and who's going to do it—no one person sits down and reads that whole proposal; they'll probably split it off into parts, maybe according to disciplines, and they'll give it to different people. They look for certain words in the RFP. They look to see if you responded to certain points and how thoroughly you seem to have responded, and they give you a grade depending on that. And when the scores are totaled up, the company that came in with the best score wins the contract, all other things being equal.

on meeting the deadline

You have to get the proposal in when it's due, and you can't allow for any possibility of its not happening. If they say it has to be delivered by twelve noon on a certain date, they mean that, and if it comes in at one second after then they can't take it; they have to reject it. So what we always did [when submitting proposals to federal government agencies in Washington, D.C.] was send three copies of the proposal. We would send one copy by overnight express; another copy would go by air; the marketing rep would take the third copy down in the trunk of his car. We had millions of dollars tied up in that proposal; we couldn't afford to miss the deadline.

Jim Hill, now retired, had a 30-year career as a proposal writer and manager for such companies as DuPont, Westinghouse, and E-Systems.

managers, lawyers, and document-production specialists. For this reason, you need to determine whether these people are available.

Usually, a project manager coordinates the writing of a large proposal. In most cases, the parts of the proposal are written at different times by different people. These components might not come together into the final package until a few days—or a few hours—before the proposal deadline.

For more information on collaboration, see Chapter 3.

The Structure of the Proposal

Most proposals follow a basic structural pattern. If the authorizing agency provides an IFB, an RFP, or a set of guidelines, follow it to the letter. If guidelines have not been supplied, or you are writing an unsolicited proposal, use the following conventional structure shown in Figure 21.4.

Summary   For any proposal of more than a few pages, provide a brief summary. Many organizations impose a length limit—for example, 250 words—and ask the writer to present the summary, single spaced, on the title page. The summary is crucial, because in many cases it will be the only item the readers study in their initial review of the proposal.

The summary covers the major elements of the proposal but devotes only a few sentences to each. To write an effective summary, first define the problem in a sentence or two. Next describe the proposed program. Then provide a brief statement of your qualifications and experience. Some organizations wish to see the completion date and the final budget figure in the summary; others prefer that this information be displayed separately on the title page along with other identifying information about the supplier and the proposed project.

For more information on summaries, see Chapters 8 and 11.

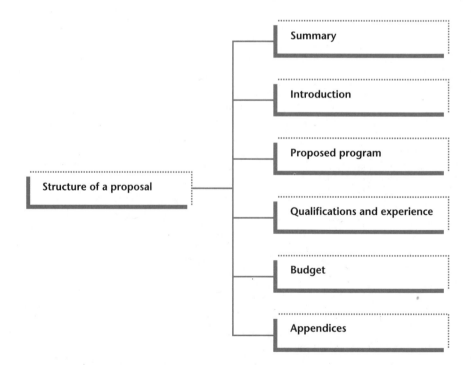

FIGURE 21.4

Structure of a Proposal

Figure 21.5 shows an effective summary taken from a proposal (Farrington, 1993) submitted by a student working for BCE, Inc., a small company that fabricates computer boards. (An additional example of a proposal is presented at the end of this chapter.)

Introduction The introduction of a proposal is like that of any other kind of document. Its purpose is to make sure the reader knows the answers to the following seven questions:

- *What is the problem or opportunity?* Be as specific as possible. Whenever you can, *quantify.* Describe the problem or opportunity in monetary terms because the proposal itself will include a budget of some sort and you want to convince your readers that spending money on what you propose is smart. Don't say that a design problem is slowing down production; say that it is costing $4,500 a day in lost productivity.
- *What is the purpose of the proposal?* Even though it might seem obvious to you, the purpose of the proposal is to describe a problem or opportunity and propose a course of action that will culminate in some deliverable. Be as specific as you can in explaining what you want to do.
- *What is the background of the problem or opportunity?* In answering this question, you probably will not be telling your readers anything they

Note: Proposals are often double spaced.

The background.

The research the author has already done.
The proposal and the deliverable.

The schedule, budget, and the writer's credentials.

FIGURE 21.5

Summary of a Proposal

Summary

BCE, Inc., uses the hand-soldering method to attach components to its PC boards (approximately 10,000 connections per day). Because of a recent increase in orders, we will be unable to meet shipping deadlines without modifying our assembly process.

Recent articles in the trade journals suggest that the easiest and fastest way to satisfy the increased demand would be to purchase a wave-solder machine.

I am proposing that I be permitted to compare wave-flow technology with hiring additional workers or subcontracting as a means of improving our productivity. After completing my research, if wave flow still appears to be the best method for reducing costs and meeting our production deadlines, I will recommend the most appropriate machine and provide you with costs and an implementation schedule.

Performing the research and writing the report should require about 35 hours over four weeks, for a total cost of $303. As a marketing major with over a year's fabrication experience at BCE, I feel qualified to perform this research effectively.

don't already know (except, perhaps, if your proposal is unsolicited). Your goal here is to show them that *you* understand the problem or opportunity: the circumstances that led to the discovery, the relationships or events that will affect the problem and its solution, and so on.

- *What are your sources of information?* Review the relevant literature, either internal reports and memos or external published articles or even books, so that your readers will understand the context of your work.
- *What is the scope of the proposal?* If appropriate, indicate what you are proposing to do, as well as what you are *not* proposing to do.
- *What is the organization of the proposal?* Indicate the organizational pattern you will use in the proposal.
- *What are the key terms that will be used in the proposal?* If you will use any new terms, the introduction is an appropriate place to define them.

Figure 21.6 is the introduction to the BCE proposal.

Proposed Program   Once you have defined the problem or opportunity, say what you want to do about it. As noted earlier, the proposed program demonstrates clearly how much work you have already done. Be specific. You won't persuade by saying that you plan to "gather the data and analyze it." How will you gather it? What tech-

Note: Proposals are often double spaced.

The background and options.

The research the writer has done.

The proposal and the deliverables.

The logic and organization of the proposal.

FIGURE 21.6

Introduction to a Proposal

Introduction

The current manufacturing process for PC boards involves hand soldering an average of 10,000 connections per day. Because of a recent increase in orders, we will be unable to meet shipping deadlines without modifying our assembly process. We have three alternatives:

- add at least one full-time employee
- subcontract some of the soldering
- purchase a single-wave solder machine

The two most recent articles on single-wave solder machines are by Lau (1991) and Woodgate (1988), who argue that the machine would greatly increase production speed while maintaining or improving product quality.

The purpose of the proposal is to request resources to investigate these three alternatives, and then present my findings in a report to you.

The report will compare the costs and benefits of the three options, focusing on cost and quality-control criteria. For the wave-solder machine, I will present the criteria necessary for selecting one, compare the available machines, and describe the steps we would need to take to purchase and set up a machine.

niques will you use to analyze it? Every word you say—or don't say—will give your readers evidence on which to base their decision. If you know your subject, the proposed program will show it. If you don't, you'll inevitably slip into meaningless generalities or include erroneous information that undermines the whole proposal.

If your project concerns a subject written about in the professional literature, show your familiarity with the scholarship by referring to the pertinent studies. However, don't just toss a bunch of references onto the page. For example, don't write, "Carruthers (1993), Harding (1994), and Vega (1993) have all researched the relationship between acid-rain levels and groundwater contamination." Rather, use the recent research to sketch in the necessary background and provide the justification for your proposed program. For instance:

> Carruthers (1993), Harding (1994), and Vega (1993) have demonstrated the relationship between acid-rain levels and groundwater contamination. None of these studies, however, included an analysis of the long-term contamination of the aquifer. The current study will consist of . . .

You might include just one reference to recent research. However, if you have researched your topic thoroughly, you might devote several paragraphs or even several pages to recent scholarship.

Whether your project calls for secondary research, primary research, or both, the important point is that the proposal will be unpersuasive if you haven't already done a substantial amount of the research. For instance, say you are writing a proposal to do research on chain saws. You are not being persuasive if you write that you are going to visit Sears, JC Penney, and Coast to Coast Hardware to see what kinds of chain saws they carry. This statement is unpersuasive for two reasons:

- You need to justify why you are going to visit those three retailers rather than all the others. Anticipate your readers' questions: Why did you choose these three retailers? Why didn't you choose more specialized tool dealers?

- You need to have visited the appropriate stores already—and done any other preliminary research. If you haven't done the homework, readers have no assurance that you will in fact do it and that it will pay off. If your supervisor has already authorized the project but then learns that none of the chain saws on the market meets your organization's needs, you will have to go back and submit a different proposal—an embarrassing move.

Unless you can show in your proposed program that you have done the research—and that the research indicates that the project will likely succeed, the reader has no reason to authorize the project.

Figure 21.7 on pages 594–595 shows the proposed program for the BCE project.

Note: Proposals are often double spaced.

Overview of this section of the proposal.

Proposed Procedure

To help you determine the best course of action for BCE, I will perform the following tasks:

1. determine BCE's soldering needs
2. evaluate the three options for meeting those needs
3. prepare a report that presents my methods and findings

The writer shows that she has already done some research and thought about the problem.

Determining BCE's Soldering Needs

To determine our current production and predict our near-term soldering needs, I have studied the 1994 Production Logs and interviewed two key personnel: Larry Abrams, Production Supervisor, and Karen Kiramidjian, Assistant Production Supervisor. According to our Production Logs and the interviews, our needs can be described by the following criteria:

- Volume: BCE is currently averaging production of approximately 100 new boards per week.
- Design/type of boards: Most of BCE's boards are less than twelve inches in diameter and less than four layers thick, with a combination of through-hole and surface-mount components.
- Physical requirements: We have an area approximately 9 × 5 feet. The production area is already outfitted with a suitable power supply, exhaust venting system, air compressor, and clean air source, should we decide to buy a machine.
- Quality: BCE has a hard-earned reputation for high quality. Any solution we choose has to enable us to maintain and enhance that reputation.

Again the writer shows the research and thinking she has already done. Notice how the writer integrates her sources.

Evaluating the Three Options for Meeting BCE's Needs

Three feasible options are available to BCE for increasing its production capacity.

- Hiring an additional employee. I have contacted the Ada County Job Bank, the BSU College of Technology, and ITT Technical Institute. This preliminary research suggests that qualified workers are available, and that the current wage would be approximately $7.50/hour. For this option to be attractive, our workload would have to be relatively steady; if our workload varies greatly, we would have one of two problems: we would still have a problem meeting production needs or we would be paying a worker for not working. I am currently reviewing the Production Logs to determine whether our soldering needs are steady or whether they fluctuate according to industry conditions.
- Subcontracting the work to an independent solder company. Woodgate (1988) lists the industry groups that I can contact to inquire about subcontracting. I showed this list to Larry and Karen, who recommended two other groups to contact to determine the availability and requirements of subcontractors. I have drafted a letter and faxed it to the groups and am now awaiting the replies. The letter solicits information about the

FIGURE 21.7

Proposed Program of a Proposal

(Figure 21.7 continued)

costs, availability, quality-control, and turnaround time of subcontractors. There are no subcontractors in the Boise area, so we would have at least a three-day shipping time added to the fabrication time.

An additional factor needs to be determined: lot size. Much of BCE's work involves small, very specialized orders. Most of the large wave-solder machines are designed to handle runs of hundreds or thousands of boards, while many of our boards are manufactured in lots of 10 or less.

- Purchasing automated solder equipment. Studying this option involves answering two questions: (1) Would acquiring such equipment be the best solution to our production problem? and (2) If so, which brand and model should we acquire, and which supplier should we contact?

Bahr (1989), Lau (1991), and Woodgate (1988) offer full descriptions of the different kinds of automated solder equipment. There are two main types of automated soldering systems: wave soldering and reflow. Bahr (1989) defines wave soldering as "a process in which printed boards are brought in contact with the surface of continuously flowing and circulating solder." The other type, reflow soldering, is defined as "a process for joining parts by tinning the mating surfaces, placing them together, heating until the solder fuses, and allowing to cool in the joined position." Reflow soldering is effective for surface-mounted components, but almost all of BCE's PC boards contain a mixture of through-hole and surface-mounted components. Therefore, the wave process would be indicated for BCE.

On the basis of my conversations with Larry and Karen and my research, I will examine the available wave-solder machines according to the following criteria:

- What kind of fluxer does it have? I will look for versatility, reliability, accuracy, failure-warning, easy cleaning, and corrosion resistance.
- Does it have air knife wipe off? This optional feature aids in removing excessive flux from the boards before they enter the preheater.
- What is the quality, reliability, etc., of the solder pot, the pumps and nozzles, conveyors and automated process controls?
- What is its provision for disposing of wastes and contaminants? Some fluxes are considered hazardous waste so we'll need to make sure we're in compliance with state, local, and EPA requirements.
- What are the projected maintenance/service costs and frequency?
- Are field reps and service technicians readily available?
- How easy is it for us to convert to an automated process? How much training would our operators need?

Preparing a Report That Presents My Methods and Findings

After completing my research, I will prepare a report that summarizes my findings in comparing and contrasting the three alternative solutions to our problem. This report should enable you to make an informed decision.

The writer has already determined what criteria she will use in evaluating the machines.

The writer describes the deliverable.

Qualifications and Experience

After you have described how you would carry out the project, demonstrate your ability to undertake it. Unless you convince your readers that you can turn an idea into action, your proposal will be unpersuasive.

The more elaborate the proposal, the more substantial the discussion of qualifications and experience has to be. For a small project, a few paragraphs describing your technical credentials and those of your coworkers will usually suffice. For larger projects, the résumés of the project leader—often called the *principal investigator*—and the other important participants should be included.

External proposals should also discuss the qualifications of the supplier's organization. Essentially similar to a discussion of personnel, this section outlines the pertinent projects the supplier has completed successfully. For example, a company bidding for a contract to build a large suspension bridge should describe other suspension bridges it has built. The discussion also focuses on the necessary equipment and facilities the company already possesses, as well as the management structure that will ensure successful completion of the project. Everyone knows that young, inexperienced persons and new firms can do excellent work. But when it comes to proposals, experience wins out almost every time.

Figure 21.8 shows the qualifications-and-experience section of the BCE proposal.

Budget

Good ideas aren't good unless they're affordable. The budget section of a proposal specifies how much the proposed program will cost.

Budgets vary greatly in scope and format. For simple internal proposals, the writer adds the budget request to the statement of the proposed program: "This study will take me about two days, at a cost of about $400" or "The variable-speed recorder currently costs $225, with a 10 percent discount on orders of five or more." For more complicated internal proposals and for all external proposals, a more explicit and complete budget is usually required.

The writer describes her practical and academic experience. Notice how she links her practical work with her academic experience in doing the kind of research she is proposing, making a persuasive argument that she can fulfill the promise of her proposal.

FIGURE 21.8

Qualifications-and-Experience Section of a Proposal

Credentials

I have over one year of electronic assembly experience at BCE, including hand soldering, rework, operating a reflow solder machine, and prepping PC boards for through-hole and surface-mount soldering. I am also in my junior year as a BSU marketing major, with a concentration in entrepreneurship. In my academic work, I have written several feasibility studies comparing capital expenditures and hiring personnel. I am familiar with basic techniques of calculating return on investment and payback periods.

Most budgets are divided into two parts: direct costs and indirect costs.

- *Direct costs* include such expenses as salaries and fringe benefits of program personnel, travel costs, and any necessary equipment, materials, and supplies.

- *Indirect costs* cover the intangible expenses that are sometimes called *overhead*. General secretarial and clerical expenses not devoted exclusively to any one project are part of the overhead, as are other operating expenses such as utilities and maintenance costs. Indirect costs are usually expressed as a percentage—ranging from less than 20 percent to more than 100 percent—of the direct expenses. In many external proposals, the client imposes a limit on the percentage of indirect costs.

Figure 21.9 shows the budget section from the BCE proposal.

Appendices Many different types of appendices might accompany a proposal. Most organizations have *boilerplate* descriptions (standard descriptions that a writer can modify or place directly into a document) of other projects they have already performed. Another popular kind of appendix is the *supporting letter*—a testimonial to the supplier's skill and integrity, written by a reputable and well-known person in the field. Two other kinds of appendices deserve special mention: the task schedule and the evaluation description.

The writer presents and justifies the costs associated with the project.

Budget

Table 1 is the budget itemization. The budget is based on my estimate of the time required for me to carry out the research, analyze the data, and prepare the report. I have included telephone and travel costs to accommodate the faxing, miscellaneous phone calls to out-of-state vendors and professional organizations, and travel to libraries.

Table 1. Budget Itemization

Category	Item	Cost
Wages	Laura Farrington, Equipment Intern, 35 hours labor @ $7.25	253.75
Miscellaneous	Long-distance telephone	30.00
	Travel	20.00
	Total:	$303.75

FIGURE 21.9

Budget Section of a Proposal

The *task schedule* is almost always drawn in one of three graphical formats:

- table
- bar chart
- network diagram

Tables

The simplest but least informative way to portray a schedule is to use a table, as shown in Figure 21.10.

Activity	Start Date	Finish Date
Design the security system	4 Oct. 94	19 Oct. 94
Research available systems	4 Oct. 94	3 Jan. 95
etc.		

FIGURE 21.10

Tabular Schedule

Although creating a table is better than writing out the information in sentences, readers cannot "see" the information; they have to read it and figure out how long each activity will last. In addition, readers cannot tell whether any of the activities are interdependent. That is, they cannot tell what would happen to the overall project schedule if one of the activities were delayed.

Bar Charts

Bar charts, also called *Gantt charts* after the early twentieth-century British civil engineer who first used them, are a distinct improvement over tables. The basic bar chart, as shown in Figure 21.11, allows readers to see how long each activ-

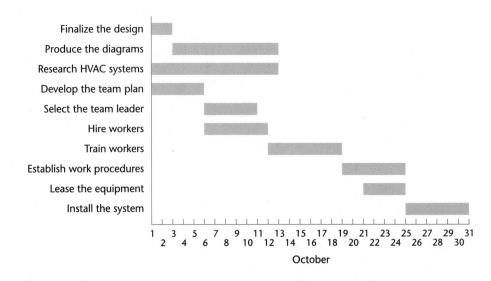

FIGURE 21.11

Bar Chart

ity will take; in addition, bar charts let readers see different activities occurring simultaneously. Like tables, however, bar charts do not indicate the interdependency of tasks.

Network Diagrams

There are many different types of network diagrams, but they all show interdependence, clearly indicating which activities must be completed before other activities can begin.

Of the different kinds of network diagrams, the most popular are PERT (Program Evaluation and Review Technique) and CPM (Critical Path Method). Both techniques use nodes and arrows. As shown in Figure 21.12, PERT and CPM use slightly different symbols.

Figure 21.13 on page 600 shows a relatively simple network diagram. As you can see, understanding a network diagram requires experience. For this reason, a network diagram might be inappropriate for most general readers. But even this brief look at scheduling graphics shows that a network diagram provides more useful information than either a table or a bar chart.

For more information on scheduling graphics, I recommend the following book:

Spinner, M. P. (1989). *Improving project management skills and techniques.* Englewood Cliffs, NJ: Prentice Hall.

Figure 21.14 on page 600 is the task schedule for the BCI proposal.

Other Appendices

Much less clear-cut than the task schedule is the description of *evaluation techniques*. In fact, the term *evaluation* means different things to different people, but in general an evaluation technique can be defined as any procedure for determining whether the proposed program is both effective and efficient. Evaluation techniques can range from simple progress reports to sophisticated statistical analyses. Some proposals provide for evaluation by an outside agent—a con-

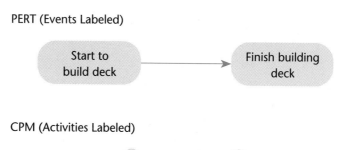

FIGURE 21.12

Symbols Used in PERT and CPM Diagrams

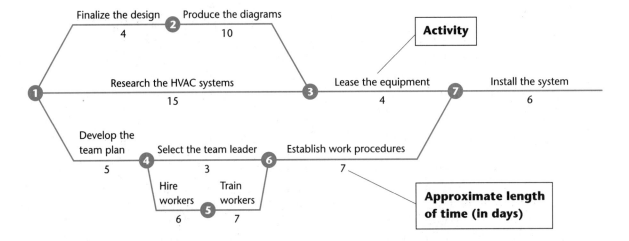

FIGURE 21.13

Network Diagram

sultant, a testing laboratory, or a university. Other proposals describe evaluation techniques that the supplier itself will perform, such as cost/benefit analyses.

The subject of evaluation techniques is complicated by the fact that some people think in terms of *quantitative evaluations*—tests of measurable quantities, such as production increases—whereas others think in terms of *qualitative evaluations*—tests of whether a proposed program is improving, say, the workmanship of a product. And some people imply both qualitative and quantitative testing when they refer to evaluations. An additional complication is that projects can be tested both while they are being carried out (*formative evaluations*) and after they have been completed (*summative evaluations*). Formative evaluations are intended to help people improve the quality of the work they are doing; summative evaluations are intended to assess the quality of a completed project.

When an RFP calls for "evaluation," experienced writers of proposals know that it's a good idea to contact the sponsoring agency's representatives to determine precisely what they mean.

Figure 21.15 is a collection of appendices that make up the description of the evaluation techniques for the BCE project.

Task Schedule

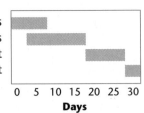

FIGURE 21.14

Gantt Chart

Figure 1. Project Schedule

Appendix A. Evaluation Techniques
I will submit a progress report at the end of week two of the project, at which time I should have completed almost all my research.

Appendix B. List of Solder Machine Vendors in North America
[Here the writer lists the names, addresses, and phone numbers of the 12 dealers she will contact.]

Appendix C. References
[Here the writer cites the sources referred to in the proposal.]

The Format of the Proposal

Proposals are presented in many different formats. Sometimes the organization receiving the proposal specifies the format. For large proposals—100 pages or longer—many organizations use a book format. As shown in Figure 21.16, each two-page unit has the same format: text on one page and graphics on the facing page. The two-page format is intended to force the writers to present material in small, easy-to-read units, with conveniently located graphics. All the information a reader needs to understand the topic is contained on the two pages.

For more information on page design, see Chapter 13.

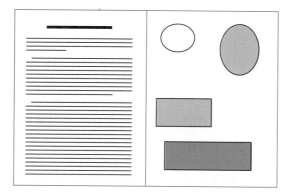

Sample Internal Proposal

Following is an internal proposal (Boully, 1993). The author was a student working as a co-op student for Autosort, a company that provides mailing services to businesses.

Note: Proposals are often double spaced.

The writer identifies the document as a proposal in the identifying information.

DATE: February 27, 1993
TO: Chuck Harding, Owner, Autosort Mail Services
FROM: Danny J. Boully, Supervisor of Autosort's Pre-Sort Mail Division
SUBJECT: Proposal to research barcoding systems for possible purchase by
             Autosort Mail Services

Purpose

This proposal recommends that I be authorized to travel to three cities to evaluate the efficiency and effectiveness of barcoders that Autosort might purchase, and then write a report presenting my findings.

The summary discusses the background, the proposed program, the schedule, the budget, the deliverable, and the writer's credentials.

Summary

The U.S. Postal Service has ruled that large business mailers such as Autosort must barcode their mail by February 1, 1995. If we purchase a barcode system in the near future, we will gain a competitive advantage over our competition in the pre-sort market. My research has shown three models of barcode machines that might fulfill Autosort's needs. I propose that I be sent to gather information on these three models by visiting the manufacturing plants that produce them and the companies that are using them. These trips can be completed in five weeks, at a cost of $3,000. My recommendation will be ready for management by April 28, 1993. On the basis of my experience with Autosort and the research on barcode systems I have already completed, I feel I am qualified to complete this project.

The introduction sketches in the background: the opportunity to turn the change in postal regulations into a business advantage. It also shows the amount of research the writer has already done. Finally, it describes the organization of the proposal.

Introduction

The U.S. Postal Service announced on March 1, 1989, that business customers must use barcoded mail by February 1, 1995, to aid in the increased automation of its mail sorting systems. Autosort must therefore purchase a barcoding machine before this deadline; not purchasing it will result in a serious loss of profit and potentially the end of Autosort's pre-sort division, which is responsible for over 60 percent of Autosort's total profit. If Autosort purchases a barcoding machine now, we will gain a significant advantage over our competition; our two major competitors have not yet made the serious capital investment to obtain one of these new machines. And when the business community is ready to purchase barcoding services, Autosort will have an established reputation in this new field.

According to Beardly (1992) and Parson (1992), there are six major manufacturers of the type of barcoding equipment that we will need to automate our mail sorting system. Because of Autosort's size and the type of customer we serve, three of the six major suppliers have the best potential to meet our needs. I propose that I be sent to evaluate the manufacturing and customer-service facilities of these three companies. An on-site inspection would allow Autosort to better gauge the potential of these new products. Also, while conducting my research, I have found mailing companies similar to ours that I can visit at the same time. If I have an opportunity to see firsthand how these pre-sort companies run their barcoding systems, we would be able to avoid potential problems when we set up our new system.

In the following sections I describe the proposed procedure for this project, the costs, my credentials, and my information sources.

Here the writer provides an overview of the proposed procedure.

Proposed Procedure

I have already completed four tasks:

- determined criteria by which to evaluate available systems
- determined which systems meet our criteria
- studied the product literature on the systems that meet our criteria
- arranged plant visits with pre-sort mailers

The writer describes how he did his preliminary research in the journals.

Determining Criteria by Which to Evaluate Available Systems

I began my initial research by studying *Business Periodicals Index* for the last five years and using the *ABI/Inform* CD-ROM. These searches yielded two recent articles on barcoding systems—Beardly (1992) and Parson (1992)—in two highly respected postal trade journals. Writing in *Postal Times,* Parson evaluated the top eight manufacturers from the perspective of the small growing company. I compared this article with Beardly's article in *Direct Mailers,* which covered barcoding systems. The two articles correlated very closely in their evaluation methods, their conclusions, and their recommendations.

The main difference between the two articles is that *Postal Times* included two very small hand-operated barcoding systems that are rated at only 1,000 pieces of mail per hour. The two articles agreed on six barcoders that might meet the needs of a growing company like ours; I did not consider other systems because Autosort has a firm upper limit of $350,000, which eliminated the rest of the potential candidates. I do not feel this will prevent Autosort from getting a high-quality system, because the systems I am considering control their costs by using innovative engineering and marketing to beat the competition (Beardly, 1992, p. 29).

I contacted these six companies for information packets, which I studied to familiarize myself with their potential and to compare the advertised claims with the findings of the two articles. The advertised claims appear to be reliable.

Here the writer describes how he derived criteria from this research and his knowledge of the company.

To evaluate these six systems for Autosort I used the needs of our company and suggestions from the two articles. The resulting criteria for evaluating the barcoding systems are that they should:

- cost less than $350,000
- process at least 20,000 pieces of mail per hour
- have at least 40 sorting bins
- use the new address computer system
- be able to be upgraded as Autosort grows

- use the new 11-digit destination barcode
- have good local customer support
- need less than two weeks of training time for the lead operator to run the system
- be designed so that most repairs can done by the lead operator to minimize down time

Determining Which Systems Meet Our Criteria

Of the six major suppliers, three do not meet even the basic criteria for Autosort's needs (see the appendix). The three manufacturers with the best potential are the following:

[Here the writer lists the names, addresses, and phone numbers of the three dealers.]

Studying the Product Literature on the Systems That Meet Our Criteria

To further prepare myself for the task of evaluating barcoding systems, I sent for the training manuals that these three companies produce. This will allow me to ask the right questions while I am on-site, and to understand the answers I receive. My goals while on-site are to see and learn as much as possible about each system, to learn how each company fills out the Postal Service paperwork, and to learn the new rules for pre-sorting barcoded mail.

Arranging Plant Visits with Pre-sort Mailers

I have also made contact with three pre-sort mailing companies, located near the three manufacturers, that own the three systems we are considering. The three pre-sort mailing companies are the following:

[Here the writer lists the three companies.]

I would like to visit the three manufacturers and the three companies that own their products. With this firsthand look at the products and frank discussions with customers, I will be in a good position to report to you on the strengths and weaknesses of the three machines.

I propose beginning my research on March 22 and completing my on-site reviews by April 8. After my on-site research is complete, I will submit a progress report with my initial findings.

I recommend that Autosort invite the sales representatives to Boise during the week of April 19 to give their presentations and bids. If I receive verified pricing information by April 26, I will be able to give a recommendation in my completion report by April 28.

The writer describes how he used the criteria to eliminate some of the options.

The writer describes how he acquired more product information.

The writer describes his proposal.

The writer presents a budget for the project.

Costs

The total expense of the research trip will be approximately $3,000. The costs break down as follows:

Airfare	$1,200
Wages	900
Hotel	450
Meals	300
Car rental	150
Total	$3,000

The writer describes his credentials.

Credentials

I have worked at Autosort since November of 1991. During this time I have been involved in nearly every aspect of our operation and have become very familiar with our needs. As pre-sort supervisor I work closely with both our employees and the USPS. This experience would allow me to anticipate potential advantages and problems as I evaluate the new barcoding systems.

In addition, I have studied the barcoding requirements for a project in a senior-level finance seminar at BSU. I have attached the seminar paper.

The writer provides a work schedule.

Task Schedule

Task	March	April
Fly to Tulsa, Oklahoma. Visit Postal Technologies and Odo's Mail House.	22 ▓ 24	
Fly to Evanston, Illinois Visit Bell and Howell and Short Stamp.	29 ▓ 1	
Fly to Stockton, California. Visit Pitney Bowes and the California State Mail Distributor.		5 ▓ 8
Prepare my progress report.	22 ▓▓▓▓▓▓▓▓▓▓▓▓ 12	
Receive bids from sales reps.		▓ 19
Prepare completion report.	22 ▓▓▓▓▓▓▓▓▓▓▓▓▓▓▓▓ 28	

Appendix

This chart was compiled from the information in Beardly (1992) and Parson (1992). See the key below. Boldface indicates a failure to meet a criterion.

The writer presents this data table in an appendix; it might also have been located in the proposed-procedure section of the proposal.

Manufacturer	Cost ($000)	Pieces (000/hr)	Number of bins	Computer	Upgrade	11 digit	Down Time (weeks)	Support	Repair
Bell and Howell	$320	36	16–40	yes	yes	yes	2	yes	yes
Franklin	$120	**10**	10	**no**	**no**	**no**	1	yes	**no**
Mason Mail	$335	36	16–40	yes	**no**	yes	9	yes	**no**
Pitney Bowes	$300	30	16–40	yes	yes	yes	4	yes	yes
Postal Technologies	$340	36	10–50	yes	yes	yes	2	no *(see repair)	total self-repair
Pre-sort Specialties	$350	30	**20–38**	**no**	yes	**no**	6	yes	**no**

*Postal Technologies will train our people at their factory to do all necessary repairs and will sell us the parts, but they do not have local salespeople.

Key:

Manufacturer: The company that produces the system evaluated on the chart.

Cost ($000): The estimated cost of the system in thousands of dollars.

Pieces (000/hr): Estimated number of letters (in thousands) the system can scan and sort in one hour.

Number of bins: The minimum and maximum number of sorting bins that can be purchased with the system. The more bins a system has, the more efficient it will be.

Computer: The system's computer should use the new address-verification program that checks each letter to make sure it is headed to a valid address. Also, the computer system itself should be high quality and upgradable.

Upgrade: The system can be upgraded as needs change.

11 digit: This is the newest improvement in barcoding. It shows the zip + 4 and indicates where the house is on that street.

Down Time: This refers to system down time, in weeks: how often the system has to be off line for repairs and maintenance.

Support: This grades the customer support that is available from the factory and service representatives.

Repair: Most repairs and maintenance on the system should be simple enough for the lead worker to perform.

References

Beardly, F. (1992, November). How to grow into big business mailings. *Direct Mailers, 24–40.*

Parson, J. (1992, July). The barcoder for your growing business. *Postal Times, 56–59, 67–70.*

The writer cites his sources.

The progress report written after the project was under way is included at the end of Chapter 22. The completion report is in Chapter 23. Marginal notes have been added.

The following checklist covers the basic elements of a proposal. Any guidelines established by the recipient of the proposal should of course take precedence over these general suggestions.

1. Does the summary provide an overview of
   ❏ the problem or the opportunity?
   ❏ the proposed program?
   ❏ your qualifications and experience?

2. Does the introduction indicate
   ❏ the problem or opportunity?
   ❏ the purpose of the proposal?
   ❏ the background of the problem or opportunity?
   ❏ the scope of the proposal?
   ❏ the organization of the proposal?
   ❏ the key terms that will be used in the proposal?

3. Does the description of the proposed program
   ❏ cite the relevant professional literature?
   ❏ provide a clear and specific plan of action?

4. Does the description of qualifications and experience clearly outline
   ❏ your relevant skills and past work?
   ❏ the skills and background of the other participants?
   ❏ your department's (or organization's) relevant equipment, facilities, and experience?

5. Is the budget
   ❏ complete?
   ❏ correct?

6. Do the appendices include the relevant supporting materials, such as a task schedule, a description of evaluation techniques, and evidence of other successful projects?

EXERCISES

Exercises 2–4 call for you to write a memo. See Chapter 16 for a discussion of memos.

1. Write a proposal for a research project that will constitute a major assignment in this course. Start by defining a technical subject that interests you. (This subject could be one that you are involved with at work or in another course.) Using abstract services and other bibliographic tools, create a bibliography of articles and books on the subject. (See Chapters 6 and 7 for discussions of choosing a topic

and finding information.) Create a reasonable real-world context. Here are three common scenarios from the business world:

• Our company uses Technology X to perform Task A. Should we instead be using Technology Y to perform Task A? For instance, our company subcontracts the writing and production of our monthly employee newsletter. Should we be producing it ourselves using desktop-publishing? What kinds of personnel are needed? What skills do they require? How much time does it take? What kind of hardware and software is required?

• Our company has decided to purchase a particular kind of tool to perform Task A. Which make and model of the tool should we purchase, and which supplier should we buy or lease it from? For instance, our company has decided to purchase 10 multimedia computers. Which brand and model should we buy, and whom should we buy them from?

• Our company does not currently perform Function X. Is it feasible to perform Function X? For instance, we do not currently offer daycare for our employees. Should we? What are the advantages and disadvantages of doing so? What are the different forms that daycare can take? How is it paid for?

Following are some additional ideas for topics.
• the need to provide Internet access to students
• the value of using the Internet to form ties with another technical-communication class on another campus
• the need for expanded opportunities for internships in your major
• the need to create an advisory board of people from industry to provide expertise about your major
• the need to raise money to keep the college's computer labs up-to-date
• the need to evaluate your major to ensure it is responsive to student needs
• the advisability of starting a campus branch of a professional organization in your field
• improving parking facilities on campus
• creating or improving organizations for minorities or for women on campus

Notice that these topics can be approached from different perspectives. For instance, the first one—

on providing Internet access to students—could be approached in several ways:

- Our college currently purchases journals but does not provide Internet access for students. Should we consider taking some money from the library's journal budget to subsidize Internet access for students?
- Our college has decided to provide Internet access for its students. How should it do so? What vendors provide such services? What are the strengths and weaknesses of each vendor?
- Our college does not offer Internet access to students. Should we make it a goal to do so? What are the advantages of doing so? The disadvantages?

2. Secure an RFP issued by a city government. What kind of information does it call for? What does the city government consider the most important attributes of a proposal? How effectively does the RFP explain to proposal writers how to make their case? Respond in a 1,000-word memo to your instructor. Attach a copy of the RFP.

3. Secure the proposal that responded successfully to the RFP described in exercise 2. In a 1,000-word

memo to your instructor, analyze the strengths of the proposal. What attributes made it successful? Would you have responded to the RFP differently? Explain. If possible, attach a copy of the proposal.

4. Interview a local professional who writes proposals. (See Chapter 6 for a discussion of interviewing.) What are the main challenges he or she faces in creating proposals? How does he or she meet those challenges? To what extent does the information in this chapter match what the proposal writer has told you? Respond in a 1,000-word memo to your instructor.

REFERENCES

Boully, D. (1993). *Proposal to research barcoding systems for possible purchase by Autosort Mail Services.* Unpublished document.

Dennis, J. (1990). *Non destructive evaluation of plasma sprayed coatings.* Unpublished document.

Farrington, L. (1993). *Proposal to research wave-solder machines for possible use at BCE.* Unpublished document.

A progress report communicates to a supervisor or sponsor the current status of a project that is not yet complete. As its name suggests, a progress report is an intermediate communication between the proposal (the argument that the project be undertaken) and the completion report (the comprehensive record of the completed project). For more information on proposals, see Chapter 21; for completion reports, Chapter 23.

Although terminology varies from one organization to another, a *progress report* usually describes a single, discrete project, such as the construction of a bridge or an investigation of excessive pollution levels in a factory's effluents. The term *status report* refers to an update on the entire range of operations of a department or division of an organization. For example, the director of the marketing division of a manufacturing company might submit a monthly status report.

Progress reports can best be understood by looking at their audience, purpose, format, schedule, and tone.

- *Audience.* Progress reports are addressed to your supervisors, sponsors, or customers.

- *Purpose.* Progress reports let you check in with your audience. Supervisors are vitally interested in the status of their projects, because they have to integrate them with other present and future commitments. Sponsors (or customers) have the same interest, plus an additional one: they want the projects to be done right—and on time—because they are paying for them.

- *Format.* A progress report can take any number of forms. A small internal project might require only brief memos, or even phone calls. A small external project might be handled with letters. For a larger, more formal project—either internal or external—a formal report usually is appropriate. Sometimes a combination of formats is used: for example, reports at the ends of quarters and memos at the end of each of the other eight months. (See Chapter 16 for a discussion of memos.)

- *Schedule.* Usually, the supervisor or sponsor establishes the schedule for submitting progress reports. A relatively short-term project—one expected to take a month to complete, for example—might require a progress report after two weeks. A more extensive project usually requires a series of progress reports submitted on a fixed schedule, such as every month or every quarter.

- *Tone.* Regardless of whether the project is proceeding smoothly or has encountered difficulties, you need to explain clearly and fully what happened and how it will affect the overall project. Your tone should be objective, neither defensive nor casual. Unless ineptitude or negligence caused the problem, you're not to blame. Regardless of what kind of news you are delivering—good, bad, or mixed—your job is the same: to provide a clear and complete account of your activities and to forecast the next stage of the project.

Progress reports are crucial because they enable managers to make informed decisions when things go wrong. What can go wrong? Problems generally fall into three categories:

- *The deliverable won't be what you thought it would be.* The *deliverable* is what you plan to deliver to your manager or sponsor at the end of the project; it could be a report or a product or service. The progress report is your opportunity to explain that even though you thought you would be able to recommend which rolling machine under $300,000 to purchase for your paper mill, you will not be able to do so. Then you explain why. Or you explain that the deliverable will not meet one of the specifications you listed in the proposal.

- *You won't meet your schedule.* Explain why you are going to be late and specify when the project will be complete.

- *You won't meet the budget.* Explain why you need more money, and specify how much more you will need.

I don't mean to sound pessimistic, but it's a fact: seldom do projects exceed their initial projections. Rarely does the deliverable exceed the specifications agreed on in the proposal. Rarely is the project completed early, and rarely does it cost less than the initial budget specified. Equipment fails; the weather doesn't cooperate; personnel change positions; prices go up. An experienced manager could list a hundred typical difficulties.

A couple of happy exceptions: sometimes prices go down, particularly in high-tech industries, as you may have noticed with personal computers. And sometimes jobs are completed early, particularly when the contract calls for a bonus for early completion and a penalty for late completion.

A note about ethics. You might be tempted to cover up the problems and hope that you can solve them before the next progress report. This course of action is unwise and unethical. It is unwise because chances are the problems will multiply, and you will have a harder time explaining why you didn't alert your readers earlier. It is unethical because by withholding bad news you are violating your readers' rights to full and complete information. Because they are the sponsors or supervisors of the project, they have a right to know how it is going.

Writing the Progress Report

Writing a progress report is similar to writing any other kind of technical-communication document in that you analyze your audience and purpose, evaluate the implications of your audience and purpose, generate ideas, and so on.

As always, you draft the body of the progress report before you draft its front matter. And as always, you will find yourself going back to do prewriting tasks—analyzing your audience, for example—after you thought you were on to the next phase of the writing process.

To draft the progress report, start with a copy of your proposal; many elements of the progress report are taken directly from it. Much of the introduction, such as the statement of the problem or opportunity that led to the project, can often be presented intact. The discussion of the work you have completed is often a restatement of a portion of the proposed program, with the future tense changed to past. You might want to reproduce your task schedule—modified to show your current status—and your bibliography.

The Structure of the Progress Report

Progress reports vary considerably in structure because of differences in format and length. Written as a one-page letter, a progress report is likely to be a series of traditional paragraphs. As a brief memo, it might also contain section headings. As a report of more than a few pages, it might contain the elements of a formal report.

Regardless of these differences, however, most progress reports consist of the five components shown in Figure 22.1.

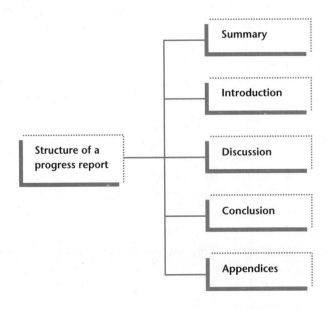

FIGURE 22.1

Structure of a Progress
Report

Summary   Progress reports of more than a few pages are likely to contain a summary section, which, like all other summaries, provides a brief overview of the discussion for those readers who do not need all the technical details. Often, the summary enumerates the accomplishments achieved during the reporting period and then comments on the current work.

Figure 22.2 is the first section of a progress report (Merrick, 1993) written by a student intern at a real-estate company. The student is studying GIS—Geographical Information Systems—a technology that might be useful for his company.

For more information on summaries, see Chapter 11.

Introduction   The introduction of a progress report is like that of any other kind of document. Its purpose is to make sure the reader knows the answers to the following five questions:

Note: Progress reports are often double spaced.

The writer identifies this as a progress report in the identifying information section.

The purpose statement describes this progress report in the context of future progress reports.

The summary reviews the progress made during this reporting period and mentions that the completion report should be submitted on schedule.

To: Arlene Marshall, Owner, M&M Real Estate Developers, Inc.
From: Doug Merrick, Technical Analyst, Technical Offices, M&M Real Estate Developers, Inc.
Subject: Midpoint progress report on the study of Geographical Information Systems uses at M&M Real Estate
Date: April 14, 1993

Purpose

This progress report covers the first month of my study of Geographical Information Systems and its applications. It covers Phase I (the library-research phase) and part of Phase II (interviewing the staff and making presentations to them).

Summary

During the first two phases of my study on Geographical Information Systems (GIS) I have been able to find and present to staff members much helpful information. Although the real-estate industry has not yet exploited the possibilities of GIS, a great deal of information is publicly available that can be of tremendous value to realtors. The major applications will probably include neighborhood information, development planning, environmental-impact analysis, and mapping. I have opened channels for useful discussion and planning with the staff, and they seem very interested in GIS applications. These early phases of the project will be beneficial to M&M Real Estate Developers; we need a thorough understanding of GIS concepts before we begin to analyze hardware capabilities and costs in Phase III of the project.

I anticipate that the completion report will be submitted on schedule.

FIGURE 22.2

Summary Section of a Progress Report

- *What is the subject of the report?* Copy the discussion of the purpose of the project from the proposal. Although the purpose has already been defined in the proposal, repeat it in each progress report for the benefit of readers who are not following the project closely.
- *What is the purpose of the report?* Even though you have probably made it clear already in the title or the subject line, identify the document as a progress report and indicate the period of time the report covers. If more than one progress report has been (or will be) submitted, place the report in the proper sequence—for example, as the third quarterly progress report.
- *What is the background of the subject?* Copy this information from the proposal.
- *What is the organization of the report?* Indicate what organizational pattern you will use in the progress report.
- *What key terms will be used in the report?* If you will introduce any new terms, define them in the introduction.

Figure 22.3 is an introduction to the GIS progress report.

Discussion   The discussion elaborates on the points listed in the summary. The discussion serves readers who want a complete picture of the team's activities during the period covered; many readers, however, will not bother to read it unless the summary highlights an unusual or unexpected development during the reporting period.

Note: Progress reports are often double spaced.

The introduction describes the progress reported here against the backdrop of the whole project. The introduction also explains the organization of the progress report.

FIGURE 22.3

Introduction to a
Progress Report

Introduction

On May 15 I was given approval to conduct a four-phase study on GIS to determine whether it might have applications for M&M Real Estate.

- Phase I of the study was to do secondary research to increase my awareness of the latest advances of this highly specialized field.
- Phase II, which is still underway, is to meet with staff members to present basic concepts of GIS and to gather ideas for our future development of GIS.
- Phase III will be to contact and visit GIS hardware distributors.
- Phase IV will be to visit agencies that are potential data sources.

This progress report presents the results of Phase I of the project and then offers comments on the future phases.

There are two basic organizational patterns for the discussion section:

- *The time pattern.* This is the simplest way to organize the discussion: past work/future work. After describing the problem, you describe all the work that has been completed in the present reporting period and then sketch in the work that remains. This scheme is easy to follow and easy to write. Some writers use a past/present/future pattern, which enables them to focus on a long or complex task that they are working on at the moment.

- *The task pattern.* If the project requires that you work on several tasks simultaneously, the task pattern is particularly effective, for it enables the writer to describe, in order, what has been accomplished on each task. Often, the task-oriented structure incorporates the past work/future work structure.

Table 22.1 shows the difference between the two organizational patterns.

TABLE 22.1	*Organizational Patterns Used in Progress Reports*
The Time Pattern	**The Task Pattern**

III. Discussion
    A. Past Work
    B. Future Work
	III. Discussion
	    A. Task I
	        1. past work
	        2. future work
	    B. Task II
	        1. past work
	        2. future work

Figure 22.4 on pages 616–618 exemplifies the standard chronological progression—from the problem to past, present, and future work. Notice, however, how the writer combines generic and specific phrases in the headings.

Conclusion A progress report is, by definition, a description of the present status of a project. The reader will receive at least one additional communication—the completion report—on the same subject. The conclusion of a progress report, therefore, is more transitional than final.

In the conclusion of a progress report, you evaluate how the project is proceeding. In the broadest sense, you have one of two messages:

- Things are going well.
- Things are not going as well as anticipated.

Note: Progress reports
are often double spaced.

The writer begins by
introducing the two sec-
tions of the Past Work
discussion.

The writer offers a full
description of the results
of his library research.

The writer describes the
background of GIS.

Past Work

The following sections discuss Phase I, which has been completed, and Phase II, which is still in progress.

Phase I: Research Work Completed

I was able to locate a number of useful publications on the subject of GIS and digital imagery. The two most helpful books on GIS concepts are the following:

- Dent, Borden D. *Cartography—Thematic Map Design.*
- Star, Jeffrey, and John Estes. *Geographic Information Systems—An Introduction.*

These books are available to the staff, in the library. I also contacted a number of government GIS users, private GIS hardware and product representatives, and various trade publications and had them send me information regarding their present GIS applications. I was not able to find any specific studies on GIS uses in real estate but did find some information in a newsletter produced by Environmental Systems Research Institute, Inc., in Redlands, California, called *ARC News.* On page 16 is the article "Title Company Implements ARC/INFO for Real Estate Information Management." The article states

> GIS technology offers the real estate industry a valuable tool for managing real estate information and maps. Using digital parcel maps and ownership information acquired from local governments, census data, environmental data, scanned photographs, building permit data, and an array of other accessible data, a GIS can be used for the following types of real estate analysis:
>
> - Investment potential
> - Neighborhood characteristics
> - Land evaluation and appraisal
> - Environmental hazards determination
> - Development planning
> - Environmental impact analysis and abatement planning
> - Asset management
> - Sales and lease management

Many of these uses have been discussed by staff members as areas where we might use GIS. Real-estate developers have only just begun to explore these uses, so the benefits are not fully known. The article further describes some software products for automating title insurance processing and mapping capabilities. I will contact the company that is developing these products during the next phase of this study.

GIS technology has advanced at a very rapid pace in the last two years due to the development of faster personal computers. The private use of GIS is in its

FIGURE 22.4

**Discussion Section of a
Progress Report**

(Figure 22.4 continued)

infancy, and objective information about this technology is limited. However, I have found government agencies to be a good source of some specific uses of GIS. City planning agencies have made significant headway in the use of GIS. Many of these uses could easily be adapted to the real-estate business, especially the use of Census Bureau Tiger files. I contacted Charles Trainor at Ada Planning Association, the Ada County planning agency located in Boise, who informed me that they have recently begun to develop the use of ARC/INFO software as their GIS system. I have an appointment to meet with him in two weeks for a demonstration of their system, which includes the use of the Census Bureau Tiger Files.

The USDA Soil Conservation Service is also using GIS technology in ways that might be of interest to us. They have current updates of soils databases for most of the land within the United States. This information is available to the public at very reasonable prices. I have obtained an informative packet from USDA Soil Conservation Service and will circulate it to staff members. I will be meeting with Dave Hoover of the Soil Conservation Service here in Boise for a demonstration of their GIS. He will allow up to eight of our staff members to participate in that demonstration. I'll post the date and time for this as soon as possible.

Other government agencies have also been developing GIS uses, and many of these are located in Boise. I have been in contact with the Forest Service, BLM, and the Department of Fish and Game and will visit them during the third phase of this study to evaluate their system and uses.

I ordered a one-year trial subscription to four GIS publications and have received the first issues of *GPS World—News and Applications of the Global Positioning System, Geo Info Systems—Applications of GIS and Related Spatial Information Technologies, GRASSCLIPPINGS—The Journal of Open Geographic Information Systems,* and *ARC News.*

GRASSCLIPPINGS is a government newsletter for the GRASS GIS software user community. This is the public-domain software I used while employed by the USDA Agricultural Research Service. The newsletter contains many articles about particular government agencies' applications of GIS. Although many of these uses are highly specialized, the concepts can be adapted by M&M Real Estate Developers.

I will be receiving many other packets of information from various agencies that I believe will be informative. Many of these will be from representatives of private development companies who will, no doubt, highly praise their own products. However, all of this information can be helpful to us. After most of this information has arrived, I will set up a display table in the library.

(Figure 22.4 continued)

The writer describes his progress on Phase II.

Phase II: Staff Interviewing Completed

I have met with staff members as a group once, and with most members personally at least once. I demonstrated the basic concepts of GIS and distributed a 10-page document explaining further concepts and uses. This information has been met with much interest by all staff members up to this point. I have been flooded by ideas, big and small, for potential uses of GIS technology. Many of these ideas are too sophisticated for the initial phases of a GIS implementation, but the following ideas are feasible:

- Querying data for land or structural characteristics
- Investment potential studies
- Basic environmental impact analysis
- Sales and Leasing management
- Area demographics analysis
- Utilities location mapping
- Traffic counts
- Information services (distance of shopping, schools, and services from properties)

Communication channels have been opened with the staff. No idea will be considered too radical, and each has been recorded for present and future reference. I have categorized each idea according to what I presently believe to be feasible now, and according to what has future potential. This brainstorming process is well documented, with an explanation of the idea, the person who suggested it, and the data needed. I will continue this process throughout Phase II, now in progress.

Future Work

After completion of the staff interviews and study of our specific GIS needs and wants, I will begin Phase III of the study—to contact GIS distributors—followed shortly by Phase IV—to research data sources.

The writer looks to the future phases of the project.

If the news is good, convey your optimism, but avoid overstatement.

OVERSTATED: We are sure the device will do all that we ask of it, and more.

REALISTIC: We expect that the device will perform well and that, in addition, it might offer some unanticipated advantages.

Beware, too, of promising early completion. Such optimistic forecasts rarely prove accurate, and it is always embarrassing to report a failure after you have promised success.

However, don't panic if the preliminary results are not as promising as you had planned, or if the project is behind schedule. Readers know that the most levelheaded and conservative proposal writers cannot anticipate all problems. As long as the original proposal was well planned and contained no wildly inaccurate computations, don't feel personally responsible. Just do your best to explain the unanticipated problems and the current status of the work. If you suspect that the results will not match earlier predictions—or that the project will require more time, personnel, or equipment—say so, clearly. If your news is bad, at least give your reader as much time as possible to deal with it effectively.

Figure 22.5 is the conclusion of the GIS progress report.

Appendices   In the appendices to the progress report, include any supporting materials that you feel your reader might wish to consult: computations, printouts, schematics, diagrams, charts, tables, or a revised task schedule. Be sure to cross-reference these appendices in the body of the report, so that the reader can consult them at the appropriate stage of the discussion.

Figure 22.6 on page 620 is an updated task schedule. The writer has taken the original task schedule from the proposal and added a screen to show the tasks that have already been completed.

The writer draws conclusions from his progress, and then indicates his anticipation that the completion report will be submitted on time.

Conclusion

Phase I has been completed successfully; I was able to gather a great deal of helpful information to further educate myself. Phase II, meeting with staff members about GIS technology and uses, is proceeding well. This phase has proven useful in bringing about intelligent conversations about GIS uses for M&M Real Estate. The knowledge we have gained has greatly increased our confidence in this technology as a useful tool. We will be able to make better-informed and wiser decisions about our company's needs for GIS and data when we begin Phase III. Specific information about GIS uses in real estate development is limited, but other specialized uses contain many concepts that can be applied to our needs.

I anticipate that the completion report will be submitted on schedule.

FIGURE 22.5

Conclusion of a Progress Report

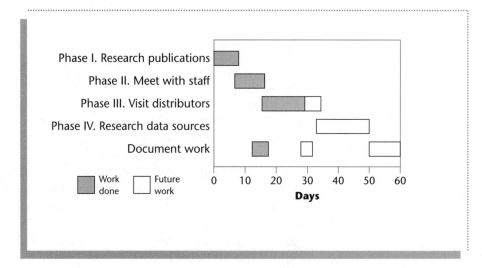

The writer uses two patterns on the bars to distinguish past work from future work.

FIGURE 22.6

Updated Task Schedule
of a Progress Report

Sample Progress Report

Following is the progress report written as a follow-up to the proposal included in Chapter 21. (The completion report on this study is included in Chapter 23.) Marginal notes have been added to the progress report.

Note: Progress reports are often double spaced.

The writer identifies this as a progress report in the subject heading.

The writer summarizes the background of the project and its present status, and then indicates that the completion report will be submitted on time.

DATE: April 11, 1993
TO: Chuck Harding, Owner, Autosort Mail Services
FROM: Danny J. Boully, Supervisor of Autosort's Pre-Sort Mail Division
SUBJECT: Progress report on the project to research barcoding systems

Purpose

This is the progress report on the project to investigate barcoding systems for possible purchase by Autosort.

Summary

The Postal Service has announced that large business mailers such as Autosort must barcode their mail by February 1, 1995. Autosort's management has decided to purchase a barcode system in the near future. This will give Autosort a strategic advantage over its competition in the pre-sort market.

There are three models of barcode machines that might fulfill Autosort's needs. I was authorized to gather information on these three models by visiting the manufacturing plants that produce them and companies that are using them. I have completed two of the three trips.

You will receive my completion report on April 28, 1993.

Introduction

The U.S. Postal Service has ruled that business customers must use bar-coded mail by February 1, 1995, to aid in the greater automation of the USPS's mail sorting systems. Autosort's management decided to purchase a barcoding system before this deadline to prevent a serious loss of profit in its pre-sort division and to comply with the new regulation before the deadline. In addition to preventing loss of profit, purchasing a system in the near future will give Autosort a strategic advantage in the Boise market. I have found three models of barcoding systems that might meet Autosort's unique needs. I have completed two of the three trips to evaluate the manufacturers of these systems and to tour pre-sort companies that are similar to our own.

With the information I collect on my trips, Autosort's management will be prepared for negotiations with the sales representatives from the three companies. After management finishes negotiations with the sales representatives and receives the final bids, I will be able to finish my completion report. If I receive the final bids by April 25, I will be able to give you my completion report by April 28. In my completion report I will recommend which system I feel best meets our needs. If there is not a clear-cut best system, I will explain the pros and cons of the two best candidates.

This progress report presents my findings from the two trips. First I describe the process by which I narrowed the list of manufacturers to three. After a narrative evaluation of the two systems, I present a decision-matrix table for the two systems I have studied.

Results of Research

My goals in this project were to gather information for the purchase of the best system for Autosort, to learn how the barcoding systems work, and to learn how to do the new USPS barcode paperwork. My research has been very successful in all three areas.

In my proposal, dated February 27, 1993, I described how I established criteria for evaluating barcoders. Following are the criteria I established. The barcoders should

- cost less than $350,000
- process at least 20,000 pieces of mail per hour
- have at least 40 sorting bins
- use the new address computer system
- be able to be upgraded as Autosort grows
- use the new 11-digit destination barcode

The writer fills in the background and the purpose of the project, and then explains the purpose and organization of the progress report.

The writer repeats elements from his proposal about establishing criteria. He includes the data table again, too.

- have good local customer support
- need less than two weeks of training time for the lead operator to run the system
- be designed so that most repairs can done by the lead operator to minimize down time

I presented the following table showing how only three of the six machines meet our criteria. This chart was compiled from the information in Beardly (1992) and Parson (1992) and my own research at Autosort. Boldface indicates a failure to meet a criterion.

Manufacturer	Cost ($000)	Pieces (000/hr)	Number of bins	Computer	Upgrade	11 digit	Down time (weeks)	Support	Repair
Bell and Howell	$320	36	16–40	yes	yes	yes	2	yes	yes
Franklin	$120	**10**	10	**no**	**no**	**no**	1	yes	**no**
Mason Mail	$335	36	16–40	yes	**no**	yes	9	yes	**no**
Pitney Bowes	$300	30	16–40	yes	yes	yes	4	yes	yes
Postal Technologies	$340	36	10–50	yes	yes	yes	2	no*(see repair)	total self-repair
Pre-sort Specialties	$350	30	**20–38**	**no**	yes	**no**	6	yes	**no**

*Postal Technologies will train our people at their factory to do all necessary repairs and will sell us the parts, but they do not have local salespeople.

Key:

Manufacturer: The company that produces the system evaluated on the chart.

Cost: The estimated cost of the system in thousands of dollars.

Pieces: Estimated number of letters (in thousands) the system can scan and sort in one hour.

Number of bins: The minimum and maximum number of sorting bins that can be purchased with the system. The more bins a system has, the more efficient it will be.

Computer: The system's computer should use the new address-verification program that checks each letter to make sure it is headed to a valid address. Also, the computer system itself should be high quality and upgradable.

Upgrade: The system can be upgraded as needs change.

11 digit: This is the newest improvement in barcoding. It shows the zip+4 and indicates where the house is on that street.

Down time (weeks): This refers to system down time (in weeks): how often the system has to be off-line for repairs and maintenance.

Support: This grades the customer support that is available from the factory and
service representatives.
Repair: Most repairs and maintenance on the system should be simple enough
for the lead worker to perform.

Here I present my narrative evaluation of the two systems I have seen in operation, followed by a decision matrix table.

Postal Technologies

I started my research trip by visiting Postal Technologies manufacturing
plant and training center in Tulsa, Oklahoma. They have very good facilities
and were very professional. Postal Technologies was originally a pre-sort and
bulk mail company like Autosort, but they decided to expand into the production of barcoding equipment.

Their model 2A has been designed to be totally repaired by the user. It is
built along a modular design, so if something goes wrong the lead worker just
unplugs that part and replaces it with a new one. The quality of the machine
appears to be high and the training the lead workers receive is the best of the
three companies. Postal Technologies will train two workers as part of the purchase price, but if we want to train more it will cost $3,000 per person for the
two-week course in Tulsa.

Odo of Odo's Mail House uses this system for all of their pre-sort and some
of their bulk mail barcoding. He has been pleased with the function of the system and has doubled his business since obtaining his barcoder. He loves the
ability to repair the machine on his own but stresses the need to keep a supply
of spare parts on hand. One disadvantage is that the manual is extremely confusing.

With Autosort's fairly high turnover rate, keeping fully trained personnel
available might be difficult.

Bell and Howell

The Bell and Howell model 1500 is very impressive. The plant foreman,
Wesly Crusher, showed me a new feature they have developed in their feed
mechanism. It will feed letters as thin as tracing paper or as thick as ¼ inch (the
maximum thickness for pre-sort mail) with only minor adjustments. The
machine can handle our thinnest mail at a rate of 35,000 pieces per hour. This is
a practical work rate that includes the normal stops and starts of a work period.
With standard size mail, the letters seem to just fly through the system. I estimate that the normal workload of a trained lead worker would be nearly 37,000
pieces of mail per hour.

Short Stamp has expanded its operations into a new building and runs mail
to the post office twice a day to meet its new volume of mail. Ben Cisco, the
owner, said that the training was conducted in their business and that the trainer
was helpful in customizing the computer system to their needs. Training was

The writer gives a full
account of his research
trips.

done during the light morning hours, and they practiced on the type of mail that Short Stamp was actually taking to the post office. During their practice runs, they identified three-digit groups that would qualify and added them to the bin system. I saw Ben and his crew handle common adjustments and routine maintenance. He said most of the problems can be handled by the crew, but if something major goes wrong, the service representative has someone on call during working hours.

The writer fills in the decision matrix for the two manufacturers he has visited.

Decision Matrix

I gave the two systems a grade of 1 to 10 on each criterion, with 1 being the lowest and 10 the highest. See the key below for an explanation of the grading system.

Manufacturer	Cost	Pieces per hr.	Number of bins	Computer	Upgrade	11 digit	Down time	Support	Repair	Total points
Bell and Howell	7	9	9	10	9	10	8	8	8	78
Pitney Bowes										
Postal Technologies	6	8	10	10	9	10	8	0* (see repair)	10	71

*Postal Technologies will train our people at their factory to do all necessary repairs and will sell us the parts, but they do not have local sales people.

Key:

Manufacturer: The company that produces the system evaluated on the chart.

Cost: This score reflects the price of the system.

Pieces per hr.: This score was reached by comparing how many letters the system actually scanned per hour with what the advertising claimed the system could do.

Number of bins: This score was reached by comparing the number of bins that are available with the system with Autosort's needs. The more bins a system has, the more efficient it will be.

Computer: This score was reached by verifying that the system uses the new address system. Because it is a yes or no answer, 0 was given for a no and 10 for a yes. Also, the computer system itself should be high quality and upgradable.

Upgrade: This score was reached by finding out how expensive and difficult it would be to expand or improve the system.

11-digit: This score was reached by verifying that the system uses the new barcoding system. Because it is a yes or no answer, 0 was given for a no and 10 for a yes.

Down time: This score was reached by comparing the maintenance schedules of the three systems. The more time the manual suggests for down time, the lower the score.

Support: This score was reached by comparing the customer support that is available from the factory and service representatives.

Repair: This score was reached by comparing how many of the normal breakdowns and how much of the maintenance can be done by the lead worker without calling the sales representative.

Total points: This is the total number of points for each system.

Future Work

The writer speculates on the future work.

I have completed all of the scheduled tasks except for visiting Pitney Bowes and writing the completion report. I recommend that Autosort invite the sales representatives to Boise during the week of April 19 to give their presentations and bids. If I receive verified pricing information by April 26, I will be able to give a recommendation in my completion report by April 28.

Updated Budget

[Here the writer provides a list of his expenses for each of the two trips.]

Updated Task Schedule

My schedule for this research has been as follows:

Task	March	April
Fly to Tulsa, Oklahoma. Visit Postal Technologies and Odo's Mail House.	22■24	
Fly to Evanston, Illinois. Visit Bell and Howell and Short Stamp.	29■1	
Fly to Stockton, California. Visit Pitney Bowes and the California State Mail Distributer.		5▒8
Prepare my progress report.	22■■■■■■11	
Receive bids from sales reps.		▒19
Prepare completion report.	22■■■■■■▒▒▒▒▒ 28	

■ = work completed
▒ = future work

References

Beardly, F. (1992, November). How to grow into big business mailings. *Direct Mailers,* pp. 24–40.

Parson, J. (1992, July). The barcoder for your growing business. *Postal Times,* pp. 56–59, 67–70.

WRITER'S CHECKLIST

1. Does the summary
   - ❏ present the major accomplishments of the period covered by the report?
   - ❏ present any necessary comments on the current work?
   - ❏ direct the reader to crucial portions of the discussion section of the progress report?

2. Does the introduction answer the following questions about the document?
   - ❏ What is the subject of the report?
   - ❏ What is the purpose of the report?
   - ❏ What is the background of the subject?
   - ❏ What is the scope of the report?
   - ❏ What is the organization of the report?
   - ❏ What key terms will be used in the report?

3. Does the discussion
   - ❏ describe the problem that motivated the project?
   - ❏ describe all the work completed during the period covered by the report?
   - ❏ describe any problems that arose, and how they were confronted?
   - ❏ describe the work remaining to be done?

4. Does the conclusion
   - ❏ accurately evaluate the progress on the project to date?
   - ❏ forecast the problems and possibilities of the future work?

5. Do the appendices include any supporting materials that substantiate the discussion?

EXERCISES

Exercises 2 and 3 call for you to write a memo. See Chapter 16 for a discussion of memos.

1. Write a progress report describing the work you are doing for the major assignment you proposed in Chapter 21.

2. Secure a progress report written for a project subsidized by a city or federal agency or a private organization. In a 1,000-word memo in your instructor, describe how the progress report differs from the model that appears in this chapter. What challenges have the writers faced in describing their progress, and how have they met those challenges? Overall, how effective is the progress report? If possible, submit a copy of the report along with your memo.

3. Interview a local professional who writes progress reports. (See Chapter 6 for a discussion of interviewing.) What process does he or she use in creating the report, and what does he or she see as the most important aspects of writing a progress report? To what extent does the information provided by the professional match the information contained in this chapter? Respond in a 1,000-word memo to your instructor.

REFERENCES

Boully, D. (1993). *Progress report on the project to research barcoding systems.* Unpublished document.

Merrick, D. (1993). *Midpoint progress report on the study of Geographical Information Systems uses at M&M Real Estate.* Unpublished document.

| CHAPTER 23 | Completion Reports |

A completion report is the final document in the series that began with the proposal (Chapter 21) and the progress report (Chapter 22). A completion report, written when the work is finished, provides a permanent record of the entire project.

Chapter 21 explained that there are two kinds of deliverables: (1) research reports and (2) goods and services. When you have completed a research project, the deliverable is the completion report itself. When you have completed a goods-and-services project, you provide two deliverables: the goods or services you promised, plus the completion report. In this case, the completion report is an argument that the product meets the specifications. It might also contain other information about the product or service you are providing.

Understanding the Functions of a Completion Report

A completion report has two basic functions:

- *Immediate documentation.* For the sponsors of the project, the report provides the necessary facts and figures, linked by narrative discussion, that enable them to understand the project. For research projects, the sponsors want to know how the researcher carried it out, what he or she found, and, most important, what those findings mean. For goods-and-services projects, the sponsors want to know as much as possible about what they have bought. All completion reports lead to a presentation of results, at least. For example, a small research project might call for the writer to determine the operating characteristics of three competing models of a piece of lab equipment. The heart of that completion report will be the results.

  Many completion reports call for the writer to go beyond the results by analyzing them and presenting conclusions. The writer of the report on the lab equipment might inform the readers which of the three machines appears to be the most appropriate for his organization's needs. And finally, many completion reports go one step further and present recommendations: suggestions about how to proceed in light of the conclusions. The writer of the lab equipment report might have been asked to recommend which of the three machines—if any—should be purchased. Conclusions and recommendations are carefully reasoned arguments based on evidence in the report.

- *Future reference.* The completion report is referred to in five common situations:

  - *When personnel change jobs.* A new employee is likely to consult reports to determine the kinds of projects the organization has completed recently. Reports are not only the best source of this information—but also they are often the only source, because the employees who participated in the project might have left the organization.

– *When the organization is contemplating a major new project.* Usually, the organization wants to determine how the new project would affect existing procedures or operations. Completion reports are the best source of this information. For example, if the owner of an office complex wants to computerize the temperature control of her buildings, she will bring in an expert to consult the reports on the electrical wiring and the heating, ventilating, and air-conditioning systems to determine whether computerization is technically feasible and economically justifiable.

– *When a similar project is being contemplated.* When an organization is considering a project that is similar to a past project, employees study the report about the past project to see if they can learn anything that will help them carry out the new project more effectively or more efficiently.

– *When the organization responds to legal or regulatory review.* Employees study the completion report when the organization needs to take legal action, and regulators study the report as part of a regulatory review.

– *When a problem develops after the project has been completed.* Employees will turn first to the project's completion report to figure out what went wrong. Analyzing a breakdown in a production line requires the technical description of the production line—the completion report written when the line was implemented.

In these five situations, completion reports are valuable long after the projects they describe have been completed.

Understanding the Types of Completion Reports

Terminology varies from organization to organization. Some organizations consider procedures manuals and policy descriptions to be completion reports because they report on the completion of projects: to establish procedures or policies. Some organizations consider a proposal to be a completion report because it completes a recognizable phase of a project.

Completion reports at the end of goods-and-services projects vary greatly in size and structure. Sometimes they are brief letters, with a set of operating instructions attached. Sometimes they are conventional reports that begin with a description of the problem or opportunity that led to the project, the methods the researchers used, and their findings—their rationale for the decisions they made.

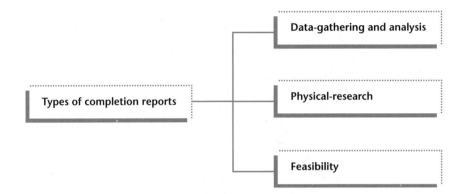

FIGURE 23.1
Types of Completion
Reports

Completion reports at the end of research projects fall into the three broad categories shown in Figure 23.1.

Data-Gathering and Analysis Reports

A data-gathering and analysis report, as the term suggests, brings information together in one place and analyzes it. For instance, if you work for a software manufacturer that provides customer support for its customers, you might commission a data-gathering and analysis report to learn why your customer-support costs are high. In the data-gathering phase, you find out exactly what your customer-support costs are, focusing on which products, and which aspects of the products, are causing your customers the most problems. For example, you might find that most problems are occurring when your customers try to perform a specific type of calculation using your database program. In the analysis phase, you use these data to figure out why this calculation is hard to perform. Data-gathering and analysis reports involve analyzing the data to determine cause-and-effect relationships (see Chapter 8). Until you understand the problem, you can't fix it.

Physical-Research Reports

A physical-research report concerns a project involving substantial empirical research carried out in a lab or in the field. Empirical research lies at the heart of the scientific method: you begin with a hypothesis, conduct experiments to test it, record your results, and determine whether your hypothesis was correct as it stands or needs to be modified or rejected. Empirical research means conducting tests and generating new information. See Chapter 7 for more information on conducting experiments.

For instance, you work for the U.S. Navy. Your supervisors are considering building the hulls of minesweeper ships from glass-reinforced plastic—GRP—instead of wood. GRP would seem to offer several advantages, but first its properties have to be established. How strong is it? How flexible? How well does it withstand temperature variations? The list of questions is long. Each question will be answered by physical research on GRP in the laboratory, on mock-ups of hulls, and even on computer simulations of the material. On the basis of this research, you will argue for or against the use of GRP.

Feasibility Reports   A feasibility report documents a study that evaluates at least two alternative courses of action. For example, should our company hire a programmer to write a program we need, or should we have an outside company write it for us? Should we expand our product line to include a new item, or should we make changes in an existing product?

A feasibility report is an argument that answers three kinds of questions:

- *Questions of possibility.* We would like to build a new rail line to link our warehouse and our retail outlet, but if we cannot raise the cash the project is not possible at this time. Even if we have the money, do we have the necessary permission from government authorities? If we do, are the soil conditions adequate for the rail link?

- *Questions of economic wisdom.* Even if we can raise the money to build the rail link, is it wise to do so? If we use up all our credit on this project, what other projects will have to be postponed or canceled? Is there a less expensive—or a less risky—way to achieve the same goals as a rail link?

- *Questions of perception.* For instance, if your company's workers have recently accepted a temporary wage freeze, they might view the rail link as an unnecessary expenditure. The truckers' union might see it as a threat to their job security. Some members of the general public also might be interested parties. Any sort of large-scale construction might affect the environment. Even though your plan might be perfectly acceptable according to its environmental-impact statement—the study required by the government—some citizens might disagree with the statement or might still oppose the project on aesthetic grounds. Whether or not you agree with the objections, going ahead with the project might create adverse publicity.

The three types of completion report—data-gathering and analysis, physical-research, and feasibility—are often interrelated. For instance, the project on GRP might begin with a data-gathering and analysis report to find out how the material is used in the industry. If the material looks promising, the navy might commission physical research on it. If it still looks promising, the navy might do a full-scale feasibility study involving economic and public-policy issues.

Writing the Completion Report

Whether your completion report is a relatively small document written by one person or a large, complex document written by numerous people, you must analyze your audience carefully before beginning to write. One factor that distinguishes the writing process for completion reports from that for other kinds of documents is the extent to which the completion report relies on previous documents: the proposal and the progress reports.

Analyzing Your Audience

The writing process used for most documents—prewriting, drafting, and revising—is effective for reports. As usual, analyzing your audience is critical because it helps you determine how much information, and what kinds of information, you need to provide. In addition, it helps you determine where to put the information. For instance, if your readers are knowledgeable and do not need much detail to understand the project, you might decide to put much of the information in appendices rather than in the body of the document. This way, the information is available for future reference, but you are not forcing anyone to read it.

Think about the following five questions before you start to write:

- *How well do your readers know your field?* The better they know your field, the less explanation you need to provide in the body of the report. If some of your readers are managers who do not know the field well, consider adding detailed explanations as appendices.

- *What is your standing within the organization?* Writers with strong technical credentials and experience do not need to justify their assertions in as much detail as writers without an extensive track record.

- *How standard is the project?* Some fields have accepted methods for carrying out projects. For example, if in assessing hurricane damage to a physical structure you always perform a visual inspection first, there is no need to explain how you performed the visual inspection or why you performed it. But if the project is unusual or unique, your readers will want to understand why you chose your approach.

- *Why are the readers reading the report?* If they merely need to understand you, you need to provide less information than if they will use your report as the basis for some further action. For instance, if some of your readers will duplicate your methods, you need to provide complete details. Again, it might be appropriate to put the complete description of the methods in an appendix.

- *Are your readers negative, neutral, or positive about the project?* The attitude of the reader might affect the organizational pattern you choose, as described in Chapter 8.

For all reports, however, you must describe and justify your actions carefully. Cite the authority for your methods. And spell out the logic you used to draw conclusions from your results. Be equally careful when explaining how you derived your recommendations from your conclusions.

Using the Proposal and Progress Reports Effectively

The fact that the completion report relies so much on the proposal and progress reports makes your job easier in one sense: much of the writing is already done. The basic elements of the introduction are already written, as is the methods section and some of the results.

However, relying on the previous documents can be a challenge as well, for your understanding of the project has developed and improved over the course of doing the work. Some of the things you wrote in the earlier document will seem to you naive, incomplete, or even wrong.

Is that a problem? No. Just tell the truth. Your progress reports might already have informed the reader that the project will not turn out as it was described in the proposal. In the completion report, explain the situation as objectively as you can. If what you need to say now is clearly at odds with what you said in an earlier document, point out and explain the discrepancy. No one expects the writer to understand the subject as fully at the start of the project as at the end—and no one expects the project to have proceeded exactly as planned.

Structuring the Completion Report

Like a proposal or a progress report, a completion report has to make sense without the author there to explain it. You can never be sure when your report will be read—or by whom. All you can be sure of is that some of your readers will be managers who are *not* technically competent in your field and who need only an overview of the project, and that others will be technical personnel who *are* competent in your field and who need detailed information. There will be yet other readers, of course—such as technical personnel in related fields—but in most cases the divergent needs of managers and of technical personnel are all you need to consider.

To accommodate these two basic types of readers, a completion report today generally contains an executive summary that precedes the body. These two elements overlap in their coverage but remain independent; each has its own beginning, middle, and end. Most readers will be interested in one of the two, but probably not in both. As a formal report, the typical completion report will contain the following standard elements:

- front matter
  - title page
  - abstract

　　　　　　　　　　– table of contents
　　　　　　　　　　– list of illustrations
　　　　　　　　　　– executive summary
　　　　　　　　　　– glossary
　　　　　　　　　　– list of symbols
　　　　　　　　• body
　　　　　　　　• back matter
　　　　　　　　　　– appendices

This chapter will concentrate on the body of a completion report; the other elements common to most formal reports are discussed in Chapter 11.

The Components of the Body of the Completion Report

The body of a typical completion report contains the five elements shown in Figure 23.2.

In the *introduction,* you explain the context of the report: the problem you were investigating, as well as your purpose. In the *methods* section, you describe what you did in carrying out the project. In the *results* section, you present the information you found or created. Like the methods section, the results section is a simple statement of facts; you do not interpret the information. In the *conclusions,* you interpret the results by explaining what they mean. And in the *recommendations,* you suggest to your readers what they should do next. This organization, which is used throughout engineering and the sciences, enables different readers to find the information they need in the report.

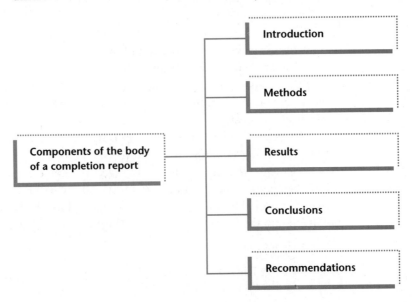

FIGURE 23.2

Components of the Body
of a Completion Report

Some writers like to draft these elements in the order in which they will be presented. These writers like to compose the introduction first because they want to be sure that they have a clear sense of direction before they draft the discussion and the findings. Other writers prefer to put off the introduction until they have completed the other elements of the body. Their reasoning is that in writing the discussion and the findings they will inevitably have to make some substantive changes; therefore, they would have to revise the introduction if they wrote it first. In either case, brainstorming and careful outlining are necessary before you begin to draft.

The Introduction

The introduction enables readers to understand the technical discussion that follows. Start by analyzing the audience and then considering the standard questions to which all introductions respond.

In analyzing your audience, think about the questions mentioned earlier in this chapter:

- How well do your readers know your field?
- What is your standing within the organization?
- How standard is the project?
- Why are the readers reading the report?
- Are your readers negative, neutral, or positive about the project?

In drafting the introduction, consider the following seven questions:

- *What is the subject of the report?* You can probably copy this from the proposal or progress reports and then modify it as necessary.

- *What is the purpose of the report?* The purpose of the report is not the purpose of the project. The purpose of the report is to present information to enable readers to understand a subject, to affect readers' attitudes toward some subject, or to enable readers to carry out a task. Be as specific as you can. See Chapter 5 for a discussion of purpose.

- *What is the background of the subject?* You can probably copy this information from a previous document. Remember, in describing the background, you are addressing readers who might not have read your previous documents and who therefore might be ignorant of the project; do not assume that they have been following along. Even those readers who have read the previous documents might need a refresher.

- *What are your sources of information?* Review the relevant literature, either internal reports and memos or external published articles or even books that help your readers understand the context of your work. Sometimes interviews are a valuable source of information.

- *What is the scope of the report?* Indicate what topics you are including, as well as what you are not including.

- *What is the organization of the report?* Indicate what organizational pattern you will use in the report, so readers can understand where you are going and why.
- *What key terms will be used in the report?* If you will introduce any new terms, the introduction is an appropriate place to define them.

For more information on introductions, see Chapter 8.

Figure 23.3 is an introduction to the body of a completion report. The subject of the report (Corder, 1993) is an investigation to determine whether the Quality Assurance (QA) lab at a microchip manufacturer should purchase a scanning electron microscope (SEM). The report is titled "QA Lab SEM Evaluation: Project Completion Report." Marginal notes have been added.

The Methods   In drafting the methods section, analyze your audience, and then provide the appropriate kind and amount of information.

To analyze your audience, think about the questions mentioned earlier. Consider your readers' knowledge of the field, their perception of you, the uniqueness of the project, and the readers' purpose in reading and their attitudes toward the project.

Note: Reports are often double spaced.

The background of the project.

An overview of the project.

The organization of the report.

Introduction

 In his memo of January 12, 1993, the Director of Quality Assurance authorized the purchase of a modern scanning electron microscope (SEM). However, the criteria for choosing a SEM were not established. In the current project I have established the criteria for evaluating the different SEMs on the market and evaluated them. This report describes the project and includes my recommendation.

 Early in March I undertook a project to recommend which SEM to add to the Quality Assurance (QA) lab. This project was undertaken in three phases.

- Phase I involved evaluating SEM requests and interviewing QA SEM lab customers. From this information I developed a set of criteria.
- Phase II involved contacting the SEM manufacturers and evaluating their systems based on our criteria.
- Phase III involved compiling the information in this report, which presents my conclusions and recommendation.

First I present my methods, followed by the results, conclusions, and recommendation.

FIGURE 23.3

Introduction to a Completion Report

TABLE 23.1	*Information in the Methods Section*	
Type of Completion Report	**Kind of Information to Provide in the Methods Section**	**Comments**
DATA-GATHERING AND ANALYSIS	Describe the following information sources you used: • the documents you studied • the people you telephoned or interviewed • the calculations or statistical procedures you performed	Explain how you carried out the project so that the readers can understand your thinking. If appropriate, justify your methods; that is, explain why you used this methodology rather than another.
PHYSICAL-RESEARCH	List the equipment and materials you used, then describe the procedure.	If your readers merely want to understand your methods, you might use traditional paragraphs. If they want to duplicate your methods, you might switch to a numbered list format. If the information is too detailed for some or most of your readers, consider moving it to an appendix. If appropriate, justify your methods.
FEASIBILITY	Describe what you did, including the following: • physical experiments • theoretical studies • site visits • interviews • library research	Feasibility studies are often quite complex and involve major decisions. Be sure to explain your methods sufficiently; otherwise, readers will question not only your methods but also your results, conclusions, and recommendations.

Table 23.1 shows the kind of information to provide in the methods section for each of the three types of completion reports.

Figure 23.4 on page 638 shows the methods section from the report on the SEM.

The Results   The results are the data you discovered or created. Present the results objectively, without commenting on them; save the interpretation of the results—the conclusion—for later. If you combine results and conclusions, your readers might be unable to follow your reasoning process. Consequently, they will not be able to tell whether your conclusions are justified by the evidence—the results.

Just as the methods section answers the question "What did you do?" the results section answers the question "What did you see?"

Note: Reports are often double spaced.

Methods

To carry out this project, I performed four tasks:

1. I studied all the internal QA records of the requests that the SEM lab has received over the last year. For each request I noted the department of the requester and the kind of information requested. For example, some requesters needed to determine the location of a fault. Others needed to determine the nature of the fault or needed an image of the fault.
2. I interviewed the 12 requesters to find out how well the lab is meeting their needs and what further information would help them do their jobs. I used a standard script (see Appendix B) that combined closed-ended and open-ended questions.
3. I wrote to the six North American SEM manufacturers (see Appendix C) to get specifications, warranties, and prices on the available systems that would fulfill the requirements I established on the basis of the requests and the interviews.
4. I analyzed the information from the SEM manufacturers and the independent reviews of the equipment in the professional journals (see References).

Notice how specifically the methods are described.

FIGURE 23.4

Methods Section of a Completion Report

The nature of the project will help you decide how to structure the results. Table 23.2 shows the kind of information typically found in the results section of a completion report.

Of course, always try to consider the needs of your readers. How much they know about the subject, what they plan to do with the report, what they anticipate your recommendation will be—these and many other factors will affect your decision on how to structure the discussion.

For instance, suppose that your company is considering installing a LAN. In the introduction you have already discussed the company's current system and its disadvantages. In the methods section, you have described how you established the criteria to apply to the available systems, as well as the research procedures you carried out. In the results section, you provide the details of each LAN you are considering.

Figure 23.5 on pages 640–642 is an excerpt from the results section of the SEM report.

Conclusions   Conclusions are the implications—the "meaning"—of the results. Drawing valid conclusions from results requires great care. Suppose, for example, that you work for a company that manufactures and sells clock radios. Your records tell you that in 1994, 2.3 percent of the clock radios your company produced were

TABLE 23.2	*Information in the Results Section*	
Type of Completion Report	**Kind of Information to Provide in the Results Section**	**Comments**
DATA-GATHERING AND ANALYSIS	• Quotations from people you interviewed and documents you studied. • The results of any statistical computations you performed.	Organize the information topically; that is, discuss all the results pertaining to one aspect of the project in one place. Do not organize the results according to the nature of the source.
PHYSICAL-RESEARCH	Data that you observed or created through inspection, calculation, or experimentation.	Present the results as a brief series of paragraphs and graphics. If in the methods section you described a series of tests, in the results section you simply report the data in the same sequence you used for the methods.
FEASIBILITY	The same kinds of information presented in data-gathering and physical-research reports: • Quotations from people you interviewed and documents you studied. • Data that you observed or created, through observation, calculation, or experimentation. • The results of any statistical computations you performed.	The comparison-and-contrast structure (see Chapter 8) is usually the most accessible. When you are evaluating a number of different alternatives, the whole-by-whole pattern, with the best alternative first, might be most appropriate. When you are evaluating only a few alternatives, the part-by-part structure might work best.

returned as defective. An analysis of company records over the previous five years yields these results:

Year	% Returned As Defective
1993	1.3
1992	1.6
1991	1.2
1990	1.4
1989	1.3

One obvious conclusion can be drawn: a 2.3 percent defective rate is much higher than the rate for any of the last five years. And that conclusion is certainly a cause for concern.

But do those results indicate that your company's clock radios are less well made than they used to be? Maybe. To reach a reasonable conclusion from these results, you must consider two other factors:

Note: Reports are often double spaced.

The writer describes his interviewing techniques.

Job Requests and Interviews

I began by collecting all SEM job requests submitted to the QA SEM lab over a period of one year. In analyzing these job requests, I found that the largest proportion (some 75%) of the job requests came from either the failure-analysis engineers in QA and product engineering, or from the R&D design engineers. Most of the requests were similar: find a specific bit or location on an individual die, and image the area to find defects or failure mechanisms.

I simultaneously began interviewing the 12 requesters. I asked them if our lab provided them with adequate images to define failure mechanisms, and if there were other functions that would make their identification of defects and locations simpler, quicker, or better in some other way.

The answers to these questions were that the images and data provided by the QA SEM lab are of sufficient resolution and contain adequate information. However, many requesters expressed a desire to have the images on the network and available for manipulation with an image processor, such as Semicaps. According to the requesters, this would accomplish three things for the engineers:

- It would allow them to include images in their reports through Harvard Graphics or other similar software.
- It would allow the engineers to make adjustments to the contrast, brightness, gamma, and other aesthetic characteristics, at their local workstation.
- It would allow the images to be archived electronically, rather than saving file cabinets full of Polaroid photos, as is currently done.

Determining Criteria

The writer explains his criteria carefully.

Using the information I obtained through the analysis of requests and the interviews of requesters, I developed a set of criteria for evaluating the various SEMs available on the market. The features I specified as criteria for the QA lab are the following:

- Turbomolecular vacuum pump (TMP)
- Lanthanum hexaboride electron gun (LaB6 Cathode)
- Digital imaging
- Solid state backscattered electron detector (BSE)
- 100 mm or larger specimen stage
- Eucentric Stage Controls
- Minimal Integral Image Processing

Turbomolecular Pump

Of the two major types of vacuum pumps used on SEMs—diffusion and turbomolecular—turbomolecular will better meet our needs.

Diffusion pumps use a heated oil vapor to remove air molecules. These pumps are quite efficient and relatively inexpensive to operate, but simple

FIGURE 23.5

Excerpt from the Results Section in the Body of a Completion Report

(Figure 23.5 continued)

errors can cause the oil to backstream into the specimen chamber and even into the column. Backstreaming oil causes many problems once inside the SEM column.

Turbomolecular pumps use a series of very high speed, rotating turbines to remove air molecules. These pumps are very dependable, very fast, and very efficient. Most important, however, they cannot backstream oil into the column; therefore, they are a much safer type of vacuum pump for a SEM.

Lanthanum Hexaboride Electron Gun (LaB6)

The lanthanum hexaboride electron gun (LaB6) is the best of the three common types of electron guns used in SEMs.

- The most common is a tungsten gun. Tungsten is the material used for filaments in light bulbs. Although the value of tungsten guns is well documented, they have problems. A tungsten filament (the consumable portion of the gun) is very inexpensive, at about $25 each, but typical lifetimes are only 20–25 hours of use.
- The LaB6 gun uses an LaB6 crystal instead of a tungsten filament. There are several advantages to this. First, LaB6 puts out more electrons at lower accelerating voltages than does tungsten. Second, LaB6 filaments last several hundreds of hours. The trade-off is that LaB6 filaments cost close to $600 each. The extended use and the enhanced signal, however, more than make up for the extra cost of replacement.
- The field-emission (FE) gun uses a slightly different technique. FE guns put out even more electrons at lower accelerating voltages than LaB6, but they are much more delicate to operate. FE guns require a higher vacuum; therefore, more vacuum pumps. FE guns are temperamental and difficult to use. They are also typically very expensive to operate, up to 50 percent more expensive. The difference in low accelerating voltage does not justify the use of an FE gun in the QA lab.

Digital Imaging

Digital imaging capability, a requirement for a new SEM, is a standard feature on most new SEMs.

Other Criteria

All modern SEMs meet our other four criteria:

- Solid state backscattered electron detector (BSE)
- 100 mm or larger specimen stage
- Eucentric stage controls
- Minimal integral image processing

(Figure 23.5 continued)

The writer explains the
next phase of his project.

Contacting SEM Manufacturers

I contacted the SEM manufacturers and asked them to provide brochures
and information on the SEMs they have available.

We had assumed that because ten years have passed since we purchased our
JEOL 840, many advances had been made in SEM technology. That assumption
proved to be false. The technology of producing images with a scanning electron beam has not changed significantly. No dramatically different columns
or detectors have been developed. The major changes include only two
additions:

- Nearly all SEMs now produce digital video images, as opposed to the
  analog imaging of our 840.
- Nearly all SEMs now include much higher degrees of computerization
  and integral image processing.

The computerization of SEMs is very much like the Semicaps software
already mentioned. The difference is that the image processors that are integral
to many SEMs are not compatible with systems we already have in use in QA.
Nearly any image processor can manipulate a digital image, and most SEMs
built today produce digital images. Therefore, "new" SEM technology is fairly
old.

- *Consumer behavior trends.* Perhaps consumers were more sensitive to
  quality in 1994 than they had been in previous years. A general increase
  in awareness of quality—or a widely reported news item about clock
  radios—might account in part or in whole for the increase in consumer
  complaints. To test this hypothesis, you would try to find out whether
  other manufacturers of similar products have experienced similar patterns of returns.
- *Your company's policy on defective clock radios.* If a new, broader policy
  was instituted in 1994, the increase in the number of returns might imply
  nothing about the quality of the product. In fact, the clock radios sold in
  1994 might even be better than the older models.

In other words, beware of drawing hasty conclusions. Examine all the relevant
information.

Just as the results section answers the question "What did you see?" the
conclusions section answers the question "What does it mean?"

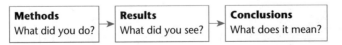

For more information on causal reasoning and on conclusions, see Chapter 8.

The conclusions section of the SEM report is shown in Figure 23.6.

Note: Reports are often double spaced.

Conclusions

Based on its technical capabilities, price, and service record, the JEOL 5100 SEM is the best option for us.

In addition to Amray and JEOL, I interviewed representatives from Hitachi, Olympus, ISI, Cambridge, and Electroscan. Each of these companies offer SEMs that meet the basic criteria I established. I have evaluated the differences that each of these companies highlights. I found none of these features to be significant for our application. In addition, our experience is quite limited with most of these companies.

On the other hand, we have considerable experience with both Amray and JEOL. We have a good idea of their reliability; we have established service contracts with both companies. Therefore, we have established relationships with their field-service engineers. In addition, we have some stock of spares from each company, many of which are applicable to their new models.

With all these factors considered, Amray and JEOL were the only manufacturers whom I invited to visit our site for more thorough discussions and to provide system quotes.

Amray quoted a model 1810 SEM. The total price from Amray is $100,000. JEOL quoted a model 5100 SEM. The total price of the JEOL system is $115,500. While the JEOL appears 15% higher in price, we have learned from experience that the Amray will require a retrofitted Eddie Fjeld stage before it is operational. The Eddie Fjeld stage suitable for the Amray model 1810 is available for $28,000, installed. Therefore, the actual price for the Amray 1810 is $128,000.

However, beyond the price difference, the SEM technicians, the equipment engineers, and I all have greater confidence in the JEOL service organization on the basis of our excellent relationship with them over the years.

The writer explains his reasoning carefully. Notice that he describes, in the last paragraph, the unanimous opinion of his colleagues.

FIGURE 23.6

Conclusions Section from the Body of a Completion Report

Recommendations

Recommendations are statements of action. Just as the conclusions section answers the question "What does it mean?" the recommendations section answers the question "What should we do now?"

Methods	**Results**	**Conclusions**	**Recommendations**
What did you do?	What did you see?	What does it mean?	What should we do?

As you draft your recommendations, consider four factors:

- *Content.* When you tell your readers what you think they should do next, be clear and specific. If the project you are describing has been unsuccessful, don't simply recommend that your readers "try some other alternatives." What other alternatives do you recommend, and why?
- *Tone.* When you recommend a new course of action, you run the risk of offending whoever formulated the earlier course. Do not write that your new direction will "correct the mistakes" that have been made recently. Instead, write that your new action "offers great promise for success." If the previous actions were not proving successful, your readers will probably already know that. A restrained, understated tone is not only more polite but also more persuasive: it indicates that you are interested only in the good of your company, not in personal rivalries.
- *Form.* If the report leads to only one recommendation, use traditional paragraphs. If the conclusion of the report leads to more than one recommendation, use a numbered list.
- *Location.* The recommendations section is always placed at the end of the body; because of its importance, however, it is often also summarized—or inserted verbatim—after the executive summary.

Figure 23.7 shows the recommendations section of the SEM report.

Note: Reports are often double spaced.

FIGURE 23.7

Recommendations Section from the Body of a Completion Report

Recommendation

I recommend that we purchase a JEOL 5100 SEM configured with the features listed in the report and on the quotation. The total system price will be $115,000. Installation will be another $1,000.

This report delivers good news: a relatively small expenditure (for equipment of this sort) will improve the capabilities of the QA engineers. Often, however, feasibility reports such as this one yield mixed news or bad news; that is, none of the available options would be an unqualified success, or none of the options would work at all. Don't feel that a negative recommendation reflects negatively on you. If the problem being studied were easy to solve, it probably would have been solved before you came along. Give the best advice you can, even if that advice is to do nothing. The last thing you want to do is to recommend a course of action that will not live up to the organization's expectations.

Sample Completion Report

The following completion report (Boully, 1993) is a follow-up to the writer's proposal (Chapter 21) and progress report (Chapter 22).

Autosort Pre-sort Mail Division
2314 Miller Street Suite C
Boise, ID 83705

April 28, 1993

Mr. Chuck Harding
Autosort Mail Services
600 S. 8th Street
Boise, ID 83705

Dear Mr. Harding:

The purpose, methods, and background of the project.

Attached is the completion report of the project to determine which barcoding system would best fulfill Autosort's needs. I compiled the information in this report by reading about barcoders in postal trade journals and by on-site inspections and tours. I inspected the manufacturing plants of three barcoding systems. I also toured three mail companies similar to Autosort that are using these systems and saw how these companies use the three brands of automated barcoding systems in their daily mail processing.

More information on the methods.

I evaluated each system according to the criteria that were developed from Autosort's needs and from suggestions found in postal trade journals. The main goal was to find a system that costs less than $350,000 and yet is still effective and reliable.

The results, conclusions, and recommendation.

All of the systems that I researched meet Autosort's requirements with good quality and acceptable price, but one system was clearly superior: Bell and Howell's model 1500. This machine not only exceeds all of the minimum criteria but also has an innovative feeding system that moves the thinnest letters through the barcoder with minimum damage. I was very impressed with the design, quality, and customer service and highly recommend that this system be purchased by Autosort.

A polite offer to provide more information.

I would like to thank you for the opportunity to do this research. If you have any questions about my recommendation or the report, I would be glad to answer them at your convenience.

Sincerely,

Danny J. Boully
Pre-sort Division Supervisor

The title indicates the
subject and purpose of
the report.

**Recommendation for Autosort's Purchase
of a Barcoding System**

Prepared for: Chuck Harding, Owner
                 Autosort Mail Services

by: Danny J. Boully
                 Pre-sort Supervisor
                 Autosort Pre-sort Mail Division

                 April 28, 1993

Note: Reports are often
double spaced.

The abstract focuses on
the technical specifica-
tions of the machines.

Abstract
"Recommendation for Autosort's Purchase of a Barcoding System"
by Danny J. Boully

The United States Postal Service has announced that all large business mailers must barcode their mail so the Postal Service may increase their use of automated sorting equipment. To maintain our pre-sort division, Autosort must purchase a barcoding system. Initial research narrowed the number of potential
systems down to three. This report presents the criteria for judging those systems, as well as the information I gathered about the three systems. After working with all three systems, I feel that all of them are high-quality machines that
will meet Autosort's minimum needs but that one system is clearly superior to
the other two: the Bell and Howell model 1500. Offering 16–40 bins, the 1500
can process 36,000 pieces per hour, its computer uses the new address-verification system, it is upgradable, and it has a good repair and support record. It will
feed letters as thin as tracing paper or as thick as 1/4 inch (the maximum thickness for pre-sort mail) with only minor adjustments. For these reasons, I recommend that Autosort purchase the Bell and Howell model 1500 to meet its barcoding needs.

i

Contents

The executive summary
concentrates on manage-
rial concerns—cost, training, maintenance—not on
technical aspects of the
project.

Executive Summary

The Postal Service has announced that large business mailers such as
Autosort must barcode their mail by February 1, 1995. Autosort's management
has decided to purchase a barcode system in the near future. This purchase will
give Autosort a strategic advantage over its competition in the pre-sort market.

Three models of barcode machines might fulfill Autosort's needs. I was sent
to gather information on these three models by visiting the manufacturing
plants that produce them and companies that are using them.

The on-site research provided all the information I needed to make an
informed recommendation as to which system Autosort should purchase: a Bell
and Howell model 1500.

At a cost of $320,000, the Bell and Howell is an expensive machine, but it is
the most sophisticated of the available machines, with an advanced feeder system that will process all the mail we currently handle, from the thinnest to the
thickest, without jamming. It is a highly reliable barcoder with an excellent service record. In addition, it does not require expensive training for our personnel. The Bell and Howell 1500 is clearly the best choice for Autosort's pre-sort
mail division.

The writer fills in the background, the purpose, the methods, and the organization of the report.

Introduction

The United States Postal Service has ruled that business customers must use barcoded mail by February 1, 1995 to aid in the greater automation of the USPS's mail sorting systems. Autosort's management decided to purchase a barcoding system before this deadline to prevent a serious loss of profit in its pre-sort division and to comply with the new regulation before the deadline. In addition to preventing loss of profit, purchasing a system in the near future will give Autosort a strategic advantage in the Boise market. My research found three barcoding systems that might meet Autosort's unique needs. I traveled to three cities to evaluate the manufacturers of these systems and to tour pre-sort companies that are similar to our own.

My goals in this project were to gather information for the purchase of the best system for Autosort, to learn how the barcoding systems work, and to learn how to do the new USPS barcode paperwork.

In this completion report I recommend the system that I believe best meets our needs. First I describe the methods I used: library research followed by the visits. Then I evaluate each of the three systems and provide a decision matrix table for the three systems. Finally, I present my conclusions and recommendation.

The writer briefly lists his working methods.

Methods

To carry out this project, I performed the following tasks:

1. Did library research on barcoders. I checked *Engineering Index* and *Business Periodicals Index* for the past five years. And I checked the two postal journals—*Direct Mailers* and *Postal Times*—for the last ten years.
2. Determined our needs on the basis of the library research; interviews with you and Bob Stipe, the foreman; and my knowledge of Autosort.
3. Visited the three leading manufacturers and three of their customers. These visits enabled me to see the equipment up close and to interview the workers.
4. Solicited and received bids from the three finalists.
5. Analyzed the information the manufacturers provided, as well as the information I gained from my trips, against our needs.

In the first section of the results, the writer presents a table he already presented in his proposal and progress report.

Results

First I present the basic data on the six leading barcoders. Then I present the criteria against which I evaluated the barcoders. Next I describe in detail the three barcoders that best meet our criteria. Finally, I present the decision matrix I used in evaluating the three barcoders.

Basic Data on Six Leading Barcoders

This chart was compiled from the information in Beardly (1992) and Parson (1992).

Manufacturer	Cost ($000)	Pieces (000/hr)	Number of bins	Computer	Upgrade	11 digit	Down time (weeks)	Support	Repair
Bell and Howell	$320	36	16–40	yes	yes	yes	2	yes	yes
Franklin	$120	10	10	no	no	no	1	yes	no
Mason Mail	$335	36	16–40	yes	no	yes	9	yes	no
Pitney Bowes	$300	30	16–40	yes	yes	yes	4	yes	yes
Postal Technologies	$340	36	10–50	yes	yes	yes	2	no* (see repair)	total self-repair
Pre-sort Specialties	$350	30	20–38	no	yes	no	6	yes	no

*Postal Technologies will train our people at their factory to do all necessary repairs and will sell us the parts, but they do not have local salespeople.

Key:
Manufacturer: The company that produces the system.
Cost: The estimated cost of the system in thousands of dollars.
Pieces: Estimated number of letters the system can scan and sort in one hour.
Number of bins: The minimum and maximum number of sorting bins that can be purchased with the system. The more bins a system has, the more efficient it will be.
Computer: The system's computer should use the new address-verification program that checks each letter to make sure it is headed to a valid address. Also, the computer system itself should be high quality and upgradable.
Upgrade: The system should be upgradable as needs change.
11 digit: This is the newest improvement in barcoding. It shows the zip+4, as well as where the house is on that street.
Down Time: This refers to system down time (in weeks): how often the system has to be off line for repairs and maintenance
Support: This grades the customer support that is available from the factory and service representatives.
Repair: Most repairs and maintenance on the system should be simple enough for the lead worker to perform.

Establishing Criteria and Eliminating Systems

To best evaluate these six systems for Autosort I used the needs of our company (as defined in your memo to me dated February 3, 1993) and suggestions from the two articles I found during my research. The resulting criteria for evaluating the barcoding systems were that they should:

The writer repeats another element: the criteria.

3

- cost less than $350,000
- process at least 20,000 pieces of mail per hour
- have at least 40 sorting bins
- use the new address computer system
- be able to be upgraded as Autosort grows
- use the new 11-digit destination barcode
- have good local customer support
- need less than two weeks of training time for the lead operator to run the system
- be designed so that most repairs can done by the lead operator to minimize down time

The first four of these criteria are absolute requirements based on our capital budget, the quantity of our business, and the demands of the new postal regulations. The remaining five criteria are desired characteristics that will help us determine the best machine to buy in case more than one meets the absolute characteristics.

Evaluations of the Three Systems

Here I evaluate the three systems that meet our absolute requirements, and then I present a decision matrix table.

Bell and Howell

The Bell and Howell model 1500 is very impressive. The plant foreman, Wesly Crusher, showed me a new feature they have developed in their feed mechanism. It will feed letters as thin as tracing paper or as thick as 1/4 inch (the maximum thickness for pre-sort mail) with only minor adjustments. The machine can handle our thinnest mail at a rate of 35,000 pieces per hour. This is a practical work rate that includes the normal stops and starts of a work period. I estimate the normal workload of a trained lead worker would be nearly 37,000 pieces of mail per hour.

Short Stamp has expanded its operations into a new building and runs mail to the post office twice a day to meet its increased volume. Ben Cisco, the owner, said that the training was conducted in their business and that the trainer was helpful in customizing the computer system to their needs. Training was done during the light morning hours, and they practiced on the type of mail that Short Stamp was actually taking to the post office. During their practice runs they identified three-digit groups that would qualify and added them to the bin system. I saw Ben Cisco and his crew handle common adjustments and routine maintenance. He said most of the problems can be handled by the crew, but if something major goes wrong, the service representative has someone on call during working hours.

The writer provides the narrative evaluations of the two systems that he presented in the progress report, then adds the evaluation of the third system.

In this completion report, the writer sequences the evaluations of the three systems from the highest-scoring system to the lowest.

4

Postal Technologies

Postal Technologies, in Tulsa, Oklahoma, was originally a pre-sort and bulk mail company like Autosort, but they decided to expand into the production of barcoding equipment.

Their model 2A has been designed to be totally repaired by the user. It is built along a modular design, so if something goes wrong the lead worker just unplugs that part and replaces it with a new one. The quality of the machine appears to be high, and the training the lead workers receive—two weeks at the factory—is the best of the three companies. Postal Technologies will train two workers as part of the purchase price, but if we want to train more it will cost $3,000 per person for the course in Tulsa.

Odo's Mail House uses this system for all of its pre-sort and some of its bulk-mail barcoding. Larry Odo, the company president, has been very pleased with the function of the system and attributes the doubling of his business to the machine's high productivity, low damage rate, and low down time. He loves the ability to repair the machine on his own but stresses the need to keep a supply of spare parts on hand. One disadvantage is that the manual is extremely confusing.

With Autosort's fairly high turnover rate, keeping fully trained personnel available might be difficult.

Pitney Bowes

Pitney Bowes is the same company we lease our meters from. They have a good machine and system, and it is the lowest priced system of the three we are considering. Their machine is rated at 36,000 pieces per hour, but only with standard letters. With the thinner billings that we send every day, the machine can run only 15–20,000 pieces per hour because they tend to fold up and jam the machine at higher speeds. The training would be done in our shop, about twice a week. The Pitney Bowes training specialist is stationed in Salt Lake and would travel here to teach us how to run the new system. Portions of the system can be upgraded, but upgrades require service calls from the factory technicians.

The California State Mail Distributor is run by Myra Kira, who is somewhat dissatisfied with the customer service for her barcoder. While I was there, a rubber roller came apart and sent fragments through most of the system. Although this problem is rare, the repair person did not arrive for three hours. Myra says this is a typical response time because she calls an 800 service number and they try to find someone who can come out to fix the problem. While they waited, Myra's people went through the system themselves and cleaned up the mess, so all the service representative had to do was replace the roller.

5

The writer completes the decision matrix by adding the third system.

When he arrived, he replaced the part, checked for additional damage, and left. The bill was the minimum service charge, which is $250 plus parts. This is very similar to the customer service we receive from Pitney Bowes when we need service for our meters.

Decision Matrix

I gave the three systems a grade of 1 to 10 on each criterion, with 1 being the lowest and 10 the highest.

Manufacturer	Value	Pieces per hr.	Number of bins	Computer	Upgrade	11 digit	Down time (weeks)	Support	Repair	Total points
Bell and Howell	7	9	9	10	9	10	8	8	8	78
Pitney Bowes	8	6	9	10	6	10	6	6	6	67
Postal Technologies	6	8	10	10	9	10	8	0* (see repair)	10	71

*Postal Technologies will train our people at their factory to do all necessary repairs and will sell us the parts, but they do not have local salespeople.

Key:

Manufacturer: The company that produces the system evaluated on the chart.

Value: This score was reached by comparing the total quality and features of the system with the price.

Pieces per hr.: This score was reached by comparing how many letters the system actually scanned per hour with what the advertising claimed the system could do.

Number of bins: This score was reached by comparing the number of bins that are available with the system with Autosort's needs. The more bins a system has, the more efficient it will be.

Computer: This score was reached by verifying that the system uses the new address system. Because it is a yes or no answer, 0 was given for a no and 10 for a yes. Also the computer system itself should be good quality and upgradable.

Upgrade: This score was reached by finding out how expensive and difficult it would be to expand or improve the system.

11 digit: This score was reached by verifying that the system uses the new barcoding system. Because it is a yes or no answer, 0 was given for a no and 10 for a yes.

Down time: This score was reached by comparing the maintenance schedules of the three systems. The more time the manual suggests for down time, the lower the score.

6

Support: This score was reached by assessing the customer support that is available from the factory and service representatives.

Repair: This score was reached by comparing how many of the normal break downs and how much of the maintenance can be done by the lead worker without calling the sales representative.

Total points: This is the total number of points for each system.

The writer presents and justifies his conclusions about the three systems.

Conclusions

Each system will perform the basic functions that Autosort requires from an automated barcoding system. All three systems will read the addresses of our customer's mail and sort that mail into bins for packaging. The Bell and Howell system scored the highest on the chart, and has the best feeding system that I worked with. Postal Technologies has an excellent system, but with Autosort's high employee turnover, I do not think the training of new personnel would be cost effective. The Pitney Bowes system is the slowest of the three I investigated, and has the lowest-quality customer support.

The writer presents and justifies his recommendation.

Recommendation

I recommend that Autosort purchase a Bell and Howell model 1500 to fulfill its need for an automated barcoding system. The training will be done in our office, which will allow us to practice on our type of mail and customize the system to meet our needs. Also, the system will be supported by a local sales representative, and it has the best feeding system of the three proposed systems. With this system, Autosort will truly become "Automated Sorting" and move into the next generation of mail sorting technology.

References

Beardly, F. (1992, November). How to grow into big business mailings. *Direct Mailers,* pp. 24–40.

Parson, J. (1992, July). The barcoder for your growing business. *Postal Times,* pp. 56–59, 67–70.

7

The following checklist applies only to the body of the completion report. The checklist pertaining to the other report elements is included in Chapter 11.

1. Does the introduction
   - ❏ explain the subject of the report?
   - ❏ explain the purpose of the report?
   - ❏ explain the background of the report?
   - ❏ describe your sources of information?
   - ❏ indicate the scope of the report?
   - ❏ explain the organization of the report?
   - ❏ define any key terms that will be used in the report?

2. Does the methods section
   - ❏ describe your methods in sufficient detail?
   - ❏ list or mention the equipment or materials?

3. Are the results presented
   - ❏ clearly?
   - ❏ objectively?
   - ❏ without interpretation?

4. Are the conclusions
   - ❏ presented clearly?
   - ❏ drawn logically from the results?

5. Are the recommendations
   - ❏ clear?
   - ❏ objective?
   - ❏ polite?
   - ❏ expressed in an appropriate form: list or paragraph?
   - ❏ placed in an appropriate location?

EXERCISES

Exercises 2 and 3 call for you to write a memo. See Chapter 16 for a discussion of memos.

1. Write the completion report for the major assignment you proposed in Chapter 21.

2. Secure a completion report written for a project subsidized by a city or federal agency, a private organization, or a university committee or task force. In a 1,000-word memo to your instructor, describe how the completion report differs from the model that appears in this chapter. What challenges have the writers faced in describing their project, and how have they met those challenges? Overall, how effective is the report? If possible, submit a copy of the report along with your memo.

3. Interview a local professional who writes completion reports. (See Chapter 6 for a discussion of interviewing.) What process does he or she use in creating the report, and what does he or she see as the most important aspects of writing a completion report? To what extent does the information provided by the professional match the information contained in this chapter? Respond in a 1,000-word memo to your instructor.

REFERENCES

Boully, D. J. (1993). *Recommendation for Autosort's purchase of a barcoding system.* Unpublished document.

Burrows, S. (1994). *Health club membership for New Beginnings clients: A comparison study.* Unpublished document.

Corder, Z. (1993). *QA Lab SEM evaluation: Project completion report.* Unpublished document.

Appendix A
Handbook

This handbook concentrates on style, punctuation, and mechanics. Where appropriate, it defines common errors directly after discussing the correct usage.

Many of the usage recommendations made here are only suggestions. If your organization or professional field has a style guide that makes different recommendations, you should of course follow it.

Also, note that this is a selective handbook. It cannot replace full-length treatments, such as the handbooks often used in composition courses.

Sentence Style

Avoid Sentence Fragments

frag

A *sentence fragment* is an incomplete sentence. Most sentence fragments are caused by one of two problems:

- a missing verb

FRAGMENT: The pressure loss caused by a worn gasket.

This example is a fragment because it lacks a verb. (The word *caused* does not function as a verb here; rather, it introduces a phrase that describes the pressure loss.)

COMPLETE SENTENCE: The pressure loss *was caused* by a worn gasket.

In this revision, *pressure loss* has a verb: *was caused*.

COMPLETE SENTENCE: We identified the pressure loss caused by a worn gasket.

In this revision, *pressure loss* becomes the object in a new main clause: *We identified the pressure loss*.

FRAGMENT: A plotting program with clipboard plotting, 3D animation, and FFTs.

COMPLETE SENTENCE: It *is* a plotting program with clipboard plotting, 3D animation, and FFTs.

COMPLETE SENTENCE: A plotting program with clipboard plotting, 3D animation, and FFTs *will be released* today.

- a dependent element used without an independent clause

FRAGMENT: The article was not accepted for publication. *Because the data could not be verified.*

Because the data could not be verified is a fragment because it lacks an independent clause: a clause that has a subject and verb and could stand alone as a sentence. It needs more information to be complete.

COMPLETE SENTENCE: The article was not accepted for publication because the data could not be verified.

In this revision, the dependent element is joined with the independent clause that precedes it.

COMPLETE SENTENCE: Because the data could not be verified, the article was not accepted for publication.

In this revision, the dependent element is followed by the independent clause.

FRAGMENT: *Delivering over 150 horsepower.* The two-passenger coupe will cost over $22,000.

COMPLETE SENTENCE: Delivering over 150 horsepower, the two-passenger coupe will cost over $22,000.

COMPLETE SENTENCE: The two-passenger coupe will deliver over 150 horse-power and cost over $22,000.

Avoid Comma Splices

cs

A *comma splice* is the error in which two independent clauses are joined, or spliced together, by a comma. Independent clauses can be linked correctly in three ways:

- by a comma and a coordinating conjunction

COMMA SPLICE: The 909 printer is our most popular model, it offers an unequaled blend of power and versatility.

CORRECT: The 909 printer is our most popular model, for it offers an unequaled blend of power and versatility.

In this case, a comma and one of the coordinating conjunctions (*and, or, nor, but, for, so,* and *yet*) link the two independent clauses. The coordinating conjunction explicitly states the relationship between the two clauses.

- by a semicolon

COMMA SPLICE: The 909 printer is our most popular model, it offers an unequaled blend of power and versatility.

CORRECT: The 909 printer is our most popular model; it offers an unequaled blend of power and versatility.

In this case, a semicolon links the two independent clauses. The semi-colon creates a somewhat more distant relationship between the two clauses than the comma-and-coordinating-conjunction link; the link remains implicit.

- by a period or other terminal punctuation

COMMA SPLICE: The 909 printer is our most popular model, it offers an unequaled blend of power and versatility.

CORRECT: The 909 printer is our most popular model. It offers an unequaled blend of power and versatility.

In this case, the two independent clauses are separate sentences. Of the three ways to punctuate the two clauses correctly, this punctuation suggests the most distant relationship between them.

Avoid Run-On Sentences

run

A run-on sentence (sometimes called a *fused sentence*) is a comma splice without the comma. In other words, two independent clauses appear without any punctuation between them. Any of the three strategies used to fix a comma splice fixes a run-on sentence.

RUN-ON SENTENCE: The 909 printer is our most popular model it offers an unequaled blend of power and versatility.

CORRECT: The 909 printer is our most popular model, for it offers an unequaled blend of power and versatility.

CORRECT: The 909 printer is our most popular model; it offers an unequaled blend of power and versatility.

CORRECT: The 909 printer is our most popular model. It offers an unequaled blend of power and versatility.

Avoid Ambiguous Pronoun Reference

ref

Pronouns must refer clearly to the words or phrases they replace. Ambiguous pronoun references can lurk in even the most innocent-looking sentences:

UNCLEAR: Remove the cell cluster from the medium and analyze it. (Analyze what, the cell cluster or the medium?)

CLEAR: Analyze the cell cluster after removing it from the medium.

CLEAR: Analyze the medium after removing the cell cluster from it.

CLEAR: Remove the cell cluster from the medium. Then analyze the cell cluster.

CLEAR: Remove the cell cluster from the medium. Then analyze the medium.

Ambiguous references can also occur when a relative pronoun such as *which,* or a subordinating conjunction such as *where,* introduces a dependent clause:

UNCLEAR: She decided to evaluate the program, which would take five months. (What would take five months, the program or the evaluation?)

CLEAR: She decided to evaluate the program, a process that would take five months. (By replacing *which* with *a process that,* the writer clearly indicates that the evaluation will take five months.)

CLEAR: She decided to evaluate the five-month program. (By using the adjective *five-month,* the writer clearly indicates that the program will take five months.)

UNCLEAR: This procedure will increase the handling of toxic materials outside the plant, where adequate safety measures can be taken. (Where can adequate safety measures be taken, inside the plant or outside?)

CLEAR: This procedure will increase the handling of toxic materials outside the plant. Because adequate safety measures can be taken only in the plant, the procedure poses risks.

CLEAR: This procedure will increase the handling of toxic materials outside the plant. Because adequate safety measures can be taken only outside the plant, the procedure will decrease safety risks.

As the last example shows, sometimes the best way to clarify an unclear reference is to split the sentence in two, eliminate the problem, and add clarifying information. Clarity is always a fundamental characteristic of good technical communication. If more words will make your writing clearer, use them.

Ambiguity can also occur at the beginning of a sentence:

UNCLEAR: Allophanate linkages are among the most important structural components of polyurethane elastomers. They act as cross-linking sites. (What act as cross-linking sites, *allophanate linkages* or *polyurethane elastomers?*)

CLEAR: Allophanate linkages, which are among the most important structural components of polyurethane elastomers, act as cross-linking sites. (The writer has changed the second sentence into a clear nonrestrictive modifier and combined it with the first sentence.)

Your job is to ensure that the reader will know exactly which word or phrase the pronoun is replacing by either restructuring the sentence or dividing it in two.

If you begin a sentence with a pronoun that might be unclear to the reader, be sure to follow it immediately with a noun that clarifies the reference.

UNCLEAR: The new parking regulations require that all employees pay for parking permits. These are on the agenda for the next senate meeting. (What are on the agenda, the *regulations* or the *permits?*)

CLEAR: The new parking regulations require that all employees pay for parking permits. These regulations are on the agenda for the next senate meeting.

Compare Items Clearly

compar

When comparing or contrasting items, make sure your sentence clearly communicates the relationship. A simple comparison between two items often causes no problems: "The X3000 has more storage than the X2500." However, don't let your reader confuse a comparison and a simple statement of fact. For example, in the sentence "Trout eat more than minnows," does the writer mean that trout don't restrict their diet to minnows or that trout eat more than minnows eat? If a comparison is intended, a second verb should be used: "Trout eat more than minnows do." And if you are introducing three items, make sure that the reader can tell which two are being compared:

AMBIGUOUS: Trout eat more algae than minnows.

CLEAR: Trout eat more algae than they do minnows.

CLEAR: Trout eat more algae than minnows do.

Beware of comparisons in which different aspects of the two items are compared:

ILLOGICAL: The resistance of the copper wiring is lower than the tin wiring.

LOGICAL: The resistance of the copper wiring is lower than that of the tin wiring.

In the illogical construction, the writer contrasts *resistance* with *tin wiring* rather than the resistance of copper with the resistance of tin. In the revision, the pronoun *that* substitutes for *resistance* in the second part of the comparison.

Use Adjectives Clearly

adj

In general, adjectives are placed before the nouns that they modify: *the plastic washer.* In technical communication, however, writers often need to use clusters of adjectives. To prevent confusion, use commas to separate coordinate adjectives, and use hyphens to link compound adjectives.

Adjectives that describe different aspects of the same noun are known as coordinate adjectives:

portable, programmable CD player

adjustable, removable housings

In these cases, the comma replaces the word *and*.

Note that sometimes an adjective is considered part of the noun it describes: *electric drill.* When one adjective modifies *electric drill,* no comma is required: *a reversible electric drill.* The addition of two or more adjectives, however, creates the traditional coordinate construction: *a two-speed, reversible electric drill.*

The phrase *two-speed* is an example of a compound adjective—one made up of two or more words. Use hyphens to link the elements in compound adjectives that precede nouns:

a *variable-angle* accessory

increased *cost-of-living* raises

The hyphens in the second example prevent the reader from momentarily misinterpreting *increased* as an adjective modifying *cost* and *living* as a participle modifying *raises.*

But note that hyphens are not used when the adjective follows the noun:

raises to cover increased *cost of living*

A long string of compound adjectives can be confusing even if hyphens are used appropriately. To ensure clarity in such a case, put the adjectives into a clause or phrase following the noun:

UNCLEAR: an *operator-initiated, default-prevention* technique

CLEAR: a technique *initiated by the operator to prevent default*

Maintain Number Agreement

Number disagreement commonly takes one of two forms in technical communication: (1) the verb disagrees in number with the subject when a prepositional phrase intervenes or (2) the pronoun disagrees in number with its referent when the latter is a collective noun.

Subject-Verb Disagreement

agr s/v

A prepositional phrase does not affect the number of the subject and the verb. The following examples show that the object of the preposition can be plural in a singular sentence, or singular in a plural sentence. (The subjects and verbs are italicized.)

INCORRECT: The *result* of the tests *are* promising.

CORRECT: The *result* of the tests *is* promising.

INCORRECT: The *results* of the test *is* promising.

CORRECT: The *results* of the test *are* promising.

Don't be misled by the fact that the object of the preposition and the verb don't sound natural together, as in *tests is* or *test are*. As long as the subject and verb agree, the sentence is correct.

Pronoun-Antecedent Disagreement

agr p/a

The problem of pronoun-antecedent disagreement occurs most often when the antecedent, or referent, is a collective noun—one that can be interpreted as either singular or plural, depending on its usage:

INCORRECT: The *company* is proud to announce a new stock option plan for *their* employees.

CORRECT: The *company* is proud to announce a new stock option plan for *its* employees.

In this example, *company* acts as a single unit; therefore, the singular verb, followed by a singular pronoun, is appropriate. When the individual members of a collective noun are emphasized, however, plural pronouns and verbs are appropriate: *The inspection team have prepared their reports.* Or *The members of the inspection team have prepared their reports.*

Use Tenses Correctly

Two tenses are commonly misused in technical communication: the present tense and the past perfect tense.

$\boxed{t}$

Present Tense

The present tense is used to describe scientific principles and recurring events.

> INCORRECT: In 1992, McKay and his coauthors argued that the atmosphere of Mars *was* salmon pink.
>
> CORRECT: In 1992, McKay and his coauthors argued that the atmosphere of Mars *is* salmon pink.

Notice that although the argument was made in the historical past—1992—the point is expressed in the present tense, because the atmosphere of Mars continues to be salmon pink. When the date of the argument is omitted, some writers express the entire sentence in the present tense.

> McKay and his coauthors *argue* that the atmosphere of Mars *is* salmon pink.

Past Perfect Tense

The past perfect tense is used to describe two events that occurred in the past, with one occurring before the other.

> We *had begun* excavation when the foreman *discovered* the burial remains.

Had begun is the past perfect tense. The excavation began before the burial remains were discovered.

> The seminar *had concluded* before I *got* a chance to talk with Dr. Tran.

Punctuation

The Period

$\boxed{.}$

Periods are used in the following instances.

1.   At the end of sentences that do not ask questions or express strong emotion:

   The lateral stress still needs to be calculated.

2.   After some abbreviations:

   M.D.

   U.S.A.

   etc.

   (For a further discussion of abbreviations, see pp. 679–680.)

3.   With decimal fractions:

   4.056

   $6.75

   75.6 percent

The Exclamation Point

!

The exclamation point is used at the end of a sentence that expresses strong emotion, such as surprise or doubt:

> The nuclear plant, which was originally expected to cost $1.6 billion, eventually cost more than $8 billion!

Because technical communication requires objectivity and a calm, understated tone, exclamation points are rarely used.

The Question Mark

?

The question mark is used at the end of a sentence that asks a direct question:

> What did the commission say about effluents?

Do not use a question mark at the end of a sentence that asks an indirect question:

> He wanted to know whether the procedure had been approved for use.

When a question mark is used within quotation marks, the quoted material takes no other end punctuation:

> "What did the commission say about effluents?" she asked.

The Comma

The comma is the most frequently used punctuation mark, as well as the one about whose usage writers most often disagree. Following are the basic uses of the comma.

- To separate the clauses of a compound sentence (one composed of two or more independent clauses) linked by a coordinating conjunction (*and, or, nor, but, so, for, yet*):

  > Both methods are acceptable, but we have found that the Simpson procedure gives better results.

  In many compound sentences, the comma is needed to prevent the reader from mistaking the subject of the second clause for an object of the verb in the first clause:

  > The RESET command affects the field access, and the SEARCH command affects the filing arrangement.

  Without the comma, the reader is likely to interpret the coordinating conjunction *and* as a simple conjunction linking *field access* and *SEARCH command.*

- To separate items in a series composed of three or more elements:

  > The manager of spare parts is responsible for ordering, stocking, and disbursing all spare parts for the entire plant.

  Despite the presence of the conjunction *and*, most technical-communication style manuals require that a comma follow the second-to-last item. The comma clarifies the separation and prevents misreading. For example, sometimes in technical communication the second-to-last item will be a compound noun containing an *and*.

CONFUSING: The report will be distributed to Operations, Research and Development and Accounting.

CLEAR: The report will be distributed to Operations, Research and Development, and Accounting.

The series comma is also useful in preventing another kind of misreading. Look at the following beginning of a sentence:

According to Johnson, Smith and Jones . . .

Smith and Jones could be part of a series:

According to Johnson, Smith, and Jones, the new technology will never live up to its potential.

Or *Smith and Jones* could be the subject of a clause:

According to Johnson, Smith and Jones have done outstanding work this quarter.

Using the series comma prevents this ambiguity.

- To separate introductory words, phrases, and clauses from the main clause of the sentence:

However, we will have to calculate the effect of the wind.

To facilitate trade, the government holds a yearly international conference.

Whether the workers like it or not, the managers have decided not to try the flextime plan.

In each of these three examples, the comma helps the reader understand the sentence. Notice in the following example how the comma actually prevents misreading:

Just as we finished eating, the rats discovered the treadmill.

The comma is optional if the introductory text is brief and cannot be misread.

CORRECT: First, let's take care of the introductions.

CORRECT: First let's take care of the introductions.

- To separate the main clause from a dependent clause:

Although most of the executive council saw nothing wrong with it, the advertising campaign was canceled.

Most PCs use green technology, even though it is relatively expensive.

- To separate nonrestrictive modifiers (parenthetical clarifications) from the rest of the sentence:

Jones, the temporary chairman, called the meeting to order.

- To separate interjections and transitional elements from the rest of the sentence:

Yes, I admit your findings are correct.

Their plans, however, have great potential.

- To separate coordinate adjectives:

The finished product was a sleek, comfortable cruiser.

The heavy, awkward trains are still being used.

The comma here takes the place of the conjunction *and*. If the adjectives are not coordinate—that is, if one of the adjectives modifies the combination of the adjective and the noun—do not use a comma:

They decided to go to the first general meeting.

- To signal that a word or phrase has been omitted from an elliptical expression:

Smithers is in charge of the accounting; Harlen, the data management; Demarest, the publicity.

In this example, the commas after *Harlen* and *Demarest* show that the phrase *is in charge of* has been omitted.

- To separate a proper noun from the rest of the sentence in direct address:

John, have you seen the purchase order from United?

What I'd like to know, Betty, is why we didn't see this problem coming.

- To introduce most quotations:

He asked, "What time were they expected?"

- To separate towns, states, and countries:

Bethlehem, Pennsylvania, is the home of Lehigh University.

He attended Lehigh University in Bethlehem, Pennsylvania, and the University of California at Berkeley.

Note the use of the comma after *Pennsylvania*.

- To set off the year in a date:

August 1, 1997, is the anticipated completion date.

Note the use of the comma after *1997*. If the month separates the date from the year, the first comma is not used; the numbers are not next to each other:

The anticipated completion date is 1 August 1997.

- To clarify numbers:

12,013,104

(European practice is to reverse the use of commas and periods in writing numbers: periods signify thousands, and commas signify decimals.)

- To separate names from professional or academic titles:

Harold Clayton, Ph.D.

Marion Fewick, CLU

Joyce Carnone, P.E.

Note that the comma also follows the title in a sentence:

Harold Clayton, Ph.D., is the featured speaker.

Common Errors in Using Commas

- The writer does not use a comma between the clauses of a compound sentence:

INCORRECT: The mixture was prepared from the two premixes and the remaining ingredients were then combined.

CORRECT: The mixture was prepared from the two premixes, and the remaining ingredients were then combined.

- The writer does not use a comma (or uses just one comma) to set off a nonrestrictive modifier:

INCORRECT: The phone line, which was installed two weeks ago had to be disconnected.

CORRECT: The phone line, which was installed two weeks ago, had to be disconnected.

See the discussion of restrictive and nonrestrictive modifiers, pages 394–396.

- The writer does not use a comma to separate introductory words, phrases, or clauses from the main clause, when misreading can occur:

INCORRECT: As President Canfield has been a great success.

CORRECT: As President, Canfield has been a great success.

- The writer does not use a comma (or uses just one comma) to set off an interjection or a transitional element:

INCORRECT: Our new statistician, however used to work for Konaire, Inc.

CORRECT: Our new statistician, however, used to work for Konaire, Inc.

- The writer joins two independent clauses with a comma, creating a comma splice (a comma used to "splice together" independent clauses not linked by a coordinating conjunction):

INCORRECT: All the motors were cleaned and dried after the water had entered, had they not been, additional damage would have occurred.

CORRECT: All the motors were cleaned and dried after the water had entered; had they not been, additional damage would have occurred.

CORRECT: All the motors were cleaned and dried after the water had entered. Had they not been, additional damage would have occurred.

For more information on comma splices, see pages 659–660.

- The writer uses unnecessary commas:

INCORRECT: Another of the many possibilities, is to use a "first in, first out" sequence.

CORRECT: Another of the many possibilities is to use a "first in, first out" sequence.

In the incorrect sentence, the comma should be removed because it separates the subject (*Another*) from the verb (*is*).

INCORRECT: The schedules that have to be updated every month are, numbers 14, 16, 21, 22, 27, and 31.

CORRECT: The schedules that have to be updated every month are numbers 14, 16, 21, 22, 27, and 31.

In the incorrect sentence, the comma separates the verb from its complement: the word or words that complete the meaning of the sentence.

INCORRECT: New and old employees who use the processed order form, do not completely understand the basis of the system.

CORRECT: New and old employees who use the processed order form do not completely understand the basis of the system.

In the incorrect sentence, a comma separates the subject and its restrictive modifier from the verb.

INCORRECT: The company has grown so big, that an informal evaluation procedure is no longer effective.

CORRECT: The company has grown so big that an informal evaluation procedure is no longer effective.

In the incorrect sentence, the comma separates the predicate adjective *big* from the clause that modifies it.

INCORRECT: Recent studies, and reports by other firms confirm our experience.

CORRECT: Recent studies and reports by other firms confirm our experience.

In the incorrect sentence, the comma separates the two elements in the compound subject.

The Semicolon
`;`

Semicolons are used in the following instances.

- To separate independent clauses not linked by a coordinating conjunction:

  The second edition of the handbook is more up-to-date; however, it is more expensive.

- To separate items in a series that already contains commas:

  The members elected three officers: Jack Resnick, president; Carol Wayshum, vice president; Ahmed Jamoogian, recording secretary.

  In this example, the semicolon acts as a "supercomma," keeping the names and titles clear.

Common Error in Using Semicolons

Sometimes writers incorrectly use a semicolon when a colon is called for:

INCORRECT: We still need one ingredient; luck.

CORRECT: We still need one ingredient: luck.

The Colon
`:`

Colons are used in the following instances.

- To introduce a word, phrase, or clause that amplifies, illustrates, or explains a general statement:

  The project team lacked one crucial member: a project leader.

  Here is the client's request: we are to provide the preliminary proposal by November 13.

  We found three substances in excessive quantities: potassium, cyanide, and asbestos.

  The week was productive: fourteen projects were completed and another dozen were initiated.

  Note that the text preceding a colon should be able to stand on its own as a main clause:

  INCORRECT: We found: potassium, cyanide, and asbestos.

  CORRECT: We found potassium, cyanide, and asbestos.

- To introduce items in a vertical list, if the sense of the introductory text would be incomplete without the list:

We found the following:
potassium
cyanide
asbestos

See Chapter 13, pages 385–389, for more information on constructing lists.

- To introduce long or formal quotations:

The president began: "In the last year . . ."

Common Error in Using Colons

Writers sometimes incorrectly use a colon to separate a verb from its complement:

INCORRECT: The tools we need are: a plane, a level, and a T-square.

CORRECT: The tools we need are a plane, a level, and a T-square.

CORRECT: We need three tools: a plane, a level, and a T-square.

The Dash

Dashes are used in the following instances.

- To set off a sudden change in thought or tone:

The committee found—can you believe this?—that the company bore full responsibility for the accident.

That's what she said—if I remember correctly.

- To emphasize a parenthetical element:

The managers' reports—all 10 of them—recommend production cutbacks for the coming year.

Arlene Kregman—the first woman elected to the board of directors—is the next scheduled speaker.

- To set off an introductory series from its explanation:

Wetsuits, weight belts, tanks—everything will have to be shipped in.

When a series *follows* the general statement, a colon replaces the dash:

Everything will have to be shipped in: wetsuits, weight belts, and tanks.

With most word-processing programs, you have to use a special character to make a dash; there is no dash key on the keyboard. Another way to make a dash is to use two uninterrupted hyphens (--). Do not add space before or after the hyphens. This form of the dash looks less professional than the special dash character.

Common Error in Using Dashes

Sometimes writers incorrectly use a dash as a "lazy" substitute for other punctuation marks:

INCORRECT: The regulations—which were issued yesterday—had been anticipated for months.

CORRECT: The regulations, which were issued yesterday, had been anticipated for months.

INCORRECT: Many candidates applied—however, only one was chosen.

CORRECT: Many candidates applied; however, only one was chosen.

Parentheses
( )

Parentheses are used in the following instances.

- To set off incidental information:

Please call me (x3104) when you get the information.

Galileo (1546–1642) is often considered the father of modern astronomy.

H. W. Fowler's *Modern English Usage* (New York: Oxford University Press, 2nd ed., 1965) is still the final arbiter.

- To enclose numbers and letters that label items listed in a sentence:

To transfer a call within the office, (1) place the party on HOLD, (2) press TRANSFER, (3) press the extension number, and (4) hang up.

Use both a left and a right parenthesis—not just a right parenthesis—in this situation.

Common Error in Using Parentheses

Sometimes writers incorrectly use parentheses instead of brackets to enclose their interruption of a quotation (see the discussion of brackets):

INCORRECT: He said, "The new manager (Farnham) is due in next week."

CORRECT: He said, "The new manager [Farnham] is due in next week."

The Hyphen
-

Hyphens are used in the following instances.

- In general, to form compound adjectives that precede nouns:

general-purpose register

meat-eating dinosaur

chain-driven saw

Note that hyphens are not used after adverbs that end in *-ly*:

newly acquired terminal

Also note that hyphens are not used when the compound adjective follows the noun:

The Woodchuck saw is chain driven.

Many organizations have their own preferences about hyphenating compound adjectives. Check to see if your organization has a preference.

- To form some compound nouns:

  vice-president

  editor-in-chief

- To form fractions and compound numbers:

  one-half

  fifty-six

- To attach some prefixes and suffixes:

  post-1945

  president-elect

- To divide a word at the end of a line:

  We will meet in the pavil-
  ion in one hour.

Whenever possible, avoid such breaks; they slow the reader down. When you do use them, check the dictionary to make sure you have divided the word *between* syllables. Do not assume that your word processor has divided the word correctly.

The Apostrophe   Apostrophes are used in the following instances.

'

- To indicate the possessive case:

  the manager's goals   the employee's credit union

  the worker's lounge   Charles's T-square

  For joint possession, add the apostrophe and the *s* to only the last noun or proper noun:

  Watson and Crick's discovery

  For separate possession, add an apostrophe and an *s* to each of the nouns or pronouns:

  Newton's and Galileo's ideas

  Make sure you do not add an apostrophe or an *s* to possessive pronouns: *his, hers, its, ours, yours, theirs.*

- To indicate the possessive case when a noun modifies a gerund:

  We were all looking forward to Bill's joining the company.

  In this case, the gerund—*joining*—is modified by the proper noun *Bill*.

- To form contractions:

  I've   shouldn't

  can't   it's

  The apostrophe usually indicates an omitted letter or letters. For example, *can't* is *can(no)t*, *it's* is *it(i)s*.

  Some organizations discourage the use of contractions; others have no preference. Find out the policy your organization follows.

- To indicate special plurals:

  three 9's

  two different JCL's

  the why's and how's of the problem

  As in the case of contractions, it is a good idea to learn the stylistic preferences of your organization. For plurals of numbers and acronyms, some style guides omit the apostrophe: *9s, JCLs.* Usage varies considerably.

Common Error in Using Apostrophes

Writers sometimes incorrectly use the contraction *it's* in place of the possessive pronoun *its*.

INCORRECT: The company does not feel that the problem is *it's* responsibility.

CORRECT: The company does not feel that the problem is *its* responsibility.

Quotation Marks

" "

Quotation marks are used in the following instances.

- To indicate titles of short works, such as articles, essays, or chapters:

  Smith's essay "Solar Heating Alternatives"

- To call attention to a word or phrase that is being used in an unusual way or in an unusual context:

  A proposal is "wired" if the sponsoring agency has already decided who will be granted the contract.

  Don't use quotation marks as a means of excusing poor word choice:

  The new director has been a real "pain."

- To indicate direct quotation, that is, the words a person has said or written:

"In the future," he said, "check with me before authorizing any large purchases."
As Breyer wrote, "Morale *is* productivity."

Do not use quotation marks to indicate indirect quotation:

INCORRECT: He said that "third-quarter profits will be up."

CORRECT: He said that third-quarter profits will be up.

CORRECT: He said, "Third-quarter profits will be up."

Related Punctuation

Note that if the sentence contains a *tag*—a phrase identifying the speaker or writer—a comma separates it from the quotation:

Wilson replied, "I'll try to fly out there tomorrow."

"I'll try to fly out there tomorrow," Wilson replied.

Informal and brief quotations require no punctuation before the quotation marks:

She said "Why?"

In the United States (unlike most other English-speaking nations), commas and periods at the end of quotations are placed within the quotation marks:

The project engineer reported, "A new factor has been added."

"A new factor has been added," the project engineer reported.

Question marks, dashes, and exclamation points, however, are placed inside the quotation marks when they apply only to the quotation and outside the quotation marks when they apply to the whole sentence:

He asked, "Did the shipment come in yet?"

Did he say, "This is the limit"?

When a punctuation mark appears inside a quotation mark at the end of a sentence, do not add another punctuation mark.

INCORRECT: Did she say, "What time is it?"?

CORRECT: Did she say, "What time is it?"

Block Quotations

When quotations reach a certain length—generally, more than four lines—switch to a block format. In typewritten manuscript, a block quotation is usually

- indented 10 spaces from the left-hand margin
- typed without quotation marks
- introduced by a complete sentence followed by a colon

(Different organizations observe their own variations on these basic rules.)

McFarland writes:

> The extent to which organisms adapt to their environment is still being charted. Many animals, we have recently learned, respond to a dry winter with an automatic birth control chemical that limits the number of young to be born that spring. This prevents mass starvation among the species in that locale.

Hollins concurs. She writes, "Biological adaptation will be a major research area during the next decade."

Mechanics

Ellipses

`. . .`

Ellipses (three spaced periods) indicate the omission of some material from a quotation. A fourth period with no space before it precedes ellipses when the sentence in the source has ended and you are omitting material that follows:

> "Send the updated report . . . as soon as you can."

> Larkin refers to the project as "an attempt . . . to clarify the issue of compulsory arbitration. . . . We do not foresee an end to the legal wrangling . . . but perhaps the report can serve as a definition of the areas of contention."

In the second example, the writer has omitted words after *attempt* and after *wrangling*. In addition, the writer uses a sentence period plus three spaced periods after *arbitration*, which ends the source's sentence, to indicate that the following sentence has been omitted.

Brackets

`[ ]`

Brackets are used in the following instances.

- To indicate words added to a quotation:

  > "He [Pearson] spoke out against the proposal."

  A better approach would be to shorten the quotation:

  > The minutes of the meeting note that Pearson "spoke out against the proposal."

- To indicate parenthetical information within parentheses:

  > (For further information, see Charles Houghton's *Civil Engineering Today* [New York: Arch Press, 1994].)

Italics

ital

Italics (or underlining) are used in the following instances.

- For words used as words:

  > In this report, the word *operator* will refer to any individual who is actually in charge of the equipment, regardless of that individual's certification.

- To indicate titles of long works (books, manuals, and so on), periodicals and newspapers, long films, long plays, and long musical works:

See Houghton's *Civil Engineering Today.*

We subscribe to the *Wall Street Journal.*

In the second example, note that the word *the* is not italicized or capitalized when the title is used in a sentence.

- To indicate the names of ships, trains, and airplanes:

The shipment is expected to arrive next week on the *Penguin.*

- To set off foreign expressions that have not become fully assimilated into English:

The speaker was guilty of *ad hominem* arguments.

Check a dictionary to determine whether the foreign expression has become assimilated.

- To emphasize words or phrases:

Do not press the ERASE key.

If your word processor does not have italic type, indicate italics by underlining:

Darwin's Origin of Species is still read today.

Numbers

num

The use of numbers varies considerably. Therefore, in choosing between words and numerals, consult the guidelines that your organization or research area follows. Many organizations use the following guidelines.

- Use numerals for technical quantities, especially if a unit of measurement is included:

3 feet   43,219 square miles

12 grams   36 hectares

- Use numerals for nontechnical quantities of 10 or more:

300 persons

12 whales

35 percent increase

- Use words for nontechnical quantities of fewer than 10:

three persons

six whales

- Use both words and numerals
  - For back-to-back numbers:

    six 3-inch screws

    fourteen 12-foot ladders

    3,012 five-piece starter units

    In general, use the numeral for the technical unit. If the nontechnical quantity would be cumbersome in words, use the numeral.

  - For round numbers over 999,999:

    14 million light-years

    $64 billion

  - For numbers in legal contracts or in documents intended for international readers:

    thirty-seven thousand dollars ($37,000)

    five (5) relays

  - For addresses:

    3801 Fifteenth Street

Special Cases

- If a number begins a sentence, use words, not numerals:

  Thirty-seven acres was the size of the lot.

  Many writers would revise the sentence to avoid this problem:

  The lot was 37 acres.

- Don't use both numerals and words in the same sentence to refer to the same unit:

  On Tuesday the attendance was 13; on Wednesday, 8.

- Write out fractions, except if they are linked to technical units:

  two-thirds of the members

  3½ hp

- Write out approximations:

  approximately ten thousand people

  about two million trees

- Use numerals for titles of figures and tables and for page numbers:

  Figure 1

  Table 13

  page 261

- Use numerals for decimals:

3.14

1,013.065

Add a zero before decimals of less than one:

0.146

0.006

- Avoid expressing months as numbers, as in *3/7/96:* in the United States, this means March 7, 1996; in many other countries, it means July 3, 1996. Use one of the following forms:

March 7, 1996

7 March 1996

- Use numerals for times if A.M. or P.M. are used:

6:10 A.M.

six o'clock

Abbreviations
ab

Abbreviations save time and space, but use them carefully; you can never be sure that your readers will understand them. Many companies and professional organizations provide lists of approved abbreviations.

Analyze your audience in determining whether and how to abbreviate. If your readers include nontechnical people unfamiliar with your field, either write out the technical terms or attach a list of abbreviations. If you are new in an organization or are writing for publication for the first time in a certain field, find out what abbreviations are commonly used. If for any reason you are unsure about whether or how to abbreviate, write out the word.

The following are general guidelines about abbreviations.

- You may make up your own abbreviations. For the first reference to the term, write it out and include the abbreviation in parentheses. In subsequent references, use the abbreviation. For long works, you might want to write out the term at the start of major units, such as chapters.

The heart of the new system is the self-loading cartridge (slc).

This technique is also useful, of course, in referring to existing abbreviations that your readers might not know:

The cathode-ray tube (CRT) is your control center.

- Most abbreviations do not take plurals:

1 lb

3 qt

- Most abbreviations in scientific writing are not followed by periods:

  lb

  cos

  dc

  If the abbreviation can be confused with another word, however, use a period:

  in.

  Fig.

- Spell out the unit if the number preceding it is spelled out or if no number precedes it:

  How many square meters is the site?

Capitalization

cap

For the most part, the conventions of capitalization in general writing apply in technical communication.

- Capitalize proper nouns, titles, trade names, places, languages, religions, and organizations:

  William Rusham

  Director of Personnel

  Quick-Fix Erasers

  Bethesda, Maryland

  Methodism

  Italian

  Society for Technical Communication

  In some organizations, job titles are not capitalized unless they refer to specific persons:

  Alfred Loggins, Director of Personnel, is interested in being considered for vice president of marketing.

- Capitalize headings and labels:

  A Proposal to Implement the Wilkins Conversion System

  Mitosis

  Table 3

  Section One

  The Problem

  Rate of Inflation, 1985-1995

  Figure 6

Appendix B
Guidelines for Speakers of English as a Second Language

English is notoriously difficult to master, and no brief guide can answer all your questions about the many eccentricities of the language. But this appendix provides a concise guide to some of the most challenging aspects of English for non-native speakers.

The advice offered here is based on two highly regarded brief books:

Ann Raimes. (1992). *Grammar troublespots: An editing guide for students* (2nd ed.). New York: St. Martin's Press.

Diana Hacker. (1995). *A writer's reference* (3rd ed.). New York: St. Martin's Press.

These two books are excellent investments.

Basic Characteristics of a Sentence

A sentence has five characteristics:

- It starts with an uppercase letter and ends with a period, a question mark, or (rarely) an exclamation point.
- It has a subject, usually a noun. The subject is what the sentence is about.
- It has a verb, which tells the reader about what happens to the subject.
- It has a standard word order. The most common sequence in English is subject-verb-object. But you can add information at different locations. For instance, here is a sentence in the basic subject-verb-object order:

We hired a consulting firm.

You can add information to the start of the sentence:

Yesterday we hired a consulting firm.

Or to the end of the sentence:

Yesterday we hired a consulting firm: *Sanderson & Associates.*

Or in the middle:

Yesterday we signed *a nontransferable contract* with a consulting firm: Sanderson & Associates.

In fact, any element of a sentence can be expanded.

- It has an independent clause: an idea that can stand alone. The following is a sentence because its idea can stand alone:

SENTENCE: The pump failed because of improper maintenance.

The following is also a sentence:

SENTENCE: The pump failed.

But the following is *not* a sentence because the thought is incomplete:

NOT A SENTENCE: Because of improper maintenance.

A result clause is required to complete this sentence:

SENTENCE: Because of improper maintenance, the pump failed.

Linking Ideas by Coordination

One way to connect ideas is by coordination. Coordination means the two ideas are roughly equal in importance. There are four main ways to coordinate ideas:

- Use a semicolon (;) to coordinate two sentences with similar structures:

  The information for bid was published last week; the proposal is due in less than a month.

- Use a comma and one of the coordinating conjunctions (*and, but, or, nor, so, for,* and *yet*) to coordinate two ideas.

  The information for bid was published last week, but the proposal is due in less than a month.

  In this example, *but* clarifies the relationship between the two clauses: you haven't been given enough time to write the proposal.

- Combine two separate sentences into one. Here are two separate sentences:

  The bridge was completed last year. The bridge already needs repairs.

  One way to combine them would be as follows:

  The bridge was completed last year and already needs repairs.

  Notice that there is no comma after *year* because the two verbs in the sentence have the same subject.

- Use transitional words and phrases within and between sentences (see Chapter 14 for more information on transitional words and phrases).

  The 486 chip has already been replaced. *As a result,* it's hard to find a 486 in a new computer.

Linking Ideas by Subordination

Two separate sentences can also be linked by subordination, that is, by deemphasizing one of them. There are two basic ways to use subordination:

- Use a subordinating word or phrase. Start with the two separate sentences about a bridge:

The bridge was completed last year. The bridge already needs repairs.

Then combine them as follows:

The bridge already needs repairs *although* it was completed last year.

In this example, *although* subordinates the clause, leaving *the bridge already needs repairs* as the independent clause. Note that the order of the sentences could be reversed:

Although it was completed last year, the bridge already needs repairs.

In general, it is easier to read a sentence if the subject (*the bridge*) appears before the pronoun that replaces it (*it*). This way, readers don't have to remember the *it* clause until they find out what the subject of the sentence is. Another way to subordinate would be as follows:

The bridge, *which* was completed last year, already needs repairs.

This version emphasizes the *already needs repairs* portion of the sentence and deemphasizes the *was completed last year* portion by putting it in a *which* clause.

- Combine two separate sentences into one.

Completed last year, the bridge already needs repairs.

In this version, *completed last year* modifies *the bridge*. The independent clause is *the bridge already needs repairs*.

Verb Tenses

|t|

Verb tenses in English can be complicated, but in general there are four kinds of time relationships that you need to understand.

- Simple past, present, and future

Yesterday we *subscribed* to a new ecology journal.

We *subscribe* to three ecology journals every year. (Meaning: We subscribe to the three journals regularly.)

We *will subscribe* to the new ecology journal next year. (Or: We *are going to subscribe* to the new ecology journal next year.)

- An action in progress at a known time

We *were updating* our directory when the power failure occurred. (Meaning: After the power was restored, we continued.)

We *are updating* our directory now.

We *will be updating* our directory tomorrow when you arrive.

- An action completed before a known time

We *had already started* to write the proposal when we got your call. (The writing began before the call.)

We *have started* to write the proposal.

We *will have started* to write the proposal by the time you arrive. (Meaning: Both events are in the future, with the writing beginning before the arrival.)

- An action in progress until a known time

    We *had been working* on the reorganization when the news of the merger was publicized. (Meaning: The work was in progress when the news of the merger was publicized.)

    We *have been working* on the reorganization for over a year. (Meaning: The work will continue into the future.)

    We *will have been working* on the reorganization for two years by the time it occurs.

-ing Forms of Verbs

-ing

English uses the *-ing* form of verbs in four major ways:

- As part of a verb in a sentence:

    We are *shipping* the materials by UPS.

- To add extra information in a sentence:

    Analyzing the sample, we discovered two anomalies.

    The sample *containing* the anomalies appears on Slide 14.

    We studied the sample, *thinking* it could be important.

- As an adjective:

    the *leaking* pipe

- As a noun:

    Writing is the best way to learn to write.

    The *-ing* form used as a noun is called a *gerund*. In the previous example, the gerund acts as the subject of the sentence. A gerund can also act as a direct object:

    The designer tried *inserting* the graphics by hand.

Infinitives

inf

Infinitive phrases are used in three main ways:

- As a noun:

    The editor's goal for the next year is *to publish the journal on schedule*.

- As an adjective:

  The company requested the right *to subcontract the project.*

- As an adverb:

  We established the schedule ahead of time *to prevent the kind of mistake we made last time.*

Helping Verbs and Main Verbs

help

There are a number of helping verbs in English. The following discussion explains four categories of helping verbs.

- Modals. There are nine modal verbs: *can, could, may, might, must, shall, should, will,* and *would.* After a modal verb, use the base form of the verb (the form of the verb used after *to* in the infinitive).

  base form
  ↓
  The system *must meet* all applicable codes.

- Forms of *do.* After a helping verb that is a form of *do—do, does,* and *did—*use the base form of the verb.

  base form
  ↓
  Do we *need* to include the figures for the recovery rate?

- Forms of *have* plus the past participle (usually the *-ed* or *-en* form of the verb). To form one of the perfect tenses (past, present, or future), use a form of *have* plus the past participle of the verb.

  PAST PERFECT: We *had written* the proposal before learning of the new RFP.

  PRESENT PERFECT: We *have written* the proposal according to the instructions in the RFP.

  FUTURE PERFECT: We *will have written* the proposal by the end of the week.

- Forms of *be.* To describe an action in progress, use a form of *be* and the present participle (the *-ing* form of the verb).

  We *are testing* the new graphics tablet.

  The company *is considering* flextime.

  To create the passive voice, use a form of *be* and the past participle.

  The piping *was installed* by the plumbing contractor.

  See the discussion of active and passive voice in Chapter 15.

Agreement between the Subject and the Verb

agr s/v

The subject and the verb in a clause or sentence must agree in number. Five major constructions are important to remember:

- Simple agreement

  The new *valve* is installed according to the manufacturer's specifications.

  The new *valves* are installed according to the manufacturer's specifications.

- Agreement when the clause or sentence contains information between the subject and the verb.

  The *result* of the tests is included in Appendix C.

  The *results* of the test are included in Appendix C.

- Agreement when the clause or sentence contains special pronouns and quantifiers. Pronouns that end in -*body* or -*one*—such as *everyone, everybody, someone, somebody, anyone, anybody, no one,* and *nobody*—are singular. In addition, quantifiers such as *something, each,* and *every* are singular.

  SINGULAR: *Everybody* is invited to the preproposal meeting.

  SINGULAR: *Each* of the members is asked to submit billable hours by the end of the month.

- Agreement when the clause or sentence contains a compound subject. In such cases, the verb must be plural.

  COMPOUND SUBJECT: *The contractor and the subcontractor* want to meet to resolve the difficulties.

- Agreement when a relative pronoun such as *who, that,* or *which* begins a clause. In such cases, make sure the verb agrees in number with the noun that the relative pronoun refers to.

  The numbers that *are* used in the formula *do* not *agree* with *the ones* we were given at the site. (*Numbers* is plural, so *that* is treated as a plural noun.)

  The *number* that *is* used in the formula *does* not *agree* with *the one* we were given at the site. (*Number* is singular, so *that* is treated as a singular noun.)

Conditions

Four main types of condition are used with the word *if* in English:

- Conditions of fact:

  If you *see* "Unrecoverable Application Error," the program *has crashed.*

  If rats *are* allowed to eat as much as they want, they *become* obese.

  For conditions of fact, use the same verb tense in both clauses. In most cases, use the present tense.

- Future prediction:

  If we *win* this contract, we *will have* to add three more engineers.

  For future prediction, use the present tense in the *if* clause. Use *will, can, should,* or *might* plus the base form of the verb in the independent clause.

- Present-future speculation:

  If I *were* president of the company, I *would be* much more aggressive.

  Notice that the present-future speculation usage suggests a condition contrary to fact. For instance, the example above implies that you are not president of the company. Use *were* in the *if* clause. Use *would, could,* and so on plus the base form of the verb in the independent clause.

- Past speculation:

  If we *had won* this contract, we *would have* had to add three more engineers.

  Notice here too that the example implies that the condition is contrary to fact. Use *had* plus the past participle in the *if* clause. Use *would have, could have,* and so on in the independent clause. You should be aware of one other point about *if* conditions. You can restructure the sentence and not use *if* at all:

  Had we won this contract, we would have had to add three more engineers.

Articles

art

Few aspects of English can be as frustrating to the nonnative speaker of English as the correct usage of the simple words *a, an,* and *the.* Although there are a few rules that you should try to learn, remember that there are many exceptions and special cases. Here are three general guidelines.

- Singular proper nouns—those that name specific persons, places, and things—do not usually take an article:

  Taiwan

  James Allenby

But plural proper nouns often do take an article:

the United States

the Allenbys

- Countable common nouns take an article:

  the microscope

  a desk

  Countable nouns are persons, places, or things that can be counted: *one microscope, two microscopes; one desk, two desks.*
  But uncountable common nouns generally do not take an article:

  overtime

  equipment

  integrity

  research

  information

  How can you be sure if a word is countable or uncountable? Unfortunately, you can't. You have to keep a list.

- Common nouns can be referred to as either specific or nonspecific. The specific form takes *the;* the nonspecific form takes either *a* or *an.*

  Our department received funding for *an experiment. The experiment* will take six months to complete.

  In the first sentence in this example, *experiment* is a nonspecific noun; in the second sentence, *experiment* is a specific noun because it has been identified in the previous sentence.

Adjectives

adj

Keep in mind three main points about adjectives in English:

- Adjectives do not take a plural form.

  a complex project

  two complex projects

- Adjectives can be placed before the nouns they modify or later in the sentence.

  The *critical* need is to reduce the drag coefficient.

  The need to reduce the drag coefficient is *critical.*

- Adjectives of one or two syllables take special endings to create the comparative and superlative forms.

Positive	*Comparative*	*Superlative*
big	bigger	biggest
heavy	heavier	heaviest

Adjectives of three or more syllables take the word *more* for the comparative form and the words *the most* for the superlative form.

Positive	*Comparative*	*Superlative*
qualified	more qualified	the most qualified
feasible	more feasible	the most feasible

Adverbs

adv

Like adjectives, adverbs are modifiers, but their placement in the sentence is somewhat more complex. Remember five points about adverbs.

- Adverbs modify verbs.

  Management terminated the project *reluctantly.*

- Adverbs also modify adjectives.

  The executive summary was *conspicuously* absent.

- Adverbs also modify other adverbs.

  The project is going *very* well.

- Adverbs that describe how an action takes place can appear in different locations in the sentence.

  Carefully the inspector examined the welds.

  The inspector *carefully* examined the welds.

  The inspector examined the welds *carefully.*

  But the adverb should not be placed between the verb and the direct object.

  INCORRECT: The inspector examined *carefully* the welds.

- Adverbs that describe the whole sentence can also be placed in different locations in the sentence.

  Apparently, the inspection was successful.

  The inspection *apparently* was successful.

  The inspection was successful, *apparently.*

Omitted Words

om

Except for imperative sentences, in which the subject *you* is understood (*Get the correct figures*), all sentences in English require a subject.

> *The company* has a policy on conflict of interest.

Do not omit the *there* or *it*.

INCORRECT:　Are three reasons for us to pursue this issue.

CORRECT:　　*There* are three reasons for us to pursue this issue.

INCORRECT:　Is important that we seek his advice.

CORRECT:　　*It* is important that we seek his advice.

See the discussion of expletives in Chapter 15.

Repeated Words

- Do not repeat the subject of a sentence.

INCORRECT:　The company we are buying from *it* does not permit us to change our order.

CORRECT:　　The company we are buying from does not permit us to change our order.

- In an adjective clause, do not repeat an object.

INCORRECT:　The technical communicator does not use the same software that we were writing in *it*.

CORRECT:　　The technical communicator does not use the same software that we were writing in.

It is not correct because *that* is the object of the preposition *in*.

In an adjective clause, do not use a second adverb.

INCORRECT:　The lab where we did the testing *there* is an excellent facility.

CORRECT:　　The lab where we did the testing is an excellent facility.

Appendix C
Commonly Misused Words

This appendix explains the proper usage of some of the most commonly misused words and phrases in technical communication. A brief list like this one cannot replace a full-length work: I recommend Harry Shaw's *Dictionary of Problem Words and Expressions* (New York: McGraw-Hill, 1987) and Theodore M. Bernstein's *The Careful Writer: A Modern Guide to English Usage* (New York: Atheneum, 1965).

The definitions in this appendix provide brief explanations and/or examples. The examples are enclosed within quotation marks.

accept, except. "We will not *accept* delivery of any items *except* those we have ordered."

adapt, adopt. Adapt means to adjust or to modify; adopt means to accept. "Management decided to adapt the quality-circle plan rather than adopt it as is."

affect, effect. *Affect* is a verb: "How will the news *affect* him?" *Effect* is most commonly a noun: "What will be the *effect* of the increase in allowable limits?" *Effect* is also (rarely) a verb meaning to bring about or cause to happen: "The new plant is expected to *effect* a change in our marketing strategy."

already, all ready. "The report had *already* been sent to the printer when the writer discovered that it was not *all ready.*"

alright, all right. *Alright* is a misspelling of *all right.*

among, between. In general, *among* is used for relationships of more than two items, *between* for only two items. "The collaboration *among* the writer, the illustrator, and the printer," but "the agreement *between* the two companies."

amount, number. *Amount* is used for noncounting items; *number* refers to counting items: "the *amount* of concrete," but "the *number* of bags of concrete."

assure, ensure, insure. *To assure* means to put someone's mind at ease: "let me *assure* you." *To ensure* and *to insure* both mean to make sure: "the new plan will *ensure* [or *insure*] good results." Some writers prefer to use *insure* only when referring to insurance: "to *insure* against fire loss."

can, may, might. *Can* refers to ability: "We *can* produce 300 chips per hour." *May* refers to permission: "*May* I telephone your references?" *Might* refers to possibility: "We *might* see further declines in PC prices this year."

compliment, complement. A *compliment* is a statement of praise: "The owner offered a gracious *compliment* to the architect on his design." *Compliment* is also a verb: "The owner graciously *complimented* the architect." A *complement* is something that fills something up or makes it complete, or something that is an appropriate counterpart: "The design is a perfect *complement* to the landscape." *Complement* is also a verb: "The design *complements* the landscape perfectly."

could of. This is not a correct phrase; it is a corruption of *could've*, the contraction of *could have*: "She *could have* mentioned the abrasion problem in the progress report."

criteria, criterion. *Criteria*, meaning standards against which something will be measured, is plural; *criterion* is singular.

data, datum. *Data* is plural; *datum* is singular. However, the distinction is fading in popular usage, although not in some scientific and engineering applications. Check to see how it is spelled in your company or field.

discreet, discrete. *Discreet* means careful and prudent: "She is a very *discreet* manager; you can confide in her." *Discrete* means separate or distinct: "The company will soon split into three *discrete* divisions."

effective, efficient. *Effective* means that the item does what is is meant to do; *efficient* also carries the sense of accomplishing the goal without using more resources than necessary. "Air Force One is an *effective* way to fly the president around, but it is not *efficient;* it costs some $40,000 per hour to fly."

either . . . or; neither . . . nor. *Either . . . or* means one of two; *neither . . . nor* means not one of two: "*Either* Jim *or* I will attend the meeting, but *neither* Bob *nor* Ahmed will."

farther, further. *Farther* refers to distance: "one mile *farther* down the road." *Further* means greater in quantity, time, or extent: "Are there any *further* questions?"

feedback. Many writers will not use *feedback* to refer to a response by a person: "Let me have your *feedback* by Friday." They limit the term to its original meaning, dealing with electricity, because a human response involves thinking (or should, anyway).

fewer, less. *Fewer* is used for counting items: "*fewer* bags of cement"; *less* is used for noncounting items: "*less* cement." It's the same distinction as between *number* and *amount*.

foreword, forward. A *foreword* is a preface, usually written by someone other than the author, introducing and praising the book. *Forward* refers to being in advance: "The company decided to move *forward* with the project."

i.e., e.g. *I.e.*, Latin for *id est*, means *that is*. *E.g.*, Latin for *exempli gratia*, means *for example*. Writers often confuse them. That's why I recommend using the English versions. Also, add commas after them: "Use the main entrance, *that is*, the one on Broadway."

imply, infer. The writer or speaker *implies;* the reader or listener *infers*.

input. People who don't like to give their *feedback* also don't like to offer their *input*.

its, it's. *Its* is the possessive pronoun: "The lab rat can't make up *its* mind." *It's* is the contraction of *it is:* "*It's* too late to apply for this year's grant." Why do people mix up these two words? Because they remember learning that possessives take apostrophes—"Bob's computer"—when they use the possessive form of it, they add the apostrophe. But *its* is a possessive pronoun, like *his, hers, theirs, ours,* and *yours,* a word specifically created to fulfill only one function: to indicate possession. It is not the possessive form of another word, and therefore it does not take an apostrophe.

-ize. Many legitimate words end in *-ize*, such as *harmonize* and *sterilize*, but many writers and readers can't stand new ones (such as *prioritize*) when there are perfectly fine words already (such as *rank*).

lay, lie. *Lay* is a transitive verb meaning to place: "*Lay* the equipment on the table." *Lie* is an intransitive verb meaning to recline: "*Lie* down on the couch." The complete conjugation of *lay* is *lay, laid, laid, laying;* of *lie,* it is *lie, lay, lain, lying*.

lead, led. *Lead* is the infinitive verb: "We want to *lead* the industry." *Led* is the past tense: "Last year we *led* the industry."

parameter. This is a mathematical term referring to a constant whose value can vary according to its application. Many writers object to the nonmathematical uses of the term, including such concepts as perimeter, scope, outline, and limit. (You guessed it: the same people who don't provide *input* or *feedback* don't use *parameter* very much either.)

phenomena, phenomenon. *Phenomena* is plural; *phenomenon* is singular.

plain, plane. *Plain* means simple and unadorned: "The new company created a very plain logo." *Plane* has several meanings: an airplane, the act of smoothing a surface, the tool used to smooth a surface, and the flat surface itself.

precede, proceed. *Precede* means to come before: "Should Figure 1 *precede* Figure 2?" *Proceed* means to move forward: "We decided to *proceed* with the project despite the setback."

shall, will. *Shall* suggests a legal obligation, particularly in a formal specification or contract: "The contractor *shall* remove all existing debris." *Will* does not suggest a legal obligation: "We *will* get in touch with you as soon as possible to schedule the job."

sight, site, cite. *Sight* refers to vision; *site* is a place; *cite* is a verb meaning to document a reference.

than, then. *Than* is a conjunction used in comparisons: "Plan A works better *than* Plan B." *Then* is an adverb referring to time: "First we went to the warehouse. *Then* we went to the plant."

their, there, they're. *Their* is the possessive pronoun: "They brought *their* equipment with them." *There* refers to a place—"We went *there* yesterday"—or in expletive expressions—"*There* are three problems we have to solve." *They're* is the contraction of *they are*.

to, too, two. *To* is used in infinitive verbs ("*to* buy a new microscope") and in expressions referring to direction ("go *to* Detroit"). *Too* means excessively: "The refrigerator is *too* big for the lab." *Two* is the number 2.

viable. This is a fine Latin word meaning able to sustain life: "*viable* cell culture" and "*viable* fetus." Many writers avoid such clichés as *viable alternative* (while they're avoiding *input* and *feedback*).

weather, whether. *Weather* refers to sunshine and temperature. *Whether* refers to alternatives. "The demonstration will be held outdoors *whether* or not the *weather* cooperates."

who's, whose. *Who's* is the contraction of *who is*. *Whose* is the possessive case of *who:* "*Whose* printer are we using?"

-wise. Don't say, "*RAMwise*, the computer has 8 MB." Instead, say "The computer has 8 MB of RAM."

Xerox. The people at Xerox become unhappy when writers ask for a xerox copy. The correct word is *photocopy; Xerox* is a copyrighted term.

your, you're. *Your* is the possessive pronoun: "Bring *your* calculator to the meeting." *You're* is the contraction of *you are*.

Appendix D
Selected Bibliography

Technical Communication

Allen, O. J., & Deming, L. H. (1994). *Publications management: Essays for professional communicators*. Amityville, NY: Baywood.

Beer, D. F. (1992). *Writing and speaking in the technology professions: A practical guide*. New York: IEEE.

Blicq, R. S. (1993). *Technically—Write! Communicating in a technological era* (4th ed.). Englewood Cliffs, NJ: Prentice Hall.

Brusaw, C. T., Alred, G. J., & Oliu, W. E. (1993). *Handbook of technical writing* (4th ed.). New York: St. Martin's, 1991.

Day, R. A. (1992). *Scientific English*. Phoenix, AZ: Oryx.

Dombrowski, P. M. (Ed.). (1994). *Humanistic aspects of technical communication*. Amityville, NY: Baywood.

Dragga, S., & Gong, G. (1989). *Editing: The design of rhetoric*. Amityville, NY: Baywood.

Mathes, J. C., & Stevenson, D. W. (1991). *Designing technical reports: Writing for audiences in organizations* (2nd ed.). Indianapolis: Bobbs-Merrill.

Pickett, N. A., & Laster, A. A. (1993). *Technical English* (6th ed.). New York: HarperCollins.

Sides, C. H. (1991). *How to write and present technical information* (2nd ed.). Phoenix, AZ: Oryx.

Also see the following journals:
IEEE Transactions on Professional Communication
Journal of Business and Technical Communication
Journal of Technical Writing and Communication
Technical Communication
Technical Communication Quarterly

Ethics

Beauchamp, T. L., & Bowie, N. E. (1993). *Ethical theory and business* (4th ed.). Englewood Cliffs, NJ: Prentice Hall.

Brockmann, R. J., & Rook, F. (Eds.). (1989). *Technical communication and ethics*. Washington, DC: Society for Technical Communication.

Velasquez, M. G. (1992). *Business ethics: Concepts and cases* (3rd ed.). Englewood Cliffs, NJ: Prentice Hall.

Collaborative Writing

Blyler, N. R., & Thralls, C. (Eds.). (1993). *Professional communication: The social perspective*. Newbury Park, CA: Sage.

Cross, G. A. (1993). *Collaboration and conflict: A contextual exploration of group writing and positive emphasis*. Cresskill, NJ: Hampton.

Ede, L., & Lunsford, A. (1990). *Singular texts/plural authors: Perspectives on collaborative writing.* Carbondale: Southern Illinois University.

Forman, J. (Ed.). (1992). *New visions of collaborative writing.* Portsmouth, NH: Boynton/Cook.

Lay, M. M., & Karis, W. M. (1991). *Collaborative writing in industry: Investigations in theory and practice.* Amityville, NY: Baywood.

Secondary Research

Lavin, M. R. (1991). *Business information: How to find it, how to use it* (2nd ed.). Phoenix, AZ: Oryx.

Malinowsky, H. R. (1994). *Reference sources in science, engineering, medicine, and agriculture.* Phoenix, AZ: Oryx.

Mount, E., & Kovacs, B. (1991). *Using science and technology information sources.* Phoenix, AZ: Oryx.

Zimmerman, D. E., & Muraski, M. L. (1994). *The elements of information gathering.* Phoenix, AZ: Oryx.

Usage and General Writing

Bush, D. W., & Campbell, C. P. (1995). *How to edit technical documents.* Phoenix, AZ: Oryx.

Corbett, E. P. J. (1990). *Classical rhetoric for the modern student* (3rd ed.). New York: Oxford University Press.

Fowler, H. W. (1987). *A dictionary of modern English usage* (2nd ed.). Rev. by Sir E. Gowers. New York: Oxford University Press.

Maggio, R. (1991). *The dictionary of bias-free usage: A guide to nondiscriminatory language.* Phoenix, AZ: Oryx.

Strunk, W., & White, E. B. (1979). *The elements of style* (3rd ed.). New York: Macmillan.

Williams, J. (1994). *Style: Ten lessons in clarity and grace* (4th ed.). New York: HarperCollins.

Style Manuals

American National Standards, Inc. (1979). *American National Standard for the preparation of scientific papers for written or oral presentation.* ANSI Z39.16—1972. New York: American National Standards Institute.

CBE Style Manual Committee. (1983). *Council of Biology Editors style manual: A guide for authors, editors, and publishers in the biological sciences* (5th ed.). Washington, DC: Council of Biology Editors.

The Chicago manual of style. (1993). (14th ed.). Chicago: University of Chicago, 1993.

Dodd, J. S. (Ed.). (1986). *The ACS style guide: A manual for authors and editors.* Washington, DC: American Chemical Society.

Li, X., & Crane, N. B. (1993). *Electronic style: A guide to citing electronic information.* Westport, CT: Meckler.

Pollack, G. (1977). *Handbook for ASM editors.* Washington, DC: American Society for Microbiology.

Publications manual of the American Psychological Association. (1994). (4th ed.). Washington, DC: American Psychological Association.

U.S. Government Printing Office style manual. (1988). (Rev. ed.). New York: Outlet.

Also, many private corporations, such as John Deere, DuPont, Ford Motor Company, General Electric, and Westinghouse, have their own style manuals.

Graphics, Design, and Multimedia

Baird, R. N., Turnbull, A. T., & McDonald, D. (1992). *The graphics of communication* (6th ed.). San Diego: Harcourt Brace.

Burke, C. (1990). *Type from the desktop: Designing with type and your computer.* Chapel Hill, NC: Ventana.

Foley, J. D. (1990). *Computer graphics: Principles and practice* (2nd ed.). Reading, MA: Addison-Wesley.

Hoffman, E. K., & Teeple, J. (1990). *Computer graphics applications: An introduction to desktop publishing and design, presentation graphics, animation.* Belmont, CA: Wadsworth.

Horton, W. (1994). *The icon book: Visual symbols for computer systems and documentation.* New York: Wiley.

Horton, W. (1991). *Illustrating computer documentation: The art of presenting information graphically on paper and online.* New York: Wiley.

Kelvin, G. V. (1992). *Illustrating for science.* New York: Watson-Guptill.

Leonard, D. C., & Dillon, P. M. (1994). *Multimedia technology from a to z.* Phoenix, AZ: Oryx.

Parker, R. C. (1993). *Looking good in print: A guide to basic design for desktop publishing* (3rd ed.). Chapel Hill, NC: Ventana.

Shushan, R., & Wright, D. (1994). *Desktop publishing by design* (3rd. ed.). Redmond, WA: Microsoft.

Talman, M. (1992). *Understanding presentation graphics.* San Francisco: Sybex.

Tufte, E. R. (1990). *Envisioning information.* Cheshire, CT: Graphics Press.

Tufte, E. R. (1983). *The visual display of quantitative information.* Cheshire, CT: Graphics Press.

White, J. V. (1990). *Color for the electronic age.* New York: Watson-Guptill.

Also see the following journals:
Graphic Arts Monthly
Graphics: USA

Technical Manuals

Barker, T. (1991). *Perspectives on software documentation: Inquiries and innovations.* Amityville, NY: Baywood.

Brockmann, R. J. (1990). *Writing better computer documentation: From paper to hypertext.* New York: Wiley-Interscience.

Crown, J. (1992). *Effective computer user documentation.* New York: Van Nostrand Reinhold.

Dumas, J. S., & Redish, J. C. (1993). *A practical guide to usability testing.* Norwood, NJ: Ablex.

Forbes, M. (1992). *Writing technical articles, speeches, and manuals* (2nd ed.). New York: Krieger.

Hackos, J. T. (1994). *Managing your documentation projects.* New York: Wiley.

Horton, W. (1990). *Designing and writing online documentation* (2nd ed.). New York: Wiley.

Lanyi, G. (1994). *Managing documentation projects in an imperfect world.* Columbus, OH: Battelle.

Price, J., & Korman, H. (1993). *How to communicate technical information: A handbook of software and hardware documentation.* Redwood City, CA: Benjamin/Cummings.

Rubin, J. (1994). *Handbook of usability testing: How to plan, design, and conduct effective tests*. New York: Wiley.

Slatkin, E. (1991). *How to write a manual*. Berkeley, CA: Ten Speed.

Weiss, E. H. (1992). *How to write usable user documentation* (2nd ed.). Phoenix, AZ: Oryx.

Wieringa, D., Moore, C., & Barnes, V. (1993). *Procedure writing: Principles and practices*. Columbus, OH: Battelle.

Woolever, K. R., & Loeb, H. M. (1994). *Writing for the computer industry*. Englewood Cliffs, NJ: Prentice Hall.

Oral Presentations   D'Arcy, J. (1992). *Technically speaking: Proven ways to make your next presentation a success*. New York: AMACOM.

Smith, T. C. (1991). *Making successful presentations: A self-teaching guide* (2nd ed.). New York: Wiley.

Proposals   Bowman, J. P., & Branchaw, B. P. (1992). *How to write proposals that produce*. Phoenix, AZ: Oryx.

Hill, J. W., & Whalen, T. (1993). *How to create and present successful government proposals*. New York: IEEE.

Lefferts, R. (1991). *Getting a grant in the 1990s: How to write successful grant proposals*. Englewood Cliffs, NJ: Prentice Hall.

Meader, R. (1991). *Guidelines for preparing proposals* (2nd ed.). Chelsea, MI: Lewis.

Miner, L. E., & Griffith, J. (1993). *Proposal planning and writing*. Phoenix, AZ: Oryx.

Stewart, R. D., & Stewart, A. L. (1992). *Proposal preparation* (2nd ed.). New York: Wiley-Interscience.

Acknowledgments

Figure 1.2: An example of technical communication. Page 5–7 of *LaserJet 4L User's manual*. Copyright 1993 by Hewlett-Packard Company. Reprinted by permission.

Chapter 1, Exercise 5: From Concord Coalition brochure. Reprinted by permission of The Concord Coalition.

Figure 4.3: Writing addressed to an expert audience. From Yazici, H., Benjamin, C., & McGlaughlin, J., "AI-based generation of production engineering labor standards." *IEEE Transactions on Engineering Management 41* (3), 302–309. © 1994 IEEE.

Figure 6.5: Guide to the use of an abstract service. From the front matter of *America: History and Life*. Copyright © 1995 by ABC-Clio, Inc. Reprinted with permission of the publisher.

Figure 8.7: Argument organized chronologically. Based on Bell, T. E., "Bicycles on a personalized basis." *IEEE Spectrum 31* (5), 20–31. © 1993 IEEE.

Figure 10.6: A process description based on a graphic. "Putting privacy in escrow," by Patrick J. Lyons. Copyright © 1994 by the New York Times Company. Reprinted with permission.

Figure 12.6: Color used to establish patterns. Jain, R. "What is multimedia, anyway?" *IEEE MultiMedia 1* (3), 3. © 1994 IEEE.

Figure 12.28: Diagram used to clarify a difficult concept. Labuz, R. *How to typeset from a word processor: An interfacing guide*. New York: R. R. Bowker, 1984. Reprinted with permission of R. R. Bowker, a Reed Reference Publishing Company. © 1984 by Reed Elsevier Inc.

Figure 12.32: A table used to communicate a maintenance schedule. From McComb, G., *Troubleshooting and repairing VCRs*, 2nd ed., p. 133. © 1991 by TAB/McGraw-Hill, Inc. Reprinted with permission of McGraw-Hill, Inc.

Figure 12.37: Sinclair Stammers/Science Photo Library.

Figure 12.38: Image Quest.

Figure 12.39: Screen shot. From Wu, J. K., & Narasimhalu, A. D., "Identifying faces using multiple retrievals." *IEEE MultiMedia 1* (2), 27–38. © 1994 IEEE.

Figure 12.40: Line drawing. From *Briggs & Stratton service and repair instructions*. Part 270962. Courtesy of Briggs & Stratton.

Chapter 13, Exercise 2: Scrabble brochure. © 1995 Milton Bradley Company. SCRABBLE® is a trademark of Hasbro, Inc. Used with permission. All rights reserved.

Figure 15.1: Screen from a style program. From *Grammatik Windows version 2.0 user's guide*. San Francisco: Reference Software International. © 1991 Novell, Inc. All Rights Reserved. Used with permission.

Figure 15.9: Information Mapping®. Horn, R. E., "Results with structured writing using the Information Mapping® writing service standards." In T. M. Duffy and R. Waller (Eds.), *Designing usable texts*. Orlando, FL: Academic Press. Reprinted with permission from Information Mapping®, Inc., Copyright, Information Mapping®, Inc., 1985. All rights reserved.

Figure 18.2: Safety information on machinery. From p. 7 of the operator's manual for 21-inch Walk Behind Rotary Mowers. Reproduced by permission of Deere & Company, © 1989. Deere & Company.

Figure 18.5: Instructions. From "Use and care guide for electric range models RF396PXY and RF396PCY." Reprinted with permission of Whirlpool Corporation.

Figure 18.6: Conventions. From *Concise user's guide: Microsoft® MS-DOS® 6 for the MS-DOS operating system*. Redmond, WA: Microsoft Corporation. 1993. Reprinted with permission from Microsoft Corporation.

Figure 18.7: Introduction to a manual. From *Using OpenScript*. Bellevue, WA: Asymetrix Corporation. © 1991 Asymetrix Corporation. All rights reserved. Reproduced with permission.

Figure 18.8: Pages from the body of a manual. *Series 935 user's manual*. Winona, MN: Watlow Controls. © 1994–95.

Figure 18.9: Page from a user's manual for a printer. Page 4–6 of *LaserJet 4L User's manual*. Copyright 1993 by Hewlett-Packard Company. Reprinted by permission.

Figure 18.11: Hypertext screens. From Hall, W., "Ending the tyranny of the button." *IEEE MultiMedia 1* (1), 60–68. © 1994 IEEE.

Chapter 18, Exercise 2: Installation instructions for sliding shower door. Copyright © 1986 by NoviAmerican, Inc., Novi, Michigan.

Index

Date Due

JUL 0 8 2004